Adaptronics and Smart Structures

Hartmut Janocha (Editor)

Adaptronics and Smart Structures

Basics, Materials, Design and Applications

Second, Revised Edition

With 432 Figures and 17 Tables

Springer

Prof. Dr.-Ing. habil. Hartmut Janocha
Universität des Saarlandes
Lehrstuhl für Prozessautomatisierung
Gebäude A5 1
66123 Saarbrücken
Germany
E-mail: janocha@lpa.uni-saarland.de

ISBN 978-3-642-09110-0 e-ISBN 978-3-540-71967-0

Springer is a part of Springer Science+Business Media

springer.com

Cover: WMXDesign GmbH, Heidelberg

Preface to the 2nd Book Edition

The coined word *adaptronics* describes technical fields that have become known internationally under the names smart materials, smart structures or intelligent systems. The term adaptronics was originally formulated by the limited liability company VDI-Technologiezentrum in Düsseldorf, Germany. In the autumn of 1991 the term was sanctioned by a board of independent experts. Initially, the term encompassed all functions of traditional control loops, which are applied to generate adaptive behaviour, i. e. adaptronic systems or structures that adapt automatically to different operating and environmental conditions. Furthermore, in contrast to conventional control loops in which each functional element is a separate component, adaptronics is characterised by multi-functional components. Thus, several application-specific functional elements are embodied in one single component (e. g. a self-sensing actuator), which is preferably integrated into the structure or the system. The intention is to build lightweight adaptive systems and structures to be as simple as possible, with the ultimate goal of reducing the material and energy resources needed for implementation and operation to an absolute minimum.

Given this background it is obvious that apart from the technical requirements for automation, modern functional materials are an essential basis for the successful design and application of adaptronic products. Today, the most well known of these materials are shape-memory alloys, magnetorheological fluids and piezoelectric materials. An old example of an adaptronic product that has been cited numerously are glasses made of photochromic glass. These glasses automatically change the light transmission depending on the surrounding light intensity by performing sensor, actuator and closed-loop control functions for transmission adaptation. Looking to other technical areas, adaptronics has great potential for application in vibration and noise reduction. Fields of application include, for instance, the automotive industry, mechanical engineering, architecture as well as the aerospace industry. Other kinds of application scenarios focus on nature trying to simulate fundamental 'vital functions' by means of adaptronics. One aspect is the ability of biological systems to recognise and automatically correct local disfunctions in their structure. Naturally, this feature is also desirable for technical systems and structures, especially in areas where safety is essential (civil structures, aircraft).

With this book the editor and the publisher have tackled the task of presenting the state of the art of this both fascinating and demanding technological-scientific field. In this 2nd. book edition the contents from the 1st. edition from 1999 have been updated and extended corresponding to the development progress. The outline, which has proven worthwhile, has been maintained: following an introduction describing the aims and the content of adaptronics, subsequent chapters present the 'scientific pillars' from the viewpoint of the various basic disciplines involved. Thereafter, important components of adaptronic structures and systems, such as actuators and sensors, are described. The remaining chapters are dedicated to applications of adaptronics in the various technological and biological/medical fields of daily life, and an outlook towards future developments concludes the book.

It is obvious that no one single person can master all the specialist knowledge involved in such a detailed and varied field as adaptronics. Thus, we recognize both a necessity and a great opportunity in bringing together, in a fundamental work, the knowledge and the experience of proven experts from across the range of adaptronic disciplines. The editor is proud of the fact that numerous experts from all over the world have supported him in performing this task. To all of these he expresses his gratitude. It will not escape the attention of the reader that, in their nuances, viewpoints about adaptronics may diverge somewhat. However, this situation is actually both attractive and stimulating. It is also hardly surprising in view of the fact that adaptronics has only begun a few years ago, to establish itself as a discipline in its own right.

With this background in mind, the editor and publisher hope that the 2nd. edition of this book will also become a useful source of information and ideas, which a large number of readers can rely on time and again. Perhaps it will help some readers to discover their interest or their vocation to actively and creatively support the field of adaptronics along its path to maturity.

Finally, the editor would like to thank his co-workers Petra Detemple, Chris May and Andreas Biehl for their untiring help in transferring the manuscripts and figures, which the contributing authors had presented in widely varied forms, into a uniform format. He also thanks the publishing house Springer-Verlag for the appealing outward design of the book.

In conclusion, the editor wants to assure the critical readership that its constructive comments about the conception, content and presentation of this book are welcome and will be taken into consideration, if possible, in future editions.

Saarbrücken, Germany
Juli 2007

Hartmut Janocha

Contents

List of Contributors

Horst Baier
Institute of Lightweight Structures,
Aerospace Department,
Faculty of Mechanical Engineering,
Technische Universität München
baier@llb.mw.tum.de

Christian Boller
The University of Sheffield,
Department of Mechanical
Engineering, Mappin Street,
Sheffield S1 3JD,
United Kingdom
c.boller@sheffield.ac.uk

Peter Boltryk
School of Engineering Sciences,
University of Southampton,
Southampton, SO17 1BJ, UK
boltryk@soton.ac.uk

William A. Bullough
Prof. William A. Bullough, Department of Mechanical Engineering,
The University of Sheffield, Mappin
Street, Sheffield, S1 3JD, UK.
Hoodgreen7@aol.com

Wenwu Cao
Department of Mathematics,
The Pennsylvania State University,
339 McAllister Bldg.,
University Park, PA 16802, USA
cao@math.psu.edu

J. David Carlson
LORD Corporation,
406 Gregson Drive,
Cary, NC 27511-6445, USA
jcarlson@lord.com

Federico Carpi
Interdepartmental Research Centre
E. Piaggio,
Faculty of Engineering,
University of Pisa,
Via Diotisalvi 2, 56126 Pisa, Italy
f.carpi@ing.unipi.it

Frank Claeyssen
CEDRAT Technologies, Zirst,
F38246 Meylan Cedex, France
Frank.Claeyssen@cedrat.com

Brian Culshaw
University of Strathclyde,
Department of Electronic
& Electrical Engineering,
204 George Street,
Glasgow G1 1XW
bc@eee.strath.ac.uk

Frank Döngi
EADS Astrium SAS,
31, rue des Cosmonautes,
31402 Toulouse Cedex 4, France
frank.doengi@astrium.eads.net

Goran Engdahl
Cedrat Recherche, Zirst,
F38246 Meylan Cedex, France
goran.engdahl@ekc.kth.se

Victor Giurgiutiu
Department of Mechanical Engineering, University of South Carolina Columbia, SC 29208, USA
giurgiut@engr.sc.edu

Martin Gurka
Neue Materialien Würzburg GmbH, Friedrich Bergius Ring 22a, 97076 Würzburg, Germany
gurka@nmwgmbh.de

Wolfgang R. Habel
Bundesanstalt für Materialforschung und -prüfung (BAM), Fachgruppe VIII.1: Mess- und Prüftechnik, Sensorik, Leiter der Arbeitsgruppe "Faseroptische Sensorik", Unter den Eichen 87, 12205 Berlin, Germany
wolfgang.habel@bam.de

Jürgen Hesselbach
TU Braunschweig, Institut für Werkzeugmaschinen und Fertigungstechnik, Langer Kamp 19b, 38106 Braunschweig
j.hesselbach@tu-bs.de

Hartmut Janocha
Universität des Saarlandes, Lehrstuhl für Prozessautomatisierung, Gebäude A5 1, D-66123 Saarbrücken, Germany
janocha@lpa.uni-saarland.de

Klaus Kuhnen
Universität des Saarlandes, Lehrstuhl für Prozessautomatisierung, Gebäude A5 1, D-66123 Saarbrücken, Germany
k.kuhnen@lpa.uni-saarland.de

Ronan Leletty
CEDRAT Technologies, Zirst, F38246 Meylan Cedex, France
Ronan.Leletty@cedrat.com

Hiroshi Matsuhisa
Dept. of Mechanical Engineering, Kyoto University, Kyoto, 520-8501, Japan
matsu@me.kyoto-u.ac.jp

Dirk Mayer
Fraunhofer Institute for Structural Durability and System Reliability, Department of Mechatronics/ Adaptronics, Bartningstr. 47, Post Office Box 100545, 64289 Darmstadt, Germany
dirk.mayer@lbf.fhg.de

Tobias Melz
Fraunhofer Institute for Structural Durability and System Reliability, Department of Mechatronics/ Adaptronics, Bartningstr. 47, Post Office Box 100545, 64289 Darmstadt, Germany
tobias.melz@lbf.fraunhofer.de

Jörg-Uwe Meyer
Head of Research, Drägerwerk AG, Moislinger Allee 53-55, D-23542 Lübeck, Germany
joerg-uwe.meyer@draeger.com

Uwe Müller
Institute of Lightweight Structures, Aerospace Department, Faculty of Mechanical Engineering, Technische Universität München
umueller@llb.mw.tum.de

Werner Nachtigall
Prof. Dr. rer. nat. Werner Nachtigall,
Zoologie, Universität
des Saarlandes, Gebäude A2 4,
66041 Saarbrücken, Germany.
gtbb@mx.uni-saarland.de

Dieter Neumann
Acteos GmbH & Co. KG,
Talhofstr. 30a, 82205 Gilching,
Germany
dieter.neumann@acteos.de

Raino Petricevic
Neue Materialien Würzburg GmbH,
Friedrich Bergius Ring 22a,
97076 Würzburg, Germany
petricevic@nmwgmbh.de

Danilo De Rossi
Interdepartmental Research Centre
E. Piaggio,
Faculty of Engineering,
University of Pisa,
Via Diotisalvi 2, 56126 Pisa, Italy
d.derossi@ing.unipi.it

Helmut Seidel
University of Saarland,
Institute for Micromechanics,
Microfluidics/Microactuators,
University Campus, Building A5 1,
P.O. Box 151150,
D-66041 Saarbrücken, Germany
seidel@lmm.uni-saarland.de

Thomas Stieglitz
Laboratory for Biomedical
Microtechnology, Department of
Microsystems Engineering,
University of Freiburg IMTEK,
Georges-Köhler-Allee 102,
D-79110 Freiburg, Germany
stieglitz@imtek.de

Martin Thomaier
Fraunhofer Institute for
Structural Durability
and System Reliability,
Department of Mechatronics/
Adaptronics,
Bartningstr. 47,
Post Office Box 100545,
64289 Darmstadt, Germany
martin.thomaier@lbf.fhg.de

Neil M. White
School of Electronics
and Computer Science,
University of Southampton,
Highfield, Southampton, SO17 1BJ,
UK
nmw@ecs.soton.ac.uk

Thomas Würtz
Universität des Saarlandes,
Lehrstuhl für
Prozessautomatisierung,
Gebäude A5 1,
D-66123 Saarbrücken, Germany
tw@zip.uni-sb.de

1 Adaptronics: A Concept for the Development of Adaptive and Multifunctional Structures

D. Neumann

1.1 What is Adaptronics?

In German-speaking areas 'adaptronics' is the comprehensive generic term for disciplines that, on an international level, are known by names such as 'smart materials', 'smart structures', 'intelligent systems' etc. The technical term adaptronics (Adaptronik) was created by the VDI Technology Centre and was submitted as a proposed name to a body of experts. Within the scope of a workshop, fourteen experts from the fields of research, development and technology management agreed on the introduction of this new technical term, along with the pertinent definition and delimitation. This was the origin of the term 'adaptronics'.

The term adaptronics designates a system (and its development process) wherein all functional elements of a conventional regulator circuit are existent and at least one element is applied in a multifunctional way. The conformity with a regulator circuit guarantees that the structure shows autonomic adaptive characteristics and can thus adapt itself to different conditions. The limits to the classic control circuit, where normally each single function is achieved through a separately built component, are fixed by the use of multifunctional elements (functional materials). These elements are decisive for making such a technically utilizable system less complex.

An adaptronic system thus is characterized by adaptability and multifunctionality. The aim is to combine the greatest possible number of application-specific functions in one single element and, if appropriate, in one specific material (see Fig. 1.1).

1.2 Examples of Adaptronic Systems

A prime example of an adaptronic system is spectacles equipped with photochromic glass. A photochromic glass which, in dependence on the external ambient brightness, darkens or lets move light through in a self-regulating manner, combines all necessary application-specific functions. It not only covers all three elements of a regulator circuit – the sensor, the actuator and the controlling unit – but also covers the shaping and optical functions as further interesting material properties. This example shows that it is possible

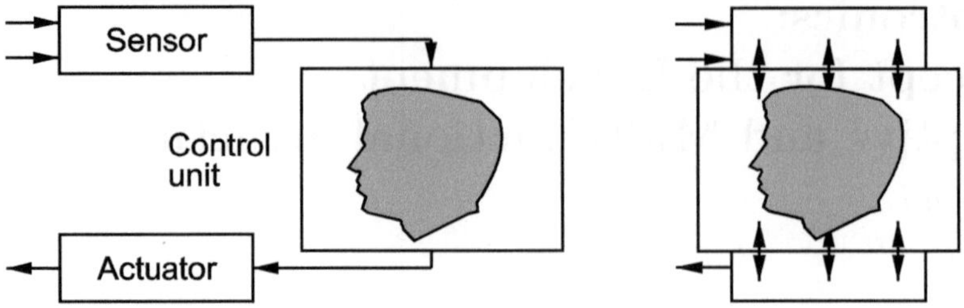

a b

Fig. 1.1. Transition from **a** a conventional system to **b** an adaptronic system

to successfully combine all functional components of a system into one single element, in this case even into one material; further external components are no longer required. The spectacles glass represents a complete functional unit.

Examples for adaptronic systems with a more distinct visionary character are window panes whose transparency automatically regulates itself or can be adjusted within seconds by pressing a button; and hydroplanes whose aerodynamic profile adapts itself to the prevailing flight conditions.

Taking an adaptronic shock absorber as an example, Fig. 1.2 shows four different levels of creating an adaptronic system. On the basic level it is first necessary to produce materials that have both suitable passive qualities and application-specific functional qualities. Depending on the specific application, passive qualities can be of a mechanical, chemical, thermal, optical or electrical nature. For instance, required characteristic features can be resistance to high and/or low temperatures, high mechanical stability, light-transmitting capacity, or good electrical conduction. Functional qualities can be structural changes, changes in the dynamic or static features, or in the chemical, electrical, thermal or optical properties. They can, among other things, manifest themselves in a change of transparency depending on the luminous intensity, in a voltage-dependent change in viscosity, or in a temperature-dependent change in dimension or shape.

The example of an adaptronic shock absorber shows how the electrorheological fluid is simultaneously used as a 'classic' absorber fluid and as an actuator (if necessary, additionally as a sensor). This use is made possible by the capacity of such fluids to change their viscosity to a vast extent in less than a second when they are influenced by an electric field.

Functional qualities can, however, only be used in terms of adaptronics if there is success in combining the specific release phenomena with the respective desired functions. What is therefore required in the conception of multifunctional elements (level II) is the release and specific use of the material-inherent options. For this purpose it is necessary to make use of release phenomena of a physical, chemical or biological nature on material in such a way that, as necessary, several effects can be combined by taking well-directed action. For example, the application of electrorheological fluids

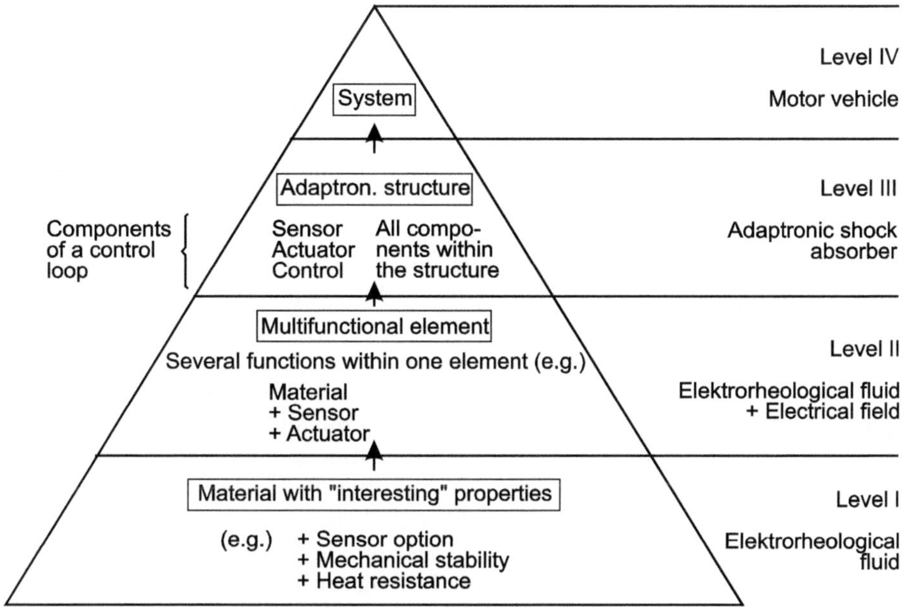

Fig. 1.2. Adaptronics: link between material and system

in an adaptronic shock absorber requires the production of an electric field, as well as the recording of a motion-dependent, variable intensity of current (i. e., use of the sensor effect). Hence, the multifunctional element does not exclusively consist of the electrorheological fluid but necessarily also of an electric voltage and field-producing electrodes.

At the structural level, multifunctional elements must be supplemented to form a complete regulator circuit, always aiming at building up a structure that is marked by minor complexity, low weight, high functional density, and economic efficiency. The successful achievement of this objective will normally depend on the degree to which the functional density is already in existence within the individual elements forming the structural components. In an ideal case – as in case of photochromic glass – all application-specific functions exist in one single element. The outcome will, however, not always be successful. For instance, the multifunctional element existing for the construction of an adaptronic shock absorber must be supplemented by a controlling mechanism, as well as by the structural components required to produce the electric field.

The system level – in the present example the entire motor vehicle – calls for the need to conceptualize during the creation of the adaptronic structure. For instance, the structural shape and damping characteristic of a shock absorber must harmonize with the overall design of a moving gear. Here again, the aim is to optimize the functionality of the entire system.

1.3 Multifunctional Elements

Functional materials constitute the essential basis of all adaptronic systems. The made-to-measure production of functional materials, wherein several functions are interlinked at a molecular level, is therefore of special importance. The more application-specific functions are combined in one single element, the bigger is the advantage in terms of an adaptronic system optimization. Multifunctionality can, however, not be a characteristic feature of an isolated element, but should always manifest itself by meeting user-specific requirements within a system interrelationships. Thus the same element can produce a decisive compression of functions in a given case (A), while it can be completely worthless in a given case (B).

Multifunctionality is by no means required to be coupled to highly sophisticated functional materials. Sometimes amazingly simple concepts lead to a problem-adjusted solution. It is, for instance, conceivable that a gas-filled balloon regulates the volume flow in a fluid flow tube in a temperature-dependent manner. The gas expands with rising temperature, whereupon the balloon reduces the uncovered tubular cross-section. If the temperature decreases, the volume flow is increased along with a smaller balloon cross-section.

This example shows that no limits are set to the users creativity. Mechanically simple solutions are often advantageous compared with high-technology concepts: they are not only more often reasonably priced, but also frequently marked out by greater functional safety. Made-to-measure solutions, however, can in most cases not fulfill their function without high-technology concepts of material scientists.

Materials represent the essential basis for all multifunctional effects. The conception of multifunctional elements is therefore mainly based on the made-to-measure production of functional materials, wherein several functions are interlinked at a molecular level. However, the fact that this is not sufficient in all cases is clearly shown by taking adaptronic shock absorbers as an example, because some effects can only be produced if several materials are combined in suitable interconnected layers or other compounds.

Functional materials, which are characterized by a high potential of functional and application options, are amongst others: shape memory elements; bimetals; electrorheological, magnetorheological, thixotropic and rheopex fluids; piezoelectric elements; electrostrictors; magnetostrictors; chemochromic, electrochromic, hydrochromic, photochromic, and thermochromic elements; and functional gels.

1.4 Fields of Technology and Application

The foregoing explanations show that a basis for adaptronic structures is created in numerous different disciplines of science. The range of applications covers various physical, but also chemical and biological technologies (see Fig. 1.3). What prove to be especially user-relevant here are the often interdisciplinary interactions, such as the physical reaction to a specific chemical stimulus or the reaction of micro-organisms to a modification of physical and/or chemical environmental parameters. Scientific disciplines, such as biophysics, biochemistry, and physical or biophysical chemistry, are of special importance here.

The scope of the application of adaptronic structures or systems can be restricted as the spectrum of influential scientific disciplines. Almost each scientific field covers applications, whose technical benefit and business management utility can be improved by realizing adaptronic concepts. While the need for efficient multifunctional materials certainly originates in the high-technology area, the scope of application is by no means exclusively confined to this field. For example, multifunctional adjusting elements of shape memory alloys are successfully applied for the automatic control of ventilation flaps in greenhouses.

However, even products resulting from highly specialized materials are only partially needed for the realization of efficient adaptronic concepts. Simple adaptive systems, with a minimal number of elements in motion, are of special importance in a surrounding field, where the protection against shortfalls is a decisive factor and where little or no well-trained staff are available for the removal of technically complex problems. The broad range of applications covers a number of areas where adaptronic concepts have been intensively pursued and partially have already been translated into concrete action. The specific interest shown in a particular line of business is a result

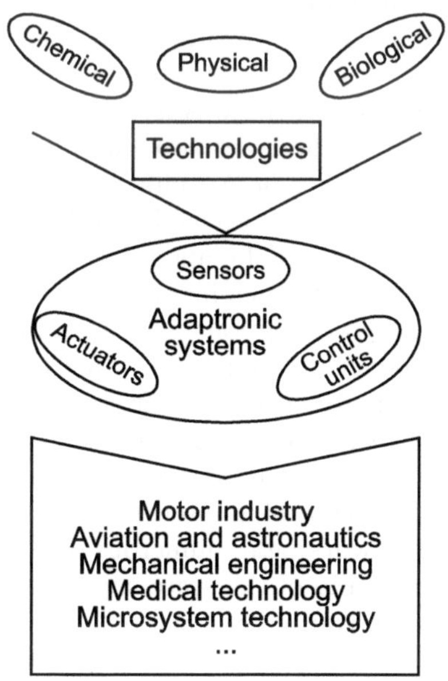

Fig. 1.3. Fields of technology and application

of special security and performance requirements. In this context the fields of aviation and astronautics hold a key position, as both aforementioned aspects are of special importance here. These fields of technology have always been marked by a generally high level of innovation, particularly as a result to their significant financial resources.

1.5 Historical Review

Adaptronics as an overall concept for the development of adaptronic structures and systems is still a young discipline, which was only able to establish itself a few years ago. On the other hand, the research in the fields of multifunctional materials and multifunctional elements, which are the basic elements of adaptronics, started much earlier. The origins of adaptronics – under a different name – go back to the early 1980s. Early progress came from the arms research sector, especially from various air forces.

In the early eighties, government-sponsored efforts were made in the United States to interlink functions, for instance integrate headlights in the outside plating of combat aircrafts. This type of integration not only aimed at the optimization of functions but also at the reduction of weight. This 'smart skin' program lasted nearly one decade, up to the early 1990s. By the mid-eighties, the US airforce likewise had started further adaptronic-oriented programs, which concentrated on the integration of sensor networks in combat aircrafts for system supervisor programs. Both the research and application aspects have considerably gained in importance in the United States, although the main fields of application are still aviation and space technology.

In Japan, the driving force behind initial developments was not the military, but mainly the civil sector. At first, however, these activities were less concentrated on the conception of systems and rather on a well-structured and broadly conceived development of multifunctional materials. In 1985, the 'New Glass Forum' came into existence as a program of Japans Ministry of International Trade and Industry (MITI), the tasks of which included the development of sensor materials with different evaluation options – for example by changing the optical, mechanical and/or chemical conduction properties of the materials. In 1987, the New Glass Forum was dismissed from MITI and a New Glass Association was established in its place. This association was joined by more than 200 enterprises from different sectors of industry and trade. From July 1987 through November 1989, far-reaching interdisciplinary discussions and harmonizations among scientists working in numerous different areas of research took place under the leadership of the state-supported Council for Aeronautics, Electronics and Other Advanced Sciences. The participants came from various sectors, such as medicine, pharmacy, engineering sciences, physics, biology and chemistry, as well as electronics and computer science. The general aim was to formulate and adopt a program for the development of made-to-measure functional materials. In 1989, a comprehensive report was delivered to the Science and Technology Agency (STA), which formed the basis for further promotional activities. Although, in Japan, the expenditures for research activities are largely borne by private enterprises, governmental institutions such as the MITI or the STA exert a significant coordinating influence, pointing the way ahead, despite their comparatively small funds for promotional measures. Within the scope of the 'Basic Technologies for Future Industries' project organized by MITI, the partial project, named 'High Performance Materials', was initiated in 1989 and was carried out up to 1996.

The first German activities in terms of an integrated approach to adaptronics were initiated in the late 1980s in the areas of aviation and space technology. The main topic within the scope of the experiments, which were initially almost exclusively carried out by the big research institutes and large groups of companies, was active vibration suppression. The interest and activities of public institutions started in 1990. The German Federal Minister of Research and Technology entrusted the VDI Technology Centre in Düsseldorf with the coordination of this topic, and initial discussions and harmonization planning took place in 1991. In autumn of that year, the VDI Technology Centre was up and running, and soon formed an expert workshop, in which fourteen reputable specialists from the fields of research and development participated. Within the scope of this event, the term adaptronics was introduced and clearly defined within the German language.

In 1992, the first government funded projects were incorporated by the German Ministry of Research and Technology in its material research program. These projects initially concentrated on the improvement of pure material functions. However, it quickly proved necessary to enlarge the basic area of materials and to develop integrated concepts for multifunctional adaptive structures or systems in terms of adaptronics. In this context the objective was the application-orientated optimization of functional materials and their functional integration in a system.

In the spring of 1993 the Ministry of Research and Technology published a study under the title 'Technologies of the 21st Century', wherein those technologies and trends were described which offered the best chance for maintaining (or even increasing) the competitiveness of German industry. In this study the field of adaptronics was emphasized as one of eight disciplines that were seem to help ensure economic growth parallel to the protection of existing resources. In the early 1994 the first system- and application-oriented projects were started, all focusing on the damping of vibrations in measurement robots.

In November 1994 a further expert workshop took place in Düsseldorf, on the occasion of which some of the main subjects within the broad and interdisciplinary field of adaptronics were thoroughly analysed. In the experts opinion during that workshop, the greatest application potential could be found in vibration and noise damping in the automobile and mechanical industries, as well as in the fields of aviation and space technology.

The aim for the future, apart from the promotion of individual pilot projects, is the further state-supported advancement of specific areas of adaptronics, which are marked by significant high-technology and application potential.

References

1. Culshaw, B.; Gardiner, P.T.; McDonach, A.: Proceedings First European Conference on Smart Structures and Materials. IOP Publishing Ltd., Bristol, GB (1992)
2. Martin, W.E.; Drechsler, K.: *Smart Materials and Structures – Present State and Future Trends.* Technische Niederschrift der Messerschmidt-Bölkow-Blohm GmbH, München (1990)
3. Neumann, D.: *Bausteine 'Intelligenter' Technik von morgen – Funktionswerkstoffe in der Adaptronik.* Wissenschaftliche Buchgesellschaft, Darmstadt (1995)
4. Newnham, R.E.: *Smart, Very Smart and Intelligent Materials.* In: MRS Bulletin, Vol. XVIII, No. 4, April (1993)
5. Rogers, C.A.: *Intelligent Material Systems – The Dawn of a New Materials Age.* In: Journ. of Intelligent Material Systems and Structures, Vol. 4, Technomic Publishing Company, Lancaster, USA (1993)
6. Science and Technology Agency (Government of Japan): *The Concept of Intelligent Materials and the Guidelines on R&D Promotion.* Tokyo, Japan (1989)
7. Takagi, T.: *A Concept of Intelligent Materials.* In: Journ. of Intelligent Material Systems and Structures, Vol. 1, Technomic Publishing Company, Lancaster, USA (1990)
8. Thomson, B.S.; Gandhi, M.V.: *Smart Materials and Structures Technologies.* An intelligence report, Technomic Publishing Company, Lancaster, USA (1990)

2 Concepts of Adaptronic Structures

V. Giurgiutiu

2.1 What are Adaptronic Structures?

Adaptronic structures (also referred to as smart materials or intelligent structures) are defined in the literature in the context of many different paradigms; however, two are prevalent. In the *technology paradigm*, adaptronic structures are seen as an 'integration of actuators, sensors, and controls with a material or structural component', see Fig. 2.1. In the *science paradigm*, adaptronic structures are 'material systems that have intelligence and life-like features integrated in the microstructure of the material in order to reduce to total mass and energy and produce an adaptive functionality'. The vision and guiding analogy of adaptronic structures is that of learning from nature and living systems in such a way as to enable man-made artifacts to have the adaptive features of autopoiesis we see throughout nature. This leads to the description of the anatomy of an adaptronic material system: actuators or motors that behave like muscles; sensors that have the functionality of the five senses (hearing, sight, smell, taste, and touch); and communication and computational networks that represent the nerves, brain, memory, and muscular control systems [1]. Although the leading analogy is that towards biological systems, it must be emphasized that adaptronic structures are designed by human beings in order to achieve human-related objective. Therefore, the system boundary of the adaptronic structures must necessarily be drawn to include the human end user.

What kind of life-like functions can we expect from adaptronic structures? Natures systems have a few general attributes that we can aspire to instill in synthetic material systems. Many of natures systems can change their properties, shape, color, and load paths to account for damage and allow for repair; and can also manage the graceful retirement of aged systems, to name a few. Engineers and scientists have developed a plethora of devices that are inspired by some of nature's capabilities; however, little has been accomplished towards realizing the integration of life-like functions at the system level to create materials systems that would be able to learn, grow, survive, and age with grace and simplicity. The survival of biological structures depends on nature's ability to balance the metabolic cost (economy of construction and maintenance) with the required mechanical properties,

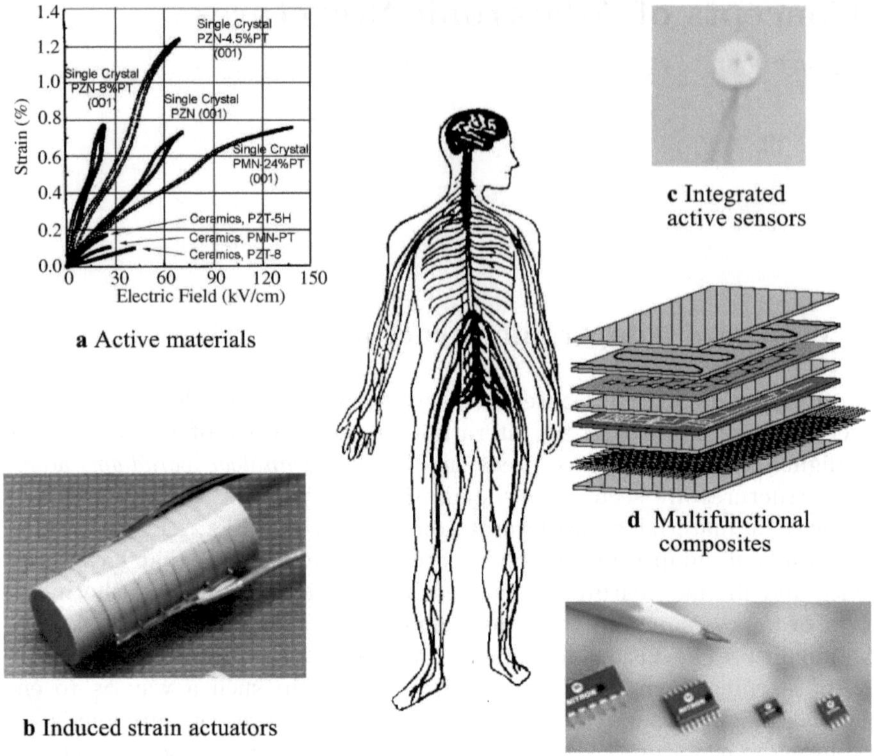

a Active materials

b Induced strain actuators

c Integrated active sensors

d Multifunctional composites

e Microcontrollers

Fig. 2.1. The bio-inspired approach to adaptronic structures: **a** active materials, **b** induced-strain actuators, **c** integrated active sensors; **d** multifunctional composites, **e** microcontrollers

such as strength, toughness, resistance to impact, etc. This balance is precisely what we aim for when we specify material and structural requirements in order to attain a design that simultaneously satisfies economic viability and mission-oriented performance. Besides, a particularly attractive feature of biological systems is their unique ability to diagnose localized damage (through a continuously distributed sensor network) and to initiate a self-repair process. Such an attribute would be a most desirable function in an adaptronic structural system.

Although present day researchers are concentrating on adaptronic structures that may seem rudimentary when compared with mammalian systems, their efforts lay the foundation for the future engineered systems. Controlling the movement of an arm is a wonderful example of the seemingly effortless task that biological creatures perform each day, but which has been quite difficult for engineers to mimic. Consider a situation in which you are sitting at a table that has one leg shorter than the others, and you wish to draw a straight line on a piece of paper that is resting on this table. Before you begin, you recognize that the table is unstable and that it will be difficult for you to draw such a line; in fact, you may have tried this task before and feel uncertain about the dynamics of the table. When you begin to draw the straight line, you will contract certain muscles to force movement of the pencil upon the paper. To draw a straight line, you normally need to contract no more than one muscle of an antagonist muscle group at a time; however, you will contract both your biceps and triceps simultaneously in an effort to better control the pencil. The biceps and triceps are antagonist muscles, meaning that they work against each other, resulting in a 'stiff' elbow joint. Activating both the biceps and the triceps is energy intensive; you are consuming a large of amount of energy to do no mechanical work (there is no work done if there is no displacement). However, stiffening the elbow joint creates a more stable control system, i. e., minimizes the influence of an unknown disturbance (the rocking motion of the table) on the output (drawing a straight line). Upon succeeding in drawing a straight line, you are asked to draw a straight line several more times on the same unsteady table. As you draw each line, you begin to formulate a sense of the dynamics of the table – and better understand the environment in which you are working – and as this occurs, you begin to conserve energy by not co-contracting the biceps and triceps to the same degree as in previous attempts. When the environment has been sufficiently sampled and you learn the dynamics of the table, your body will try to conserve as much energy as possible and tend towards no co-contraction of muscles. If, however, someone wanders in the room and creates a disturbance in your task, e. g., bumps your arm or the table, then you will once again co-contract your muscles to again increase the accuracy.

The *classical engineering approach* to this same task would be to formulate mathematical models for the table dynamics, the mechanism that draws the line, the interaction between the table surface and the paper and the paper and the pen, and any other aspect of the problem that would seem important to an engineer. Using these models, a deterministic plan or control algorithm would be developed to control the movement of the pen upon the paper while calculating what is expected to happen to the unstable surface when the pen creates a force at various locations. The engineer would then measure the response of the table and the straightness of the line. Once implemented, this algorithm would perform the same function at each and every time – it never gets any better, and it never gets any worse. It uses the same amount of energy at each and every time. In all likelihood, the mechanisms to be used would be conceptually different from those used in the human arm. Most robots that mimic arm motion use a rotary motor at the joint and do not have co-contraction capabilities. This basic difference in algorithm and architecture highlights one of the fundamental deficiencies of todays robot systems as compared to biological systems. When a robot arm in a manufacturing plant or the arm of the space shuttle moves quickly, the robot arm vibrates because of the sudden deceleration. The human arm can generally out-perform a robot in this type of combined slewing (moving from one position to another) and vibration control. The arm will use only the muscles needed to quickly perform the slewing motion, and then use co-contraction to stiffen the structure and reduce any vibration that might be caused by decelerating the mass of the arm and the payload that it may be carrying.

The *adaptronic approach* would be one that would borrow directly from the biological world. Materials that behave more or less like muscles can be used in adaptronic structures and are called *induced strain actuators*. When energy is applied to the actuators, they attempt to expand/contract and work against any load that is applied to them. The actuators are typically bonded to the surface of a structure, or embedded within the material. This means that the artificial muscles must now work against the inherent structural impedance of the component, just as human muscles are parallel to the skeletal structure or bone. However, whereas the arm has discrete joints about which rotation occurs, the adaptronic structure may be a continuum, thereby necessitating a distributed actuation system. For example, the tip motion of a beam will not occur by rotating the beam about a joint but by inducing its deformation by means of induced strain actuators placed on the beam.

A basic premise of adaptronic structures is the intelligent use of energy transduction principles. In a conventional design, a structure would be calculated to resist the worst-case scenario. This usually results in gross over design. A ladder designed for the worst-case scenario would be, 99% of the time, *too strong* and *too heavy* for what is being used for. However, an adaptronic ladder would be designed much lighter, and, through the energy transduction, would be able to modify its behavior to cover its utility envelope. For example, an adaptronic ladder that is overloaded could use electrical energy to stiffen or strengthen itself while alerting the user that the normal loading capacity is being exceeded. The overload response should also be based upon the actual 'life experience' of the ladder to account for aging or a damaged rung; therefore, the ladder would determine its current state of health and use this information in assessing when it has been overloaded. At some point in time, the ladder will graciously announce its retirement, as it can no longer perform even minimal tasks.

2.2 Construction of Adaptronic Structures

Adaptronic structures are complex systems displaying motion, sensing, and artificial intelligence functions synergistically to duplicate life-like functions. In line with the bio-inspired approach, we will consider in turn the actuators (artificial muscles), the sensors (artificial senses) and the microcontroller-artificial intelligence network (artificial nerves, brain, and mind).

2.2.1 Artificial Muscles: Actuators

Materials that allow an intelligent or smart structure to adapt to its environment are known as actuators. These materials have the ability to change the shape, stiffness, position, natural frequency, damping, friction, fluid flow rate, and other mechanical characteristics of adaptronic structures in response to changes in temperature, electric field, or magnetic field. The most common actuator materials are shape memory alloys, piezoelectric materials, magnetostrictive materials, electrorheological fluids, and magnetorheological fluids [2]. Actuators with these materials will be described in detail in Sects. 6.2 to 6.6; therefore you will find only a brief overview below.

Shape memory alloys (SMA) undergo solid-to-solid martensitic phase transformations, which allow them to exhibit large, recoverable strains [3]. Nickel-titanium, also known as nitinol (Ni for nickel, Ti for titanium, and nol for Naval Ordnance Lab), are high-performance shape memory alloy actuator materials exhibiting strains of up to 8% by heating the SMA above its phase transformation temperature – a temperature which can be altered by changing the composition of the alloy.

Nitinol wires embedded in composite materials yield adaptive composite structures with muscle similarities. They have been shown to display large bending deformation when activated (Fig. 2.2). In addition to applying forces or changing the shape of the structure, the Nitinol wires can be used to change the modal characteristics of the composite by changing the stiffness or state of stress in the structure. Photoelastic damage control experiments have shown that embedded Nitinol actuators can also be used to reduce stress concentrations in notched tensile coupons by creating localized compressive stresses.

Piezoelectric materials can enact deformation and mechanical forces in response to an applied voltage. Rather than undergoing a phase transformation, piezoelectric materials change shape when their electrical dipoles spontaneously align in electric fields, causing deformation of the crystal structure.

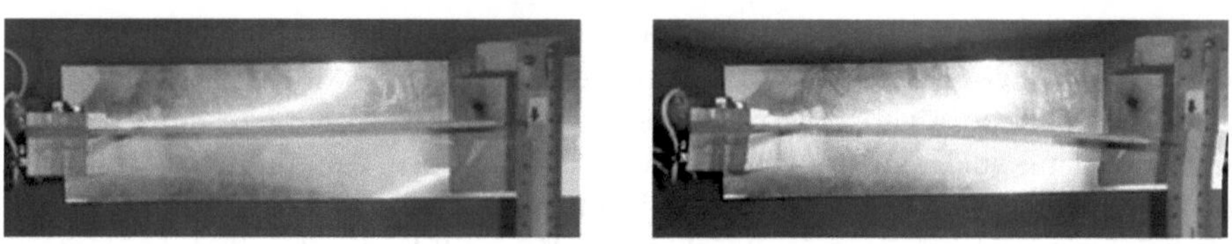

a b

Fig. 2.2. Polymeric composite with embedded Nitinol wires displaying large bending deformation when activated: **a** beam configuration before activation, **b** deflected beam after SMA activation

Maximum strains of over 10^{-3} are now possible at kHz frequencies. When these small deformations are constrained, large mechanical forces, energy and power densities are generated. Examples of systems using piezoelectric actuators are: optical tracking devices, magnetic heads, adaptive optical systems, micropositioners for robots, ink jet printers, and speakers. Recent research has focused on using piezoelectric actuators with sophisticated control systems in adaptronic structures to perform active acoustic attenuation, active structural damping, and active damage control.

In contrast with linear piezoelectricity, the electrostrictive response is quadratic in electric field. Hence, the direction of the electrostriction does not switch as the polarity of the electric field is switched.

Magnetostrictive actuator materials are similar to piezoelectric materials, but respond to magnetic, rather than electric, fields. When placed in a magnetic field, the magnetic domains in a magnetostrictor rotate until they are aligned with the field, resulting in expansion of the material. Magnetostrictive material response is basically quadratic in magnetic field, i. e., the magnetostrictive response does not change sign when the magnetic field is reversed. However, the nonlinear magnetostrictive behavior can be linearized about an operating point through the application of a bias magnetic field. In this case, piezomagnetic behavior, in which response reversal accompanies field reversal, can be obtained.

Active fluids can also act as actuators in adaptronic structures. Electrorheological (ER) and magnetorheological (MR) fluids experience reversible changes in rheological properties (viscosity, plasticity, and elasticity) when subjected to electric and magnetic fields, respectively. These fluids contain micron-sized particles which form chains when placed in an electric or magnetic field, resulting in increases in apparent viscosity of up to several orders of magnitude. These fluids can be used to make simple hydraulic valves which contain no moving parts. Other applications include tunable dampers, vibration isolation systems, clutches, brakes, other frictional devices, and robot arms.

2.2.2 Artificial Nerves: Sensors

One of the critical functions instilled in adaptronic structures is that of sensing. Vibration detection and dampening, acoustic attenuation, intelligent processing, damage detection and control are just a few examples. Sensing capabilities can be given to structures by externally attaching sensors or by incorporating such sensors within the structure during manufacturing. Some of the sensing materials used for this purpose include optical fibers, piezoelectric materials, 'tagging' particles, etc. You will find a detailed description of the corresponding sensors in Sects. 7.2 and 7.3, thus there is only a brief overview here.

Piezoelectric materials have found widespread use as sensors in adaptronic structures [4] (see Sect. 7.3). Piezoelectric ceramics and polymers produce measurable electrical charges and voltages in response to mechanical stress. Because of the brittle nature of ceramics, piezoelectric polymers [5], such as polyvinylidene fluoride (PVDF), are more often used for sensing of flexible structures. PVDF can be formed in thin films and bonded to many surfaces. Uniaxial films, which are electrically poled in one direction, can measure stresses along one axis, while biaxial films can measure stresses in a plane. The sensitivity of PVDF films to pressure changes has been utilized in tactile sensors that can read the Braille alphabet and distinguish different grades of sandpaper. Tactile sensors with ultra-thin (200 ... 300 μm) PVDF films have been proposed for use in robotics. A skin-like sensor that replicates the temperature and pressure sensing capabilities of human skin can be used in different modes to detect edges, corners, and geometric features or to distinguish between different grades of fabric. The pyroelectric effect, which allows piezoelectric polymers to sense temperature, also limits their use to lower temperature ranges. Piezoelectric composite materials have been developed to overcome the brittleness of piezoelectric ceramics and the temperature limitations of piezoelectric polymers. Flexible composite sensors containing piezoelectric ceramic rods in a polymer-based matrix [6] have been widely used in hydrophones and medical ultrasonic transducers with improved sensitivity and mechanical performance over the original piezoelectric ceramics. Polymers containing piezoelectric powders have also been investigated for use as sensing materials. Piezoelectric paint and coatings are being developed that can be applied to complex shapes to provide information about the state of stress and health of the underlying structure.

Sensing with optical fibers can be done either extrinsically or intrinsically [7]. When used extrinsically, the optical fiber does not act as a sensor; it merely transmits light. An example of an extrinsic fiber optic sensor is a position sensor which uses the fiber to collect light from a source. Breaks in the light beam are used to accurately determine the position of a work piece in robotics applications. Security systems also use this technique to detect intruders. Displacement sensing can be achieved using the Sagnac, Mach-Zehnder, and Fabry-Perot interferometer sensors (see Sect. 7.2). Intrinsic sensing relies on changes in the light transmission characteristics of the optical fiber. The use of optical fibers to perform intrinsic sensing in smart structures has known an accelerated development in recent years in line with similar developments in the use of optical fiber for data transmission and communications. Fiber Bragg grating sensors are among the most common intrinsic optical fiber sensors. Strain sensors, temperature sensors, liquid level sensors, pressure sensors, humidity sensors, have been demonstrated. Fiber optic smart structures for aerospace, automotive, and civil infrastructure monitoring have been developed. Recent advances in fiber optic sensing include optical frequency domain reflectometry for high density multiplexing of multi-axis fiber Bragg grating sensors. These sensors allow the reading of strains at many locations with a single fiber connection [8].

However, fiber optics sensors cannot perform an active interrogation of the structure; they can only passively record various structural parameters such as loads, strains, environment, vibrations, acoustic emission from cracks, and the like.

2.2.3 Intelligence: Signal Processing, Communication, and Controls

Tremendous efforts have been invested in developing theories, simulations, and hardware implementations for machinery control. Modern control approaches include adaptive control, neural networks and probabilistic control, to name only a few. However, the intelligence features that the adaptronic materials community is trying to create have constraints that the engineering world has never experienced before, but that the biological world seems to accept with simplicity and grace. Namely, the tremendous number of sensors, actuators, and their associated power sources compels us to supersede the conventional central processor architecture whereby every piece of sensor and actuator information must be stored and manipulated electronically.

Norbert Wiener defined *cybernetics* as the science of communication and control in animals and machines. Nature has used natural selection to develop alternative architectural solutions that compensate for its quite restrictive and far-from-robust material selection; likewise, natural selection has evolved towards more and more elaborate cybernetic architectures to facilitate signal processing, complex communication, and advanced memory via biological constructs. The electro-bio-chemical devices that we refer to as neurons are not nearly as fast as our silicon devices; however, nature has developed a wonderful way of processing information that allows rather complex tasks to be performed with amazing speed. The key appears to be a hierarchical architecture in which signal processing and the resulting action can take place at levels below and far removed from the central processor, the brain. Removing your hand from a hot stove to prevent getting burned (damage to the system) need only be processed locally, i. e., in the spinal cord; whereas the less automatic behaviors are organized by successively higher centers within the brain. The information that you have touched a hot surface reaches the brain much later than the reflex action of contracting muscles in the arm and fingers to get away from it. This hierarchical approach not only yields control systems that are time-efficient, but yields systems that are fault-tolerant as well. Reliability is a critical factor in reducing energy costs. A failed system is a tremendous waste of resources and energy; in a biological system, the control subsystem is as important, if not more important, than the structural components in assuring a biological system that has a longer lifespan than any one of its components.

These very important concepts are of paramount importance in the design of adaptronic structures. It is essential to have a hierarchical and distributed control architecture, in which many functions can be delegated to the lower control levels, while the central processor would retain the general systemic and strategic functions. In such a concept, decisions that affect only a local substructure (such as the reflex reaction to a local stimulus or change in operating conditions) will be taken by the local controllers. Whereas, actions that require collaborative contributions from all the structural components, such as a configurational change in response to mission change, will be coordinated from a central location. Recent advances in embedded microcontrollers, digital signal processors (DSP), and field-programmable gate arrays (FPGA) make such a distributed architecture quite possible. To make such a system robust and autonomous, the issue of power supply independence should be addressed. Embedded power harvesting systems, having the capability of recharging their energy supply by scavenging environmental energy sources, have received increased attention in recent years and are likely to be essential building blocks in adaptronic structures.

2.2.4 Adaptive Algorithms for Smart Structures Control

The control systems to be used in adaptronic structure will be able to learn, then change based upon need; they will also be able to anticipate a need, and to correct a mistake. The architecture of control systems will remain an important element in the future manifestations of adaptronic structures, for it is the computational hardware and the processing algorithms that will determine how complex our systems can become – how many sensors we can utilize – and how many actuators we can use to effect change. Will all control systems be neural networks and modeled after biological systems? No. The same paradigm we use to design the material systems or structures is used to design the control system – the design that will reduce the mass and energy needs of the system to enable it to perform its adaptive functions.

Implementation of control algorithms in smart structures architecture is subject to attentive scrutiny. Conventional application of classical control algorithms is only the first step in this process. Much better results are obtained if modern adaptive control is used, such that the resulting smart structure can react to changes in the problem-definition parameters. Actual structural designs are very complex, nonlinear in behavior, and subject to load spectra that may be substantially modified during the structures service life. Under such adverse situations, the resulting uncertainty in the controlled plant dynamics is sufficient to make 'high-performance goals unreachable and closed-loop instability a likely result' [9]. To address this problem, at least three adaptive control approaches are advocated:

(a) adaptive signal processing methods; (b) model reference adaptive control (MRAC); and (c) self tuning regulators (STR). Though different in detail, all three aim at same goal, i. e., to eliminate the effect of variations in disturbance signature and plant dynamics on the smart structures performance. The topic of controllers in adaptronics, will be covered in detail in Chap. 4.

2.3 Application Examples

2.3.1 Solid State Actuation and Morphing Structures

Solid-state actuation signifies the use of the induced-strain effect present in active materials to achieve actuation without any moving parts, i. e., in a solid-state manner. Already, solid-state actuation has found niche application in the aerospace industry. The aero-servo-elastic control of vibrations and flutter with solid-state actuated flaps, tabs, vanes, etc. for helicopter rotor blades and aircraft wings is currently being experimented on.

The design with induced-strain actuators must take into consideration their specific characteristics. Induced-strain actuators can develop large forces but only a small finite stroke that must be judiciously used to achieve the design goals. By displacement-amplification (see Section 6.2), a tradeoff between stroke and force is obtained. Mitigation of the input/output requirements and induced-strain actuation capabilities is done during the design cycle. The mitigation of the input/output requirements and induced-strain actuation capabilities during the design cycle is presented schematically in Fig. 2.3.

Induced-strain Actuation for Aeroelastic and Vibration Control

Aeroelastic and vibration control technology allows flight vehicles to operate beyond the traditional flutter boundaries, improves ride qualities, and minimizes vibration fatigue damage. Conventional active flutter and vibration control technology relies on the use of aerodynamic control surfaces operated by servo-hydraulic actuators. In this conventional configuration, the

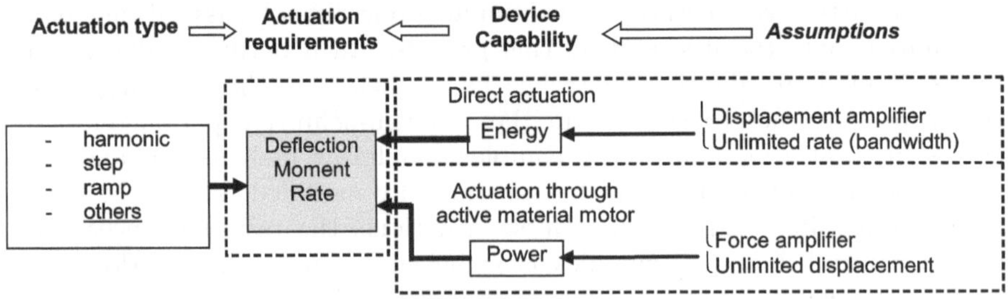

Fig. 2.3. Mitigation of the input/output requirements and induced-strain actuation capabilities

flutter and vibration suppression algorithms are implemented through the servo-valve/hydraulic actuator. Though widely used, conventional technologies for active control of flutter and vibrations have many limitations, such as: (a) multiple energy conversions (mechanical, hydraulic, electrical); (b) large numbers of parts, i. e., potential failure sites; (c) high vulnerability of the hydraulic pipes network. In contrast, active-materials technologies offer direct conversion of electrical energy to high-frequency linear motion. The application of active-materials to adaptive structural control, vibration suppression, and flutter prevention opens new and exciting technological opportunities. Helicopter applications of induced-strain actuation have received extensive attention since conventional actuation solutions (hydraulics and electric motors) are very difficult to implement for on-blade actuation. Induced-strain appears as a viable alternative. Two ways of rotor-blade induced-strain actuation have been investigated: (a) discrete actuation of a servo-aerodynamic control surface (flap, tab, blade-tip, etc.) to generate localized aerodynamic forces; and (b) distributed induced-strain actuation resulting in a continuous twisting of the blade. The former concept is easier to implement on existing structures, and hence it is amenable to structural retrofitting. However, by still dealing with discrete actuation surfaces, it is only an evolutionary rather than revolutionary change to the present state of the art. The latter concept is more revolutionary, since it removes structural discontinuities and results in better and more efficient aerodynamics.

Induced-strain Actuation of Helicopter Blades. A sustained program for full-scale implementation of smart materials actuation is under way at Boeing (Mesa). The program is called smart material actuated rotor technology (SMART). The development effort included design, fabrication, and component testing of rotor blades, trailing edge flaps, piezoelectric actuators, switching power amplifiers, and the data/power system [10]. Simulations and model scale wind tunnel tests have shown that this system can provide 80% vibration reduction, 10 dB noise reduction for a helicopter passing overhead, and substantial aerodynamic performance gains. Whirl tower testing of a 10.4 m diameter rotor demonstrated the functionality, robustness, and required authority of the active flap system. The actuator demonstrated excellent performance during bench testing and has accumulated over 60 million cycles under a spectrum of loading conditions. The flaps showed excellent authority with oscillatory thrust greater than 10% of the steady baseline thrust. Various flap actuation frequency sweeps were run to investigate the dynamics of the rotor and the flap system. Limited closed loop tests used hub accelerations and hub loads for feedback. Proving the integration, robust operation, and authority of the flap system were the key objectives met by the whirl tower test. This success depended on tailoring the piezoelectric materials and actuator to the application and meeting actuator/blade integration requirements (Fig. 2.4).

Induced-strain Actuation of Fixed-Wing Aircraft. The feasibility of using active piezoelectric control to alleviate vertical tail buffeting was inves-

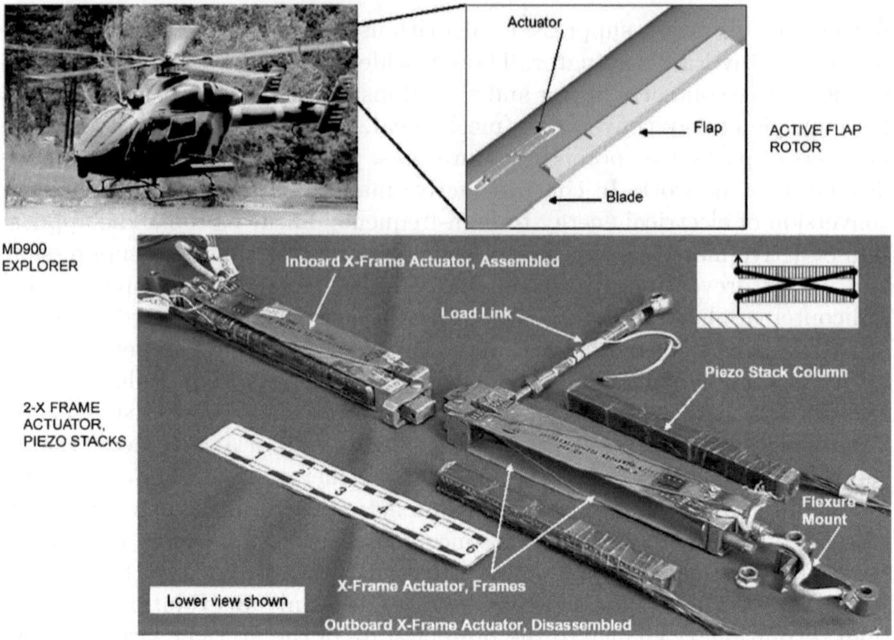

Fig. 2.4. MD 900 helicopter hingeless blade displaying the planned trim tab for in-flight tracking and active control flap for noise and vibration reduction [10]

tigated under the actively controlled response of buffet affected tails (ACROBAT) program [11]. Tail buffeting is a significant concern from fatigue and maintenance standpoints. During the ACROBAT program, active materials solutions to buffet problems were studied on 1/6-scale rigid full-span model of the F/A-18 aircraft tested in the Langley transonic dynamics tunnel (TDT). The piezoelectric wafer actuators were placed in opposing pairs on both surfaces of the vertical tails. The port vertical tail was equipped with surface-bonded piezoelectric wafer actuators, while the starboard vertical tail had an active rudder and other aerodynamic devices. Buffeting alleviation control laws aimed at reducing the fin tip acceleration were imposed (Fig. 2.5a). The tunnel was run at atmospheric pressure and 4.5 m/sec airspeed. The F/A-18 model was tested at up to 37° angles of attack. Constant-gain active control of the piezoelectric wafer actuators resulted in reduction of the root bending moment (Fig. 2.5b). The power spectral density of the root strains at the vertical-tail first bending resonance was reduced by as much as 60%, while the corresponding root mean square (rms) values were reduced by up to 19%. In achieving these results, both active rudder and piezoelectric actuators seem to be similarly effective.

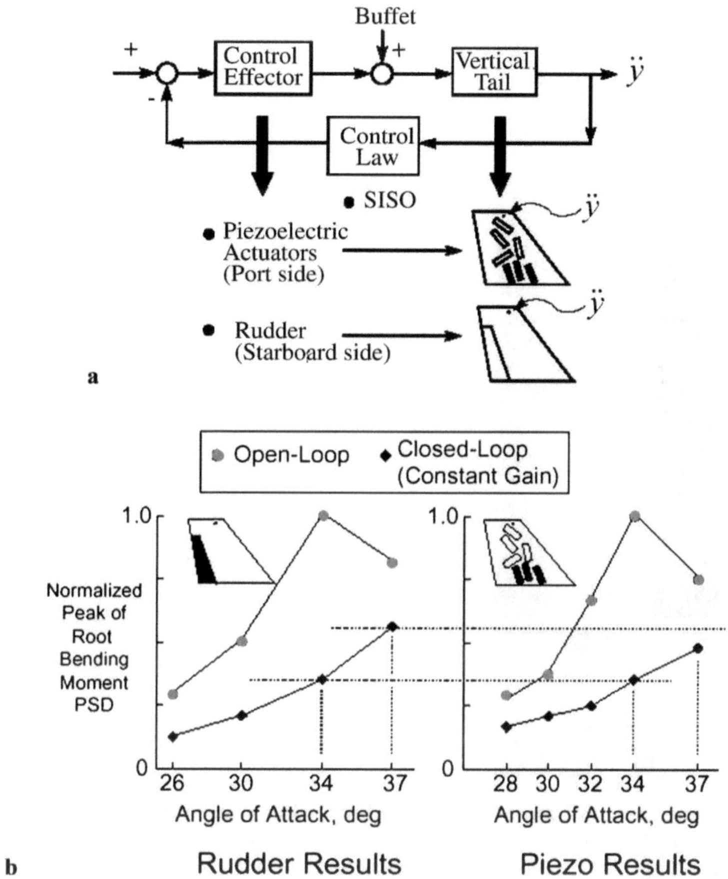

Fig. 2.5. ACROBAT tail buffet alleviation experiments: **a** single-input single-output (SISO) control law design for active rudder and piezoelectric wafers excitation, **b** power spectrum density (PSD) peak values for the root bending moment at the first bending resonance [11]

Morphing Structures

A recent example of an actuation-intensive adaptronic structure is the morphing aircraft program. Morphing aircraft refers to the use of large shape changes to effect planform change and/or for flight control [12]. Early examples are the Wright Flyer, which used wing twist for flight control, and the F-14, which changes its wing sweep to capitalize on two distinct flight regimes. Unlike past efforts, current efforts in morphing aircraft focus on multiple, large planform changes in sweep, wing extension, wing folding, etc. and in camber, twist, and asymmetric planform changes for flight control motivated by predator birds such as a hawk [13]. This bio-inspired direction for morphing aircraft structures has lead to numerous research projects span-

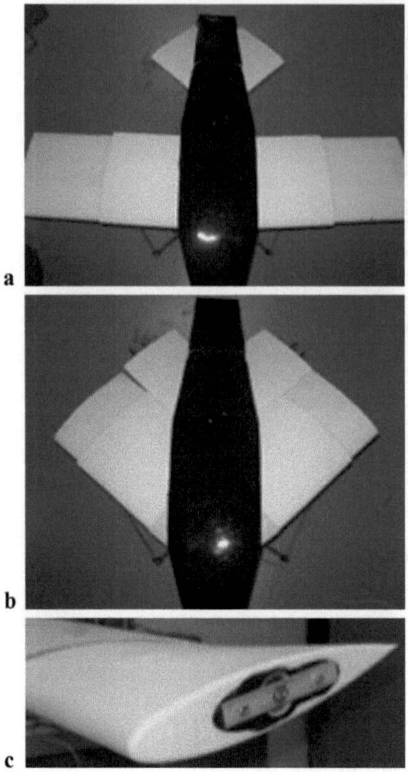

Fig. 2.6. 'Morphious', the Virginia Tech morphing wing wind tunnel simulator: **a** cruise configuration, **b** attack configuration, **c** wing twist (Photos courtesy of the designer David A. Neal, III)

ning flight dynamics, aerodynamics, structural mechanics, and control. The most common motivating example is the desire to have an unmanned aircraft that can morph from a long aspect ratio, straight winged plane for efficient loitering flight into a highly maneuverable short, swept wing aircraft that is effective in attack (Fig. 2.6). The second common example is the design of high altitude long endurance (HALE) aircraft that can take off and land on their own. Extremely long, highly flexible wingspans are required for long endurance and such wings tend to hit the ground during take off and landing. A morphing solution would be to fold or otherwise morph such wings into shapes more favorable for take off and landing.

2.3.2 Structural Health Monitoring and Self-Repairing Structures

Structural health monitoring (*SHM*), *condition-based maintenance* (*CBM*) and *birth-to-retirement* refer to the capability of using sensors throughout the life or an adaptronic structure to monitor its state of health and act ac-

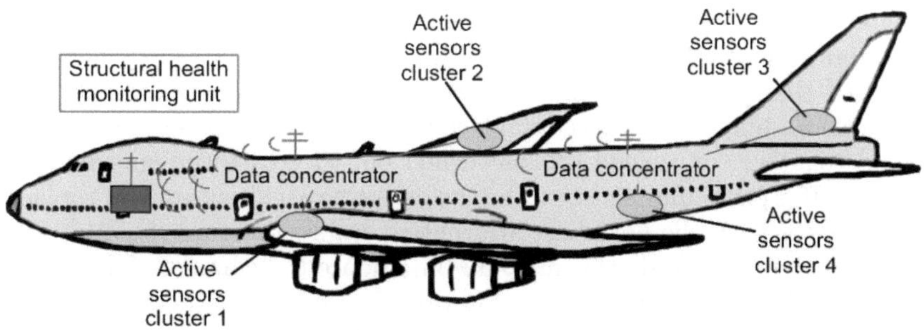

Fig. 2.7. Concept of an aging aircraft instrumented with active sensors for structural health monitoring

cordingly. The sensors would record the way the manufacturing process was implemented, and would remember the *pristine state* of the structure. At the same time, the sensory output will be used to optimize the fabrication process and ensure quality consistency. The network of sensors embedded in the adaptronic structure will be then used to monitor the structural behavior throughout its life (Fig. 2.7). A structural health bulletin will be produced on demand and life history of the structure will be gathered in the database. If needed, active measures will be taken to control and reverse the evolution of structural damage or modify the structures behavior or performance to elude damage. These sensors will monitor the structural aging process and will determine when the artifact should be repaired or even graciously retired. Thus, scheduled maintenance will be replaced by *need-based maintenance*, with associate savings in the life-cycle costs and increase in the structural safety and equipment availability. Piezoelectric materials offer the capability of performing active structural health monitoring, i. e., actively interrogating the structure with ultrasonic waves to detect damage such as cracks, de-bonding, delaminations, etc. [14]. Recently, various nondestructive evaluation (NDE) methods have been successfully demonstrated with permanently attached piezoelectric wafer active sensors (PWAS) [15]. It is predictable that in the not so distant future, adaptronic structures will be permanently equipped with an embedded NDE system that will allow on-demand structural interrogation to assess the state of structural damage, perform a structural diagnostic, issue a structural health bulletin, and even perform a prognosis of the future structural performance and remaining structural life.

Will adaptronic structures eliminate all catastrophic failures? No. Not any more than trees will stop falling in hurricane winds or birds will no longer tumble when they hit glass windows. But adaptronic structures will enable man-made inanimate objects to become more natural and life-like. The future of adaptronic structures lies in developing a system with the ability to interface and interact with the network of sensors, actuators, and controls. This interaction will allow the user/designer/builder to design a system to perform the function desired with the generic enabling system within the host material. An example can be postulated by focusing on one aspect of the material system, the sensor system. In this scenario, a sensor network is built into the system with many more sensors than are needed by any one application, but by means of adaptive architecture, these sensors can be connected together, or turned off, or turned on, to create the specific system desired. If a particular sensor fails, the adaptive architecture will replace the failed sensor with the next best alternative and reconfigure the interconnections and the control algorithm to accommodate this change. The sensor network, therefore, could look like the detail of a silicon microchip in which numerous sensors are spread about a polymeric sheet that can be used as the structural ply of a composite laminate. The sensor sheet can be produced by photolithography techniques, which are much like making a Xerox copy, for fractions of a cent per sensor and can be mass-produced. Similar 'pictures' can be painted for the other components of the system. It seems likely that a system with large arrays of sensors and actuators within a host will require three-dimensional interconnections between the power modulation devices, the control processors, and the sensors and actuators; technology that has been developed and refined, once again, by the silicon community.

Self-repair and *self-healing* is another bio inspired capability highly desirable in adaptronic structures. Once damage has been identified by the structural health monitoring system, a mechanism could be triggered to initiate a self-repair process that will restore, at least partially, the initial structural performance. This mechanism can be either an external action triggered by the SHM system, or an automatic response initiated by the adaptive material itself. An example of the latter is the self-healing composites that have been recently studied for various applications. Inspired by biological systems in which damage triggers an autonomic healing response, such polymer composite materials can 'heal' themselves when cracks develop.

The self-healing material developed at the University of Illinois, Urbana-Champaign, USA [16] considers epoxy matrix composites incorporating microcapsules of a 'healing agent' that is released upon crack intrusion. Polymerization of the healing agent is triggered by contact with an embedded catalyst. The addition of healing microcapsules can significantly toughen the neat epoxy and implicitly the composite, as long as the cracks are matrix related (such as delaminations and disbonds). Figure 2.8a presents the natural self-healing process taking place in animal bone: the internal bleeding is accompanied by the formation of a fibrin clot and then by an unorganized fiber mesh. Calcification converts the resulting fibro cartilage into fibrous bone and, eventually lamellar bone. The corresponding process developed in thermosetting composites is illustrated in Fig. 2.8b: when the crack propagating through the polymer encounters a microcapsule containing the healing agent, a self-repair process is initiated. The healing agent inside the capsule spreads out to fill the crack and becomes polymerized in contact with the catalyst agents dispersed throughout the polymeric matrix. Subsequently, the healing

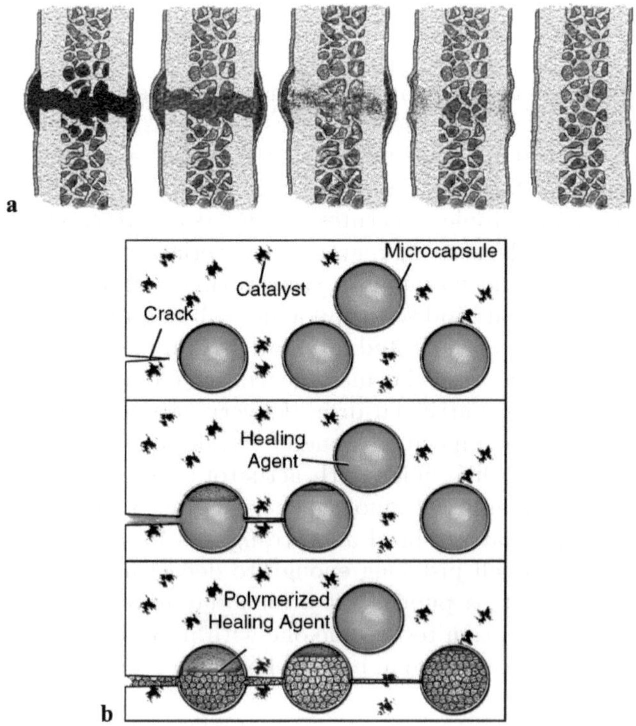

Fig. 2.8. Self healing concepts: **a** biological self-healing in animal bones: Internal bleeding, forming of fibrin clot, development of fibro cartilage and its calcification, conversion to fibrous bone and eventually lamellar bone; **b** self-healing in a thermosetting polymer [16]

is completed and the crack growth is arrested. Once healed, the self-repairing polymer has been shown to recover as much as 90% of its virgin fracture toughness [16]. Similar research is being currently conducted in Europe and Japan. As an alternative to microcapsules, researchers at the University of Bristol in the UK have studied the use of hollow fibers containing the healing resin and the catalyst.

The health monitoring and the self-repair capabilities are essential attributes of adaptronic structures and their importance cannot be over emphasized. Such capabilities are essential for maintaining our aging infrastructure and historical constructions that could be enhanced with health monitoring capabilities and external self-repair or strengthening mechanisms during upgrade/retrofitting. In addition, structural health monitoring attributes and self-repairing capabilities could be design ab initio into new structures and engineered materials, thus bringing them even closer to the adaptronic ideal.

2.4 Future Adaptronic Structures

The adaptronic structures revolution to date has focused upon learning how to use energy as a structural component, how to make structures behave like natures systems, how to make structures that are 'soft', and how to better utilize the materials around us. New compositions of matter will begin influencing the manifestations of adaptronic structures. Scientists and researchers who are developing new materials, sensory materials, materials with actuator capabilities, energy storage and modulation devices that will allow the integrated system to be autonomous and self-supporting will add fuel to this movement. Adaptronic structures are first and foremost hybrid material systems. The sensors, actuators, and artificial intelligence are reduced to the microstructure, be it nano level for artificial drug delivery systems, micron level for advanced fiber reinforced composites, or meter level for civil engineering constructions. Some may look like fluids with actuators that cannot be seen by the naked eye, but can manipulate molecules with grace and agility; others may look like materials that are hard and strong and in a moment, upon demand, can behave like a jell just long enough to deflect and absorb energy as a karate expert reacts to a punch. Yet others may have the mass of small mountains, but the perception to become one with nature to ensure the safety of the delicate and intricate human beings they have been designed to protect.

Nastic structures are a new type of bio inspired adaptive structures based on the principles of plant nastic motions that have recently started to be studied [16]. Biological nastic motion is what causes plants to angle their stems so that their leaves face light sources and flower pedals to open. Plant motor

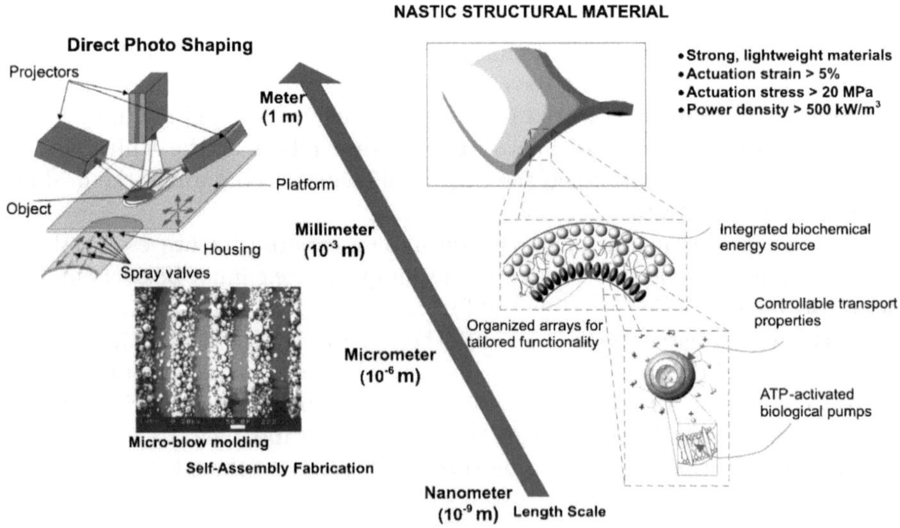

Fig. 2.9. The concept of nastic structures [17]

cells can be considered the muscles of biological systems, and the process of nastic motion the driving force. When biochemical reactions cause water to flow into or out of the plant motor cells, cellular volume change and overall tissue deformation is achieved. When the plant tissue undergoes non-uniform elongation from increased osmotic pressure or shrinkage from a decrease in pressure, the tissue will have bending deflection. Nastic structures will be capable of achieving controllable deformation and shape change through internal microactuation that functions on principles found in the biological process of nastic motion (Fig. 2.9). Nastic structures utilize localized changes in hydraulic pressure to control shape change in the material. In the current Nastic Structures program, localized pressure change is controlled by varying the concentration gradient across lipid bilayers that incorporate ion pumps. Ion pumps are used to control the transport of charge and fluid across the lipid bilayer for the purpose of controlling hydraulic pressure in a closed cavity.

Adaptronic structures may start to affect our lives even in the near future as they are being introduced commercially; but the most lasting impact will be that the philosophy of engineering design will begin to change. Engineers of the future will not have to add mass and cost to a structure to assure safety in structures that are used outside their initially intended envelope. Engineers will not have to learn from structural failures, but will be able to learn from the life experiences of the structure. Not only will adaptronic structures be of great utility to the consumer, they will have an even more profound influence on science and engineering. They will allow the silent systems we create to inform us, to enlighten us, to educate us of the physics, science, and interaction of the environment on our designs.

References

1. Giurgiutiu, V.; Lyshevski, S.E.: *Micromechatronics: Modeling, Analysis, and Design with MATLAB.* CRC Press, 856 pages, ISBN 084931593X (2004)
2. Giurgiutiu, V.: *Actuators and Smart Structures.* In: Encyclopedia of Vibrations, S.G. Braun (Editor-in-Chief), ISBN 0-12-227085-1, Academic (2001), pp. 58–81
3. Bank, R.: *Shape Memory Effects in Alloys.* p. 537. Plenum, New York (1975)
4. Chang, F.-K.: *Built-In Damage Diagnostics for Composite Structures.* Proc. 10th Int. Conf. on Composite Structures (ICCM-10), Vol. 5, Whistler, B.C., Canada, August 14–18 (1995), pp. 283–289
5. Lovinger, A.J.: *Ferroelectric Polymers.* Science 220 (1983), pp. 1115–1121
6. Smith, J.: *The Role of Piezocomposites in Ultrasonic Transducers.* Proc. IEEE Ultrasonics Symp. (1989), pp. 755–766
7. Udd, E. (Ed.): *Fiber Optic Smart Structures.* Wiley, New York (1995)
8. Kreger, S.; Calvert, S.; Udd, E.: *Optical Frequency Domain Reflectometry for High Density Multiplexing of Multi-Axis Fiber Bragg Gratings.* Proc. OFS-16, Nara, Japan (2003), p. 526
9. Clark, R.L.; Saunders, W.R.; Gibbs, G.: *Adaptive Structures – Dynamics and Control.* Wiley (1998)

10. Straub, F.K.; Kennedy, D.K.; Stemple, A.D.; Anand, V.R.; Birchette, T.S.: *Development and Whirl Tower Test of the SMART Active Flap Rotor.* Smart Structures and Materials 2004: Industrial and Commercial Applications of Smart Structures Technologies, Eric H. Anderson (Ed.), Proc. SPIE Vol. 5388 (2004), pp. 202–212
11. Moses, R.W.: *Vertical Tail Buffeting Alleviation Using Piezoelectric Actuators – Some Results of the Actively Controlled Response Of Buffet-Affected Tails (ACROBAT) Program.* SPIE Symp. on Smart Structures and Materials, Industrial and Commercial Applications of Smart Structures Technologies, SPIE Vol. 3044, San Diego, California, March 4–6 (1997), pp. 87–98
12. Bowman, J.; Sanders, B.; Weisshaar, T.: *Evaluating the Impact of Morphing Technologies on Aircraft Performance.* AIAA Paper 2002–1631, April (2002)
13. Bae, J.S.; Siegler, T.M.; Inman, D.J.: *Aerodynamic and Static Aeroelastic Characteristics of a Variable-Span Morphing Wing.* AIAA J. Aircraft, Vol. 42, No. 2. (2005), pp. 528–534
14. Giurgiutiu, V.; Cuc, A.: *Embedded Nondestructive Evaluation for Structural Health Monitoring, Damage Detection, and Failure Prevention.* Shock and Vibration Digest, Sage Pub., Vol. 37, No. 2, March (2005), pp. 83–105
15. Giurgiutiu, V.: *Embedded Ultrasonics NDE with Piezoelectric Wafer Active Sensors.* Journal Instrumentation, Mesure, Metrologie, Lavoisier Pub., Paris, France, RS series 12M, Vol. 3, No. 3–4 (2003), pp. 149–180
16. Brown, E.N.; Sottos, N.R.; White, S.R.: *Fracture Testing of Self-Healing Polymer Composites.* Experimental Mechanics, Vol. 42, No. 4 (2002), pp. 372–379
17. Leo, D.; Sundaresan, V.B.; Tan, H.; Cuppoletti, J.: *Investigation on High Energy Density Materials Utilizing Biological Transport Mechanisms.* ASME-IMECE2005-60714 (2004)

3 Multifunctional Materials: The Basis for Adaptronics

W. Cao

Two of the three components in adaptronic structures, i. e., sensors and actuators, are made of single phase or composite functional materials. In order to design better adaptronic structures, it is necessary to know a little more about these functional materials and to understand their functional origin, which will allow us to use them more efficiently and help us design and fabricate new and better functional materials for adaptronic structures.

3.1 What are Functional Materials?

Functional materials are materials that can perform certain functions when triggered by environmental changes, such as stress, electric field, magnetic field, and temperature variations, or when stimulated by control signals, such as electric or magnetic signals from a control center. The difference between a device and a functional material is that a functional material will preserve the same functional property when its volume is subdivided, while a device is usually made of many different components and will fail to function when the components are disintegrated. Functional materials may be categorized into two groups: passive and active functional materials.

The signature of passive functional materials is the appearance of anomalies, such as maxima, minima, or singularities, in at least one of their physical quantities. For crystal systems, such anomalies are often associated with a structural phase transition and are usually limited in a finite temperature range. The large amplitude change of a particular physical property in prescribed environmental conditions can be used to perform certain functions. Examples of such passive functional materials include positive temperature coefficient materials (PTC) [1], superconducting materials, and partially stabilized tetragonal ZrO_2 [2,3]. The resistivity of a doped $BaTiO_3$ can change more than four orders of magnitude immediately above the paraelectric-ferroelectric phase transition, making it a good material for thermistors. The tetragonal-monoclinic phase transition in ZrO_2 can produce up to 6% volume expansion, which can help to stop crack propagations in ceramics. There are many passive functional materials that can perform certain functions using their physical anomalies, including voltage dependent resistors (VDR), carbon fiber-polymer composite near the percolation limit, etc.

Active functional materials are those materials that can convert energy from one form to the other. Good examples of functional materials include piezoelectric materials, magnetostrictive materials, piezomagnetic materials, electrostrictive materials, and shape memory alloys (see Chap. 6). These materials can produce a large response to external stimuli without having to have physical anomalies. The basic energy forms that can be interchanged via active functional materials are: thermal energy, electric energy, magnetic energy, and mechanical energy. The energy can be either in a static form such as electrostatic energy inside a capacitor, or in a dynamical form such as electromagnetic and mechanical waves. Active functional materials, particularly piezoelectric and magnetoelectric materials, are primary materials used for most of the adaptronic structures because electric control signals are very convenient to generate. Active functional materials are sometimes called multifunctional materials since most of them have several functional properties due to cross coupling effects.

Although the definition of functional materials is not so stringent in general, it is critical that the property variation in these materials must be sufficiently large in amplitude. For example, thermal expansion alone is too small to be utilized for any control purpose; therefore, materials with normal thermal expansion properties do not qualify as functional materials. It is very important to understand the fundamental principles that make these materials functional, which can help us to use them properly and to inspire us to create better multifunctional materials based on the same physical principles.

There are many natural functional materials that have been widely used in our daily life. Many composite materials with enhanced functional properties have also been created so that the amount of functional material categories is growing fast. As most of the control systems are driven electronically, ferroelectric materials are naturally one of the best functional materials for adaptronics applications. In this chapter, we will use ferroelectric materials as examples to explain some of the fundamental physics that produce these marvellous functional properties. Three design philosophies will be given at the end of the chapter to provide general guidance in the innovation of better functional materials for adaptronic applications.

3.2 Basic Principles of Functional Materials

As defined in Chap. 1, adaptronic structures are designed to perform all three functions: sensing, control and actuation. Roughly speaking, an adaptronic structure is a primitive replica of a biological body. Multifunctional materials are essential components of an adaptronic structure in which each component must be able to communicate with others. Only a limited number of natural materials can meet the high demand of adaptronics. Therefore, understanding the physical principles of functional materials is very important, which could help us use these basic principles to engineer composite materials with enhanced functionality and/or to create new functional materials. The quality of functional materials may be measured in terms of their responsiveness and agility. The former measures the degree of response, while the latter specifies the speed of response.

3.2.1 Phase Transitions and Anomalies

High responsiveness is often found in a stability edge of a physical property, or near a structural phase transition. The commonly referred phase transitions are thermally driven structural instabilities in which ionic displacements/rearrangements occur in the crystal structure at a critical temperature T_c. In other words, at T_c the crystal structure of the high temperature phase becomes unstable and the ions will form a new crystal structure with lower crystal symmetry below T_c. As a signature of structural phase transitions, at least one physical quantity vanishes, or appears, or becomes discontinuous. Phase transitions can be induced by temperature or field changes, and are the origin of anomalous responses in many crystalline systems that are considered passive functional materials.

Figure 3.1 is an illustration of the ionic displacement pattern in the cubic to tetragonal phase transition in $BaTiO_3$ when cooling the crystal from a temperature higher than $T_c = 130\,°C$ to room temperature. Figure 3.1a is the cubic perovskite structure of the paraelectric phase. While cooling through the phase transition temperature T_c, oxygen anions move down (note: oxygen atoms on the top and bottom faces shift more than the oxygen atoms on the side faces) and the Ti-cations move up relative to the Ba frame as shown in Fig. 3.1b, forming an upward dipole in each unit cell [4]. Associated with the formation of the electric dipole, the unit cell is also elongated along the poling direction, reflecting a strong coupling between the electric dipole formation and crystal structure distortion. The symmetry of the crystal changes from a cubic $m\bar{3}m$ to tetragonal 4 mm. This phase transition is accompanied by a dielectric anomaly [5].

Ferroelectric materials are multifunctional materials with many useful functional properties. In addition, the ferroelectric phase transition can pro-

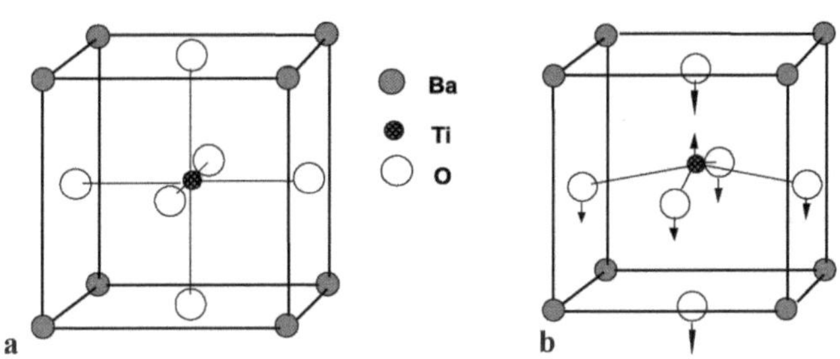

Fig. 3.1. Illustration of the ionic rearrangement in the **a** cubic to **b** tetragonal ferroelectric phase transition in $BaTiO_3$

vide useful functional properties when the material is chemically engineered to increase its electrical conductivity. In a doped ceramic $BaTiO_3$, grain boundaries can create Schottky barriers which couple with the dielectric anomaly to produce a strong PTC effect. PTC resistors have been widely used as thermistors to regulate the temperature limit in many heating devices.

Physical property anomalies may also be induced by field variations rather than temperature changes. For example, the drastic resistance change in ZnO at a critical electric field level is the basis for varistors that are used for voltage surge protections. The characteristic electric current-voltage curve for a voltage dependent resistance (VDR) material is shown in Fig. 3.2. The varistor has very high resistance at low voltage but becomes a good conductor when the voltage exceeds a critical value. When it is put in parallel with an electric device, such as a computer, a TV, etc, it will provide a bypass for the current so as to protect the device when a voltage surge occurs (for example, when there is thunderstorm).

The explanation for this anomalous behavior of ZnO ceramics is the creation of paired Schottky barriers at the grain boundaries as illustrated in Fig. 3.3. The intergrain layer (IGL) can act as acceptors to draw electrons from the semiconducting ZnO grains near the IGL region, so that this region will be positively charged. Schottky barriers are then formed at the interface between grain boundary layer and grains. The paired Schottky barriers provide high resistance to current flow in either direction. At a low electric field, the barrier for the electron flow is too high to produce good conduction. Only a small fraction of thermally activated electrons can pass through the barriers to provide very low current. At a high field level, the electron potential is raised high enough to allow the electrons to overcome the forward biased barrier and tunnel through the grain boundary to produce a surge of current. The reverse direction electron flow is the same due to the symmetry of the paired Schottky barriers so that the I–V curve is antisymmetric.

As passive functional materials are based on anomalies, the criteria for good functional materials are very different from that for common materials.

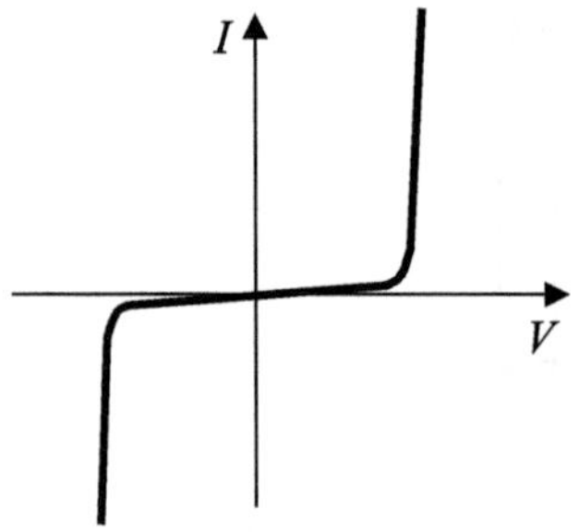

Fig. 3.2. Typical current-voltage curve for a varistor

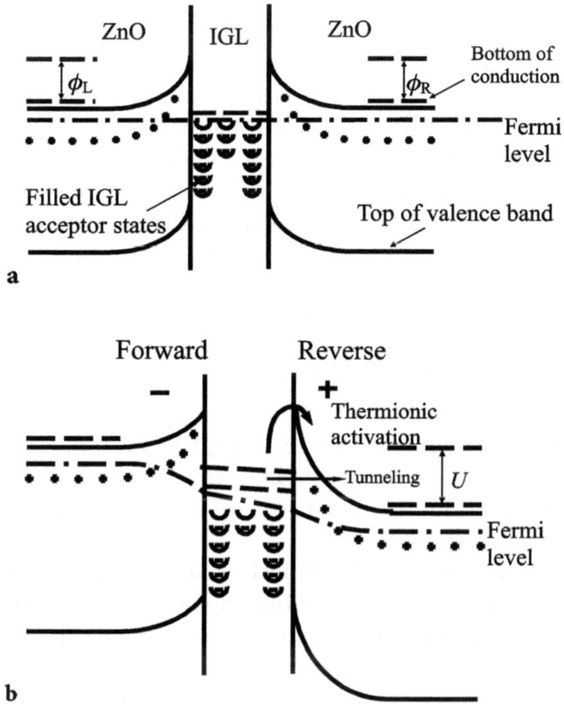

Fig. 3.3. Proposed electronic structure at a junction between semiconducting ZnO grains: **a** no voltage applied, **b** with applied voltage (after A.J. Moulson and J.M. Herbert [1])

These anomalies would be considered disastrous for many common materials since they signify breakdowns and instabilities. Such anomalies are, however, essential for constructing some of the adaptronic structures because they can provide clear signals to indicate the operating limits and can also respond in large amplitude to mend the damages caused by sudden environmental changes.

3.2.2 Microscopic, Mesoscopic, Macroscopic Phenomena and Symmetries

Most adaptronic structures are used, or are intended to be used in macroscopic devices. In a single domain crystal system, macroscopic properties are simply the statistical average of microscopic properties of each unit cell. For most functional materials, however, such a simple average fails due to nonlocal interactions and the additional mesoscopic structures created at the intermediate length scale, such as domain patterns in single crystal systems and grain microstructures in ceramics. These nonlocal interactions and mesoscale structures often produce very strong extra enhancement to the functional properties if properly utilized. Therefore, in order to grasp the whole picture of these functionalities, one must study basic principles at different length scales.

Electron band structures control the electrical conductivity and the structural stability at the microscopic level. Based on electronic band structures, inorganic materials can be classified as conductors, semiconductors and insulators. Modification of these band structures through doping of foreign elements into a crystal could change these band structures and even produce conductivity anomalies. Electronic structures also determine the stability of crystal structures. Instabilities may be created in crystal structures at designed temperatures by altering the electronic structures using chemical doping methods. Many functional materials contain elements of mixed valences, i. e., the element can have two or more different valences while forming a compound. Doping of these mixed-valence elements, such as transition d-block elements in the periodic table and the lanthanides (Eu, Yb, Ce, Pr, Tb, etc), often enhances the functionality of the material [6]. The length scale for this level of functional property manipulation is in the scale of a few angstroms, i. e., the unit cell level or below.

The next level of structures determining the functionalities of materials is the so-called microstructures, such as domains, domain walls, grains, and grain boundaries. In ferroelectric ceramics, for example, contributions to the functional properties from domain wall movements could be as high as 70% of the total functional effect [7]. For shape memory alloys, the super elasticity and shape memory effects originate from domain reorientations and/or from the creation and annihilation of domains. As mentioned above, grain boundaries play a key role in the formation of paired Schottky barriers in PTC and VDR materials. The conduction anomalies found in PTC and VDR do not even exist in a single crystal system. The length scale for these mesoscopic structures is of the order of a few to a few tens of nanometers.

The formation of domain patterns during a phase transition from a high symmetry phase to a low symmetry phase is a reflection of the system trying to recover those lost symmetries. The number of domain states or variants in the low temperature phase is equal to the ratio of the number of operations in the high and low symmetry groups. There are 230 space groups and 32 point groups describing the symmetry operations allowed in crystal structures [8]. The point groups refer to those symmetry operations without translation operations, including rotation, mirror reflection and inversion. The representation of these 32 point groups and their graphic representations are listed in Table 3.1. Although, macroscopically, we often treat many systems as isotropic, i. e., having a spherical symmetry, the highest symmetry allowed in a crystal structure is cubic $m\bar{3}m$. Structural phase transition is allowed only when the symmetry group of the low temperature phase is the subgroup of the high temperature phase. Transitions may also happen between subgroup symmetries of the same parent group, although they may not have direct group-subgroup relationship, such as between tetragonal and rhombohedral symmetries in $BaTiO_3$ and in $Pb(Zr_xTi_{1-x})O_3$ (PZT) solid solutions.

Table 3.1. The 32 point groups and the symbols of the symmetry groups. The *upper left corner* and the *lower right corner* in each cell list the Schoenflies and international symbols, respectively

CUBIC	HEXAGONAL	TETRAGONAL	TRIGONAL	ORTHO-RHOMBIC	MONOCLINIC	TRICLINIC
O_h m3m	C_6 6	C_4 4	C_3 3	C_{2v} 2mm	C_2 2	C_1 1
O 432	C_{6v} 6mm	C_{4v} 4mm	C_{3v} 3m	D_2 222	C_{2h} 2/m	S_2 $\bar{1}$
T_h m3	C_{6h} 6/m	C_{4h} 4/m	S_6 3	D_{2h} 2/mmm	C_{1h} m	
T_d $\bar{4}$3m	C_{3h} $\bar{6}$	S_4 $\bar{4}$	D_3 32			
T 23	D_6 622	D_4 422	D_{3d} $\bar{3}$m			
	D_{6h} 6/mmm	D_{4h} 4/mmm				
	D_{3h} $\bar{6}$2m	D_{2d} $\bar{4}$2m				

At a structural phase transition, there are several equivalent choices (variants) for the high symmetry phase to transform. For example, two variants exist in a ferroelastic tetragonal 4/mmm to orthorhombic 2/mmm transition, representing the elongated axis in the x- or y-directions, respectively. The situation is illustrated in Fig. 3.4, which is the unit cell projection on the x–y

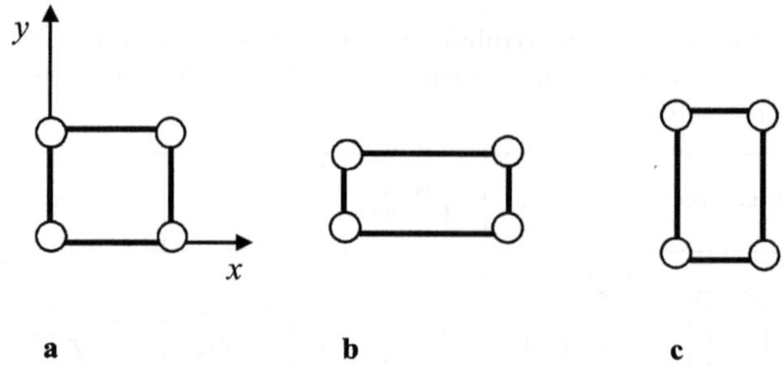

Fig. 3.4. A 2-D illustration of the two possible low temperature states in a tetragonal 4/mmm to orthorhombic 2/mmm ferroelastic phase. **a** Unit cell of the high temperature phase, **b** orthorhombic phase with the elongation along the x-direction, **c** another orthorhombic phase with the elongation along the y-direction

plane. As the two low temperature variants are energetically degenerate, they have equal chance to be formed at the phase transition. As a result, there will be a mixture of these two domain states in the low temperature phase. If the transformation was originated from a single crystal system, these two kinds of domains can form 90° twins that maintain the atomic coherency across the domain boundaries (domain walls). The domain wall orientation can be either in [110] or [1$\bar{1}$0] in this case. The two sets of twins could also coexist to form more complex domain patterns.

Twinning provides a new functional mechanism for easy shape deformation via the movement of domain walls. If the low temperature states are polarized, domain wall movements cause the polar vector to rotate in the region swept by the moving domain wall. This situation is illustrated in Fig. 3.5 for a ferroelectric twin. Under an upward electric field, the domain wall moves to the left. At the same time, the whole region II in the right hand side of the wall moves up relative to region I. The dipoles in region II are switched to more favorable positions by the external field and the global shape change caused by the domain wall movement could be substantial as shown in the figure. The switching of these dipoles in region II gives an extrinsic contribution to the dielectric susceptibility while the shape deformation caused by the wall movement contributes extrinsically to the macroscopic piezoelectric effect.

Domain walls are a special kind of defect. They create localized stress gradients and/or electric (magnetic) field gradients [9] that can strongly interact with other defects, such as dislocations, vacancies and aliovalent dopants. This interesting feature of domain walls enables us to control domain patterns and domain wall densities through different chemical doping strategies.

It is a common practice to dope aliovalent ions (non-stoichiometric doping) to create multivalence and/or vacancies in the material so that domain walls could interact with them. The charged defects created by doping can either pin the domain walls or make the walls more mobile. This method has proven effective to enhance the mesoscale functionality in some ferro-

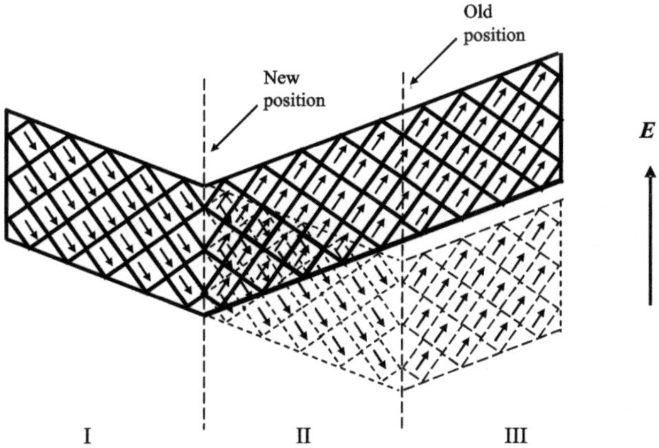

Fig. 3.5. Domain wall movement in a ferroelectric twin structure under an external electric field ***E***

electric materials. For example, the La or Nb doped PZTs have much larger piezoelectric and dielectric properties than those of the non-doped PZTs.

Inhomogeneous stresses produced by localized defects may induce local phase transitions above the normal phase transition temperature T_c, causing the material to have mixed low and high symmetry phases in certain temperature regions. Such a two-phase mixture is usually very sensitive to external fields or stresses since the phase change among the mixture becomes barrierless even for a first order phase transition [10].

The formation of domain structures and the available variants in the low symmetry phase is dictated by the crystal symmetry of the high temperature phase. However, because domain patterns may produce new symmetries at the mesoscopic scale, it is the global symmetry, not the local symmetry, which controls the macroscopic functionality of the material. Therefore, at the macroscopic level, one can make composite structures of designed average symmetries to produce better functional properties.

3.2.3 Energy Conversion

Energy conversion between different energy forms is the primary base of many adaptronic structures. Active functional materials must be used for such purposes. For each energy form, there is a set of generalized conjugate variables (their product has the dimension of energy density) consisting of a generalized force and a generalized displacement. They can be scalars, vectors or tensors. If one kind of generalized force can produce a displacement other then its own conjugate, then the material has the ability to convert energy from one form to the other, and is called an active functional material. Again, crystal symmetries dictate if some of the energy conversions are allowed in a particular crystal structure.

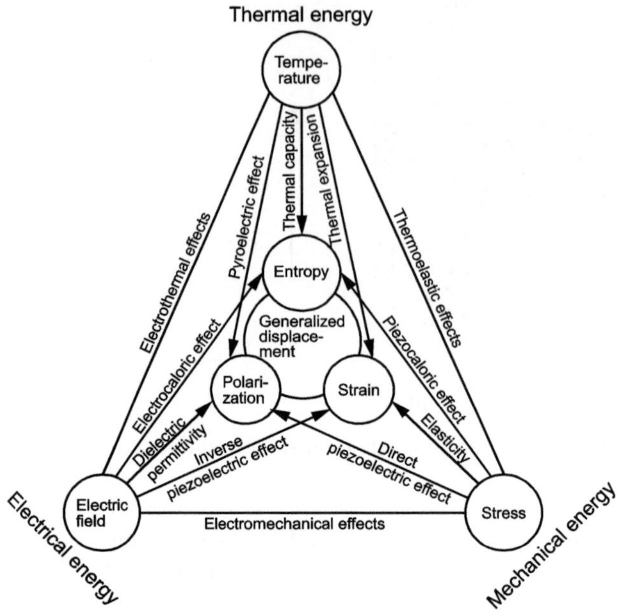

Fig. 3.6. Phenomena occurring in a ferroelectric material that can convert the three forms of energies from one to the other

There are many phenomena in nature reflecting these energy conversion effects. Figure 3.6 illustrates such energy conversion phenomena that can occur in a ferroelectric material. There are three energy forms listed, i. e., thermal, electrical and mechanical energies. The generalized force and displacement pairs corresponding to these three energy forms are: temperature and entropy, electric field and electric displacement, stress and strain. Cross coupling among different physical quantities in the three types of energy forms could be linear or nonlinear depending on the nature of the material. For example, an electric field can generate mechanical strain through the linear inverse piezoelectric effect and the nonlinear electrostrictive effect:

$$S_\lambda = d_{k\lambda} E_k \,, \quad (i, j, k = 1, 2, 3; \ \lambda = 1, 2, 3, 4, 5, 6) \tag{3.1}$$

$$S_\lambda = M_{ij\lambda} E_i E_j \,. \tag{3.2}$$

Here S_λ are the elastic strain components in Voigt notation, $d_{k\lambda}$ are the piezoelectric coefficients and $M_{ij\lambda}$ are the electrostrictive coefficients.

Some of the energy conversion effects can be two-way effects. For example, the piezoelectric effect was defined based on the crystals ability to convert stress into electric charge,

$$D_i = d_{i\lambda} T_\lambda \,, \quad (i = 1, 2, 3; \ \lambda = 1, 2, 3, 4, 5, 6) \tag{3.3}$$

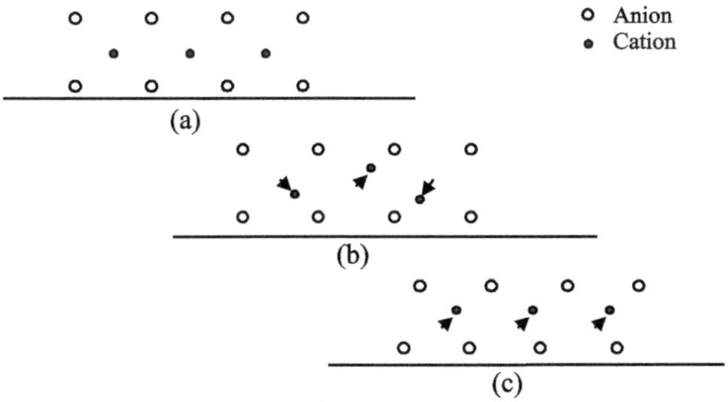

Fig. 3.7. 2-D illustration of the formation of dipoles and shear deformation during a ferroelectric phase transition from square symmetry to rhombic symmetry

where T_λ are the stress tensor components in Voigt notation, and the piezoelectric coefficients $d_{i\lambda}$ are the same as those in (3.1). Hence the piezoelectric effect is a two-way effect. The electrostriction, on the other hand, is a one-way effect due to its nonlinear nature. Theoretically speaking, the same M-coefficient as in (3.2) can be used to describe the combined stress and electric field effect on the electric displacement,

$$D_{\mathrm{i}} = 2M_{\mathrm{ij}\lambda}E_{\mathrm{j}}T_\lambda\ , \quad (i,j,k = 1,2,3;\ \lambda = 1,2,3,4,5,6)\,. \tag{3.4}$$

However, since the above equation describes a mixed phenomenon for which both the electric field and stress must be nonzero, pure stress could not generate charge through this effect when E is zero.

The fundamental principle of the cross coupling in the energy domain is illustrated in Fig. 3.7 using a simple two-dimensional lattice. Figure 3.7a is a binary compound consisting of negative ions (anions) sitting at the corners and positive ions (cations) sitting at the centers of the square lattice. Assuming the square symmetry lattice goes through a ferroelectric phase transition to become rhombic symmetry lattice, there are two kinds of ionic rearrangements involved as shown in the figure. The first kind is the formation of dipoles through the shift of the cations along the diagonal directions as shown in Fig. 3.7b. There are four variants in the low symmetry phase and the dipoles in different unit cells may or may not be aligned. The second kind is a shear deformation of the anion lattice frame to accommodate the ionic shifts as shown in Fig. 3.7b. One can see (Fig. 3.7c) that the dipole formation pushes the frame to deform, and in return, the shear deformation of the frame helps create an ordering of the dipoles. This interdependency between the ordering of dipoles and deformation strain is the fundamental principle of electromechanical coupling.

3.3 Examples of Functional Materials

In order to make the above concepts correlate to real materials, specific examples are given below to further explain some principles of functional properties in different materials. For convenience, we now discuss these functional materials based on their responsive nature, i. e., based on their potential application categories.

3.3.1 Thermal Responsive Materials

Thermally responsive functional materials can be produced in the vicinity of phase transitions. For example, the tetragonal-monoclinic phase transition in ZrO_2 can produce as large as 6% volume strain, which can be used for material toughening. Shown in Fig. 3.8 is an enlarged view at a crack tip in a partially stabilized tetragonal ZrO_2 system. The crack produces tensional stress at the crack tip, which can induce the partially stabilized tetragonal phase ZrO_2 to transform into monoclinic martensitic phase (darker shaded interior region). The large volume expansion from the induced phase transformation helps reduce the stress concentration near the crack tip to stop the crack propagation. The volume expanded martensitic phase forms twins to fit the boundary condition as indicated in the figure.

Temperature could also induce large resistivity change in doped $BaTiO_3$ mentioned above. The fundamental principle of the PTC material is the coupling of the Schottky barriers at the grain boundaries to the ferroelectric phase transition. The potential barrier similar to that plotted in Fig. 3.3 will be short circuited in the ferroelectric state due to the presence of charges at

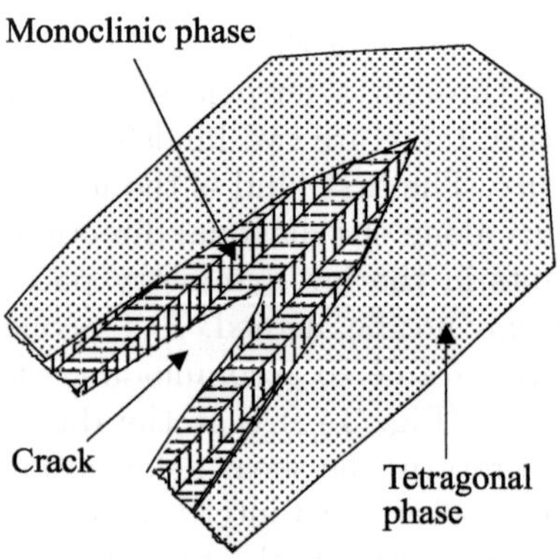

Fig. 3.8. Magnified view at a crack tip where the partially stabilized ZrO_2 transformed to monoclinic phase. The twin pattern represents the transformed martensite phase

the grain boundaries. Above the Curie point T_c, the conductivity is proportional to the Boltzmann factor $\exp(-\phi/kT)$, with the height of the barrier, ϕ, approximately given by [1]:

$$\phi = \frac{e^2 N_s^2}{8\epsilon n} , \tag{3.5}$$

where N_s is the surface density of acceptor states near the boundary, e is the electron charge, n is the volume density of donor states in the grain and ϵ is the dielectric permittivity.

Above the ferroelectric phase transition temperature T_c, the dielectric constant obeys the Curie-Weiss law: $\epsilon = C/(T - \theta)$, where C is the Curie constant and θ is the Curie-Weiss temperature, (note: $\theta = T_c$ for a second order phase transition and $\theta < T_c$ for a first order phase transition), therefore, the resistivity above the transition temperature may be written as [1]:

$$R_{gb} \propto \exp\left\{\frac{e^2 N_s^2}{8nkC}\left(1 - \frac{\theta}{T}\right)\right\} , \quad T > \theta . \tag{3.6}$$

The fast decrease of the permittivity with temperature immediately above T_c drives the resistivity to increase exponentially, producing several orders of magnitude changes to the resistivity in a temperature range of a few tens of degrees. In other words, the resistivity is super sensitive to temperature in this temperature range.

Shape memory alloys, such as Ni–Ti (Nitinol), Ni–Ti–Cu and Ni–Ti–Fe, etc., can recover their original shapes in the austenite phase from a large deformation in the martensite phase upon heating back to the austenite phase (see Sect. 6.4). This process is demonstrated in Fig. 3.9 by assuming only two variants in the low temperature martensite phase. Figure 3.9a is the high temperature austenite phase with a perfect rectangular shape. When the system is cooled through the phase transition, a shear deformation occurs and the two martensite variants will co-exist to form twin structures. A twin structure is formed between domain states 1 and 2 as shown in Fig. 3.9b. The twinning of the two domains requires no defects at the domain wall and the atomic coherency is preserved cross the wall. Now, if a tensional stress is applied as shown in Fig. 3.9c, the degeneracy of the two domain states is lifted so that one type of domain grows at the expense of the other.

New domains of type 1 may also be generated through nucleation process to speed up the domain switching process. This domain switching may continue until the unfavored domains (type 2 as shown in Fig. 3.9b) are driven out of the system. If the applied stress is compressive as shown in Fig. 3.9d, some domains are annihilated. The presence of domain walls makes the shape deformation very easy in the martensite phase. Upon heating, all shapes in Fig. 3.9b–d go back to the same shape as in Fig. 3.9a. In other words, the shape in the high temperature phase is ‘remembered’. This shape memory

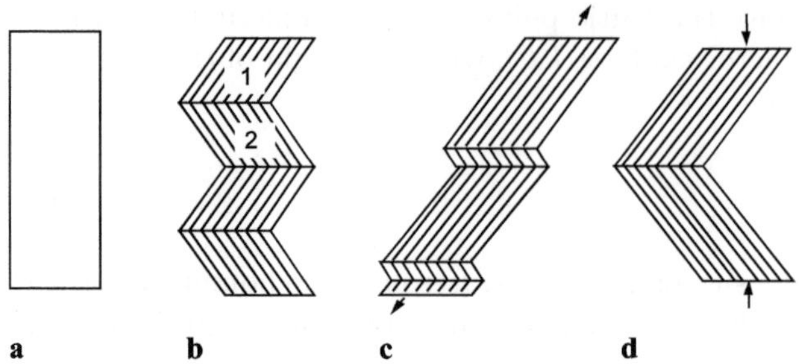

Fig. 3.9. Illustration of shape memory effect

effect extends the elastic limit of the alloy with the help of temperature and the phase transition.

Shape memory effect is directly related to the domain pattern and the interaction of domain walls with defects. How accommodating the shape of martensite depends on the available number of low temperature variants. Generally speaking, the more variants available the easier it is for the martensite to deform into arbitrary shapes. A cubic-monoclinic transition can generate up to 24 low temperature variants, therefore, such a martensite can deform into more complex and elegant shapes than the one shown in Fig. 3.9 without breaking-up the atomic coherency. Due to the drastic change of mechanical strength above and below the martensite phase transition, shape memory effect can also be used to make shape memory alloy engines that can convert thermal energy into mechanical energy [10].

Another important thermal responsive material is the pyroelectric materialthat can directly convert thermal energy into electric energy. The pyroelectric effect is a manifestation of the existence of polarization in the material. The change of the polarization amplitude with temperature generates electric charges at the sample surface where the polarization terminates. Again, the pyroelectric effect is strongest near the ferroelectric phase transition temperature because polarization changes more drastically in the vicinity of T_c. Pyroelectric materials are widely used as infrared sensors for the remote control of electronic devices and for making night vision devices.

3.3.2 Materials Responsive to Electric, Magnetic and Stress Fields

If the adaptronic structure requires temperature stability, active functional materials must be used since they can have a flat temperature response away from the phase transition and are controllable with external fields. Most materials in this category are ferroic materials, i. e., ferroelectric, ferromagnetic and ferroelastic materials.

Piezoelectric and electrostrictive materials are materials having the ability to convert electric energy into mechanical energy. The effect is called piezoelectric if the generated surface charge density is linearly proportional to the applied stress. The piezoelectric effect is reversible. The physical origin of piezoelectricity comes from the noninversion symmetry of ionic arrangement in the crystal structure. Without inversion symmetry, the anions and cations in a crystal shift in an asymmetric fashion under stress to produce a dipole moment. In fact, 20 out of the 21 non-central symmetric crystal point groups allow piezoelectricity to exist except the cubic class of 432 (see Table 3.1). The term polarization refers to the volume average of dipole moments and is measured as charge per area. For a finite system in static equilibrium the polarization projection onto a surface of the material is equal to the surface (bond) charge density.

It is important to recognize that a useful piezoelectric effect is defined macroscopically. Each unit cell has to contribute constructively in order for the macroscopic effect to occur. It is the global symmetry that determines the macroscopic piezoelectric effect. For example, a piezoelectric ceramic containing randomly oriented crystal grains has no piezoelectric effect even though the symmetry of each unit cell allows piezoelectricity. A net polarization in the material is a sufficient but not a necessary condition for the presence of piezoelectricity; for example, quartz is one of the popular piezocrystals without polarization. The existence of a polarization, however, does make the piezoelectric effect much more pronounced. In fact, the best piezoelectric materials are all ferroelectric materials. Most importantly, the hydrostatic piezoelectric effect belongs uniquely to polar materials.

Figure 3.10a shows the polarization arrangement in a ceramic system (note: domains were not explicitly drawn in here, the arrows only represent the net polarization in each grain). The macroscopic piezoelectric effect is zero due to the cancellation of oppositely polarized grains. If the ceramic material is ferroelectric, it can be made piezoelectric by aligning the polarization of different grains using an external electric field through the domain switching process. A net polarization may be produced along the field direction as illustrated in Fig. 3.10b. As the electromechanical characteristic of piezoelectric effect is reversible, piezoelectric materials can be used for both sensing and actuation functions (see Chaps. 6 and 7).

Electrostriction can generate mechanical deformation that is independent of the polarity of the electric field. It exists in almost all materials but is usually too weak for any practical use. However, it can be very large in electrostrictive materials, such as lead magnesium niobate (PMN) systems [11]. The nonlinearity often works to the advantage in such systems since it can produce tunable functional properties. As the effect is nonreversible, electrostrictive materials are better for actuator applications. Unlike the piezoelectric effect, electrostriction can even exist in systems with center symmetry. Electrostrictive materials become piezoelectric under a dc bias field.

A magnetic field is similar to electric field in many aspects, but it has its own distinctive nature. Materials responding to a magnetic field are another important category of functional materials since the change of magnetic properties can easily be converted into electric signals and vice versa.

The magnetic moment of ions is produced by the spins of unpaired electrons. These magnetic moments are randomly oriented in the paramagnetic phase. The system becomes ferromagnetic through an order–disorder phase transition that can align these magnetic moments. Unlike the ferroelectric case, the amplitude of each magnetic moment is fixed and the coupling to the lattice structure is usually much weaker compared to that of ferroelectric materials. The ordering may appear in the form of antiferromagnetic, ferrimagnetic or ferromagnetic as illustrated in Fig. 3.11. In a ferrimagnetic state, the spins are only partially aligned or having different amplitude in an antiparallel configuration.

If the magnetic spins are strongly coupled to the lattice structure, the system shows a magnetoelastic effect similar to the case analyzed in Fig. 3.7. The piezomagnetic effect is allowed in terms of crystal symmetry in many

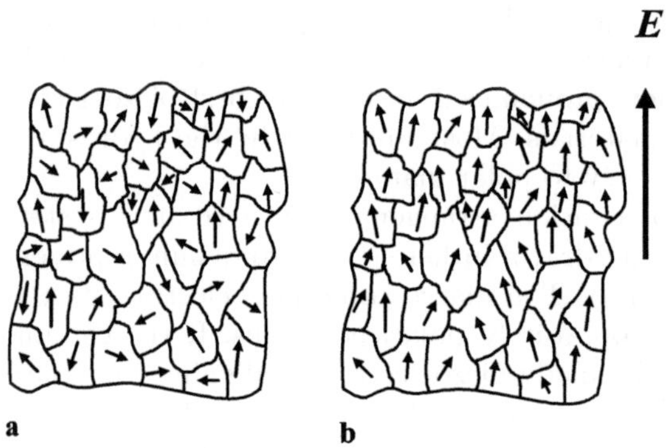

Fig. 3.10. Polarization distribution in a polar ceramic system. **a** Random orientation before poling and **b** after poling by an external electric field $\boldsymbol{E}$

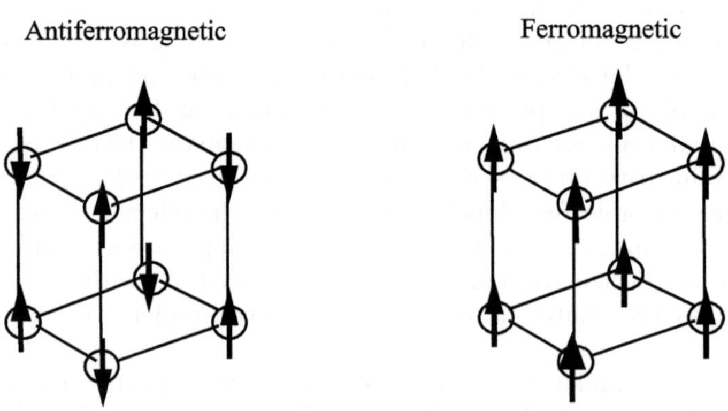

Fig. 3.11. Spin arrangements in an antiferromagnetic and a ferromagnetic system

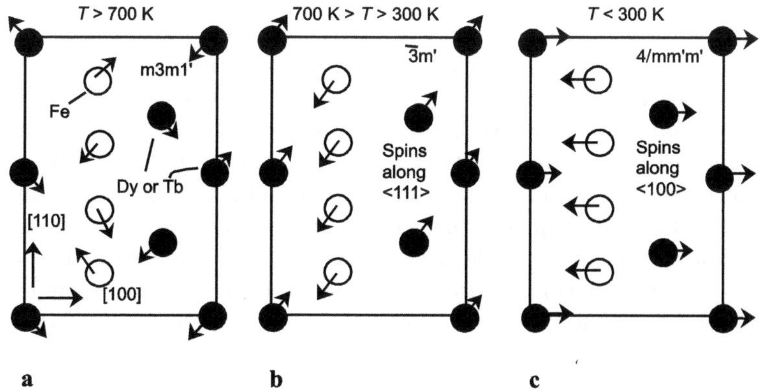

Fig. 3.12. Spin orientations in all three states of $Tb_{0.3}Dy_{0.7}Fe_2$. **a** paramagnetic phase, **b** rhombohedral ferrimagnetic phase, and **c** tetragonal ferrimagnetic phase (after R.E. Newnham [13])

systems; however, it is usually too small to be useful for any control purpose. The nonlinear effect, magnetostriction, however, can be quite large in certain systems.

For example, $Tb_{0.3}Dy_{0.7}Fe_2$ (Terfenol-D), can generate a strain level as high as 10^{-3} at room temperature [12] (see Sect. 6.3). Above 700 °C, the crystal has a cubic symmetry with a C_{15} structure in which the rare-earth atoms form a diamond-like lattice. It is paramagnetic in the cubic phase, in which the spins are randomly oriented. At T_c, (<700 °C, see the phase diagram to be discussed below), it transforms into a rhombohedral ferrimagnetic phase with the spins parallel to the <111> direction. The strong antiferromagnetic coupling between the spins of irons and the rare-earth atoms prevents the spins to align perfectly in one direction. With further cooling, $Tb_{0.3}Dy_{0.7}Fe_2$ goes through another phase transition that changes the orientation of the spins from <111> to <100> so that the system becomes a tetragonal ferrimagnetic. All three cases are shown in Fig. 3.12, which is the cross section plane of [100] and [110]. The arrows indicate the spin orientations in each of the three states.

3.4 Increased Functionality Through Material Engineering

Different strategies have been developed to engineer better functional materials for adaptronic applications. At the microscopic level, chemical mixing of different compounds may create new and better functional materials or may shift the transition temperature closer to the operating temperature so as to maximize the functionality. Experience tells us that better functional materials are mostly mixed compounds or solid solutions rather than single-phase materials. At the mesoscopic level, defects can be introduced to influence the formation of domain structures or to change the grain boundary properties. Many doping strategies have been developed which can greatly enhance the mobility of domain walls and to change the characteristics of the conductivity. One recently developed strategy of making engineered domains in relaxor based ferroelectric materials has been proven effective in improving the piezoelectric and dielectric properties. At the macroscopic level, composites of different materials can be made through structural engineering. These composites can be multifunctional and are also tunable compared to single-phase materials. Below, we will introduce a few design strategies for creating better multifunctional materials.

3.4.1 Morphotropic Phase Boundary

Some isostructural compounds can be atomistically mixed to form a solid solution with much enhanced functional properties. A solid solution without solubility gap is called a complete solid solution in which two or more compounds can be mixed in any proportion to form a single-phase material. The PZT system is a good example of such a complete solid solution. As shown in the phase diagram Fig. 3.13, the solid solution of $(1-x)PbZrO_3$-$xPbTiO_3$ has a cubic perovskite structure in the paraelectric phase and will go through a ferroelectric phase transition to become either rhombohedral or tetragonal ferroelectric depending on the composition. The near vertical line in the middle of the phase diagram specifies a compositional boundary separating the tetragonal and rhombohedral phases. This boundary is called the morphotropic phase boundary (MPB), at which the two ferroelectric phases are energetically degenerate. At room temperature, this composition corresponds to a Ti/Zr ratio of 48/52.

The reason for this MPB composition having superior functional properties is due to the fact that there are more variants in the ferroelectric phase and the energy barrier height between different states becomes lower. As mentioned above, the symmetry change produced by phase transition plays a vital role in determining the functionality of materials. Generally speaking, the more variants generated at the phase transition, the better is the functionality. For the PZT case, the high temperature phase has a cubic symmetry with point group $m\bar{3}m$, which has 48 symmetry operations. The tetragonal 4 mm and the rhombohedral 3 m symmetry groups in the ferroelectric phase contain 8 and 6 symmetry operations, respectively, so that the number of variants for the tetragonal phase is 6 and for the rhombohedral phase is 8. This situation is illustrated in the phase diagram Fig. 3.13. Depending on the composition, the dipoles formed at the phase transition in each unit cell may point to any of the six faces of the cube to form the tetragonal phase or any of the eight corners of the cube to form the rhombohedral phase. At the MPB composition, the two low temperature phases are energetically degenerate so that all 14 variants are accessible. This provides a unique situation that gives

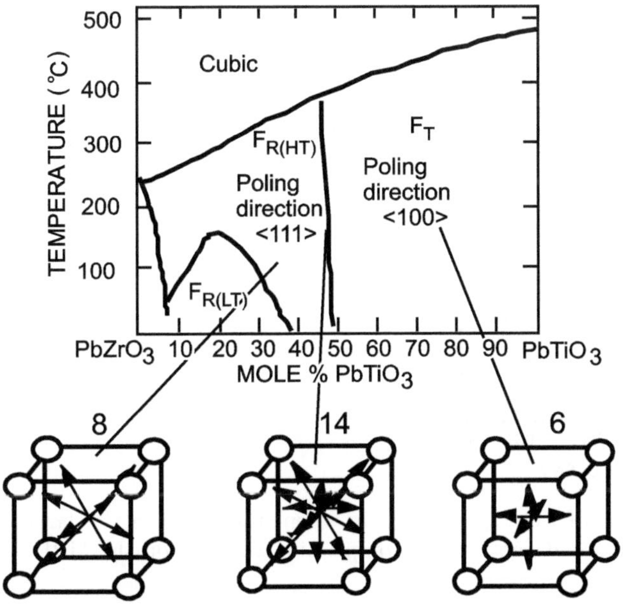

Fig. 3.13. Phase diagram of $Pb(Zr_{1-x}Ti_x)O_3$ and the illustration of available number of variants in the ferroelectric phase

the most variants in the low temperature ferroelectric phase to allow better poling of the PZT ceramic. The energy barrier for polarization rotation is also greatly reduced at the MPB so that large polarization changes can be produced with relatively small fields.

Another similar example is the Terfenol-D system. The $(1-x)TbFe_2$-$xDyFe_2$ binary alloy is a resemblance of the PZT solid solution system.

There exists a similar compositional boundary between the rhombohedral and tetragonal phases. At room temperature, the best magnetostrictive effect is given by the alloy with a Tb/Dy ratio of 30/70, which falls on the MPB line as indicated by the arrow in Fig. 3.14.

In both cases, the best functional properties are found in those compositions on the MPB. Due to the energetic degeneracy of the two structural phases on the MPB, the system cannot decide which structure to take so that both phases may co-exist. Any external stimulus could tip the balance of the situation. Therefore, this uncertainty effectively creates high responsiveness to external fields, leading to enhanced functional properties of the material. These kind of methods to enhance functional properties of materials are guided by the following design philosophy:

Design Philosophy 1: introducing instabilities into the system to create more responsive materials.

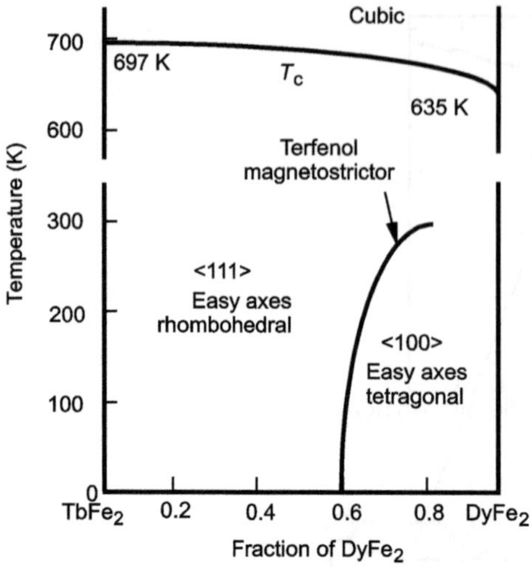

Fig. 3.14. Phase diagram of $(Tb_{1-x}Dy_x)Fe_2$. At room temperature the largest magnetostrictive effect occurs at the $x = 0.7$ composition which is on the MPB as pointed by the *arrow*

3.4.2 Domain Engineering

Material engineering in the mesoscopic level is the manipulation of domain structures and their mobility in order to increase the extrinsic effects. Aliovalent doping (charged point defects) in a ferroelectric system can create strong localized fields that may either facilitate or hinder the movements of domain walls. This method has been used to improve the piezoelectric properties in soft and hard PZT systems as mentioned above. Defects, including dislocations and point defects, can also be rearranged to accommodate the stress field generated by the formation of domain walls. For shape memory alloys, such defect alignment associated with domain patterns can provide reverse shape memory effect. Aligning these defects may take a few rounds of thermal cycling passing through the phase transition, i. e., the alloy must be trained to remember the exact locations of the domain walls formed in the martensite phase. This training process is to correlate defects with domain patterns.

The second type of domain structure manipulation is to create disorder in an ordered system using physical means other than chemical doping. Single crystal $Pb(Zn_{1/3}Nb_{2/3})O_3$-$PbTiO_3$ (PZN-PT) and $Pb(Mg_{1/3}Nb_{2/3})O_3$-$PbTiO_3$ (PMN-PT) solid solution systems have created some excitement recently in the transducer and actuator communities due to their over 90% electromechanical coupling coefficient k_{33} (compared to 68% for PZT) and very large piezoelectric coefficient ($d_{33} > 2000\,\mathrm{pC/N}$) [14–16]. Although these materials had been discovered in 1969 [17], they did not generate enough interest because they cannot retain high remnant polarization along the threefold polar axis.

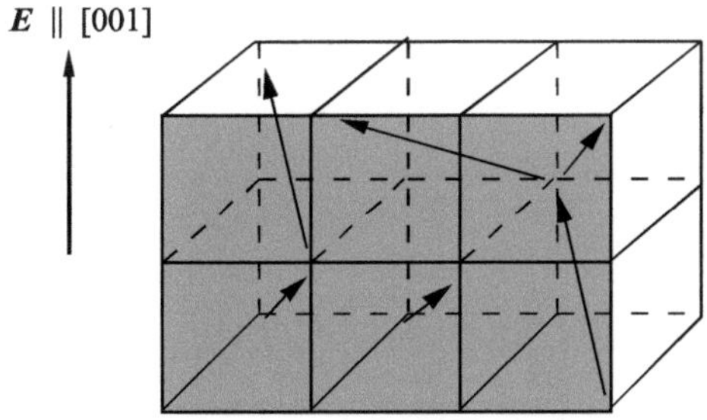

Fig. 3.15. Illustration of misorientational poling in PZN-PT single crystal system. The field is applied along [001] and the dipoles in each unit cell are pointing to the four upper corners along body diagonals

Moreover, the piezoelectric d_{33} coefficient in the single domain state is not very impressive. The crystal symmetry of these ferroelectric crystals is rhombohedral 3m with the dipoles in each unit cell pointing to the <111> (along body diagonals) of the original cubic cells. It was found that the system could sustain large polarization if the poling field is applied along <100> (one of the normal directions of the cubic cell). After poling, each unit cell has a dipole moment along four of the <111> directions in the upper half space as shown in Fig. 3.15. The polarization projections onto the directions perpendicular to the field direction are randomly oriented so that the global symmetry of the multidomain system (macroscopic average) is pseudo-tetragonal. Strong elastic interactions among neighboring cells help stabilize the poled multidomain configuration. Such nonpolar direction poling produces a new domain pattern symmetry in the macroscopic sense that is totally different from the original crystal symmetry. As the multidomain state has a higher energy compared to the ground state, it is much more responsive or unstable under external stimuli. These kind of methods to enhance functional properties of materials are guided by the following design philosophy:

Design Philosophy 2: create order in a disordered system, such as alignment of random defects in martensite to produce reverse shape memory effect, and/or create disorder in an ordered system, such as non-polar direction poling of PZN-PT and PMN-PT single crystals to increase responsiveness of the system to an external field.

3.4.3 Functional Composites

Composite engineering is to put several different materials together in certain configurations. This can be done from few nanometers up to tens of millimeters. Composite engineering allows us to use non-functional materials to enhance functional materials, or to use different functional materials to make new functional composite or multifunctional composite materials.

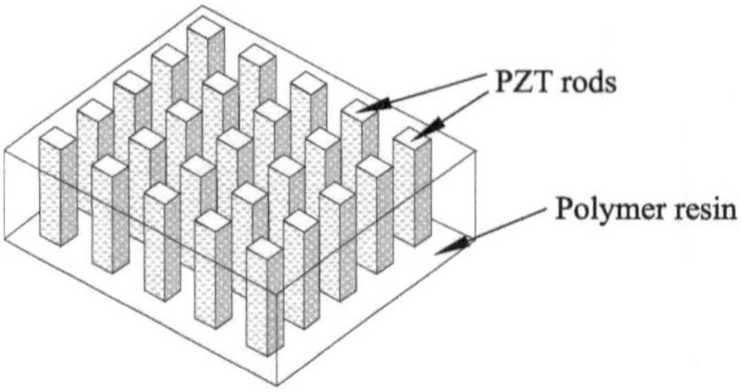

Fig. 3.16. Piezoelectric PZT-polymer composite

The constructive enhancement concept in a composite is analogous to a ballet dance in which the male dancer uses his strength to help the female dancer rotate faster on her toes than she could ever do by herself. In some occasions, he also lifts her to the air to help her 'fly' to a new height that she can never accomplish by her own ability. The beauty here is to form a complementary team so that the merits of individuals can be constructively combined and enhanced.

Shown in Fig. 3.16 is a 1–3 piezoelectric composite with PZT ceramic rods embedded in a polymer resin. This structure is now widely used in medical ultrasonic transducers because the polymer helps reducing the acoustic impedance mismatch between human body and the PZT so that energy transmission becomes more efficient. The load on the polymer phase can be transferred to the ceramic so that the effective load on the ceramic is enhanced, which produces higher electric signal when it is used as stress sensor. This composite structure also gives a much higher figure of merit for hydrophone applications [18].

The hydrostatic piezoelectric coefficient is defined as $d_{\mathrm{h}} = d_{33}+2d_{31}$. Here d_{33} represents the ability of the material to generate charges on the surface normal to the polarization under stress, and d_{31} measures the ability of the material to generate charges on the same surface by a stress perpendicular to the poling direction. Under a constant electric field, the relationship between the electric displacement D and the hydrostatic pressure P_{h} is given by $D = d_{\mathrm{h}}P_{\mathrm{h}}$. This d_{h} value is usually small due to the opposite signs of d_{33} and d_{31}. For PZTs, the d_{h} value is $20\ldots60\,\mathrm{pC/N}$, which is an order of magnitude smaller than d_{33}. The mechanism to enhance the hydrostatic effect in the 1–3 composite is to transfer the stress acting on the polymer to the ceramic rods via shear coupling at the ceramic-polymer interface [19]. This coupling effectively amplifies the pressure on the ceramic rods along the poling direction while leaving the lateral pressure unchanged. As a result, the effective $\bar{d}_{\mathrm{h}}$ value (we use an overbar symbol to represent macroscopic average) is enlarged through the enhanced effective $\bar{d}_{33}$.

The flextensional moonie structure shown in Fig. 3.17 is another good example of using this re-directing force strategy. Through a metal cap, the

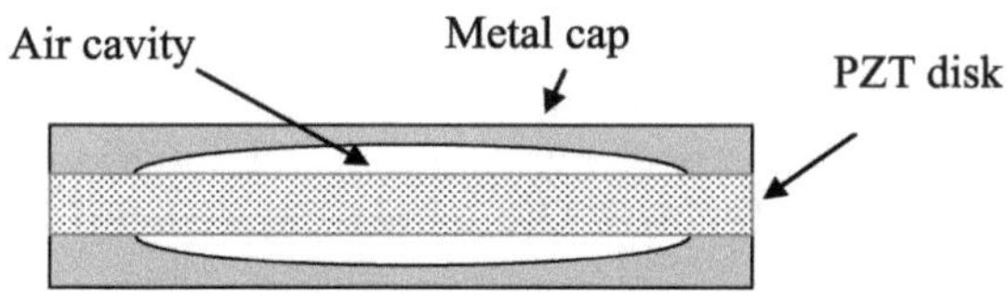

Fig. 3.17. Cross-section of a moonie transducer [20]

normal pressure applied to the top and bottom surfaces of the structure is converted to a force that has large radial component acting on the outer ring of the PZT disk. The radial component of this force counters the d_{31} effect and the normal component of this force enhances the d_{33} effect at the contact area. For a small diameter cavity, the main contribution to d_h is the effectively enhanced $\bar{d}_{33}$. While for a large diameter cavity, the main contribution is $\bar{d}_{31}$ since the contact ring area becomes very small and the cavity area does not contribute to the effective $\bar{d}_{33}$. In this case, the redirected force in the radial direction is much larger than the force produced by the pressure applied to the side of the PZT disk, so that the effective $\bar{d}_h$ value can be very large. For actuator applications, the radial contraction of the disk will be converted to a much larger normal displacement at the center region of the metal cap. This displacement adds to the displacement produced by the d_{33} so that the effective $\bar{d}_{33}$ could be increased by an order of magnitude [20].

There are many other multifunctional materials being created, for example, piezoelectric-piezomagnetic composite, magnetoelectric-piezoelectric composite, etc. They can respond to several different types of external fields and perform multiple functions.

In general, using composite scheme to enhance functional properties of materials or creating multifunctional materials is guided by the following design philosophy:

Design Philosophy 3: use nonfunctional materials to enhance the ability of functional materials through a redirecting force scheme, and make multifunctional composites using constructive integration of different functional materials.

3.5 Summary

In this chapter, we have discussed some fundamental principles of functional materials using a few examples. Nature has provided us with many functional materials, but at the same time, also puts some limitations on these materials both in terms of availability and the magnitude of functionality. The objective of material engineering is to break these natural limits and to invent new composite materials that can better meet new technological challenges. Following the above mentioned design philosophies, advanced functional materials with multifunctional properties can be developed through innovative engineering at all three length scales, i. e., microscopic, mesoscopic and macroscopic levels. With the rapid improvement of modern processing techniques and the creative imagination of scientists, we can expect more and better functional materials to be developed in the future to meet the increasing demand in adaptronics.

References

1. Moulson, A.J.; Herbert, J.M.: *Electroceramics: Materials, Properties, Applications.* Chapman & Hall, London, ISBN 0412294907 (1990)
2. Evans, A.G.: *Science and Technology of Zirconia II, Advances in Ceramics.* Proc. 2nd Int. Conf. on the Science and Technology of Zirconia; N. Claussen, M. Rühle, and A.H. Heuer (Ed.), Amer. Ceram. Soc., Vol. 12 (1984), pp. 193–212
3. Muddle, B.C.; Hannink, R.H.J.: *Crystallography of the Tetragonal to Monoclinic Transformation in MgO-Partially-Stabilized Zirconia.* J. Amer. Ceram. Soc., Vol. 69, No. 7 (1986) pp. 547–555
4. Shirane, G.; Pepinsky, R.; Frazer, D.C.: *X-ray and Neutron Diffraction Study of Ferroelectric* $PbTiO_2$*. Acta Crystallographica.* Int. Union of Crystallography, Vol. 9, No. 2 (1956), pp. 131–140
5. Merz, W.J.: *The Electric and Optical Behavior of* $BaTiO_3$ *Single-Domain Crystals.* Phys. Rev., Vol. 76, No. 8 (1949), pp. 1221–1225
6. Wang, Z.L.; Kang, Z.C.: *Functional and Smart Materials: Structural Evolution and Structure Analysis.* Plenum, New York, ISBN 0306456516 (1998)
7. Luchaninov, A.G.; Shil'nikov, A.V.; Shuvolov, L.A.; Shipkova, I.JU.: *The Domain Processes and Piezoeffect in Polycrystalline Ferroelectrics.* Ferroelectrics, Vol. 98 (1989), pp. 123–126
8. Ashcroft, N.W.; Mermin, N.D.: *Solid State Physics.* Holt, Rinehart and Winston, ISBN 0030839939 (1976)
9. Cao, W.; Cross, L.E.: *Theory of Tetragonal Twin Structures in Ferroelectric Perovskites with a First-order Phase Transition.* Phys. Rev. B, Vol. 44, No. 1 (1991), pp. 5–12
10. Cao, W.; Krumhansl, J.A.; Gooding, R.: *Defect-induced Heterogeneous Transformations and Thermal Growth in Athermal Martensite.* Phys. Rev. B, Vol. 41 (1990), pp. 11319–11327
11. Newnham, R.E.: *Composite Electroceramics.* Annual Review of Materials Science, Vol. 16 (1986), pp. 47–68
12. Clark, A.E.: *Ferromagnetic Materials: a Handbook on the Properties of Magnetically Ordered Substances.* Wohlfarth, E.P. (Ed.); North-Holland, Vol. 1, ISBN 0444898530 (1980)
13. Newnham, R.E.: *Molecular Mechanisms in Smart Materials.* Materials Research Soc. Bulletin, Vol. 22, No. 5 (1997), pp. 20–34
14. Park, S.E.; Shrout, T.: *Relaxor Based Ferroelectric Single Crystals for Electromechanical Actuators.* Materials Research Innovations, Vol. 1, No. 1 (1997), pp. 20–25

15. Yin, J.; Jiang, B.; Cao, W.: *Elastic, Piezoelectric, and Dielectric Properties of* $0.955Pb(Zn_{1/3}\ Nb_{2/3})O_3$*-*$0.045PbTiO_3$ *Single Crystal with Designed Multidomains.* IEEE Trans. on Ultrasonics, Ferroelectrics, and Frequency Control, Vol. 47, No. 1 (Jan. 2000), pp. 285–291
16. Zhang, R.; Jiang, B.; Cao, W.: *Elastic, Piezoelectric, and Dielectric Properties of Multidomain* $0.67Pb(Mg_{1/3}Nb_{2/3})O_3$*-*$0.33PbTiO_3$ *Single Crystals.* J. Appl. Phys., Vol. 90 (2001), pp. 3471–3475
17. Nomura, S.; Takahashi, T.; Yokomizo, Y.: *Ferroelectric Properties in the System* $Pb(Zn_{1/3}Nb_{2/3})O_3$*-*$PbTiO_3$. J. Physical Society of Japan, Vol. 27 (1969), p. 262
18. Skinner, D.P.; Newnham, R.E.; and Cross, L.E.: *Flexible Composite Transducers.* Materials Research Bulletin, Vol. 13, No. 6 (1978), pp. 599–607
19. Cao, W.; Zhang, Q.; Cross, L.E.: *Theoretical Study on the Static Performance of Piezoelectric Ceramic-polymer Composites with 1-3 Connectivity.* J. Applied Physics, Vol. 72 (1992), pp. 5814–5821
20. Xu, Q.C.; Yoshikawa, S.; Belck, J.R.; Newnham, R.E.: *Piezoelectric Composites with High Sensitivity and High Capacitance for Use at High Pressures.* IEEE Transactions on Ultrasonics, Ferroelectrics, and Frequency Control, Vol. 38, No. 6 (1991), pp. 634–639

4 Controllers in Adaptronics

V. Rao, R. Damle, S. Sana

4.1 Introduction

In recent years, control of smart structures has become an important component of multidisciplinary research into vibration suppression. The design of controllers for smart structures is a challenging problem because of the presence of nonlinearities in the structural system and actuators, limited availability of control force, and nonavailability of accurate mathematical models. In this study, adaptive and robust control algorithms are being investigated for designing active controllers for smart structures. Both conventional and neural network-based adaptive controllers have been designed and implemented on smart structure test articles. In addition, a neural-network based optimizing control algorithm with on-line adaptation capabilities has been developed that can incorporate nonlinearities in the smart structural system, accommodate the limited control effort and adapt on-line to time-varying dynamical properties. In this algorithm the control signal is computed iteratively while minimizing a linear quadratic (LQ) performance index with additional weighting on the control increments.

A central goal of research into robust control is to develop control algorithms for time-varying systems, nonlinear systems and systems with unknown parameters [1–6]. These controllers have the ability to adjust controller gains for multiple operating points. The adaptive control techniques have been extensively employed for designing controllers for various industrial systems. One of the objectives of this research is to investigate the applicability of adaptive and robust control algorithms for smart structures. When the desired performance of an unknown plant with respect to an input signal can be specified in the form of a linear or a nonlinear differential equation (or difference equation), stable control can be achieved using model reference adaptive control (MRAC) techniques. The idea behind MRAC is to use the output error between the plant and a specified reference-model to adjust the controller parameters. There are two basic approaches to MRAC. When the controller parameters $\boldsymbol{\theta}(k)$ are directly adjusted to reduce some norm of the output error between the reference model and the plant, it is called direct control. In indirect control, the parameters of the plant are estimated as the elements of a vector $\hat{\boldsymbol{p}}(k)$ at each instant k, and the parameter vector $\boldsymbol{\theta}(k)$

of the controller is chosen assuming that $\hat{\boldsymbol{p}}(k)$ represents the true value of the plant parameter vector $\boldsymbol{p}(k)$. Both the direct control and indirect control algorithms have been implemented on the smart structure, resulting in the following model.

Having successfully implemented conventional MRAC techniques, the next logical step was to try to incorporate the MRAC techniques into a neural network-based adaptive control system. The ability of multilayered neural networks to approximate linear as well as nonlinear functions is well documented and has found extensive application in the area of system identification and adaptive control. The noise-rejection properties of neural networks makes them particularly useful in smart structure applications. Adaptive control schemes require only limited a priori knowledge about the system to be controlled. The methodology also involves identification of the plant model, followed by adaptation of the controller parameters based on a continuously updated plant model. These properties of adaptive control methods makes neural networks ideally suited for both identification and control aspects [7–11].

A major problem in implementing neural network-based MRAC is translating the output error between the plant and the reference model to an error in the controller output, which can then be used to update the neural controller weights. One recently proposed solution to this problem is based on a constrained iterative inversion of a neural model of the forward dynamics of the plant [12]. This technique predicts the actual and desired output errors to calculate the necessary control signal at the next time instant. The algorithm has shown promise in that it offers a degree of robustness and generates a smooth control. It is from this iterative inversion process that the update method described herein is derived. We use the neural identification model to find the instantaneous derivative of the unknown plant at one instant in time. The derivative is then used iteratively to search the input space of the system to find the input $u^*(k)$ that would have resulted in the correct system output. The control signal error $e_{\mathrm{u}}(k) = u^*(k) - u(k)$ can then be used with a static backpropagation algorithm [10] to update the weights of the neural controller.

For the implementation of MRAC algorithms, we propose to investigate the use of neural networks in order to identify a linear model of a system with the objective of adjusting the parameters of a neural controller to reflect the changes in the plant parameters. This method would be particularly useful when the parameters of the plant change considerably with changes in its operating conditions.

A neural network-based eigensystem realization algorithm (ERA) [13] has been utilized to generate a mathematical model of the structural system. For smart structure applications, the size of such networks becomes very large. Therefore, we have developed an adaptive neuron-activation function and an accelerated adaptive learning-rate algorithm, which together significantly reduce the learning time of a neural network. The models obtained by these identification techniques are compared with that obtained from the swept sinewave testing and curve fitting methods [13, 14].

The remainder of this chapter is arranged as follows. A brief description of the two smart structure test articles used to evaluate the adaptive control algorithms is given in Sect. 4.2. Section 4.3 includes the outlines of the conventional model-reference adaptive control techniques and their experimental closed-loop performances on the cantilever beam smart structure test article. The neural network-based model-reference adaptive control algorithm and the neural network-based optimizing controller with on-line adaptation have been introduced in Sect. 4.4. The adaptive neuron activation function and an on-line adaptive control algorithm for the neural network-based model-reference adaptive control algorithm are also described in this latter section. The design of robust controllers for structural systems is presented in Sect. 4.5.

4.2 Description of the Test Articles

To demonstrate some of the capabilities of adaptive control using neural networks on smart structures and to determine the limitations imposed by hardware realization, we have designed and fabricated an experimental test article. The smart structure test article was an aluminum cantilever beam with shape memory actuators, strain-gauge sensors, signal-processing circuits and digital controllers. A schematic diagram of the cantilever beam is shown in Fig. 4.1. The system is a single input-single output (SISO) system with one actuator and one sensor.

The neural network-based control algorithm described in Sect. 4.4 is tested using simulation studies on a cantilever plate system with PZT actuators and PVDF film sensors. A top-view line diagram of the plate structure is shown in Fig. 4.2.

The PVDF film sensors are shaped to measure the displacement and velocity at the free end of the plate [15]. The output of the PVDF film sensor is buffered through a high-pass filter for an output in the range of $\pm 1\,\mathrm{V}$ for

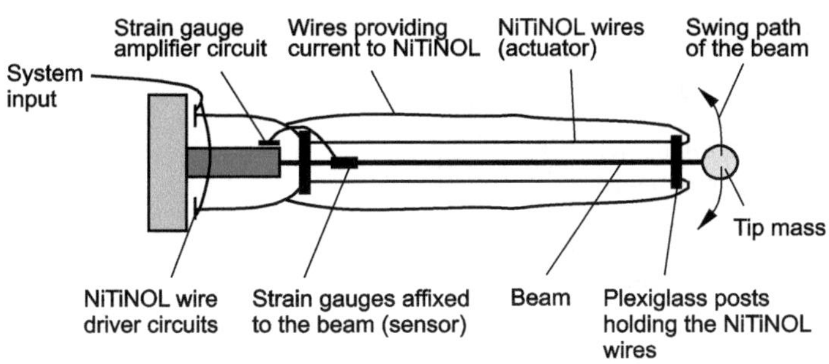

Fig. 4.1. Schematic of cantilever beam test article

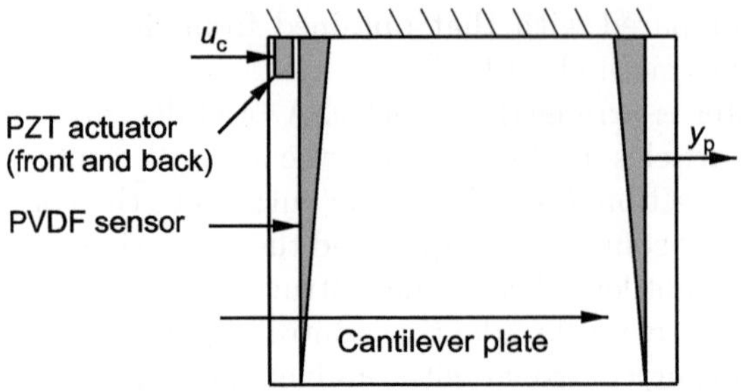

Fig. 4.2. Top view of the plate system

a nominal tip displacement of 0.5 inches. The PZT actuators are driven by a high-voltage amplifier such that the control input is in the range of $\pm 5\,\mathrm{V}$ and uses the full linear operating range of the PZT.

4.3 Conventional Model-Reference Adaptive Control Techniques

For many years, there have basically been two distinct methods for finding the solution of the adaptive control problem [2]. These are direct and indirect control methods. When the controller parameters $\boldsymbol{\theta}(k)$ are directly adjusted to reduce some norm of the output error between the reference model and the plant, this is called direct control or implicit identification. In indirect control, also referred to as explicit identification, the parameters of the plant are estimated as the elements of a vector $\hat{\boldsymbol{p}}(k)$ at each instant k, and the parameter vector $\boldsymbol{\theta}(k)$ of the controller is chosen assuming that $\hat{\boldsymbol{p}}(k)$ represents the true value of the plant parameter vector $\boldsymbol{p}$. Figures 4.3 and 4.4 respectively show the direct and indirect model-reference adaptive control structures for a linear time invariant (LTI) plant. It is important to note that in both cases efforts have to be made to probe the system to determine its behaviour because control action is being taken based on the most recent in-

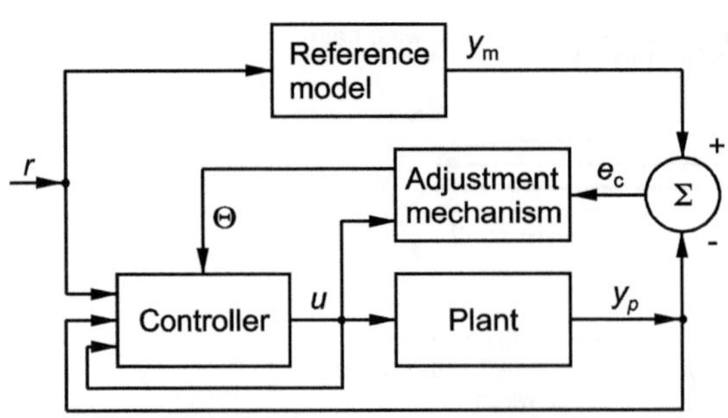

Fig. 4.3. Direct model-reference adaptive control structure

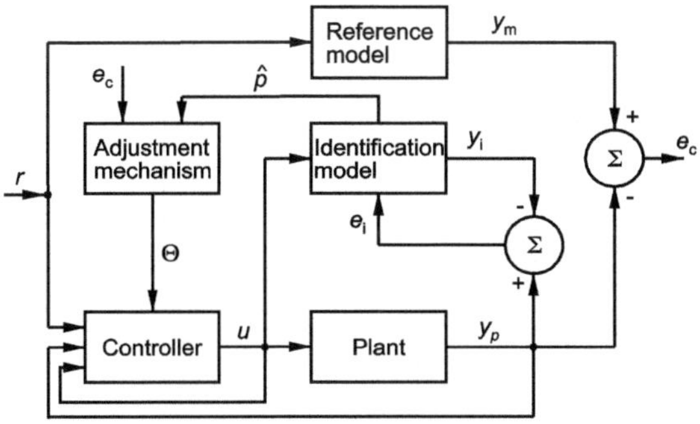

Fig. 4.4. Indirect model-reference adaptive control structure

formation available. The input to the process is therefore used simultaneously for both identification and control purposes. However, not every estimation scheme followed by a suitable control action will result in optimal or even stable behaviour of the overall system; therefore, considerable care must be taken in blending estimation and control schemes in order to achieve the desired performance [2].

4.3.1 Experimental Results

Direct Model-Reference Adaptive Control

The first controller implemented on the structure was the direct MRAC shown in Fig. 4.5. This gives a basis for comparison between direct and indirect control. Figure 4.6a shows a plot of the open-loop response envelope, the desired response envelope, and the closed-loop response achieved. As can be seen, the closed-loop system adapts to the reference-model response until the deadband is reached (after approximately 11 s), at which point adaptation is turned off. The deadband is inherent to the Nitinol wire actuators.

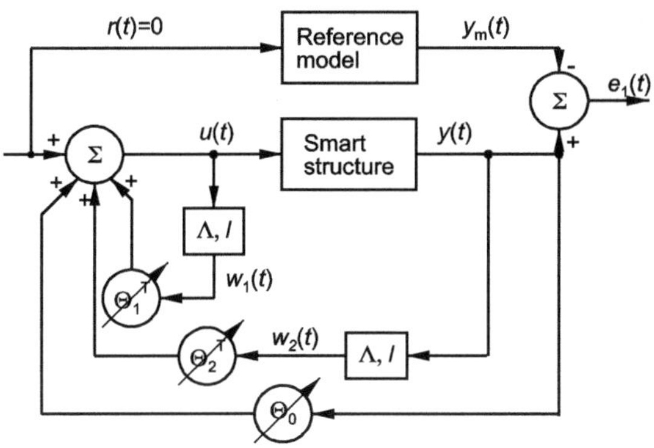

Fig. 4.5. Direct MRAC structure for smart structures

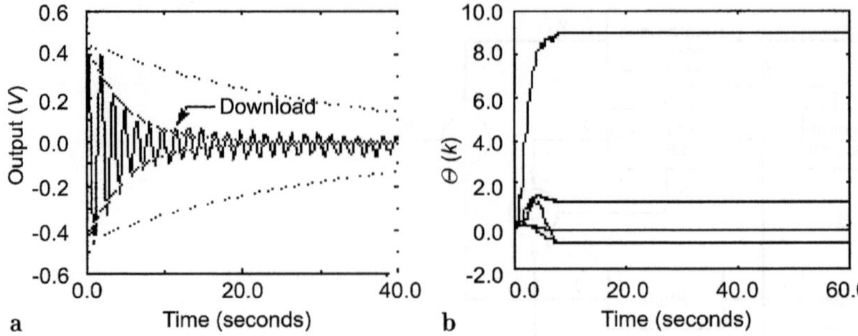

Fig. 4.6. Time-response comparison and evolution for a closed-loop system (direct MRAC): **a** variation of output with time, **b** variation of $\boldsymbol{\theta}(k)$ with time

In Fig. 4.6b, the control parameter vector $\boldsymbol{\theta}(k)$ has stabilized after about 8 s and before the deadband is reached.

The final values of the controller parameters are given in Table 4.1.

Table 4.1. Final values of the controller gains (direct MRAC)

$\boldsymbol{\theta}(k)$	Final value
$\boldsymbol{\theta}_1(k)$	$\left[-0.78069\ 0.75716\right]^{\mathrm{T}}$
$\boldsymbol{\theta}_0(k)$	8.92846
$\boldsymbol{\theta}_2(k)$	$\left[1.18814\ -0.16468\right]^{\mathrm{T}}$

Indirect Model-Reference Adaptive Control

Next, an indirect MRAC was implemented on the structure, as shown in Fig. 4.7. Figure 4.8a shows a plot of the open-loop response envelope, the desired response envelope, and the losed-loop response and Fig. 4.8b shows the time evolution of the control parameter vector. Again, the parameters converge after about 8 s with the deadband reached by 11 s.

The final values of the controller parameters are given in Table 4.2.

Table 4.2. Controller parameters of indirect MRAC

$\boldsymbol{\theta}(k)$	Final value
$\boldsymbol{\theta}_1(k)$	$\left[0.98276\ -1.01170\right]^{\mathrm{T}}$
$\boldsymbol{\theta}_0(k)$	9.36202
$\boldsymbol{\theta}_2(k)$	$\left[1.61385\ -0.14307\right]^{\mathrm{T}}$

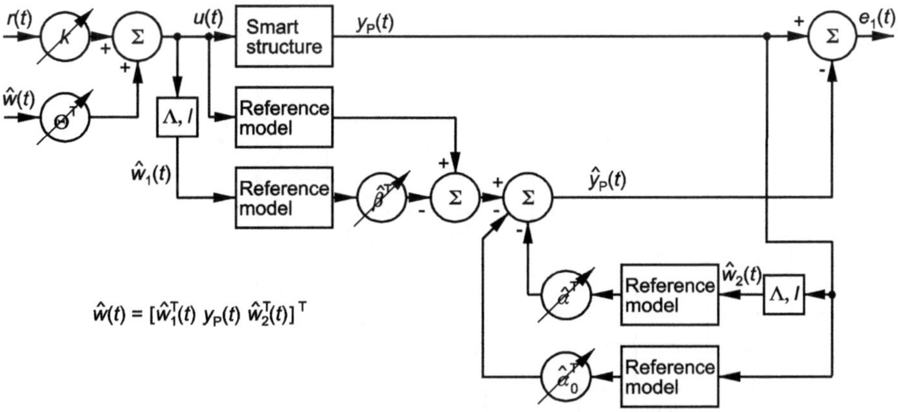

Fig. 4.7. Indirect MRAC regulator for smart structures

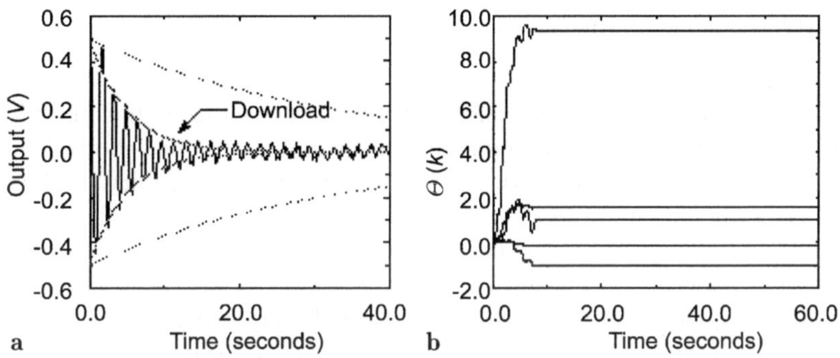

Fig. 4.8. Time-response comparison and evolution for a closed-loop system (indirect MRAC): **a** variation of output with time, **b** variation of $\boldsymbol{\theta}(k)$ with time

4.4 Adaptive Control Using Neural Networks

4.4.1 Neural Network-Based Model Reference Adaptive Control

After successful implementation of conventional model-reference adaptive controllers on smart structures, the next logical step was to investigate the possibility of using a neural network for adaptive control implementations. The linear and nonlinear mapping properties of neural networks have been extensively utilized in the design of multilayered feed-forward neural networks for the implementation of adaptive control algorithms [10].

A schematic diagram of the neural network-based adaptive control technique is shown in Fig. 4.9. A neural network identification model is trained using a static backpropagation algorithm to generate $\hat{\boldsymbol{y}}_{\rm p}(k+1)$, given past values of $\boldsymbol{y}$ and $\boldsymbol{u}$. The identification error is then used to update the weights of the neural identification model. The control error is used to update the

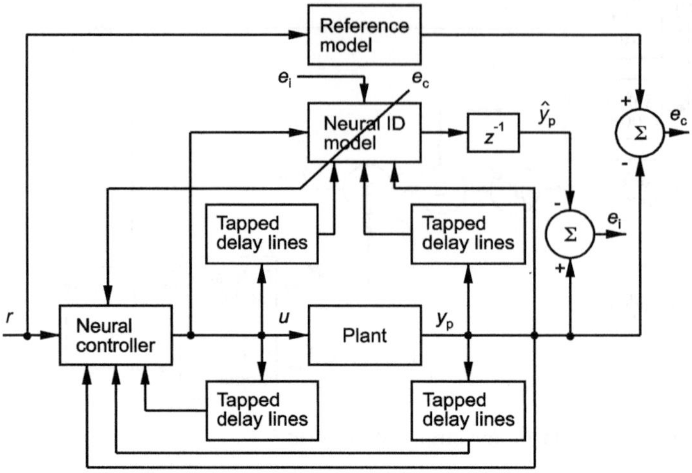

Fig. 4.9. MRAC using neural networks

weights of the neurocontrollers. Narendra [2] has demonstrated that closed-loop systems may result in unbounded solutions even if the plant is bounded-input and bounded-output stable. In order to avoid such instability, he has suggested that sufficient identification should be made before control is initiated. He has also suggested that the update rate of the identification and controller weights should be chosen carefully. Hoskins et al. [12] have presented a control optimization using a constrained iterative inversion process in order to dynamically search the input space of the identification process. This process provides stability and robustness measures for neural network-based adaptive control systems. We have utilized this technique for designing on-line adaptive algorithms; we have also developed a method for directly deriving a state-variable model using a multilayered neural network. These models are useful in generating adaptation data for neural controllers.

In the neural network-based adaptive control scheme, a neurocontroller is trained to approximate an inverse model of the plant. We have introduced an adaptive activation function for increasing the training rate of the neural controller, and the proposed function is described in this section.

Adaptive Activation Function

In order to train a neural controller, a multilayered network with linear activation functions was initially considered. During the training process, a large sum-squared error occurred due to the unbounded nature of the linear activation function that caused a floating point overflow. To avoid the floating point overflow we used the hyperbolic tangent activation functions in the hidden layers of the network. The network was unable to identify the forward

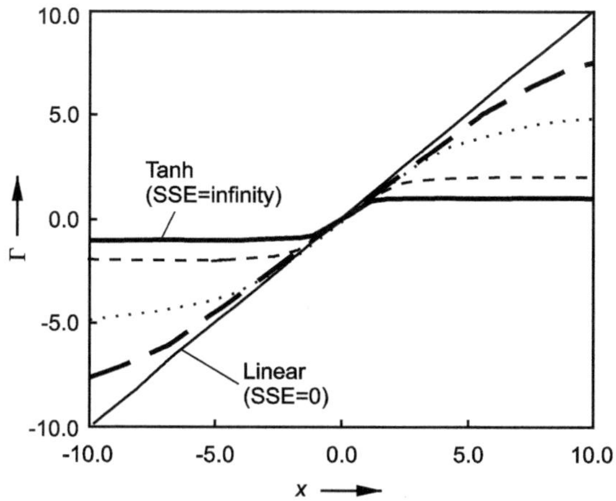

Fig. 4.10. Adaptive activation function

dynamics of the controller. To overcome this problem, we are proposing an activation function which adapts its shape depending upon the sum-squared error, as shown in Fig. 4.10.

The proposed adaptive activation function is governed by the equation

$$\Gamma(x) = \left(\frac{s+c}{s+1}\right)\tanh\left(\frac{s+1}{s+c}x\right), \tag{4.1}$$

where s is the sum-squared error over the previous time period and c is an arbitrary constant. The transition from a hyperbolic tangent to a linear function is shown in Fig. 4.10.

The function has the properties of

$$\begin{aligned} &\Gamma(x) \to \tanh(x) && \text{as} \quad s \gg c\,, \quad \text{and} \\ &\Gamma(x) \to c \cdot \tanh\left(\frac{x}{c}\right) && \text{as} \quad s \ll c\,. \end{aligned} \tag{4.2}$$

When the constant c is chosen large enough, the adaptive activation function can be replaced with a linear activation for implementation with no retraining needed. This procedure allows for a one-stage training session of the neural network.

For practical reasons when using the backpropagation training algorithm, it is convenient to be able to express the derivative of an activation function in terms of the activation function itself. The derivative of the adaptive activation function can also be expressed in the form

$$\frac{\mathrm{d}\Gamma(x)}{\mathrm{d}x} = 1 - [\Gamma(x)]^2\,. \tag{4.3}$$

The proposed activation function was successfully implemented in the training algorithm. The adaptive activation function is also feasible for hardware implementation. Specifically, the Intel i80170 electronically trainable artificial neural network (ETANN) chip [16] has an external voltage that controls the slope of the activation function. The control level could easily be made a function of the sum-squared error during training and held at the last sum-squared error achieved.

On-Line Adaptive Control Algorithm

A neural network-based model reference adaptive control scheme for nonlinear plants is presented in this section.

Let a system be described by a nonlinear difference equation

$$y_{\mathrm{p}}(k+1) = f[\boldsymbol{Y}_{\mathrm{k,n}}(k)] + g[\boldsymbol{U}_{\mathrm{k,m}}(k)] \,, \tag{4.4}$$

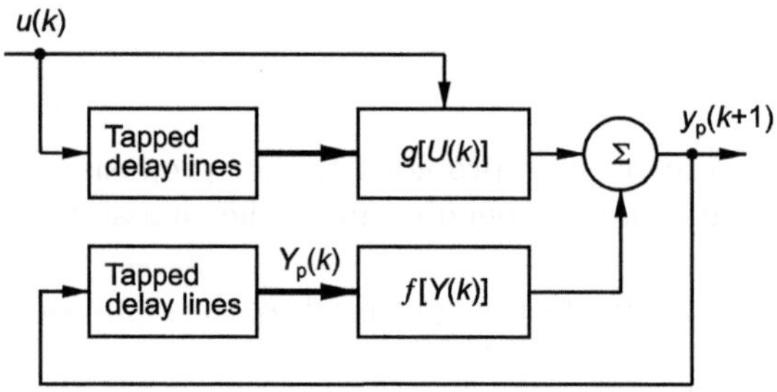

Fig. 4.11. Identification scheme for plant

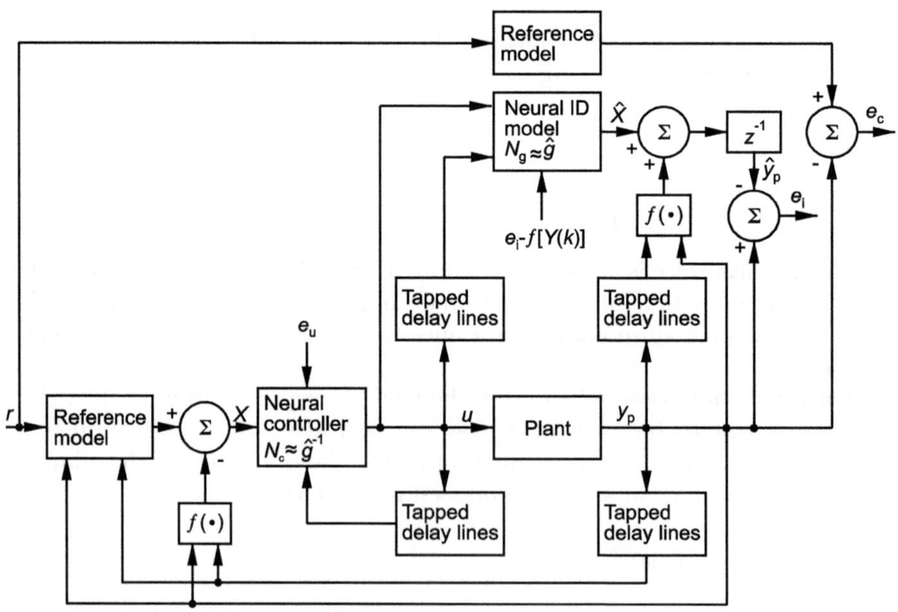

Fig. 4.12. Neural network MRAC block diagram

where f and g are both nonlinear functions in $\boldsymbol{y}$ and $\boldsymbol{u}$, respectively. This model requires two neural networks to identify the plant, one for each nonlinear function, as shown in Fig. 4.11.

For simplicity, let us assume that the function f is linear and g is nonlinear. Then a series parallel neural identification model will have the form

$$\hat{\boldsymbol{y}}_{\mathrm{p}}(k+1) = \hat{f}[\boldsymbol{y}_{\mathrm{k,n}}(k)] + N_{\mathrm{g}}[\boldsymbol{u}_{\mathrm{k,m}}(k)] , \tag{4.5}$$

where the reference model is represented by

$$\boldsymbol{y}_{\mathrm{mm}}(k+1) = f[\boldsymbol{y}_{\mathrm{k,n}}(k), \boldsymbol{R}_{\mathrm{k,m}}(k)] . \tag{4.6}$$

The desired control signal $\boldsymbol{u}(k)$ can be computed by

$$\boldsymbol{u}(k) = \hat{g}^{-1}[-\hat{f}[\boldsymbol{y}_{\mathrm{k,n}}(k)] + f[\boldsymbol{y}_{\mathrm{k,n}}(k), \boldsymbol{R}_{\mathrm{k,m}}(k)]] . \tag{4.7}$$

The schematic diagram of the model reference adaptive control system is shown in Fig. 4.12.

4.4.2 Neural Network-Based Optimizing Controller With On-Line Adaptation

In this section, a neural network-based design methodology is developed that utilizes the adaptability of neural networks to compensate for the time varying dynamical properties of smart structures. This formulation is designed to be implemented using the ETANN chip and also allows the designer to directly incorporate all the a priori information about the system that may be available. An important feature of this formulation is that it relies only on the experimental input/output data of the system for the design. The ability of neural networks to map nonlinear systems allows this formulation to be extended to incorporate nonlinearity in structural systems.

A functional block diagram of the controller is shown in Fig. 4.13, where the structural system can be represented by

$$\boldsymbol{y}_{\mathrm{p}}(k+1) = \Phi(\boldsymbol{y}_{\mathrm{p}}(k), \boldsymbol{u}_{\mathrm{c}}(k)) , \tag{4.8}$$

where Φ can be a linear or a nonlinear function.

The neural network in the controller block diagram has a model IV architecture with one hidden layer, as shown in Fig. 4.14. It is pretrained to the dynamics of the smart structural system using experimental input/output data. As shown in Fig. 4.15, the input vector to the network consists of $n+1$ samples of the plant input and $m+1$ samples of the plant output. The hidden and output layers have P and 1 neurons respectively.

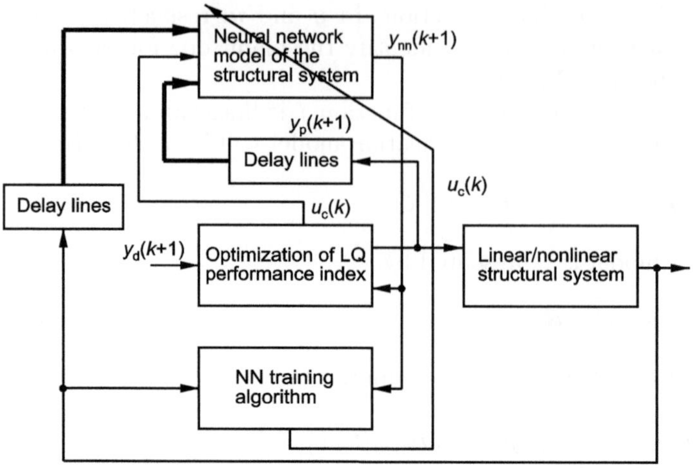

Fig. 4.13. Neural network-based controller block diagram

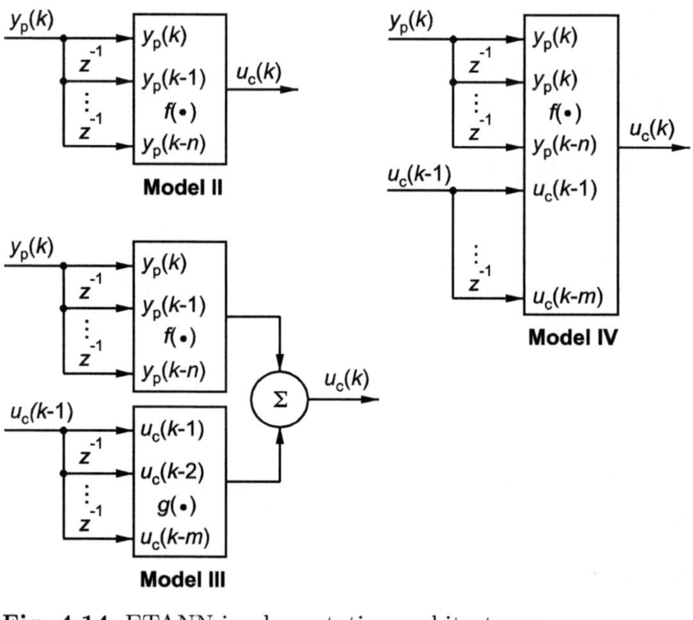

Fig. 4.14. ETANN implementation architectures

The activation function of the neurons in the hidden layer is the adaptive activation function (4.1). Models II and III are alternative neural network architectures that can be used to model a dynamical system. Model III is similar to model IV except for the additional external adder and separate network for the plant input and output parts. Model III can be used to implement high-order dynamical system models using hardware neural networks like the ETANN.

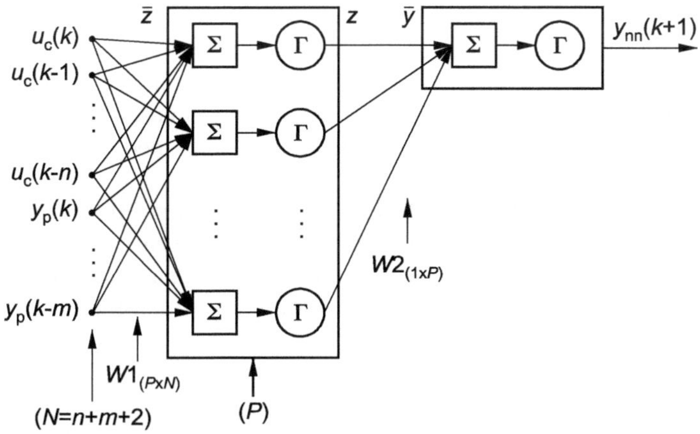

Fig. 4.15. Neural network architecture

The feed-forward equation of the network in Fig. 4.15 can be written as follows. Defining

$$W1_1 = \begin{bmatrix} W1_{11} \\ W1_{21} \\ \cdots \\ W1_{P1} \end{bmatrix}, \quad W1_2 = \begin{bmatrix} W1_{12} & \cdots & W1_{1N} \\ W1_{22} & \cdots & \cdots \\ \cdots & \cdots & \cdots \\ W1 & \cdots & W1_{PN} \end{bmatrix}, \quad \text{and}$$

$$u_2 = \begin{bmatrix} u_c(k-1) \\ \cdots \\ u_c(k-n) \\ y_p(k) \\ \cdots \\ y_p(k-m) \end{bmatrix}, \tag{4.9}$$

the outputs of each of the layers can be written as

$$z = \Gamma(W1_1 \cdot u_c(k) + W1_2 u_2) \quad \text{and} \tag{4.10}$$

$$y_{nn}(k+1) = \Gamma(W2 \cdot \Gamma(W1_1 \cdot u_c(k) + W1_2 \cdot u_2)) \,. \tag{4.11}$$

In the optimization block, the control input applied to the smart structural system is obtained by minimizing a generalized linear quadratic (LQ) performance index with weights on the control moves. The performance index is given by

$$\underset{u_c(k)}{\text{Min} J} = \frac{1}{2} E^{\mathrm{T}} Q E + \frac{1}{2} \Delta u^{\mathrm{T}} R \Delta u \,, \tag{4.12}$$

under the constraint given by (4.11). The error E is given by $E = y_{nn}(k+1) - y_d(k+1)$ and the control movement Δu is given by $\Delta u = u_c(k) - u_c(k-1)$.

Q is symmetric positive semi-definite and R is symmetric positive definite. The desired output y_{d} can either be a constant (regulator problem) or varying (tracking problem). The existence of weights on the control moves alleviates the problem of requiring large sampling times when non-minimum phase zeros exist in plants even in linear unconstrained optimization [17].

In addition to the constraint given by (4.11), any a priori information about the system or the sensors and actuators can be incorporated as additional constraints. Some of the commonly known constraints, such as control effort limits, actuator bandwidth limits and structural bandwidth limits, can be described by

$$\begin{aligned} \Delta u_{\mathrm{L}} \leq \ \Delta u \ \leq \Delta u_{\mathrm{H}} \quad & \text{actuator bandwidth limits} \\ u_{\mathrm{L}} \ \leq u_{\mathrm{c}}(k) \leq \ u_{\mathrm{H}} \quad & \text{control effort limits} \\ \Delta y_{\mathrm{L}} \leq \Delta y(k) \leq \Delta y_{\mathrm{H}} \quad & \text{structural bandwidth limits}\,. \end{aligned} \tag{4.13}$$

Since this study is restricted to a structural system that is operated in its linear region, the adaptive activation functions approximate to a linear function after sufficient training. Therefore the general nonlinear optimization problem given by (4.11)–(4.13) can be simplified for a linear case. After the neural network is sufficiently trained, (4.11) can be written as

$$y_{\mathrm{nn}}(k+1) = W_1 \cdot u_{\mathrm{c}}(k) + C_1\,, \tag{4.14}$$

where $W_1 = W2 \cdot W1_1$ and $C_1 = W2 \cdot W1_2 \cdot u_2$.

Substitution of (4.14) in the error equation above yields

$$E = y_{\mathrm{p}}(k+1) - y_{\mathrm{d}}(k+1)\,, \quad \text{or} \tag{4.15}$$

$$E = W_1 \cdot u_{\mathrm{c}}(k) + C_2 \tag{4.16}$$

where $C_2 = C_1 - y_{\mathrm{d}}(k+1)$. The control move equations can then be written as

$$\Delta u = u_{\mathrm{c}}(k) - u_{\mathrm{c}}(k-1) = C_2 - T_1\,, \tag{4.17}$$

where $T_1 = u_{\mathrm{c}}(k-1)$.

For a single input-single output system, the LQ performance index (4.12) can be written as

$$\begin{aligned} J &= \frac{1}{2}QE^2 + \frac{1}{2}R(\Delta u)^2\,, \quad \text{or} \\ J &= \frac{1}{2}(W_1^2 \cdot Q + R)u_{\mathrm{c}}^2(k) + (2W_1 \cdot C_2 \cdot Q - 2T_1 \cdot R)u_{\mathrm{c}}(k) \\ &\quad + \frac{1}{2}C_2^2 \cdot Q + \frac{1}{2}R \cdot T_1^2\,. \end{aligned} \tag{4.18}$$

This optimization problem can be solved for $u_{\mathrm{c}}(k)$ using any of the standard optimization algorithms [18].

4.5 Robust Controllers for Structural Systems

Design of controllers for smart structures requires accurate modeling of the system. Two main approaches traditionally used for obtaining models are analytical techniques and identification based on experimental data. Both of these approaches have advantages and disadvantages. The advantage of analytical modeling is that the models developed are physically intuitive and will help in the control system design. Most of the time the mathematical models developed are dependent on approximate representation of the physical phenomenon. The accuracy will depend on the complexity of the model and the assumed physical parameters incorporated in the model. Euler-Bernoulli beam model is an example of analytical models used for structural systems such as cantilever beams etc. As the size and complexity of the structural system becomes larger, analytical modeling becomes difficult in which case approximate analytical modeling methods such as finite element methods (FEM) are used. From the discussion above it is clear that the analytical modeling is prone to modeling errors due to the inaccurate physical parameters and approximation in the modeling process.

In contrast to analytical modeling, identification methods are not dependent on the physical structure of the systems but are solely data dependent. Hence the models so obtained are prone to problems such as noise and errors in the measurements, inadequate information content in the input/output data, limited wordlengths of the data acquisition system, and phase delays introduced by the aliasing and reconstruction filters etc. But, with proper care accurate models of the system including the affects of the actuators, sensors and interface electronics can be developed. In addition to the errors described above, departure of the models from the physical system characteristics can occur due to changes in the environmental conditions or operating conditions and degradation of the system due to use, ageing and other detrimental affects. The aggregate errors in the modeling are termed as uncertainty in the control literature and robust control methods are available to incorporate the effect of the uncertainties in the design. The robust control methods need some kind of quantification and representation of the manner in which the uncertainty affects the models. Due to inherent trade-off in the size of the uncertainty and the performance achieved by the control system, it is necessary that the uncertainty be represented as compactly as possible utilizing the manner in which the uncertainty affects the nominal model. Thus uncertainty is categorized as unstructured and structured uncertainty. In smart structural models both kinds of uncertainties are present with the unstructured uncertainty arising from the unmodeled or neglected dynamics and the structured uncertainty arising from the physical parameter variations and modeling errors in the nominal model. Linear fractional representations (LFRs) are widely used to describe the interaction of the uncertainty and the nominal models.

4.5.1 Uncertainty Modeling

Balas [19] developed a procedure to obtain nominal models and uncertainty representation for a multi input multi output flexible structure. Single input-multiple output (SIMO) models for each of the actuator are developed using a curve fitting method based on Chebyshev polynomials. The authors then develop an ad hoc model reduction technique based on a prior knowledge of the physical system to remove the additional dynamics obtained by combining multiple SIMO models. Based on the frequency response error between model and observed frequency response the authors generate uncertainty representation for the unmodeled dynamics. Campbell et al. [20–22] have developed a comprehensive uncertainty modeling procedure for structural systems. Their approach combines analytical modeling and identification techniques in order to retain the advantages of both the approaches. In this procedure a discrete extended Kalman filtering approach is used to estimate the modal parameters (natural frequencies, damping ratios) in a FEA model (finite element analysis, FEA) representation for the structure. Identification is performed based on several data sets obtaining the parameters corresponding to different conditions representing the errors due to noise, and variations in the operating conditions. From the set of estimated parameters, bounds and nominal values of the modal parameters are obtained which can incorporated in the modal representation of the structural system obtained from the FEA model. From the experience of the authors of the application of the procedure on the Middeck active control experiment (MACE) structure, it was found that the modeshape variations are unreliable and contributed to most of the conservativeness in their designs which prompted them to discard this variation in their future designs.

Boulet et al. [23] have considered the incorporation of structural uncertainties in coprime factorization models for structural systems. The uncertainties due to natural frequencies, damping ratios and modal gains are lumped together as unstructured uncertainties in the left coprime factors of the system normalized by low order weighting functions. The authors have successfully applied this method and designed H_∞ controllers for a large flexible space structure experiment. However, because of the individual weightings on the uncertainty due to each mode, the order of the controller design is higher than the original system model. Cockburn and Morton [24] have developed an algorithm to obtain a minimal order LFR of a system with polynomial parametric uncertainty. This method, coined by the authors as structured tree decomposition, decomposes the original polynomial matrix of the system into sums and products of simple factors named as leaves. The leaves are polynomial matrices with minimal LFRs. The LFR for the original system can be obtained by combining the LFRs for the leaves. Thus, this provides a general method of obtaining a minimal LFR for any system in which the parametric uncertainty appears polynomial. Because of the simple operations, the method can be automated.

Smith et al. [25] have approximated variations in natural frequencies and damping ratios as circular disks around the nominal eigenvalues of the structural system. These circular disks are then approximated as a complex unstructured uncertainty block weighted by a diagonal matrix containing radii of the disks of eigenvalue variations of the nominal frequencies. Because of the rectangular nature of the variations in eigenvalues due to variations in natural frequencies and damping ratios, this uncertainty representation included more plants than those for the specified variations. This will lead to conservative designs. To reduce this effect the authors modified the nominal eigenvalues corresponding to the nominal modes such that the uncertainties could be accommodated with uncertainty disks with minimum possible radii. Because of the approximation of the variations in the eigenvalues as an unstructured uncertainty, the resulting designs are conservative. The authors apply an H_∞/μ synthesis approach, but could only achieve robustness to only 1% variation in the damping ratios and 0.1% variation in the natural frequencies. In a similar procedure, Lashlee et al. [26] have formulated natural frequency variations in the smart structural models as an LFR with structured real parametric uncertainties and applied mixed H_2/H_∞ controller design procedure for designing robust controllers.

Butler [27] formulated a LFR for smart structures based on measurement errors during the identification process. Based on this uncertainty model, mixed H_2/H_∞ controllers [29] were designed incorporating actuator saturation.

4.5.2 Robust Control Design Methods

Balas and Doyle [28] formulate the problem of disturbance rejection problem for a prototype space structural system. They used a structured singular value μ synthesis approach considering uncertainties due to unmodeled dynamics and equivalent uncertainty formulations of the performance requirements on actuator limits, disturbance rejection and sensor noise by choosing appropriate weightings.

Balas and Young [29] have considered the design problem of the NASA Langley Minimast structure for disturbance rejection performance. Uncertainties due to actuator variations, unmodeled dynamics and natural frequency and damping ratio variations for modes in the controller bandwidth are considered in the design. They used two different uncertainty representations to represent the natural frequency variations in the natural frequency and damping ratios, the first one being the complex structured uncertainty, and the second one involves real-parametric structured uncertainty. In their designs the authors used the complex μ synthesis procedure based on D–K iteration, on the complex structured uncertainty model while using the less conservative real μ analysis to verify the designs.

Joshi and Kelkar [30], have developed an iterative procedure by combining LQG type synthesis with robustness and performance analysis to design controllers to reduce the vibrations due to flexible aeroelastic modes in a supersonic aircraft. The controller design utilizes a model including only a few significant modes of interest. The rest of the modes are considered as uncertainty. In the analysis iteration the robustness and performance of the controller is tested and the design iterations are continued until the desired performance and robustness are achieved. The drawback of this procedure is that because of the non-intuitive nature of the adjustment of weighting functions in LQG design, expertise is needed to achieve satisfactory designs within a lesser number of iterations. How et al. [31, 32] have used synthesis techniques based on Popov stability analysis for the control of the Middek active control experiment (MACE) system. Uncertainties due to natural frequency uncertainty is formulated as structured real parametric uncertainty. The synthesis method is based on a Quasi-Newton optimization procedure that is computationally intensive. The authors show by numerical examples that the Popov stability condition is less conservative than the complex μ synthesis procedure.

4.6 Summary

In this study, adaptive control algorithms have been utilized for designing active controllers for smart structure test articles. Adaptive control schemes require only a limited a priori knowledge about the system in order to be controlled. The availability of limited control force and inherent deadband and saturation effects of shape memory actuators are incorporated in the selection of the reference model. The vibration suppression properties of smart structures were successfully demonstrated by implementing the conventional model reference adaptive controllers on the smart structure test articles. The controller parameters converged to steady state values within 8 s for both direct and indirect MRACs.

Various neural network-based adaptive control techniques were discussed in this study. A major problem in implementing neural network-based MRACs is the translation of the output error between the plant and the reference model so as to train the neural controller. A technique called iterative inversion, which inverts the neural identification model of the plant for calculating neural controller gains, has been used. Due to the real-time computer hardware limitations, the performance of neural network-based adaptive control systems is verified using simulation studies only. These results show that neural-network based MRACs can be designed and implemented on smart structures.

A neural network-based control algorithm based on a LQ performance index which can be implemented using the ETANN chip has been developed. This formulation incorporates a priori information about the structural system. Information such as limits on the control effort and limits on the bandwidths of the sensors and actuators can be incorporated in this formulation. The on-line adaptability property of the ETANN chip-based neural network is also utilized to adapt the controller to time-varying structural systems. The capabilities of this algorithm have been demonstrated on the smart plate system through simulation studies. The ability of neural networks to map nonlinear dynamics as well as linear dynamics makes the control algorithm valid for control of smart structural systems with nonlinearities.

References

1. Aström, K.; Wittenmark, B.: *Adaptive Control.* Addison-Wesley, Reading, MA (1989), pp. 105–156
2. Narendra, K.; Annaswamy, A.: *Stable Adaptive Control.* Prentice Hall, Englewood Cliffs, NJ (1989), pp. 21–28, pp. 182–232, pp. 318–345
3. Narendra, K.: *Adaptive Control of Dynamical Systems.* In: 'Handbook of Intelligent Control: Neural, Fuzzy and Adaptive Approaches', White, D.; Sofge, D., Van Nostrand Reinhold (Eds.), New York, NY (1992)
4. Narendra, K.; Duarte, M.: *Combined Direct and Indirect Adaptive Control of Plants with a Relative Degree Greater than One.* Technical Report #8715., Center for Systems Science, Yale University, New Haven, CT (November 1987)
5. Isermann, R.: *Digital Control Systems.* Springer-Verlag, Vol. 1: 'Fundamentals, Deterministic Control'; 2nd rev. ed. (1989); Vol. 2: 'Stochastic Control, Adaptive Control Multivariable Control, Adaptive Control, Applications'; 2nd rev. ed. (1991)
6. Rao, V.; Damle, R.; Tebbe, C.; Kern, F.: *The Adaptive Control of Smart Structures using Neural Networks.* Smart Materials and Structures, No. 3 (1994), pp. 354–366
7. Chen, F.; Khalil, H.K.: *Adaptive Control of Nonlinear Systems using Neural Networks – A Dead-Zone Approach.* Proc. Amer. Control Conf. (1990), pp. 667–672
8. Chen, F.: *Adaptive Control of Nonlinear Systems using Neural Networks.* A Ph.D. Dissertation, Dept. Elec. Eng., Michigan State University (1990)
9. Tzirkel-Hancock, E.; Fallside, F.: *Stable Control of Nonlinear Systems using Neural Networks.* Tech. Report CUED/F-INFENG/TR.81, Cambridge University. Eng. Dept. (July 1991)
10. Narendra, K.; Parthasarathy, K.: *Identification and Control of Dynamical Systems Using Neural Networks.* IEEE Trans. Neural Networks (March 1990), pp. 4–27
11. Hoskins, D.A.: *Neural Network Based Model-Reference Adaptive Control.* Ph. D. Dissertation, University of Washington, UMI Dissertation Services, Ann Arbor, MI (1990)
12. Hoskins, D.A.; Hwang, J.N.; Vagners, J.: *Iterative Inversion of Neural Networks and Its Application to Adaptive Control.* IEEE Trans. Neural Networks (March 1992), pp. 292–301
13. Damle, R.; Lashlee, R.; Rao, V.; Kern, F.: *Identification and Robust Control of Smart Structures using Artificial Neural Networks.* Int. J. Smart Struct. Materials, vol. 3, no. 1 (March 1994), pp. 35–46

14. Lashlee, R.; Butler, R.; Rao, V.; Kern, F.: *Robust Control of Flexible Structures Using Multiple Shape Memory Alloy Actuators.* North American Conf. on Smart Structures and Materials, Albuquerque, NM (Feb. 1993)
15. Butler, R.; Rao, S.V.: *Identification and control of two-dimensional smart structures using distributed sensors.* Proc. North American Conf. on Smart Structures and Materials, San Diego, CA, SPIE 2442 (March 1995), pp. 58–68
16. *Intel 80170NX Electrically Trainable Analog Neural Network Data Book.* (June 1991)
17. Garcia, C.E.; Morari, M.: *Internal Model Control – Multivariable Control Law Computation and Tuning.* Industrial Engineering Chemical Process Design and Development, 24 (1985), pp. 484–494
18. *OPTIMIZATION TOOLBOX Users Guide.* The MathWorks Inc. (November 1990)
19. Balas, G.J.; Doyle, J.C.: *Identification of flexible structures for robust control.* Proc. Amer. Control Conf., 3 (1989), pp. 2566–2571
20. Campbell, M.E.: *Identification and parameter estimation for control design.* IFAC 13th Triennial World Congress (1996), pp. 209–214
21. Campbell, M.E.; Crawley, E.F.: *Development of Structural Uncertainty Models.* J. Guidance, Control and Dynamics, 20, no. 5 (1997) pp. 841–849
22. Campbell, M.E.; Grocott, S.C.O.: *Parametric uncertainty model for control design and analysis.* IEEE Trans. Control Systems Technol., 7, no. 1 (1999), pp. 85–96
23. Boulet, B.; Francis, B.A.; Hughes, PC.; Hong, T.: *Uncertainty modeling and experiments in* H_∞ *control of large flexible space structures.* IEEE Trans. on Control Systems Technol., 5, no. 5 (1997), pp. 504–519
24. Cockburn, J.C.; Morton, B.G.: *Linear fractional representations of uncertain systems.* Automatica, 30, no. 7 (1997), pp. 1263–1271
25. Smith, R.S.; Chu, C.-C.; Fanson, J.L.: *The design of H controllers for an experimental non-collocated flexible structure problem.* IEEE Trans. on Control Systems Technol., 2, no. 2 (June 1994), pp. 101–109
26. Lashlee, R.; Rao, V.S.; Kern, F.J.: *Mixed* H_2/H_∞ *Optimal Control of Smart Structures.* Proc. 33rd Conf. on Decision and Control, Lake Buena Vista, FL (1994), pp. 115–119
27. Butler, R.; Rao, V.S.; Sana, S.: *Design of Robust Controllers for Smart Structural Systems with Actuator Saturation.* J. of Intelligent Material Systems and Structures, 8, no. 9 (1997), pp. 721–811
28. Balas, G.J.; Doyle, J.C.: *Control of lightly damped, flexible modes in the controller crossover region.* J. of Guidance, Control & Dynamics, 17, no. 2 (1994), pp. 370–377
29. Balas, G.J.; Young, P.M.: *Control design for variations in structural natural frequencies.* J. of Guidance, Control & Dynamics, 18, no. 2 (1995), pp. 325–332
30. Joshi, S.M.; Kelkar, A.G.: *Inner loop control of supersonic aircraft in the presence of aeroelastic modes.* IEEE Trans. on control systems technol., 6, no. 6 (1998), pp. 730–739
31. How, J.P.; Hall, S.R.; Haddad, W.M.: *Robust Controllers for the Middeck Active Control Experiment using Popov Controller Synthesis.* IEEE Trans. on Control System Technol., 2, no. 2 (1994), pp. 73–87
32. How, J.P.; Collins, E.G.; Haddad, W.M.: *Optimal Popov controller analysis and synthesis for systems with real parameter uncertainties.* IEEE Trans. on Control Systems Technol., 4, no. 2 (1996), pp. 200–207

5 Simulation of Adaptronic Systems

H. Baier, F. Döngi, U. Müller

5.1 Introduction

In an adaptronic system the system response is observed via sensors in order to control and enhance the performance via integrated actuators which are being properly triggered by controllers. Adaptronic systems are usually dynamic systems with time-varying states subjected to external disturbances, and they 'adapt' to these disturbances in order to deliver the required performance. For the simulation of such adaptronic systems, control and system theory together with proper modelling of the plant are to be applied. Plant models might be nonlinear or linear models. They usually have to be parameterised for design studies and for final system optimisation. In the following, the focus will be on linearised, time-continuous descriptions of adaptronic mechanical systems and structures. Since related discretised models are usually quite large, proper model reduction techniques for integrated simulation of controller and plant have to be applied.

A general overview in Sect. 5.2 about the simulation of adaptronic (mechanical) systems is followed by a discussion of steps to be taken towards a mathematical model of an adaptronic structure in Sect. 5.3. Once a mathematical model of the adaptronic system has been derived and implemented numerically, analysis and simulations have to be carried out to characterise its dynamic behaviour. A survey of related methods and algorithms is given in Sect. 5.4. Simulation goals such as stability, performance and robustness are discussed, especially for the case of actively controlled structures. The modelling and simulation process is also demonstrated by a practical example in Sect. 5.5, while Sect. 5.6 gives an outlook on adaptronic system optimisation techniques.

5.2 Related Elements of System Theory

5.2.1 Linear and Nonlinear Systems

Most dynamic systems exhibit nonlinear characteristics to some extent, mainly due to strong variations in response quantities, such as large displacements or large strain, leading to material nonlinearities. Some smart materials such as electrostrictive and shape memory alloys which are often used in actuators of adaptronic systems, imply nonlinear constitutive behaviour which then requires special effort with respect to modelling and simulation techniques (see for example [1, 2]). Assuming that nonlinear effects are either small or proper linearization, e. g. via a Taylor series expansion around a chosen state in the system, can be carried out, then linear theory can be applied. Consider for example the system dynamic behaviour described by a set of n first-order nonlinear differential equations

$$\dot{\boldsymbol{x}} = f(\boldsymbol{x}, \boldsymbol{u}) \,, \tag{5.1}$$

where $\boldsymbol{x}$ and $\boldsymbol{u}$ denote state variables and external influences on the system, respectively. The dot symbolises differentiation with respect to time. A linearised representation around an equilibrium state $\boldsymbol{x} = 0$, $\boldsymbol{u} = 0$ is given by

$$\dot{\boldsymbol{x}} = \frac{\partial}{\partial \boldsymbol{x}} f(\boldsymbol{x}, \boldsymbol{u}) \bigg|_{\boldsymbol{x}=0, \boldsymbol{u}=0} \cdot \boldsymbol{x} + \frac{\partial}{\partial \boldsymbol{x}} f(\boldsymbol{x}, \boldsymbol{u}) \bigg|_{\boldsymbol{x}=0, \boldsymbol{u}=0} \cdot \boldsymbol{u} \,. \tag{5.2}$$

5.2.2 State-Space Representation

For coupled simulation of a dynamic system with second order differential equations together with its control part, the transformation to a set of first order of differential equations into the so called state space representation is desirable in order to simplify the solution process. This has to be achieved by a duplication of the number of equations.

The state-space representation of a linear or linearized system consists of the system equation

$$\dot{\boldsymbol{x}} = \boldsymbol{A}\boldsymbol{x} + \boldsymbol{B}\boldsymbol{u} \tag{5.3}$$

which is connected with the output equation

$$\boldsymbol{y} = \boldsymbol{C}\boldsymbol{x} + \boldsymbol{D}\boldsymbol{u} \,. \tag{5.4}$$

The system's dynamic response variables such as displacements and velocities are contained in the state vector $\boldsymbol{x}(n \times 1)$. Physical quantities that exert excitations on the system (e. g. external forces and actuator forces) are collected in an input vector $\boldsymbol{u}(p \times 1)$, and measured quantities (sensor signals) in an output vector $\boldsymbol{y}(q \times 1)$. For actively controlled adaptronic systems, the task is to generate a suitable input $\boldsymbol{u}(t)$ from a given output $\boldsymbol{y}(t)$ such that the system exhibits desirable dynamic behaviour.

The matrix $\boldsymbol{A}(n \times n)$ is called the state or system matrix, which comprises the properties of the adaptronic (controlled) plant. The input matrix $\boldsymbol{B}(n \times p)$ maps the excitation and control forces to the relevant degrees of freedom of the plant model, while the output matrix $\boldsymbol{C}(q \times n)$ relates the state vector with measured responses. The feed through matrix $\boldsymbol{D}(q \times n)$

of the system is zero except for cases where the input quantities (actuator forces and moments) have a direct influence on the sensor measurements. For example, this happens in the case of active struts based on integrated strain actuators in a truss structure, where sensors measure the displacement, strain or force in the strut and the strain induced by the actuator directly influences the sensor signal [3].

5.2.3 Controllability and Observability

The efficiency and proper positioning of actuators and sensors in adaptronic systems can be analysed using the concepts of controllability and observability. To make the basic ideas more clear, adaptronic structures are taken as an example. Loosely speaking, controllability and observability also mean that the actuator force and sensor vectors are not orthogonal and preferably parallel to the relevant vector (e. g. natural mode) or state to be controlled or observed.

The dynamic behaviour of structural systems can be characterised in terms of natural frequencies and modes, including possible rigid-body modes in multi-body systems. If the natural modes of a system are supposed to be actively controlled using actuators and sensors, these elements must be able to influence and sense, respectively, the appropriate modal oscillations. If a mode cannot be detected by a given sensor, it is not observable. Analogously, a pin force actuator located in a node of a mode shape is unable to excite this mode, which is then said to be not controllable.

If an adaptronic system is modelled in a state-space description (5.3), (5.4), its observability and controllability can be determined numerically by various methods. A common way is to compute the eigenvalues of the controllability and observability Gramians

$$\boldsymbol{P} = \int_0^\infty \mathrm{e}^{\boldsymbol{A}t}\boldsymbol{B}\boldsymbol{B}^{\mathrm{T}}\mathrm{e}^{\boldsymbol{A}^{\mathrm{T}}t}\mathrm{d}t\,, \quad \boldsymbol{Q} = \int_0^\infty \mathrm{e}^{\boldsymbol{A}^{\mathrm{T}}t}\boldsymbol{C}^{\mathrm{T}}\boldsymbol{C}\mathrm{e}^{\boldsymbol{A}t}\mathrm{d}t\,. \tag{5.5}$$

$\boldsymbol{P}$ and $\boldsymbol{Q}$ possess real non-negative eigenvalues. Large eigenvalues indicate good controllability and observability, respectively, while very small or zero eigenvalues correspond to non-controllable and non-observable states, respectively. Every linear time-invariant system ((5.3), (5.4)) can be transformed into its balanced realisation [4]. For collocated actuators and sensors $\boldsymbol{P}$ equals $\boldsymbol{Q}$, with the Hankel singular values σ_{k}:

$$\sigma_{\mathrm{k}} = \sqrt{\lambda_{\mathrm{k}}} \tag{5.6}$$

with λ_{k} being the eigenvalues of $\boldsymbol{PQ}$. The Hankel values can be applied to check for both controllability and observability simultaneously. For lightly damped adaptronic structures, i. e. those where damping has a small influence on eigenvalues and modes, the Hankel singular values can be determined from modal data [6] which makes numerical application quite fast and efficient. Complementary to this, proper interpretation of different simulation results together with engineering insight shall also lead to proper actuator and sensor positions in order to achieve controllability and observability. In addition, technical properties of actuators and sensors have to be considered as well. For example, the actuators have to provide sufficient stroke within relevant frequency bands ('control authority'), and sensors have to be sufficiently accurate and stable over time.

5.2.4 Stability

An important condition for a controlled dynamic system is its stability. The notion of stability implies that, after a bounded disturbance, the state variables of the system remain bounded, i. e. they stay within a defined space around a selected state (or approach this state asymptotically). In stable systems finite inputs lead to finite outputs. A mathematically more rigorous definition is given by the Lyapunov condition [1].

Controllers for adaptronic systems can be designed based on general proofs of stability, as in the case of collocated dissipative controllers, or based on a mathematical model of the system. In the latter case, it is often important to represent the dynamics of a system very accurately because the stability and performance of the controller can only be checked with the mathematical model in the first place. Discrepancies between the dynamic behaviour of the mathematical model and the real adaptronic system may lead to loss of performance and even instability when the controller is finally implemented with the real system (see Sect. 5.4.2). This then would have to be corrected by sometimes time-consuming adjustment of the controller parameters to the actual plant properties and behaviour, if possible at all.

The stability of a controlled dynamic system is said to be robust if the controller designed using a mathematical model stabilizes the real system in spite of modelling errors and/or parameter changes in the adaptronic system. A similar definition holds for the robustness of performance.

5.2.5 Alternative System Representations

An equivalent representation of a state-space system ((5.3), (5.4)) is the Laplace transform transfer function description,

$$\boldsymbol{G}(s) = \boldsymbol{C}(s\boldsymbol{I} - \boldsymbol{A})^{-1}\boldsymbol{B} + \boldsymbol{D}\ , \tag{5.7}$$

where s is a complex variable. The elements of matrix $\boldsymbol{G}(s)$ are transfer functions $n(s)/d(s)$ with nominator and denominator polynomials $n(s)$ and $d(s)$, respectively. The common denominator of these transfer functions is the characteristic polynomial of $\boldsymbol{G}(s)$, its roots are equivalent to the eigenvalues of the state matrix $\boldsymbol{A}$. Poles and zeros of the system are evident if the transfer-function matrix $\boldsymbol{G}(s)$ is transformed into its Smith-MacMillan form [7]. Variations of the transfer-function matrix representation are the zero-pole-gain and partial fraction models, where the transfer functions in (s) are factorised into nominator/denominator factors and partial fractions, respectively. Such transfer functions are also quite often used complementary to or instead of the state space representation since due to their output-input-relation they give a good and direct insight into the system behaviour.

5.3 Modelling of Adaptronic Structures

In order to make the discussion of simulation of adaptronic systems more concrete, this chapter concentrates on adaptronic or smart structural systems. They consist of the structure as a dynamic system combined with integrated, multifunctional (i. e. load-bearing) smart materials such as piezoelectric or magnetostrictive materials. If these material types are used as actuator and/or sensor materials, a linearisation of the system equations may easily be found. From the modelling point of view, the situation is much more complex in the case of electrostrictive materials or shape memory alloys [2] that exhibit highly nonlinear constitutive behaviour, or in the case of smart polymer gels [8] that imply coupling between the mechanics of large displacements, electro-diffusion processes and chemical reactions (see Chap. 6).

Figure 5.1 outlines the modelling and simulation process for adaptronic structures. Starting from proper structural modelling with the establishment of the equations of motion including excitations as well as actuator and sensor dynamics, the resulting full order model often has to be significantly reduced for investigating the adaptronic structure's performance as well as the influence of different design and controller parameters. The essential steps of this process are discussed in more detail in the following sections and are also demonstrated with the practical example in Sect. 5.5.

5.3.1 Basic Equations of Structural Mechanics

Consider a linear elastic continuum, which may consist of passive and active, i. e. adaptronic, elements. The dynamic equilibrium of the structure can be formulated using the principle of virtual displacements [9] including inertia loads. To express the internal strain energy in terms of displacement variables, the kinematics of the structure has to be considered. Various types of mechanical structures, e. g. beams, plates, or shells, are defined by kinematics relations and constraints.

Finally, stress $\boldsymbol{\sigma}$ and strain $\boldsymbol{\epsilon}$ are related to each other by the constitutive law of the material. For passive, linear elastic materials the generalized Hooke's law is a valid approximation:

$$\boldsymbol{\sigma} = \boldsymbol{E}\,\boldsymbol{\epsilon}\,. \tag{5.8}$$

Here, $\boldsymbol{E}$ denotes the elasticity tensor of the material. Examples of constitutive laws for smart materials are given in the subsequent section.

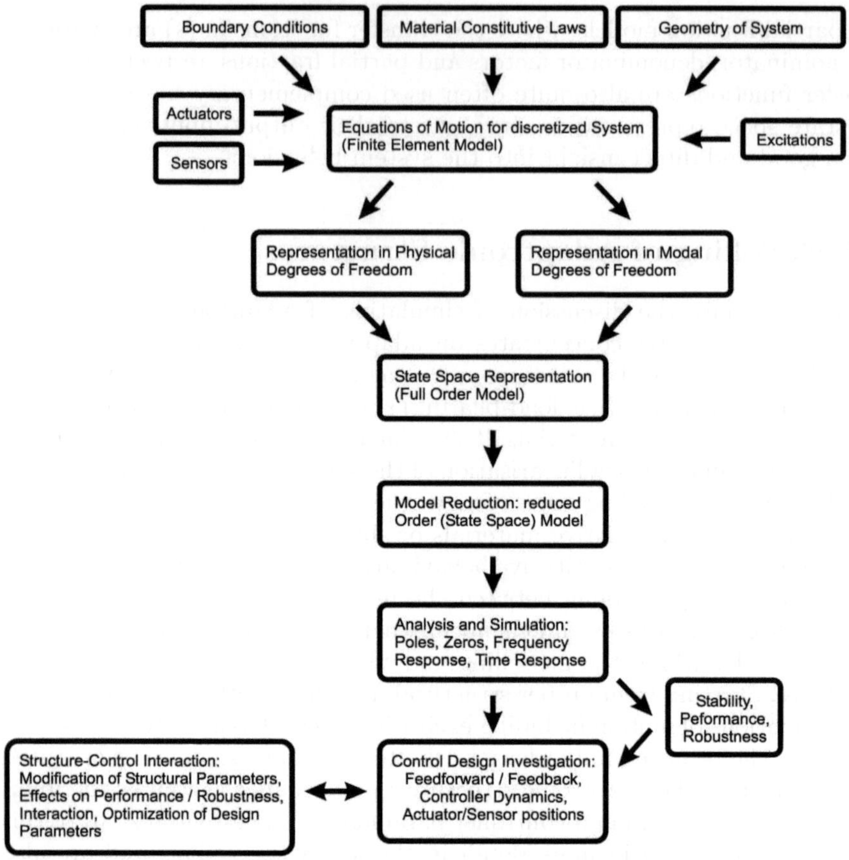

Fig. 5.1. Modelling and simulation process for adaptronic structures

In case of thermally induced strain $\epsilon_{\mathrm{T}} = \alpha_{\mathrm{T}} \Delta T$ this relation extends to

$$\boldsymbol{\sigma} = \boldsymbol{E}(\boldsymbol{\epsilon} - \boldsymbol{\alpha}_{\mathrm{T}} \Delta T) \tag{5.9}$$

with $\boldsymbol{\alpha}_{\mathrm{T}}$ containing the coefficients of thermal expansion of the material under consideration and ΔT characterising the temperature change related to a stress-free state. The equations of dynamic equilibrium, kinematics, and constitutive behaviour are combined in the variational formulation on which the discretisation using the finite element method (FEM) is based (see Sect. 5.3.3).

5.3.2 Constitutive Laws of Smart Materials

In the case of smart materials, Hooke's law has to be substituted or amended by a constitutive law that couples the mechanical properties of the material with other physical properties such as electric, magnetic, or thermal entities.

Among the large variety of smart materials discussed today, piezoelectric and magnetostrictive materials can be described by linearised constitutive laws that are given below. Other widely used material types, such as electrostrictive or shape memory materials, exhibit strongly nonlinear behaviour, the modelling of which may become quite demanding.

Piezoelectric Materials. In the constitutive law of piezoelectrics, a coupling between strain $\boldsymbol{\epsilon}$, stress $\boldsymbol{\sigma}$, electric displacement $\tilde{\boldsymbol{D}}$ and electric field $\tilde{\boldsymbol{E}}$ exists as follows:

$$\boldsymbol{\sigma} = \boldsymbol{E}(\boldsymbol{\epsilon} - \boldsymbol{d}\tilde{\boldsymbol{E}}) \tag{5.10}$$

$$\tilde{\boldsymbol{D}} = \boldsymbol{d}\boldsymbol{\sigma} + \tilde{\boldsymbol{\epsilon}}\tilde{\boldsymbol{E}} \; . \tag{5.11}$$

Here, $\boldsymbol{d}$ and $\tilde{\boldsymbol{\epsilon}}$ are the matrices of piezoelectric coupling and dielectric constants, respectively.

Magnetostrictive Materials. Substituting magneto-mechanics for electro-mechanics, mechanical strain $\boldsymbol{\epsilon}$ and stress $\boldsymbol{\sigma}$ are coupled with magnetic field intensity $\tilde{\boldsymbol{H}}$ and flux density $\tilde{\boldsymbol{B}}$ as follows:

$$\boldsymbol{\sigma} = \boldsymbol{E}(\boldsymbol{\epsilon} - \boldsymbol{d}_{\mathrm{m}}^{\mathrm{T}}\tilde{\boldsymbol{H}}) \; , \tag{5.12}$$

$$\tilde{\boldsymbol{B}} = \boldsymbol{d}_{\mathrm{m}}^{\mathrm{T}}\boldsymbol{\sigma} + \tilde{\boldsymbol{\mu}}^{\mathrm{T}}\tilde{\boldsymbol{H}} \; . \tag{5.13}$$

Here, $\boldsymbol{d}_{\mathrm{m}}$ and $\tilde{\boldsymbol{\mu}}$ denote the magnetostrictive coupling and free permeability matrices.

A comparison of the constitutive (5.10) and (5.12) with the stress-strain (5.9) shows a favourable analogy. This is often used to model smart materials as a part of an adaptronic structure e. g. by substituting $\boldsymbol{\alpha}_{\mathrm{T}}$ by analogous parts of the constitutive equations of the smart material in the finite element model of the adaptronic structure (see also below).

5.3.3 Finite Element Modelling

In the domain of structural mechanics, the finite element method (FEM) is a widespread and powerful tool for numerical analysis of complex structures (see for instance [9]). A large number of commercial and also public domain codes exist. FE codes based on the principle of virtual displacements model the spatial distribution of displacements using so called test or interpolation functions for the displacement field with discrete displacements at finite element nodal points as unknowns to be determined. In this manner the FEM reduces the continuous formulation of the system dynamics to a discrete set of differential equations for specified nodal degrees of freedom. A full coupling between mechanical and electrical or magnetic properties, respectively, would require the introduction of additional degrees of freedom to the system. In most formulations for piezoelectrics, the electric potential is considered at element nodes. Modelling of magnetic fields leads to field intensity degrees of freedom.

While only some of the FE codes include the fully coupled constitutive laws of smart materials, many formulations exist however for piezoelectric materials [10].

As an approximation, the electric or magnetic degrees of freedom can be neglected if the influence of the mechanical properties on these entities is fairly weak. In general this is the case if large mechanical structures with only a small percentage of adaptronic elements are considered, e. g. shell structures with piezoceramic patches. If a standard FE code is used, piezoelectric or magnetostrictive elements can be modelled with the thermo-elastic analogy mentioned above, where coefficients of thermal expansion are substituted by piezoelectric or magnetostrictive coupling coefficients. However, the approximation must not be made, if single actuators, such as piezoelectric stacks or magnetostrictive rods, are to be analysed in detail. From a dynamics point of view, the approximation error can be characterised as an underestimation of the system's natural frequencies.

5.3.4 Equations of Motion

Application of the FEM to structural dynamics leads to the discrete equations of motion of an adaptronic structure:

$$\boldsymbol{M}\ddot{\boldsymbol{q}} + \boldsymbol{D}\dot{\boldsymbol{q}} + \boldsymbol{K}\boldsymbol{q} = \boldsymbol{F}\boldsymbol{u} \, . \tag{5.14}$$

Here, $\boldsymbol{M}$, $\boldsymbol{D}$, and $\boldsymbol{K}$ denote the mass, damping, and stiffness matrices, respectively. In the case of full coupling for piezoelectric or magnetostrictive material elements in the structure, the vector $\boldsymbol{q}$ of degrees of freedom initially comprises both nodal displacements and electric potentials or magnetic field intensities, respectively. In general, electromagnetic processes are much faster than mechanical vibrations, so that they may be assumed as being quasi-static in the above equation. As a consequence, electric or magnetic fields only contribute to the stiffness of the system, and a static condensation [9] of the corresponding electric or magnetic degrees of freedom can be carried out. Only the mechanical degrees of freedom remain. The full coupling is represented in an electro- or magnetomechanical stiffness matrix. In (5.14) the term $\boldsymbol{F}\boldsymbol{u}$ denotes the actuator influence on the structures. The input variable $\boldsymbol{u}$ may represent externally applied actuator voltages (piezoelectric) or currents (magnetostrictive actuators) and $\boldsymbol{F}$ the corresponding influence matrix. Note that static condensation of the electric or magnetic degrees of freedom leads to changes in $\boldsymbol{F}$ in addition to those in $\boldsymbol{K}$.

The equations of motion (5.14) can be transformed into the state equations of a state-space system description (5.3, 5.4) if the displacements $\boldsymbol{q}$ and velocities $\dot{\boldsymbol{q}}$ are chosen as state variables:

$$\dot{\boldsymbol{x}} = \begin{bmatrix} \dot{\boldsymbol{q}} \\ \ddot{\boldsymbol{q}} \end{bmatrix} = \begin{bmatrix} \boldsymbol{0} & \boldsymbol{I} \\ -\boldsymbol{M}^{-1}\boldsymbol{K} & -\boldsymbol{M}^{-1}\boldsymbol{D} \end{bmatrix} \begin{bmatrix} \boldsymbol{q} \\ \dot{\boldsymbol{q}} \end{bmatrix} + \begin{bmatrix} \boldsymbol{0} \\ \boldsymbol{M}^{-1}\boldsymbol{F} \end{bmatrix} \boldsymbol{u} \tag{5.15}$$

or

$$\dot{\boldsymbol{x}} = \boldsymbol{A}\boldsymbol{x} + \boldsymbol{B}\boldsymbol{u} \; . \tag{5.16}$$

A comparison of both systems of equations shows that the state or system matrix is determined by the plant properties $\boldsymbol{K}$, $\boldsymbol{M}$ and $\boldsymbol{D}$.

Alternatively, modal amplitudes and velocities can be chosen as state variables, leading to a desirable decoupling of the state differential equations.

5.3.5 Sensor Equations

In the case of actively controlled structural dynamics, sensors may measure a variety of signals, such as accelerations (accelerometers), displacements (Hall sensors, capacitive sensors, laser interferometers, etc.), forces (force transducers), or – typically for adaptronic structures – strain or strain velocities (strain gauges, piezoelectric sensors, etc.). Most of these cases can be represented by the following sensor equation:

$$\boldsymbol{y} = \begin{bmatrix} \boldsymbol{C}_1 \; \boldsymbol{C}_2 \end{bmatrix} \begin{bmatrix} \boldsymbol{q} \\ \dot{\boldsymbol{q}} \end{bmatrix} + \boldsymbol{D}\boldsymbol{u} = \boldsymbol{C}\,\boldsymbol{x} + \boldsymbol{D}\boldsymbol{u} \; . \tag{5.17}$$

As mentioned in Sect. 5.2.2, the feedthrough term $\boldsymbol{D}\boldsymbol{u}$ becomes important, for example in the case of active struts in truss structures where piezoelectric stacks are placed in series with force transducers.

5.3.6 Model Reduction Techniques

Structural models obtained by using FEM codes are, in general, much too large for the application of control design tools. Complex structures are commonly represented by tens if not hundreds of thousand nodal degrees of freedom, whereas control design methodologies and analysis tools are often restricted to only several tens or hundred degrees of freedom. This discrepancy highlights the reason why a large variety of model reduction techniques have been developed.

In the case of model reduction for linear elastic, actively controlled structures, a comprehensive survey is given by Craig and Su [6]. It is often advantageous to transform such systems into modal space before reduction, control design, simulation and analysis are carried out. Reduction is then performed by selection of natural modes which

- lie in the frequency range of control;
- are strongly controllable and observable with the chosen actuator and sensor configuration; and
- substantially contribute to undesirable structural motion in case of disturbances.

These criteria may be expressed numerically using Hankel singular values, or balance gains [11] if the system inputs and outputs are chosen appropriately. The method is known as 'balanced reduction'. In the case of lightly-damped adaptronic structures, balanced reduction techniques can be applied based on modal data which makes this methodology numerically fast and efficient (see Sect. 5.2.3). Frequency-weighted versions [12] have been developed to account for critical frequency ranges. Another commonly used technique computes modal costs [13]. Methods based on Ritz and Krylov vector projections are advantageous with respect to representation of quasi-static system behaviour, but decoupling of the equations of motion is no longer feasible. This implies the risk of severe dynamic spillover (see Sect. 5.4.2). If the control objective is active damping, quasi-static modelling errors are not critical. Therefore, modal representations are often preferred.

5.4 Analysis of Adaptronic Systems and Structures

Numerical analysis and simulation of adaptronic systems can be performed in the time or in the frequency domain depending on the representation of the system in the state space or as a matrix of transfer functions. In addition to performance criteria, important goals are stability and robustness of an adaptronic system. In the case of adaptronic structures, performance criteria are often given in terms of allowable static and dynamic errors relating to structural shape if subjected to specified disturbances. Many applications also involve limits in energy consumption and actuator stroke or force, which must be checked in time-history simulations. A comprehensive introduction on the different aspects and their interaction can be found in [14]. Current research in the field is for instance presented in [15] and [16].

5.4.1 Stability Analysis

Every dynamic adaptronic system must be checked for stability in the case of disturbances. For linear elastic adaptronic structures, asymptotic stability as defined in Sect. 5.2.4 is guaranteed if the poles (or eigenvalues) of the closed-loop active system lie in the left complex half-plane, i. e. if they have negative real parts. More stringent stability criteria, such as the generalized Nyquist criterion [7], also consider the zeros of the adaptronic system.

In the case of nonlinear systems that cannot be reduced to a linearized system, stability is much more difficult to assess. Lyapunov's direct method [1] requires a suitable energy function to be found. Often, only numerical time integration gives an indication of the dynamic behaviour and stability that cannot be proven otherwise.

5.4.2 Spillover

Structural control systems must be designed using rather small-scale models but are applied in the real structure with a theoretically infinite number of eigenmodes. Unwanted interaction or energy flow from the control system to neglected but excitable structural modes may occur and lead to loss of performance or even instability. This effect is known as spillover [17].

Three different types of spillover can be defined:

- The actuators influence structural modes that have not been represented in the mathematical model used for control design. This type is known as *control spillover.*
- The sensors produce signals with contributions from neglected structural modes. If this type, known as *observation spillover*, coincides with control spillover in the case of observer-based state feedback control, destabilization of the closed-loop system may be the consequence.
- In case the equations of motion used as a basis for model reduction are not decoupled, coupling terms between selected and neglected degrees of freedom exist. They imply *dynamic spillover*, which may lead to instability of the closed-loop system even if no observer is involved in the design.

The notion of spillover is important with respect to neglected structural modes. Other modelling errors include parametric uncertainties, which are more difficult to model and may have a substantial impact on the stability and performance of the closed-loop system.

5.4.3 Numerical Time Integration

In many cases, stability, performance and robustness are difficult to check with general criteria. Numerical time integration of the state-space model is often used to investigate the dynamic behaviour of an adaptronic system. For the simulation of adaptronic structures without control feedback loops, it can be advantageous to use direct time integration schemes to solve the second-order equations of motion (5.14). Examples of widespread numerical integration algorithms are the Houbolt, Wilson, and Newark schemes [9]. They exhibit good performance for linear structural dynamics problems. If a large range of structural eigenfrequencies has to be covered, however, very small time steps are required in order to guarantee a stable solution. Modal decoupling of the equations of motion substantially reduces the required computation time and allows for model reduction based on modal selection (see Sect. 5.3.6).

The existence of control feedback loops, especially with actuator, sensor, or observer dynamics, makes the application of direct time integration schemes difficult. Implicit and explicit schemes based on the first-order state space differential (5.15) are preferred in this case. A large variety of algorithms exist, among them the well-known Runge-Kutta schemes with modifications for step size control [18].

5.5 Application Example

A typical practical example for adaptronic systems or systems with adaptronic subsystems are large and high precision astronomical telescopes as shown in Fig. 5.2. This example is from [19].

For example, their optical mirrors and their large support structures should have minimum deviation from their ideal shape in the sub μm and even the nm range under dynamic (wind, micro-seismics,. . .) and quasi-static (e. g. thermal) loads. In addition to that, influences on the active optics control system (AOCS) have to be taken into account for the smaller mirrors in the optical chain. From that point of view such systems have some analogy to very high precision machinery and manipulation systems. Typical diameters of the main mirrors of current telescopes are in the order of 5 to 8 m, while for newer concepts in planning – such as the Overwhelmingly Large Telescope (OWL) of the European Southern Observatory (ESO) – this might go up to the order of 50 m for a segmented mirror.

Adaptronics and control can be implemented at different points or subsystems. In order to evaluate possible concepts from a system point of view, an overall end to end model is established. The elements of such an integrated simulation model are shown in Fig. 5.3. In the left part the reduced structure (dynamics) model together with control laws are used for assessing the effects of active damping introduced by adaptronics as described below. The right part contains the optical submodel together with the telescope drives.

Fig. 5.2. Overwhelmingly Large Telescope (OWL) of the European Southern Observatory (ESO)

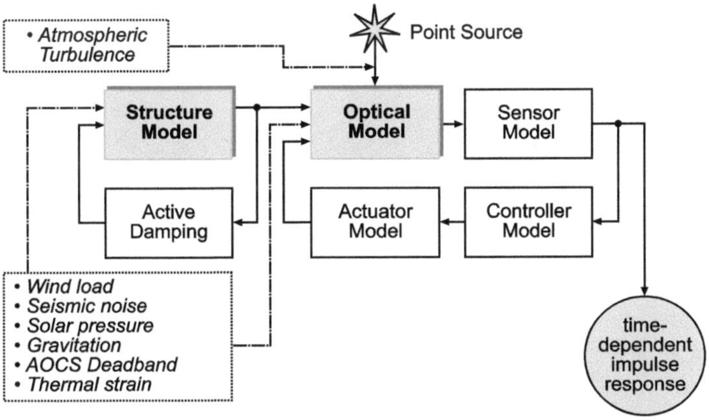

Fig. 5.3. End to end model of a large astronomical telescope

Atmospheric turbulence – which causes the stars to appear to twinkle to the human eye – is also considered and has to be also compensated for via adaptronic means.

A representative full state finite element dynamic model of OWL is outlined in Fig. 5.4. It comprises the main and secondary mirror and their support structure including the interface to the ground. A typical result for transfer functions from reduced models with 1000 states and 25 states or considered modes are given in the Fig. 5.5. The transfer function describes the movement of the secondary mirror when subjected to wind loads in the y-direction. As can be seen, the drastically reduced model still covers the

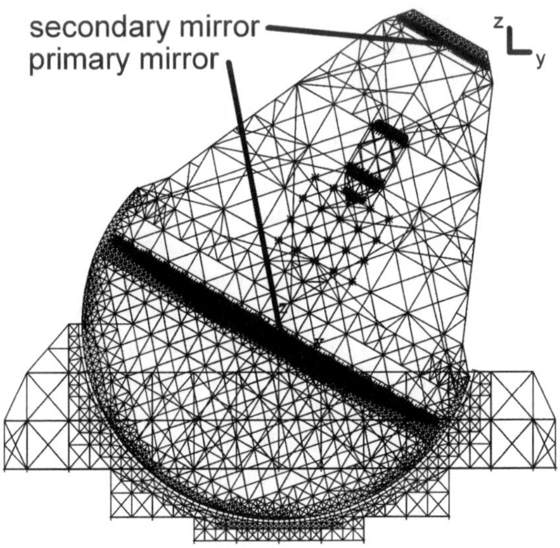

Fig. 5.4. Finite-element model of OWL

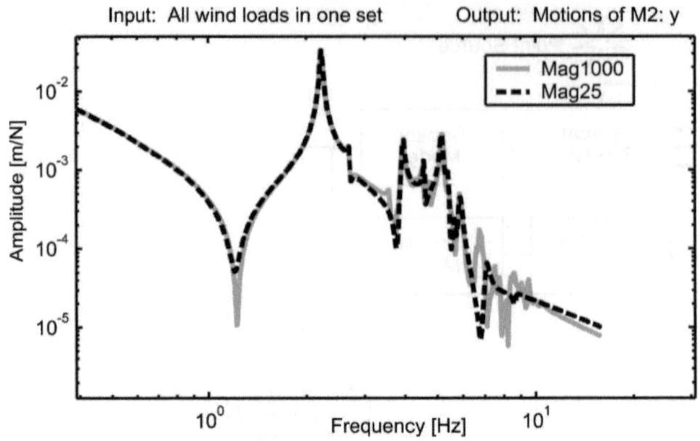

Fig. 5.5. Transfer functions of the motions of the secondary mirror M2 due to wind load on OWL (for 1000 modes and for 25 modes)

dynamic system behaviour of up to about 10 Hz. Since frequency spectra of relevant wind loads are in the range from 1 Hz to a maximum of 10 Hz, there is still some margin available for the reduced model.

Adaptronics to be included starts with quasi-static shape control (compensating gravity and thermal loads) of the main mirror with electromagnetic high force actuators attached to or integrated into its rear, and goes up to very high frequency control (in the order of 1000 Hz) of small mirrors at the end of the optical chain (not shown in the model of Fig. 5.4). These small mirrors typically have a diameter of 10 to 30 cm with integrated piezo actuators for compensation of high frequency disturbances with low force and displacement amplitudes in the µm range.

A further option for active damping is the low frequency control (typically 1 to 10 Hz) via actuators integrated into the struts of the large supporting structure of the secondary mirror starting from main mirror up to the top including the secondary mirror. Simulation of achievable active damping has shown that significant levels can be achieved only by proper positioning and also a significant number (in the order of 50 and more) of such active struts. Alternatively, four inertia or proof mass actuators placed at the top of the supporting truss have shown to be more effective for active damping in this upper structural part. This becomes obvious from the root locus curve for a relevant vibration mode given in Fig. 5.6. As a control law a velocity feedback controller is used. It can be seen that by proper gain factors stability is obtained and damping levels of roughly 10% of critical damping can be achieved (no passive structural damping assumed). For the theoretical limit case of 100% of critical damping no vibration could be excited at all (aperiodic limit case). Irrespective of the calculated achievable damping values, their technical implementation is still challenging for example when considering the required actuators and their proper integration together with their power lines etc.

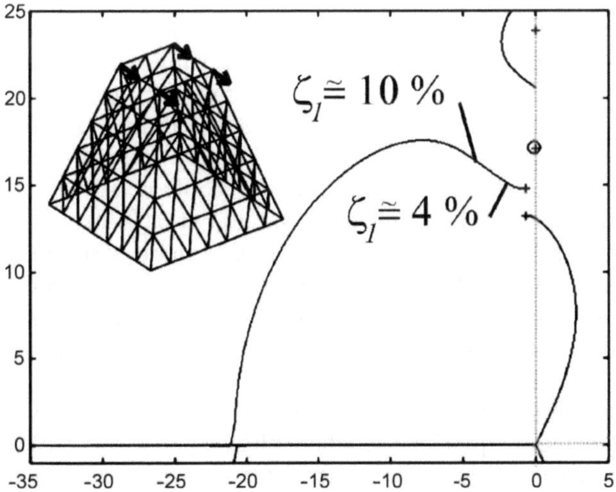

Fig. 5.6. Root locus curve (real and imaginary axis) for active damping via proof mass actuators in the secondary mirror support truss (ζ_1: percentage of critical damping)

5.6 Optimization of Adaptronic Systems

Modelling and simulation for applications also implies proper model parametrization followed by parameter studies in order to determine proper 'design variables' both of the plant or structure and of the controller including actuator positions etc. In the case of quantifiable goals and requirements, this process can be formalized via (nonlinear) optimization problems and solution processes as will be briefly addressed in the following.

5.6.1 Problem Statements

Though there exists a multitude of different possible problem statements, depending on the different technical tasks, a typical design optimization problem, with a combined mechanical (superscript m) and control subsystem (superscript c), is the following nonlinear (and usually non-convex) optimization problem:

$$
\begin{aligned}
&\text{Minimize} && f_1(\boldsymbol{v},\boldsymbol{y}) + f_2(\boldsymbol{v},\boldsymbol{y}) + \dots \\
&\text{such that} && g_{\mathrm{k}}(\boldsymbol{v},\boldsymbol{y}) \geq 0\,, \\
& && s_{\mathrm{i}}^{\mathrm{m}}(\boldsymbol{v}^{\mathrm{m}},\boldsymbol{v}^{\mathrm{c}},\boldsymbol{y}^{\mathrm{m}},\boldsymbol{y}^{\mathrm{c}}) = 0 \qquad i = 1,\dots,q^{\mathrm{m}} \\
& && s_{\mathrm{i}}^{\mathrm{c}}(\boldsymbol{v}^{\mathrm{m}},\boldsymbol{v}^{\mathrm{c}},\boldsymbol{y}^{\mathrm{m}},\boldsymbol{y}^{\mathrm{c}}) = 0 \qquad i = 1,\dots,q^{\mathrm{c}} \\
&\text{with} && \boldsymbol{v} = (\boldsymbol{v}^{\mathrm{m}},\boldsymbol{v}^{\mathrm{c}})^{\mathrm{T}} \\
& && \boldsymbol{y} = (\boldsymbol{y}^{\mathrm{m}},\boldsymbol{y}^{\mathrm{c}})^{\mathrm{T}}\,.
\end{aligned}
\tag{5.18}
$$

The design or optimization variables $\boldsymbol{v}^{\mathrm{m}}$, $\boldsymbol{v}^{\mathrm{c}}$ are to be determined such that a set of objective functions $f_1, f_2 \ldots$ is minimized while constraints g_{k} on the design variables and system response variables $\boldsymbol{y}^{\mathrm{m}}$, $\boldsymbol{y}^{\mathrm{c}}$ are to be satisfied. Typical design variables $\boldsymbol{v}^{\mathrm{m}}$ are structural stiffness properties, typical control variables $\boldsymbol{v}^{\mathrm{c}}$ are gain factors and actuator/sensor positions, while objectives are related to structural and control subsystem mass, required power, time integral of response values, etc. Constraints often are put on design variables directly, for example where structural stiffness or actuator forces must not exceed given bounds and indirectly via constraints on response quantities, for example displacements or accelerations at specific points on a structure or limits on its eigenfrequencies.

Mathematically, the coupling between the mechanical and control subsystem mainly occurs in the system equations s, where the response quantities $\boldsymbol{y}^{\mathrm{m}}$ (e. g. displacement vector) and $\boldsymbol{y}^{\mathrm{c}}$ (e. g. control forces) are determined depending on (the actual values of) the design variables $\boldsymbol{v}^{\mathrm{m}}$ and $\boldsymbol{v}^{\mathrm{c}}$. These system equations often are the state-space representation as discussed in previous sections, where in the case of adaptronic structures the equations of motion and vibration are involved. So, all the remarks on modal representation, condensation, etc. apply, including proper parameterisation in the design variables.

5.6.2 Solution Techniques

A solution technique for the optimisation problem first of all requires an appropriate overall strategy to deal with the coupled structural (plant) and optimum control problems. There are different options available, such as:

- treating the problem as fully coupled and solving for both the $\boldsymbol{v}^{\mathrm{m}}$ and $\boldsymbol{v}^{\mathrm{c}}$ simultaneously;
- using a decomposition or nested approach, where an optimal structural design with constraints for achieving good controller performance is carried out first, followed by optimal control design with optional side constraints to consider structural requirements, and then eventually followed by optimal structural design, etc.; or
- heuristic decomposition methods.

Treating the problem as fully coupled is in principle the most desirable approach, but it might be difficult to carry out for large complex problems. Depending on the type of technical problem, coupling might be weak which allows for separate determination of structural and control parameters, possibly with some additional iteration loops. Therefore a decomposition might be worthwhile, where the subsystems are treated separately without sacrificing too much of the overall optimal system performance. An improved approximation $\boldsymbol{v}^{(k+1)}$ for the optimum solution as obtained from the kth approximation $\boldsymbol{v}^{(k)}$ is obtained from

$$\boldsymbol{v}^{(k+1)} = \boldsymbol{v}^{(k)} + \Delta\boldsymbol{v}^{(k)} \; . \tag{5.19}$$

The change vector $\Delta\boldsymbol{v}^{(k)}$ is determined by a nonlinear optimisation algorithm such that the objectives in step $k+1$ are improved and the constraints are (better) satisfied compared with those at the previous step k. Optimisation algorithms range from mathematical techniques with and without the need for derivatives to evolutionary and genetic algorithms. While the latter usually needs a considerable number of optimisation steps, they are more general e. g. for handling several objectives or discrete variables. Irrespective of the type of optimisation algorithm to be chosen, the plant or structural model has to be properly condensed to a form which still contains the design variables in a parameterized manner.

5.7 Software Tools for Adaptronic Structure Simulation

A brief overview on different software tools related to the simulation of adaptronic systems is given. Since for the core tasks there are several tools which are continuously improved, actual comparisons are difficult and also depend on the specific criteria relevant for each of the application cases. So the overview should be considered as representative but not necessarily as complete.

5.7.1 Solution Techniques

For static and (structural) dynamic analysis, for determination of eigenfrequencies and eigenmodes, several different commercial tools exist such as NASTRAN, ABAQUS or ANSYS. Some of them are also able to handle actuators and piezoelectric materials, and also to carry out some types of model reduction techniques. Nevertheless, specific techniques might have to be established by the user via accessing the modal data base. These data are then also used to set up a modal or otherwise condensed state-space representation possibly including specific actuator and sensor models. A description of the transformation of finite-element models from ANSYS to dynamic models in state space form in MATLAB can be found in [20].

5.7.2 Control Design and Simulation Tools

Among the software for control design and system simulation MATLAB together with SIMULINK is a widespread tool. In particular, MATLAB includes a large variety of toolboxes for control design (standard, non-linear, robust control, etc.) and system identification (see Sect. 5.7.3). SIMULINK offers the option to graphically design and simulate dynamic systems as block diagrams without any additional programming. These tools, however, are restricted to medium-/small-scale problems, so that reduction of large-scale FE models is necessary.

5.7.3 System Identification Tools

Identification of the dynamic behaviour of adaptronic structures may be performed in the framework of modal testing (experimental modal analysis) or in a more control-oriented fashion known as system identification. In the former case, commercially available software packages can be used. They offer a variety of data acquisition and processing capabilities (modal analysis, frequency response functions, etc.) combined with comfortable graphical user interfaces.

For all of the tools mentioned, proper application requires the knowledge of the physical and modelling background together with that on the steps mentioned in this chapter, and engineering insight into the adaptronic system to be developed.

References

1. Slotine, J.-J.E.; Li, W.: *Applied nonlinear control.* Prentice-Hall, Englewood Cliffs, NJ, USA (1991)
2. Boyd, J.G.; Lagoudas, D.C.: *Thermomechanical response of shape memory composites.* J. Intelligent Material Systems and Structures, 5 (1994), pp. 333–346
3. Preumont, A.; Dufour, J.-P.; Malkian, C.: *Active damping by a local force feedback with piezoelectric actuators.* AIAA J. Guidance, Control, and Dynamics, 15 (1992), pp. 390–395
4. Moore, B.C.: *Principal component analysis in linear systems: controllability, observability, and model reduction.* IEEE Trans. Autom. Contr., AC-26 (1981), pp. 17–32
5. Gawronski, W.K: *Advanced Structural Dynamics and Active Control of Structures.* Springer Verlag New York, USA (2004)
6. Craig, R.R. Jr.; Su, T.-J.: *A review of model reduction methods for structural control design.* Proc. 1st Conf. Dynamics and Control of Flexible Structures in Space, Cranfield, UK (1990)
7. Maciejowski, J.M.: *Multivariable feedback design.* Addison-Wesley, Wokingham, UK (1989)
8. Shahinpoor, M.: *Continuum electromechanics of ionic polymeric gels as artificial muscles for robotic applications.* Smart Materials and Structures, 3 (1994), pp. 367–372
9. Bathe, K.-J.: *Finite element procedures in engineering analysis.* Prentice-Hall, Englewood Cliffs, NJ (1995)
10. Hwang, W.-S.; Park, H.C.: *Finite element modelling of piezoelectric sensors and actuators.* AIAA J., 31 (1993), pp. 930–937

11. Gregory, C.Z. Jr.: *Reduction of large flexible spacecraft models using internal balancing theory.* AIAA J. Guidance, Control, and Dynamics, 7 (1984), pp. 725–732
12. Al-Saggaf, U.M.: *On model reduction and control of discrete time systems.* Ph.D. dissertation, Inform. Syst. Lab., Dept. Electr. Eng., Stanford University (1986)
13. Skelton, R.E.; Hughes, P.C.: *Modal cost analysis for linear matrix-second-order systems.* Trans. ASME, J. Dynamic Systems, Measurement, and Control, 102 (1980), pp. 151–158
14. Preumont, A.: *Vibration Control of Active Structures, an Introduction.* 2nd ed., Kluwer, Dordrecht, NL (2003)
15. Ulbrich, H.; Günthner, W.: *Vibration Control of Nonlinear Mechanisms and Structures.* Proc. IUTAM Symp. München 2005, Springer Verlag (2005)
16. Lindner, D.K. (ed.): *Smart Structures and Materials 2006: Modeling, Signal Processing and Control.* Proc. SPIE, Vol. 6166, USA (2006)
17. Czajkowsky, E.A.; Preumont, A.; Haftka, R.T.: *Spillover stabilization of large space structures.* AIAA J. Guidance, Control, and Dynamics, 13 (1990), pp. 1000–1007
18. Gear, C.W.: *Numerical initial value problems in ordinary differential equations.* Prentice-Hall, Englewood Cliffs, NJ, USA (1971)
19. Baier, H.; Müller, U.C.: *Simulation of Adaptronic Structures.* Automatisierungstechnik, Vol. 54 (6), Oldenbourg Wissenschaftsverlag, Munich, Germany (2006), pp. 270–275
20. Hatch, M.R.: *Vibration Simulation using MATLAB and ANSYS.* Chapman and Hall/CRC, Boca Raton, FL, USA (2001)

6 Actuators in Adaptronics

6.1 The Role of Actuators in Adaptronic Systems

H. Janocha

Actuators are applied extensively in all spheres of our environment. They can be found in CD players and cameras, washing machines, heating and air-conditioning systems, machining equipment, automobiles, boats and aircraft and even respiratory equipment and artificial limbs. Actuators are also essential components in adaptronic systems, see Chap. 1. The actuators presented in Sects. 6.2 to 6.8, which are based on the transducer properties of new or improved materials, are particularly interesting for adaptronics: so-called self-sensing actuators can be implemented on the basis of multifunctional materials, which simultaneously feature sensory and actuator properties. These multifunctional components shall be described in more detail in Sect. 6.9; Sect. 6.10 will deal with amplifier concepts for driving energy converters, an often neglected subarea of actuators.

6.1.1 What is an Actuator?

An actuator is a functional element which connects the information processing part of an electronic control system with a technical or nontechnical part, e. g. biological, process. Actuators can be used to control the flow of energy, mass or volume. The output of an actuator is energy or power, often available in the form of a mechanical working capacity 'force times displacement'. The actuator input is always driven by very low electrical power, ideally without any power consumption, with currents and voltages which are, if possible, microelectronically (e. g. TTL) compatible [1].

An actuators functional structure can be described by introducing the elementary functional components of an energy controller and an energy converter (see Fig. 6.1). The output variable of an energy controller is the energy provided by an auxiliary power supply which is controlled via the input variable as it is typically done with transistors and valves (see Fig. 6.1a). An energy converters input and output variables are energies. In the case of current transformers and torque converters these two energies are of the same

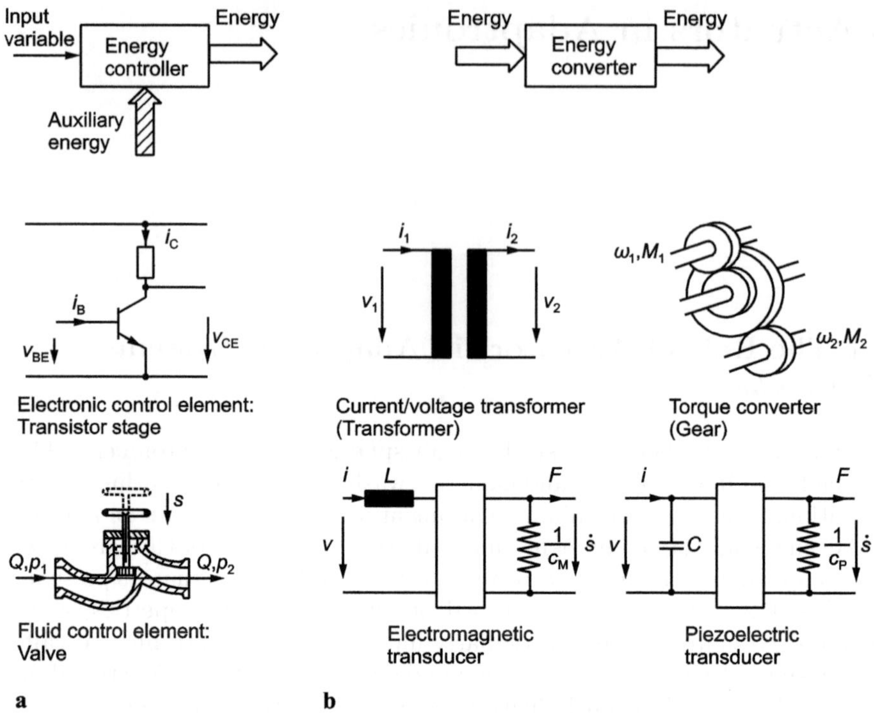

Fig. 6.1. Elementary functional components of an actuator. **a** Energy controller, **b** energy converter

kind, whereas the input and output variables of electromagnetic and piezoelectric transducers are different (see Fig. 6.1b).

As an actuator is supposed to control flows of matter and energy, an actuator must contain at least one energy controller. This is why actuators are usually a series connection of energy controllers and energy converters. The common understanding, however, leaves out one important property of actuators, and that is their controllability with a low power electrical signal. Subsequently, the term actuator refers often only to the energy converter, whereas the energy controller is called a power amplifier or a power circuit. These are not standardized but are accepted and used by the global scientific community.

For further reference, see the German DIN standard 19226 Regelungstechnik und Steuerungstechnik (closed and open loop control). Figure 6.2 describes a control system according to this DIN standard with the official translation of the technical terms. Within the actuator ('Steller'), the controller output variable y_C is turned into the manipulated variable y ('Stellgröße') which is used to drive the final controlling element ('Stellglied'). This final controlling element will influence the flow of matter and/or energy. Subsequently, the actuator definitions mentioned above are much closer to the

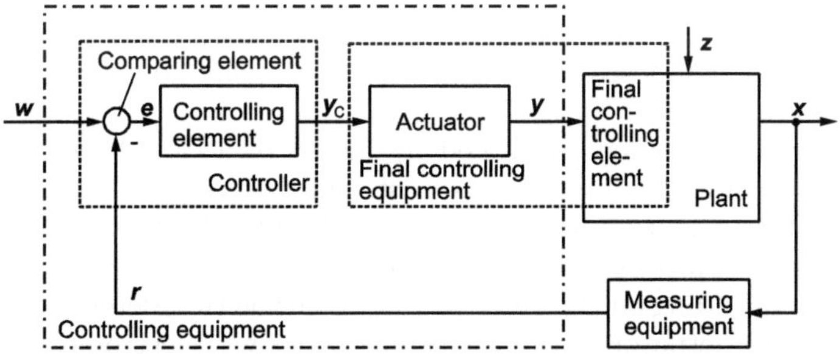

Fig. 6.2. Typical block diagram of a closed loop control system (terms as defined in DIN 19226)

DIN standard final controlling equipment ('Stelleinrichtung') and final controlling element ('Stellglied'). It is worth noting that the term actuator used in Fig. 6.2 conflicts with the actuator definition presented above which shall serve as the basis for this chapter.

6.1.2 Actuator as a System Component

Many controlling tasks that are required in the natural and artificial environment can be described with an open loop control chain, as shown in Fig. 6.3. The focus is placed on operations and processes that must be modified to achieve a certain goal. This is where actuators come into play. Their input signals are microelectronically compatible and are produced by the electronic controls inside of the information processing part of the control system. The electronic controls are often decentrally arranged and can therefore be assigned to the individual processes with respect to location and function. They are usually program controlled and can be implemented by means of

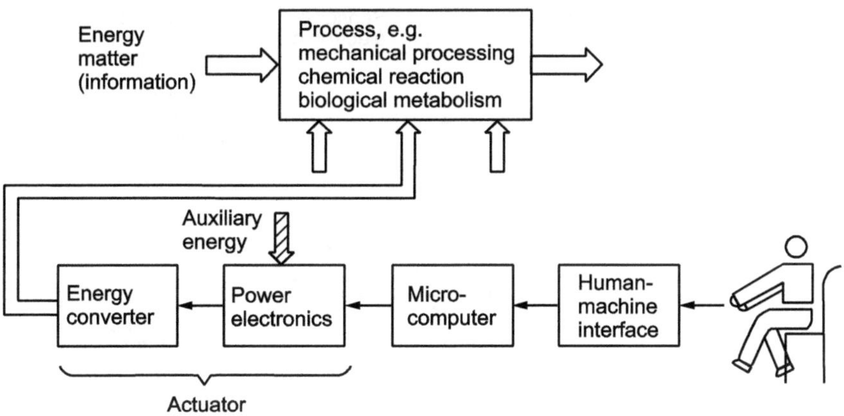

Fig. 6.3. Open loop control of processes

a personal computer. The user may modify the process via a so called human-machine interface (HMI), composed, in the simplest case, of an alphanumeric keypad and a computer monitor.

Automated processes are often controlled by means of a closed control loop (see Fig. 6.4). One of its key functions consists of measuring the characteristic process variables which are then preprocessed and fed into the control processor. The control processor compares the measured values with the given set values and, depending on the difference between the two, determines the control signal for the actuator or the corresponding power electronics by means of control algorithms in accordance with a control strategy which has been installed in the computer. The process specific parameters of any available process information the control processor might utilize, for instance a mathematical model, are determined by the control processor during an identification cycle. These parameters are the fundamentals of a controller synthesis within the computer. On a higher automated level, the controller adapts autonomously to the process-related changes of the parameters, e. g. due to wear: adaptive control, AC.

The symmetric system arrangement in Fig. 6.4 shows phenomenologically the duality of sensor and actuator technology in the field of automation engineering. It is interesting to note that an actuator alone features all the properties in terms of structure and function which comprise a complete control system including sensors and a signal processing part. A good example is the piezoelectric actuator whose displacement is detected by strain gauges

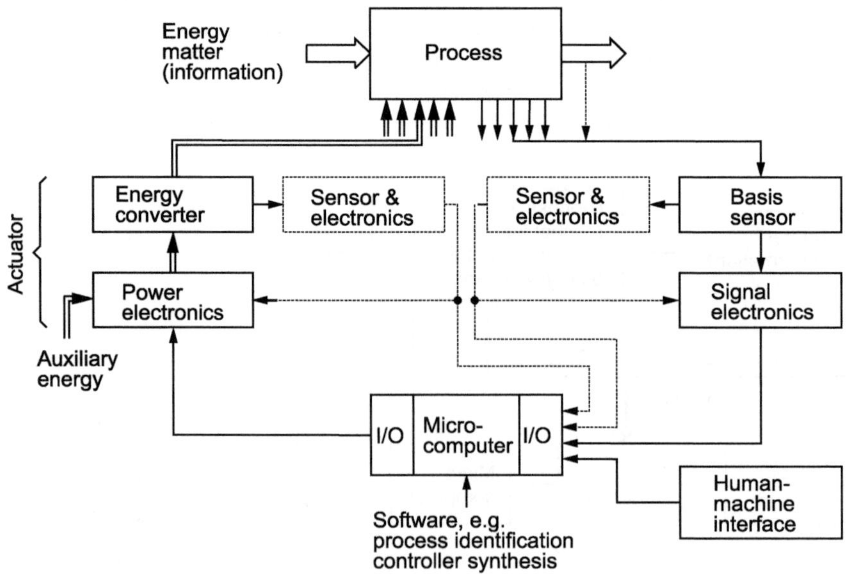

Fig. 6.4. Closed loop control of processes

which are mounted directly to the piezo crystal in order to eliminate – analogous to the methods for compensating error signals from basis sensors witch detect the process variables of interest – temporary or technology-related imperfections of the actuator such as temperature dependency, non-linearity or hysteresis of the output-input characteristic (see Sect. 6.1.4 Intelligent Solid-State Actuator). This multifunctionality is also a property of adaptronic systems.

It is possible to achieve an even higher degree of multifunctionality when multifunctional materials are being used. This shall be illustrated with the following example: actively controlling structural geometry is a typical task performed by adaptronic systems. Piezoelectric stacks, for instance, are used

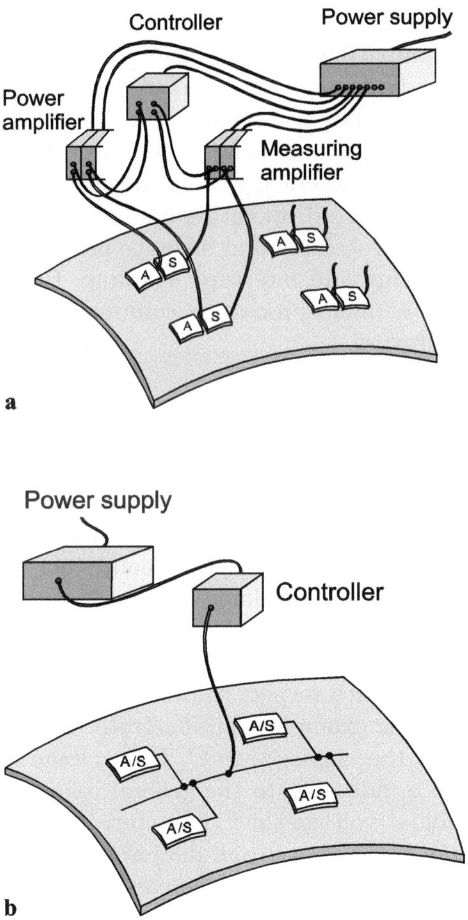

a

b

Fig. 6.5. Controlling of surface structures. **a** With standard actuator-sensor configurations (A: actuator, S: sensor), **b** with linked self-sensing actuators (A/S: adaptronic actuator-sensor module)

as active braces in truss structures, while piezoelectric flexural transducers deform surface structures such as plates and shells (see Fig. 6.5). In this particular application, the piezoelectric transducer can perform its actuator function and make use of its sensor properties at the same time. These self-sensing actuators (see Sect. 6.1.4) allow the implementation of smart structures. Their operation now involves far less devices and installation effort (compare Fig. 6.5a with b). Beyond that, the fact that actuator and sensor properties are collocated proves advantageous for the design and the operation of the controller, as controllers algorithms with simpler stability criteria can be implemented (e. g. PPF controller [2]).

Treating an actuator as a system component automatically raises the question regarding the type of its interfaces. The output or process interface can vary just as greatly as the range of actuator applications and is determined finally by the particular application. The actuators input interface, described above as microelectronically compatible, is much easier to describe. Researchers have agreed on certain standards allowing them to connect an actuator to any control processor with a standardized interface. As actuators are often included in real-time system concepts, the control processor must process the required user programs in time or practically simultaneously. Ordinary personal computers (PC) with a standard operating system usually cannot accomplish this task, in contrast to processors that have the necessary properties such as timesharing, multitasking and interrupt handling. However, it is possible to upgrade a PC to a micro-processing computer with commercially available hardware and software.

6.1.3 Power Electronics

Actuators are usually a series connection of energy controllers (power electronics) and energy converters. Subsequently, the system components strongly influence each other and depend on each other. This can be seen clearly in energy converters which – from an electrical point of view – mainly act as a reactive load (capacitance, inductance) at the amplifier output. Reactive electrical elements are accumulators of electrical or magnetic energy, which cannot be charged or discharged arbitrarily quickly due to fundamental physical laws. This has, of course, consequences with respect to the requirements a power amplifier needs to fulfil, as a simple example shall illustrate.

Suppose a piezoelectric actuator has the capacitance C. If a voltage is applied, the actuator stores the charge q, adhering to the general relationship $q = Cu$. For the case of a sinusoidal voltage-time characteristic with the angular frequency ω, the peak value of the charge and discharge current is $\hat{I} = \omega C \hat{U}$ (to simplify matters, we shall assume that the capacitance C remains constant). According to this equation, the current demand will increase with growing actuator dynamics (increasing ω). The application of piezo elements with large C, e. g. multilayer actuators (see Sect. 6.2), even increases this tendency.

Increasing the amplifier output power naturally raises the question regarding the degree of efficiency of energy transfer between the controller and the converter. The example of the harmonically operated piezo actuator leads us to the following universal approach for improving the degree of efficiency.

If the piezo element is charged within a half period (operation cycle = expansion of the piezo ceramic), it will be discharged during the next half period. The energy flowing back during discharge will either be converted into thermal energy (and therefore be lost), or can be temporarily stored in a convenient electrical component making it available to the piezo converter during the next operation cycle. It is clear that this type of energy recovery will improve the degree of efficiency of a series connection between an energy controller and an energy converter.

The higher the actuator output power required by the user, and/or the more actuators applied in an adaptronic system (distributed actuators in smart structures), the more relevant become the aspects we are discussing here. This topic is strongly linked with the question of whether it would be wiser to drive the converter with an analogue amplifier – good output signal quality, moderate degree of efficiency – or with a switching amplifier – moderate output signal quality, high degree of efficiency. Since both amplifier types have their specific strengths and weaknesses, which may become relevant depending on the application at hand, we shall deal with the topic of energy controllers in more detail in Sect. 6.10 (Power Amplifiers for Actuators).

6.1.4 'Intelligent' and Self-Sensing Actuators

The concepts of 'intelligent' and self-sensing actuators mentioned in Sect. 6.1.2 are exemplified below with solid-state actuators. The potential of both concepts is especially easy to recognize and to compare when described in terms of system theory. We will start with the conventional actuator.

The conventional actuator consists of the sub-systems feedforward controller, power electronics and solid-state transducer (see Fig. 6.6). By means of the desired displacement s_d, the feedforward controller consisting of a linear static transfer characteristic with a constant k_s produces an electrical input signal X_i for the power electronics. The power electronics generates the energy carrying output quantity X for the solid-state transducer from the information carrying electrical input signal X_i. The solid-state transducer transforms the electrical energy quantity X into a displacement s against a force F.

However, even in quasi-static operation the actual displacement and desired displacement usually do not correspond. Internal imperfections such as complex hysteretic nonlinearities described by the operator Γ_A in Fig. 6.6 and external influences such as load reactions via the surrounding mechanical

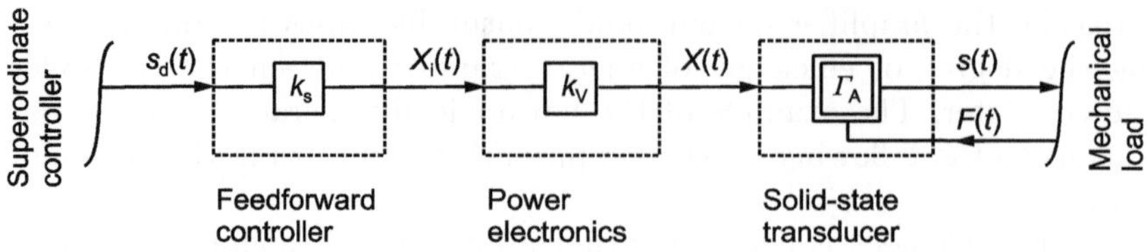

Fig. 6.6. Conventional actuator (k_V: transfer factor of the power amplifier)

structure are the main reasons for the deviation between the desired and actual values. The former imperfection provokes ambiguities between the input and output of the transducer; the latter one causes an additional deviation in the actual displacement from the desired value due to the finite stiffness of the solid-state transducer.

'Intelligent' Solid-State Actuator

According to general usage, solid-state actuators are called intelligent when their transfer characteristic is determined by a functionally allocated and electronically integrated intelligence, if necessary, with sensor support. Such intelligent actuators can recognize deviations from the desired transfer characteristic, which result from the hysteretic nonlinearities as well as from load feedback, and correct them automatically. The position controlled actuator in Fig. 6.7a is an example of such an actuator type. With this principle, the compensation of internal imperfections and external disturbances is achieved by a linear controller G_C, which receives information about the actuator out-

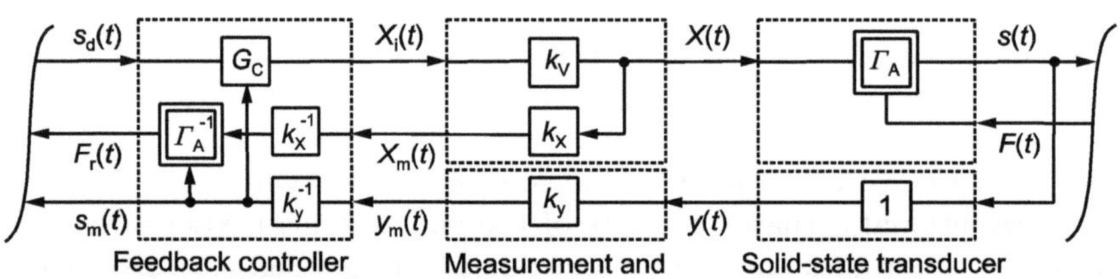

a

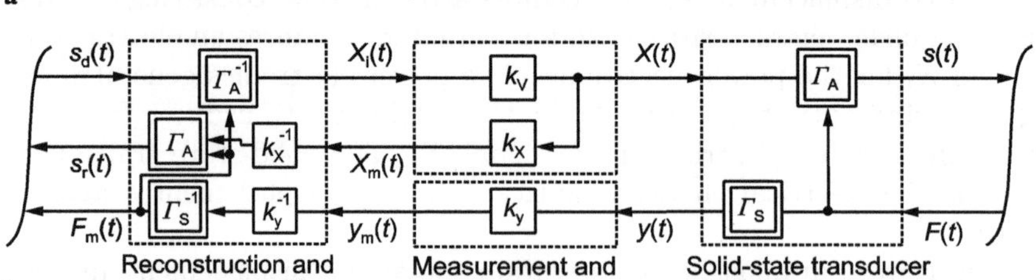

b

Fig. 6.7. Concept of 'intelligent' solid-state actuators. **a** With separated sensor, **b** with integrated sensor (k_x, k_y: Transfer factors of the sensor to measure the electric driving quantity X and the dual electric quantity y)

put from an external displacement sensor. With the reconstruction of force by means of the inverse filter Γ_A^{-1}, it is possible in this case to give feedback about the actuators current load situation to the superior control system. An electrical circuit for the measurement of the electrical quantity X is necessary for the implementation of this additional function. Such an electrical measurement circuit can be an element of the power electronics.

The actuator concept in Fig. 6.7b is sometimes used with piezoelectric transducers. It has clearly a higher measure of integration. In this case, some of the stacks ceramic disks are used as sensors in order to measure the force, whereas the major part of the stack operates purely as an actuator. For the accurate measurement of the force, the hysteretic transfer characteristic of the integrated sensor must be compensated within the electronic signal processing part by an inverse filter Γ_S^{-1}. In this case, the displacement can be reconstructed with the filter Γ_A from the electrical quantity X and the measured force F. Hysteretic nonlinearities and mechanical loading resulting during actuator operation can be compensated by implementing the inverse filter Γ_A^{-1}

Self-Sensing Solid-State Actuator

The self-sensing solid-state actuator shown in Fig. 6.8 has the highest measure of integration. However, its bidirectional function requires also the most complex mathematical and electronic signal processing unit. Characteristic of self-sensing actuators is the simultaneous utilization of actuator and sensor properties of the active material. In contrast to the intelligent concepts of Fig. 6.7, they have power electronics which contains the electronic circuits for measuring the given electrical quantity X and the dual electrical quantity y carrying the sensory information. The central function of the signal processing unit, which is responsible for the bidirectional function, is in this case the linearization and decoupling of both sensor and actuator operation.

In particular, the decoupling of both sensor and actuator operation for force and displacement reconstruction according to Fig. 6.8 is the main difference in the intelligent actuator concepts depicted in Fig. 6.7. In the case of self-sensing actuators the output y of the sensory path is strongly influenced by the driving quantity X of the solid-state transducer and must be

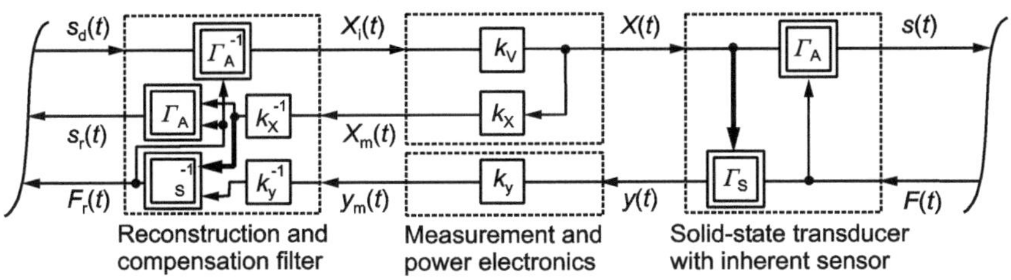

Fig. 6.8. Concept of self-sensing solid-state actuators

regarded as an external disturbance for the sensor operation. This is shown in the right hand block in Fig. 6.8. In intelligent actuators the output y of the sensor path is not influenced by the driving quantity X of the solid-state transducer, and a model-based decoupling of the sensor and actuator operation is not necessary.

The topic of intelligent actuators and self-sensing actuators will gain growing importance for adaptronic applications, e. g. in relation to structurally integrated electrical actuators. Therefore, we will look at them in more detail from a theoretical system point of view in Sect. 6.9.

6.1.5 Actuator Design

As in most technical fields, actuators are increasingly designed with the help of computers. The actuator and its surrounding are simulated as a mathematical model by means of commercially available software. Such models are fundamental for the simulation of the system response characteristic in each specific case. In this way, it is possible to find out about all the important properties of the system even before the actuator is built, and the actuators relevant parameters can be optimized to achieve the desired values. This designing strategy is exemplified below with an auxiliary mass damper which is able to withdraw kinetic energy from a host vibrating system.

Such vibration absorbers are used for instance in the automotive and aerospace industries where the vibration inclination of the car bodies or fuselages has to be attenuated. Within the scope of a first rough model the mechanical structure at the place of maximal vibration is described by the effective base mass m_1 which is excited by an unknown disturbing force F_1 causing undesirable vibrations (see Fig. 6.9). F_1 is thus a consequence of the interaction between m_1 and the remainder of the mechanical structure which is excited by externally or internally acting forces at other points. The task of the vibration absorber is to displace the auxiliary mass m_2 in such a way as to generate a secondary force $F_2 = m_2 \cdot a_2$ that will compensate the primary force F_1 and thus counteract the excitation of mass m_1.

When the force F_1 is narrow band, attenuation can be achieved with a passive vibration absorber which has to be tuned to the disturbance fre-

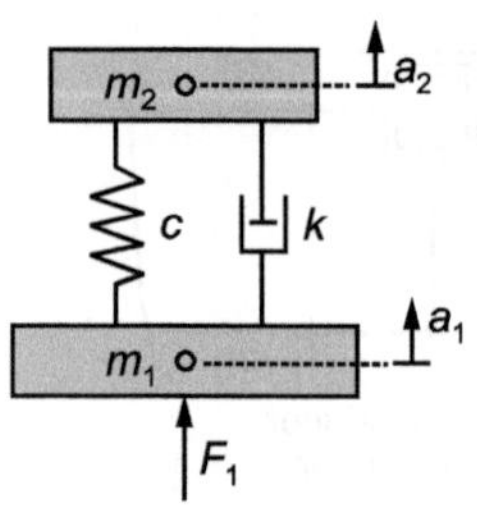

Fig. 6.9. Vibration attenuation using a passive vibration absorber

quency through its parameters stiffness c, damping constant k and mass m_2. In contrast an attenuation of broadband disturbances requires the use of an active vibration absorber whose mass is coupled to the vibrating main system via an electrically controllable interface. From the formulation of the dynamic balance of forces for the mass m_1 it follows that the acceleration a_1 of m_1 is a measure for the compensation effect of the active vibration absorber. Thus the aim of this damper principle is to displace the mass m_2 through an appropriate feedback of the acceleration a_1 in such a way that the resulting force F_2 will compensate the disturbing force F_1 and thus nullify the base acceleration a_1.

The starting point of the following specific example is a vibrating structure being stimulated to vibrate by imbalances within the rotating parts. The vibration has been dampened by a passive vibration absorber whose resonance frequency is tuned to the fundamental frequency of the vibration at 100 Hz. The disturbing force F_1 affecting the passive vibration absorber shows in addition to the 30 N value at 100 Hz other noteworthy values of 20 N and 10 N lying at 200 Hz and 300 Hz that cannot be compensated for due to the narrow-band damping characteristic of the passive vibration absorber. Now this task will be undertaken by an active piezoelectric vibration absorber.

The principle structure of the active vibration absorber corresponds approximately to the structure of the passive vibration absorber shown in Fig. 6.9 whereby the passive elastic material between m_1 and m_2 has been replaced by a piezoelectric actuator and a displacement amplification system to increase the achievable displacement of m_2. The amplification system is given in this example by elastic joints, similar to those illustrated in Fig. 6.9.

The mathematical model of the mechanical actuator system can be developed directly from the CAD design drawing by means of commercial FEM software tools, e. g. ANSYS® [3]. This model is fundamental for the calculational modal analysis which serves to find out the systems natural frequencies. Figure 6.10 shows the FEM model of the active piezo absorber and gives an impression of the third vibration mode of the structure which is used in this example for the vibration absorption.

The active compensation of the disturbing force F_1 can now be achieved by a suitable feedback of the measured base acceleration a_1 to the input of the high-voltage source for the piezo actuator. Based on a signal flow diagram, which is always to be developed by the designer, for the functionality of the force compensation will be investigated on the computer with support of an appropriate dynamic simulation and analysis software system, for example MATLAB® [4].

Figure 6.11 illustrates several results of this simulation. The frequency response in Fig. 6.11a shows the band rejection filter characteristic required for the compensation of the force F_1 lying between about 70 Hz and 329 Hz. In Fig. 6.11b the effect of the closed-loop force compensation is illustrated within the time domain, over the time interval of 0 s… 0.4 s. During the interval

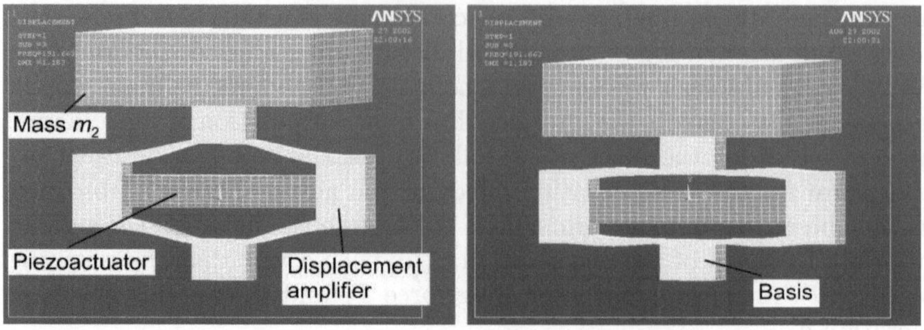

Fig. 6.10. Third vibration mode of the absorber structure analysed using ANSYS®

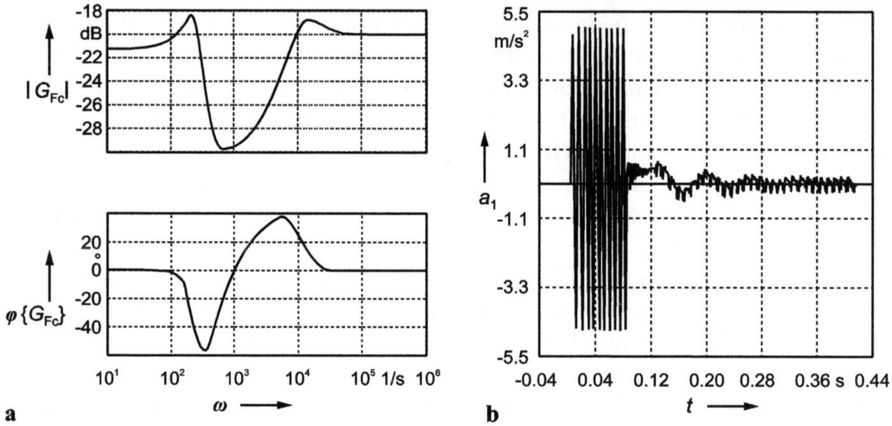

Fig. 6.11. Active vibration absorber. **a** Amplitude and phase response, **b** compensation effect within the time domain (G_{Fc}: disturbance frequency response)

0 s... 0.1 s the controller is idle, so that the vibration absorber operates passively. The maximum amplitude of the acceleration a_1 emerging due to the excitation by F_1 amounts in this operating state to about $5\,\mathrm{m/s^2}$. The controller is switched on at $t = 0.1$ s which excites the dynamics of the whole system.

This is indicated by a rapidly decaying high-frequency vibration corresponding to the second peak in the amplitude response shown in Fig. 6.11a. The high-frequency vibration is superimposed by a slower decaying low-frequency vibration corresponding to the first peak in the amplitude response. After the decay of all transient processes only the acceleration emerging due to the continued disturbance F_1 is still visible. The maximum amplitude of the acceleration a_1 at steady state is approx. $0.25\,\mathrm{m/s^2}$. Thus the force affecting the base mass m_1 can be reduced by a factor of 20.

This analysis software naturally also allows the user, for instance, to test and optimize the stability of the vibration absorber to avoid unpleasant surprises after the prototype has been built.

6.2 Piezoelectric Actuators

R. Letty, F. Claeyssen

6.2.1 Physical Effect

Certain crystals, such as quartz, feature a physical relationship between mechanical force and electric charge. When the crystal lattice ions are elastically shifted relative to one another due to an external force, an electric polarization can be detected by means of metallic electrodes on the surface. This so-called piezoelectric effect was first scientifically explained by the brothers Jacques and Pierre Curie in 1880 and forms the basis for piezo sensors (see Sect. 7.3). The effect is reversible and is then called reciprocal or inverse piezoelectric effect. If, for instance, an electric voltage is applied to a disc shaped piezo crystal, the thickness of the crystal changes due to the reciprocal piezoelectric effect. It is this property that is made use of in actuators.

Describing analytically the piezo effect by the linear state (6.1) and (6.2), the electric displacement density $\boldsymbol{D}$ and the mechanical strain $\boldsymbol{S}$ are combined with the mechanical stress $\boldsymbol{T}$ and the electrical field strength $\boldsymbol{E}$:

$$\boldsymbol{D} = \boldsymbol{d}\boldsymbol{T} + \boldsymbol{\epsilon}^{\mathrm{T}}\boldsymbol{E} \tag{6.1}$$

$$\boldsymbol{S} = \boldsymbol{s}^{\mathrm{E}}\boldsymbol{T} + \boldsymbol{d}_{\mathrm{t}}\boldsymbol{E}\ . \tag{6.2}$$

In this system of equations the piezoelectric charge constant $\boldsymbol{d}$ indicates the intensity of the piezo effect; $\boldsymbol{\epsilon}^{\mathrm{T}}$ is the dielectric constant for constant $\boldsymbol{T}$ and $\boldsymbol{s}^{\mathrm{E}}$ is the elastic compliance for constant $\boldsymbol{E}$; $\boldsymbol{d}_{\mathrm{t}}$ is the transpose of matrix $\boldsymbol{d}$. The mentioned parameters are tensors of the first to fourth order. A simplification is possible by using the symmetry properties of tensors. Usually, the Cartesian coordinate system in Fig. 6.12a is used, with axis 3 pointing in the direction of polarization of the piezo substance (see below) [5, 6].

All material dependent parameters can be described by matrices, whose elements are marked with double indices. In $\boldsymbol{d}$, the first index marks the orientation of $\boldsymbol{E}$, the second the direction of $\boldsymbol{S}$. The examples in Fig. 6.12b and c are based on the condition that the field strength works in the direction of the polarization. The resulting elongation in Fig. 6.12b points as well in direction 3 (longitudinal effect), in Fig. 6.12c however, it works in direction 1 (transversal effect). These two characteristics of the piezoelectric effect are quantified by means of the piezo constants d_{33} and d_{31}.

It is common to summarize all matrix elements in so-called coupling matrices. From the coefficients in the coupling matrix it is possible to determine an important parameter of piezo materials, the coupling coefficient k. For the coupling coefficient of the longitudinal effect k_{33} applies for instance

$$k_{33} = \frac{d_{33}}{\sqrt{s_{33}^{\mathrm{E}}\epsilon_{33}^{\mathrm{T}}}}\ . \tag{6.3}$$

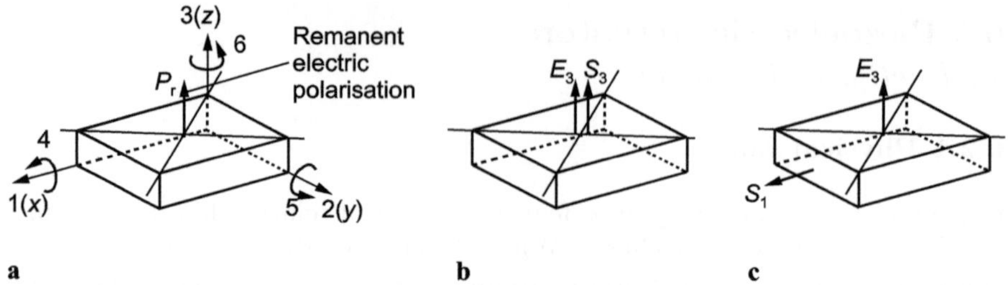

Fig. 6.12. Definition of the axes in piezo materials. **a** The digits 4, 5 and 6 indicate the shear on the axes 1, 2 and 3; **b** longitudinal (d_{33}) effect, **c** transversal (d_{31}) effect

Since k^2 corresponds to the ratio of stored mechanical energy to absorbed electrical energy, achieving actuators with high elongation efficiency requires substances with a large k.

In ferroelectric materials one must add to the linear piezo effect according to the (6.1) and (6.2) an elongation that depends on the square of the electric field strength. This elongation share is negligibly small in the traditional materials, but it can be increased systematically in order to reach the strength of the linear piezo effect. This so-called electrostrictive effect is independent of the polarity of the control voltage, and the corresponding diagram $S(E)$ shows a very small hysteresis. The effect is long-term stable (no creep, easily reproducible), however, the operational range of temperature is limited to about 30 K, and the effect is not reversible. The electrostrictive effect is presently of less significance for use in transducers.

6.2.2 Materials

Piezoelectric materials can be grouped into the class of natural crystals, such as quartz or tourmaline, into one of polymers, such as polyvinylidene fluoride (PVDF) or that of polycrystalline ceramics.

For the production of actuators, sintered ceramics are mainly used, especially lead-zirconate-titanate (PZT) compounds. After sintering, the domains of a ceramic body (i. e., the regions consisting of crystallites of uniform dipole orientation) will show a statistically distributed orientation, i. e., the macroscopic body is isotropic and has no piezoelectric properties. Only when a strong electrical dc field is applied, the dipole regions become almost completely arranged (polarization). After switching off the polarization field, this arrangement remains to a large extent, that is, the ceramic body features a remanent polarization P_r, combined with a permanent elongation S_r of the body (see Fig. 6.13).

PZT ceramics are chemically inactive and can cope with high mechanical loading, but are also brittle and therefore difficult to process. The permissible compressive stress is considerably higher than the tensile stress. This is why the elements need to be pre-stressed when extensive tensile stress is

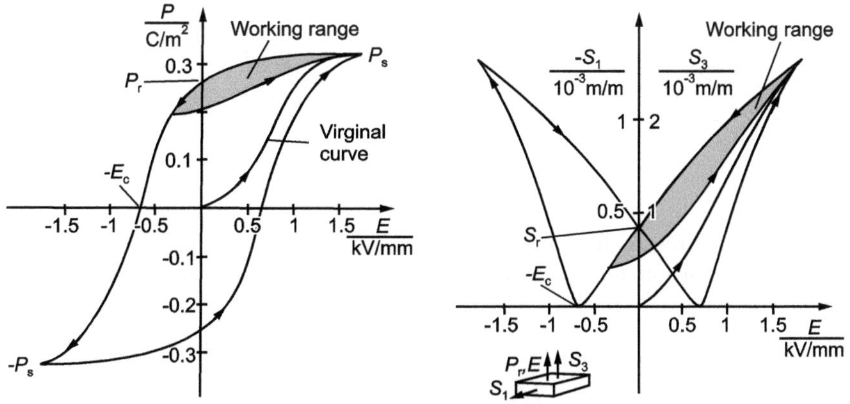

Fig. 6.13. Diagram $P(E)$ and $S(E)$ for a typical piezoceramic for $T = 0$. The actuators operation cycle starts at point $E = 0$, S_r (derived from [5])

applied. PZT ceramics belong to the group of ferroelectric materials which feature a hysteretic behaviour shown in the diagram $P(E)$ in Fig. 6.13. Due to the relation $P = D - \epsilon_0 E$ (P: electric polarization) and $D = \epsilon E$, the two diagrams $P(E)$ and $D(E)$ differ m erely by the term $\epsilon_0 E$. For actuator operation the diagram $S(E)$ of the polarized ceramic, the so-called butterfly trajectory shown in Fig. 6.13 (right hand side) is crucial. The maximum achievable strain is limited by the saturation and the repolarization. Precautions must be taken in order to avoid depolarization during actuator operation due to electrical, thermal and mechanical overload. Piezoceramics, for instance, gradually loose their piezoelectric properties even at operating temperatures far below the Curie temperature (depending on the material 120... 500 °C, for multilayer ceramics (see below) 80... 220 °C). Under certain applications when the inverse operating voltage is applied, it may not exceed 20% of the rated voltage, or depolarization may occur.

Piezoceramic elements are mainly available as plates or discs with a quadratic, circular or ring-shaped profile and a length from 0.3 upto several millimeters long, with or without metal electrodes. Most are designed to make use of the longitudinal effect (see Fig. 6.14a), which is due to the high d_{33} value, which is the strongest effect. When making use of the transversal effect the actuator stroke depends also on the dimensions of the material, whereby the influence of the quotient s/l on stiffness and elongation is oppositional (see Fig. 6.14b).

Since the 1980s, multilayer ceramics have grown more important. The so-called green and several tens of micrometers thick ceramic foil is cut into pieces and then coated with an electrode paste, similar to multilayer capacitors. The pieces are then placed on top of each other, pressed and sintered. They form a kind of monolithic object that is used as a finished transducer or as a basis for producing stacks (see Fig. 6.15). Multilayer ceramics reach the maximum permissible field strength at a driving voltage of about 100 V (low

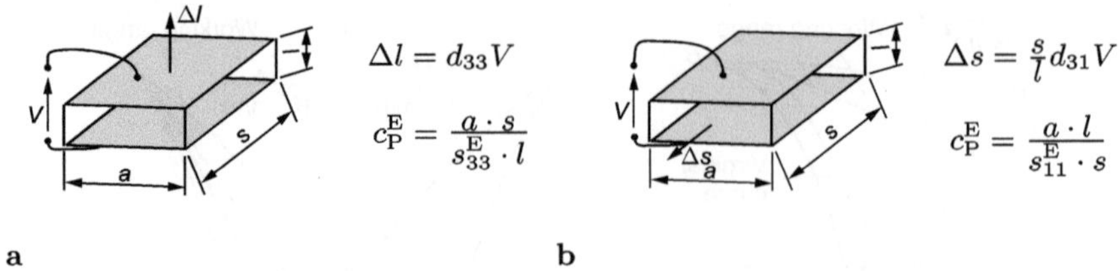

Fig. 6.14. Inverse piezo effect in polarized ceramics. Voltage V is applied in the direction of polarization P. **a** Longitudinal effect, **b** transversal effect (c_P^E stiffness of the piezo material for constant field strength E)

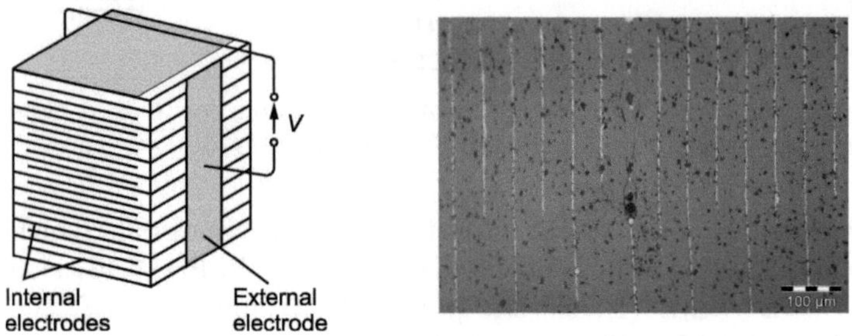

Fig. 6.15. Basic structure of a stack comprised of multilayer piezoceramic (MLA) and a section through a component

voltage actuators), and achieve therefore the same elongations as ordinary (so-called high voltage) piezoceramics do for a driving voltage in the kilovolt range.

Apart from that, piezoelectric polymers are available as foils with a thickness on the order of several tens of micrometers. Such polymers have been known of since 1924; but a major milestone was marked with the discovery of the strong piezo effect in polyvinylidene fluoride (PVDF) in 1969. Piezoelectric PVDF films are produced by mechanically drawing the material and polarizing it in order to form a useful transducer material. The drawing techniques include extrusion and stretching, and while processing the film the material is subjected to a strong electrical polarization field. Typical for PVDF piezo constants are $d_{33} \approx -30\,\text{pC/N}$ and $d_{31} > d_{32} > 0$; the coefficient of coupling k_{33} is about 0.2, and the Curie temperature is near 110 °C.

Recently, polymer foils made for example of polypropylene (PP) have become known with enclosed, lens-shaped vapor locks with dimensions in the micrometer scale, forming a kind of foam structure. Upon applying a high polarization voltage, electrical charges with opposite polarity are produced on opposing bubble walls resulting in a piezoelectric behavior. While the d_{33} values of PVDF foils are clearly below the values of piezoceramics, the values can be much higher for PP foils.

For applications in the field of microactuators, very thin piezoelectric films are preferably implemented with the help of sputter technologies. Frequently used materials include ZnO, ZnS and AlN. These are placed on appropriate substrates, for instance, in the form of beams and membranes, whereby it is also possible to produce multilayer designs. A strong anisotropy of the expansion rate leads to a distinct orientation of the polycrystalline layers, so that the piezoelectric values may reach approximately the values of polarized ceramics under optimal precipitation.

6.2.3 Design of Piezoelectric Transducers

The user can either build a piezo transducer from piezoceramics that are available on the market, or he may benefit from the broad range of avoidable standardized and cased transducers.

Stack Translator (Stacked Design)

The high voltage stack translator is the work horse of piezo actuators. Furthermore, it lends itself to explaining the construction and properties of piezoelectric actuators.

Structure. The active part of the transducer consists, for instance, of many 0.3 to 1 mm thin ceramic discs that are mounted with metal electrodes, e. g. made of nickel or copper, for applying the operating voltage. The discs are stacked up in pairs of opposing polarization and glued together. Highly insulating materials seal the stack against external electrical influences. In other designs – the so-called low-voltage actuators – the multilayer ceramics described above are used.

Figure 6.16 features the electric parallel connection and the mechanical series connection of the stack. Its displacement is the sum of the single element elongations Δl. The applied field and the achieved elongation are in line with the polarization, that is, the piezo constant d_{33} is used (longitudinal effect). The transducer can also handle tensile forces, if prestressed with a slotted cylinder casing as shown in Fig. 6.16 or with an anti-fatigue bolt, as is commonly done.

Static and Dynamic Behaviour. The static diagram $S(E)$ in Fig. 6.17 holds for no-load operation ($T = 0$ in (6.2)). The addend $s^{\mathrm{E}}T$ in (6.2) takes into account the loaded piezo transducers elastic deformation. Two cases are distinguished:

- The load is constant, e. g. weight F_{G}. In this case, the entire diagram is shifted by

$$s^{\mathrm{E}}T = -F_{\mathrm{G}}/c_{\mathrm{P}}^{\mathrm{E}} \,. \tag{6.4}$$

  The spring constant $c_{\mathrm{P}}^{\mathrm{E}}$ follows from the (6.2), if $E = 0$ (see Fig. 6.17). As long as the maximum permissible load is not exceeded, the original no-load expansion of the piezo substance holds (see Fig. 6.17a).

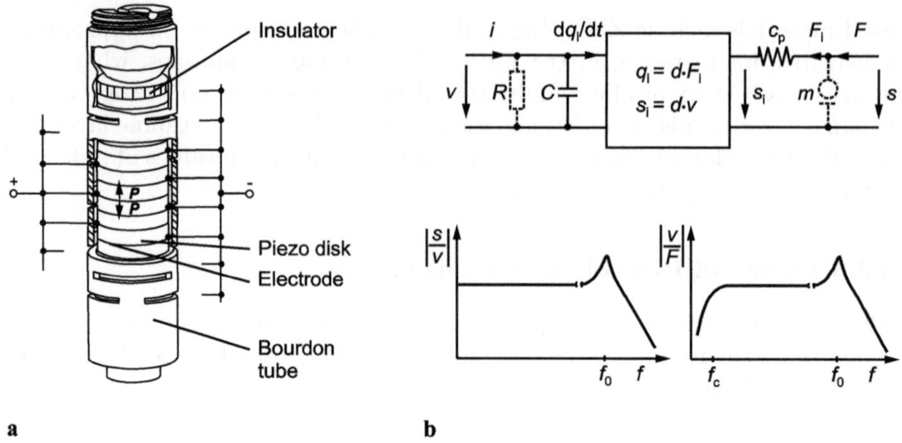

Fig. 6.16. Piezoelectric stack translator. **a** Structure, **b** electromechanical equivalent circuit and amplitude responses of the actuator and sensor transfer behaviour in small signal operation (derived from [5])

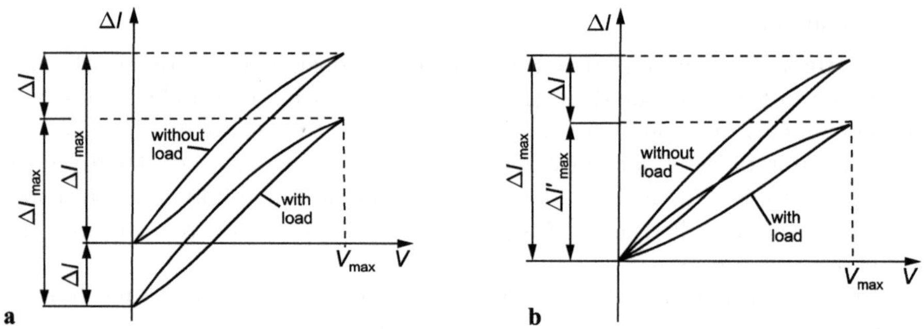

Fig. 6.17. Static displacement characteristic of a stack translator. **a** Constant load, **b** load that depends on the displacement

– The load is dependent upon the displacement, e. g. spring force $F_F = -c_F \Delta l'$. In this case, the origin of the diagram does not move, but the maximally achievable elongation is reduced by the factor $c_P/(c_P + c_F)$ (see Fig. 6.17b). In the extreme case $c_F \to \infty$ (fixed clamp support of the transducer), the transducer achieves its maximal force, the so-called clamping force or blocking force which also follows from (6.2), if $S = 0$.

Equations (6.1) and (6.2) show that an ideal piezoelectric transducer input can be considered as an electric capacitor with the capacitance C and its output as a mechanical spring with the stiffness c_P. This is illustrated in Fig. 6.14b for the d_{33} transducer, but the description holds in principle for all piezo transducers. Since C is in reality always lossy and c_P always has a mass, the amplitude response $|v/F|$ (sensory operation) has an electrically determined lower cut-off frequency f_c and a mechanical eigenfrequency f_0. When operated as an actuator, the electrical input is a voltage, that is, C is constantly recharged, so that f_c has no effect on the amplitude response $|s/v|$, as shown in Fig. 6.16b.

Laminar Translator (Laminar Design)

In contrast to the stack design, the laminar design is based on the piezo constant d_{31} and the transversal effect. The greater the quotient s/l of the piezoelectric element (see Fig. 6.14b), the bigger the effect. This leads to strip shaped elements with low stiffness. Therefore, several layers of strips are piled up, similar to the stack design, and form a so-called laminate for improving the mechanical stability. Since the transversal effect is applied, the result are flat transducers which shorten proportionally to the applied voltage, as d_{31} is negative.

Bending Elements

Bending elements feature the transversal effect as well. They can consist, for instance, of a PZT ceramic mounted onto a piece of spring metal (monomorph). If the length of the ceramic is altered while the length of the metal core stays the same, the element bends in order to compensate the different behaviour, and is therefore phenomenologically quite similar to the thermo-bimetal.

Similarly, it is possible to connect two thin ceramic strips one of which shortens while the other expands (bimorph). One can distinguish between two designs: in the series bimorph, the polarization of the two piezo layers

Fig. 6.18. Bimorph piezoelectric actuators (courtesy of NOLIAC [8])

is inversely arranged, while it is codirectional in the parallel bimorph (see Fig. 6.18). Compared to stack translators, bending elements feature a greater deflection, lower stiffness, smaller blocking force and lower eigenfrequency.

Shear Elements

Recently, Physik Instrumente [7] started offering a line of actuators based on the strong d_{15}-effect (shear effect). According to the definitions in Fig. 6.12a, the quantities E and S work along the axes 1 and 5, i. e. upon applying a voltage the piezo element experiences a shearing motion about its axis 2. Making use of this effect, the end surfaces of block-shaped elements without casing (cross sections of 3×3 to $16 \times 16\,\mathrm{mm}^2$) are shifted by up to $10\,\mu\mathrm{m}$ with respect to each other, while the shearing loads are limited to 300 N. By stacking two such elements, a x–y positioner can be created. Adding a third piezoceramic element based on the d_{33}-effect results in a 3-axis positioning system.

6.2.4 Piezoelectric Transducer With Displacement Amplification

In piezoelectric transducers with displacement amplification the achieved deflection is increased by constructive means. The stiffness of such a design decreases with the square of the displacement amplification ratio and is therefore much smaller than in the stack design.

This kind of transducer used for displacements of up to 1 mm with forces of several tens of Newtons is achieved, for instance, with elastic joints or hinges. These elastic hinges transform small angular alterations into parallel movements free of backlash. Figure 6.19 illustrates the principle.

The highly elastic material region of the displacement amplifier in Fig. 6.19a is locally concentrated, while the designs in Fig. 6.19b and c make use of the global elastic behaviour of metallic materials. The so-called moonie transducer in Fig. 6.19b consists of a piezoelectric disk sandwiched between two metal end caps. The ceramic is poled in the thickness direction and uses the d_{31} mode. In this way the small radial displacement of the disk is transformed into a much longer axial displacement normal to the surface of the

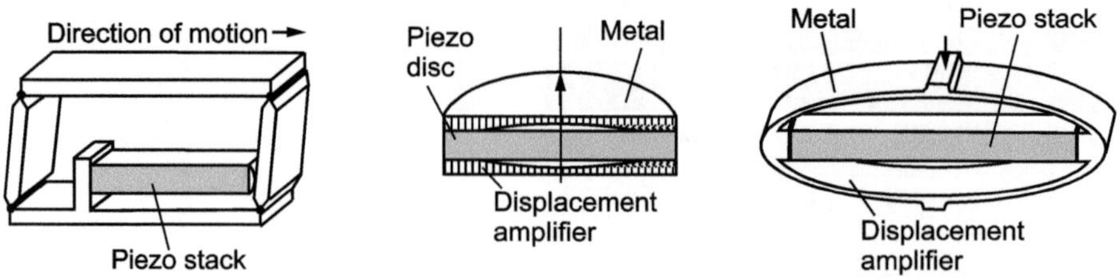

a b c

Fig. 6.19. Mechanical displacement amplification. **a** Implementation with elastic hinges, **b** moonie transducer, **c** amplified piezo actuator, APA (derived from [5])

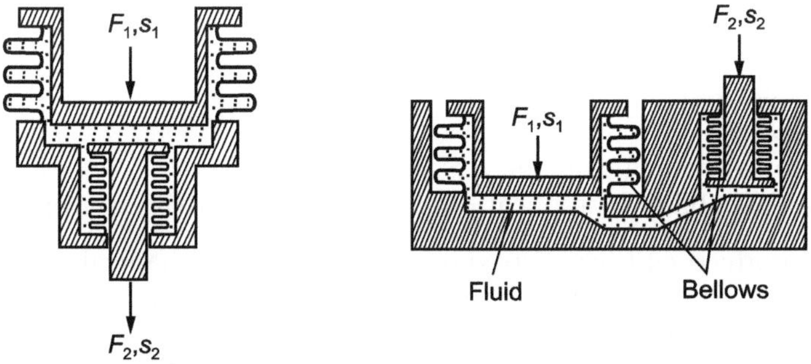

Fig. 6.20. Hydrostatic displacement amplification (derived from [5])

caps. The moonie design is very simple and its manufacture can easily be automated. It generates moderate forces and displacements, filling the gap between bimorph and multilayer actuators [9]. Figure 6.19c shows a related design in which the piezo stack and subsequently the d_{33}-mode are used. The advantages of these APAs (amplified piezo actuator) are their relatively high displacements combined with its large forces and compact size along the active axis [10].

Figure 6.20 shows an entirely different solution. A hydraulic force-displacement transformer functions according to the two-piston hydraulic principle. Leak-free operation is achieved in the presented design through the use of two folding bellows of different effective diameters. This special constructive design keeps the enclosed oil volume small increasing the stiffness of the whole design and minimizing the amount of error due to thermal fluid expansion [11].

With the above introduced principle, it is usually possible to implement an amplification factor of up to 10. Greater values are constructively possible but quickly lead to a worsening of the dynamic behaviour of the entire system.

6.2.5 Piezoelectric Motors

Piezoelectric motors use friction between a mobile part (guide, rotor) and a vibrating part (stator) in order to create motion. The vibrations of the stator are generated by piezoactive materials. The vibrations of the contact points of the stator are such that the trajectory of these points is elliptical. Using friction forces, this vibration drives the mobile part, which is pressed against the stator with a static pre-load. In unloaded conditions, the tangential speed of the mobile part is almost equal to the tangential velocity of vibration of the stator (which is the time derivative of the tangential component of displacement of the stator).

The advantages of such mechanisms are:

- large holding force or torque at rest, without power supply;

- a large actuating force or torque at low speed;
- potential for silent operation;
- nonmagnetic behaviour;
- short time response;
- very good micropositioning capability;
- high integration capability in application, including direct drive concepts.

Piezomotors can produce elliptical motions either at the mechanical resonance (leading to ultrasonic motors) or in quasistatic (leading to stepping piezoelectric motors, so-called Inchworm®) [14].

The use of this motor in direct drive means that the complete function is obtained without any additional gear mechanism (for speed reduction, or for converting rotation in translation). Optics is probably the domain where the use of the piezoelectric motors is the most advanced. The most famous example, is the Canon camera, which includes an auto focus zoom based on a piezoelectric ultrasonic motor (USM) since 1992 (Fig. 6.21) [12].

Several other concept have been developed since then; few of them have found industrial applications. The motor from Elliptec is using a multilayer component, encased in a structure to couple two flexural modes of the beam (Fig. 6.22a) [13]. The stator includes a play recovering mechanism in the form of a spring that:

- applies the preload force between the vibrating stator and the moving member;
- guides the stator;
- decouples the vibrations in the stator from the ground.

Such a vibrating stator can be implemented in various ways (Fig. 6.22b).

Several concepts of quasistatic motors exists as well. One of them is using at least one pair of amplified piezo actuators. The basic working principle of the Cedrat stepping piezomotor concept is illustrated through a simplified linear model based on a pair of APAs (Fig. 6.23). The displacements and forces produced by the APAs are transferred to the slider or the rotor by

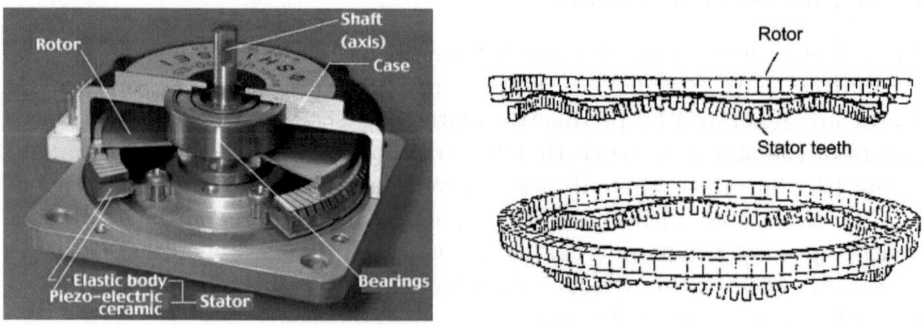

Fig. 6.21. Resonant travelling wave ultrasonic motor

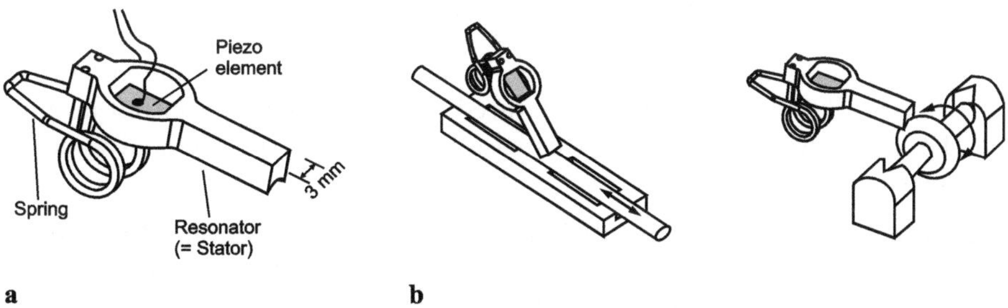

Fig. 6.22. Elliptec motor. **a** Basic structure, **b** examples of application (by [13])

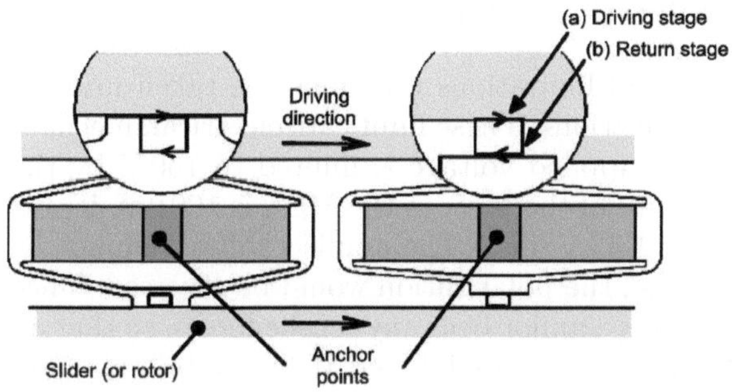

Fig. 6.23. Piezoelectric stepping motor principle [15]

friction. At least one pair of APAs is used in the following conditions: held by their centre, the APAs are actuated in opposite phase.

The motion sequence is in fact not so far from the human walking, each APA working as one leg and whose contact top would be one feet. However, the displacement sequence which produces one step is simplified in the sense that the tops are only actuated with series of pure normal or tangential displacements. During one displacement step, each APA alternatively takes part to drive the slider during a driving stage (a) by friction whereas the other APA returns backward once released from the slider (b). Both the required normal and tangential displacement can be easily obtained at the tops of the APAs with the appropriated voltage supply of its pair of piezoceramics (MLA):

- the same additional voltage supply produces a normal displacement;
- an opposite additional voltage supply produces a tangential displacement.

This piezomotor concept displays two distinct modes of running which are easily combined successively to reach the targeted position:

- a coarse mode through the above described stepping principle. In this driving mode the stroke is not limited and one linear displacement step can vary from 1 to 10 µm in length versus the voltage level applied.

– a fine mode to increase the precision positioning after a coarse approach. In this mode, the pair of APAs is driven in phase and actuated so that a tangential displacement is produced. The total stroke centred around the non powered state is in this case limited to about the equivalent of one coarse step.

This principle can be implemented in rotating and linear motors. Due to their low speed and high positioning accuracy, quasistic motors find applications in scientific and semiconductor applications.

6.2.6 Limitations of Piezoelectric Actuators

Piezoelectric actuators have several limitations that must be taken into account to properly design the applications. These limits are electrical, mechanical and thermal. The maximum applied voltage is limited to 150 V by the insulating layer. Since the thickness of the layer in the MLA is 100 µm, it corresponds to an electrical field of 1.5 kV/mm. The applied voltage cannot be decreased under −30 V. Otherwise, the polarization would be reversed. Since MLAs are laminated materials, they cannot bear any tensile forces, so that all the piezo actuators are mechanically preloaded. Since MLA is a brittle material, bending or twisting moments must be avoided as much as possible, even during the mounting procedure, especially for direct piezo actuators (DPAs). Tensile forces during dynamic operations or switched operations must also be avoided.

For designing purpose, multiplayer piezoceramic are considered as linear. Indeed, the hysteresis is in the range of 10 ... 15%, meaning that a closed loop if often required. Moreover, under a high voltage, a repoling process (corresponding to the drift) occurs and could range upto 10%.

In static operations, the lifetime is mainly limited by the humidity, which penetrates through the external insulation layer and leads to a leakage current increase. A larger leakage current can lead to an electrical breakdown. Due to the dielectric and mechanical losses, the piezoelectric actuator warms up under continuous excitation. Losses are mainly non-linear and depend on the excitation frequency, the voltage amplitude and the humidity. To avoid a depoling effect of the ceramic, the temperature in the actuator should be monitored to ensure that it stays well below the ceramics Curie temperature. So a typical range of temperatures is −40 °C to 80 °C.

This results in that the duty cycle of the piezoelectric actuator in dynamic operation is limited by the thermal behaviour.

There are currently a lot of research into materials capable of producing MLAs displaying higher working temperatures (up to 140 °C). Similarly, the standard MLAs work at low temperature and have already been tested in liquid nitrogen (77 K, −196 °C): at this low temperature, their displacement is only one third of that obtained at room temperature.

Provided the self heating and the tensile forces are prevented, the amplified piezo actuators do not show any fatigue effect. For example, a test of an amplified piezo actuator under full scale pulse (0 ... 150 V) with a driving frequency of 600 Hz, had a continuous duration of 6 months. It shows the ability of the actuator to operate for 10^{10} cycles.

The thermo-mechanics may be an issue in the case of a fine positioning application over a large range of temperature: the PZT in the multilayer technique display various coefficient of thermal expansion, CTE (as a function of some construction details). Standard amplified piezo actuators displays fairly large CTE due to some thermal mismatch between the piezo component and the shell material. There are some possibilities to cancel this CTE in the application:

- a large CTE material that compensates the low CTE from the piezo component may be added in the mechanism;
- a symmetric arrangement implemented in push-pull operation is insensitive to the CTE.

6.2.7 Example Applications of Piezoelectric Actuator Used in Adaptronics

There are many possibilities when controlling piezo actuators, which depends on the applications and the foreseen command. This section aims at covering many different applications involving a closed loop. Combining piezoelectric actuators with smart electronics can lead to numerous adaptronics applications.

Open Loop Applications

Open loop operations with high accuracy remain possible if the behaviour of the piezo actuator (hysteresis, drift effect) is well known, and if the command applied to the piezo is known [16]. Two examples have been recently investigated in active optics:

- dynamic refocusing of a laser extended cavity for a LIDAR [17] or optical delay line;
- a mechanism for CCD microscanning.

For these two applications, the command is repetitive; therefore, the drift and hysteresis can be anticipated through a feed-forward correction, which remains dependant on the temperature and the voltage. A typical command including a pre-shaper sent to the piezo actuator is shown in Fig. 6.24. The command anticipates the drift effect during the plateau. The command amplitude is a function of the temperature. This approach is simple (it does not need any position sensor) but requires a calibration effort.

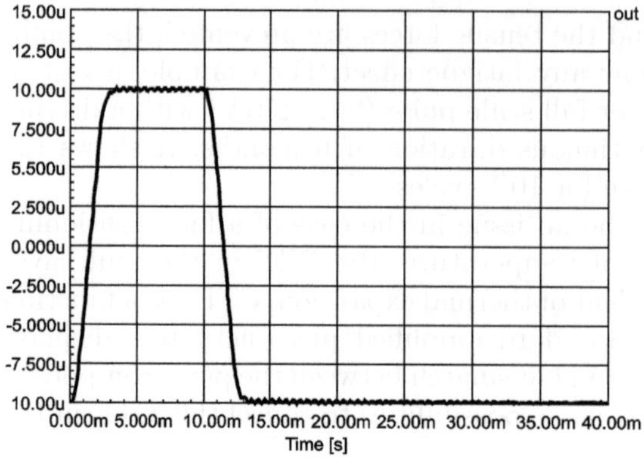

Fig. 6.24. Example of an open loop command minimizing the overshoot (rising time 2.7 ms, accuracy during the plateau: $\pm 0.2\,\mu m$)

Acceleration Closed Loop

To achieve vibration damping, the piezo actuator can be combined with an accelerometer [18]. A first solution consists of using a piezoelectric actuated proof mass damper (Fig. 6.25), in which the compliance of the proof mass corresponds to the piezoelectric compliance. The force provided by the piezo actuator is $F = N \cdot V$, where N is the force factor and V the applied voltage. This method is generally adapted to high frequency mode (e. g. 100 ... 400 Hz), as it remains difficult to build a piezo proof mass (PPM) at low frequency.

Alternatively, the piezo actuator can act in parallel to the structure and is controlled through an accelerometer on the structure. Similarly to position control, high order vibration modes can greatly influence the stability of the loop. In Fig. 6.26, it can be seen that the structure reacts under a disturbance force at $t = 0.1$ s and is quickly damped at $t = 0.5$ s, when the closed loop is switched on.

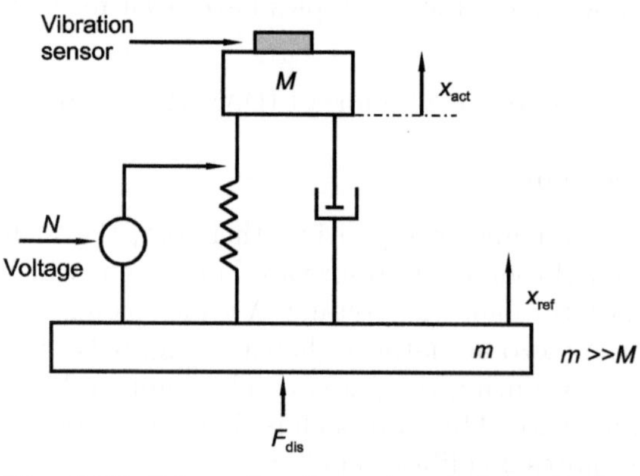

Fig. 6.25. Schematic of a system to be actively damped

a

b

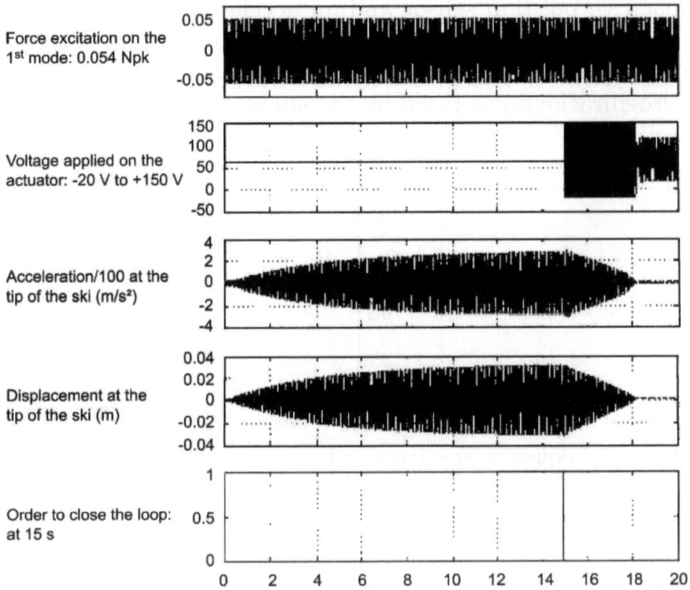

Fig. 6.26. Active damping of a ski. **a** Implementation of the APA120ML with its mechanism, **b** time diagram of the closed loop

Among several applications in space and machine tools, a nice application has been developed using this concept: the active damping of a ski [19]. The first flexural vibration mode of the ski occurring at 14 Hz is actively damped (the initial quality factor of 100 is decreased down to 10) through a piezo actuator and an accelerometer (Fig. 6.26a). A special filter is necessary to avoid instabilities coming from the high order vibration modes. The piezo actuator is mounted in front of the shoe and the accelerometer is mounted at the top of the ski. This implementation is an important step to the adaptronic application, in which it is foreseen to adapt the quality factor of the ski as a function of the snow hardness.

Combined Loops

Position and acceleration closed loops can be also combined in a single controller. This application may find application in space optics where image multiplexing and microvibration isolation can be achieved with the same piezo mechanism; it has been modelled and tested at Cedrat Technologies.

A typical case arrives with the reaction wheel,whose perturbations frequency is dependent on the rotation speed (Fig. 6.27): it is therefore necessary to have a broadband active control of vibrations.

The Fig. 6.28 shows the experiments consisting of a platform including a piezo actuator moving a guided payload through flexural springs, monitored though a capacitive position sensor and an accelerometer. This platform is shaken with a solid-state (magnetostrictive) transducer. The purpose of the controller is to accurately position the payload and remove (at the payload level) the microvibrations generated by the shaker.

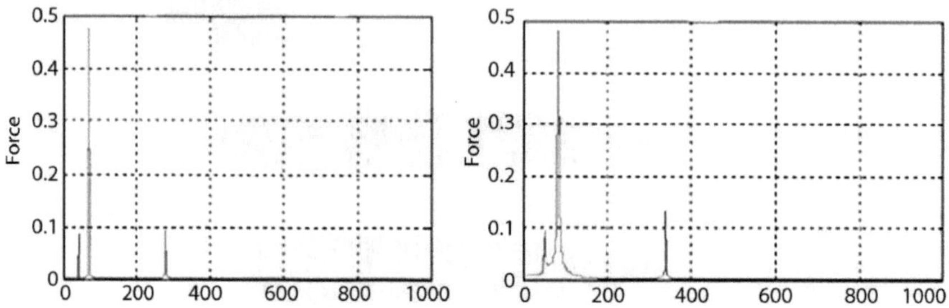

Fig. 6.27. Typical frequency spectrums of a perturbation force coming from a spacecrafts reaction wheel

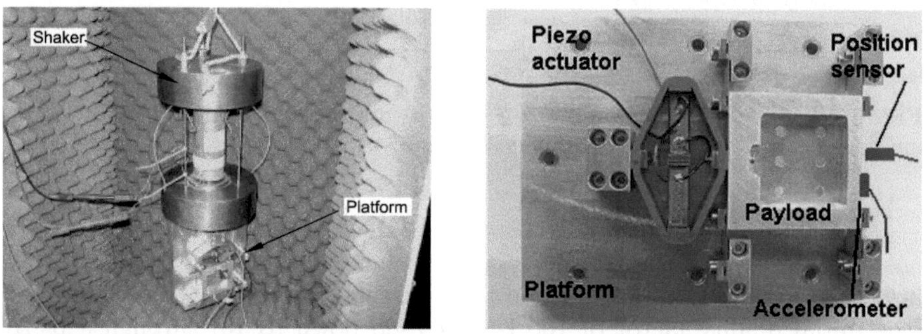

Fig. 6.28. View of the piezo actuator and its payload equipped with a position sensor and an accelerometer – the piezo actuator (*right*) is excited with a magnetostrictive actuator (*left*)

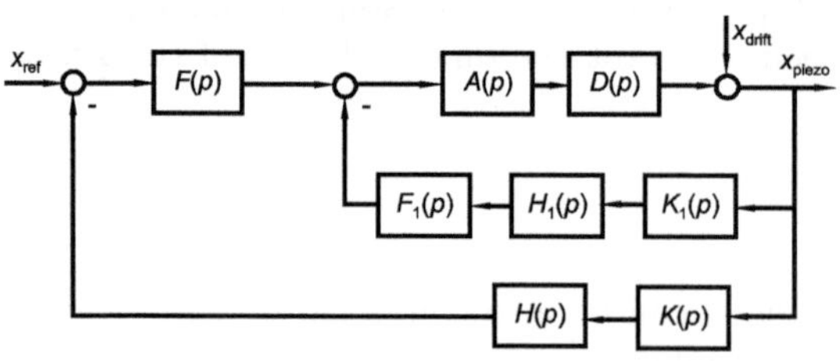

Fig. 6.29. Block diagram including the position and the acceleration closed loops

The block diagram depicted on Fig. 6.29 has been realised with analogue boards: the first step corresponds to the tests of the filtering cells in an open loop. As a first step, one checks that the pilot and the measurement accelerometer give a correct response under an excitation of the shaker. As a second step, one checks that the filtering cell $F_1(p)$ is correct. As a third step, one also checks that the filtering cell $H_1(p)$ allows isolation of the acceleration loop from a position order.

In this block-diagram (in which p is the variable of the Laplace transform):

- x_{ref} is the command for the position;
- x_{drift} is the drift of the position resulting either from a disturbance force or non linear behaviour of the piezo actuator;
- $D(p)$ is the transfer function of the piezo actuator and the payload;
- $A(p)$ is the transfer function of the power linear amplifier, including its current limitation;
- $F(p)$ is the transfer function of the position corrector;
- $H(p)$is the transfer function of the lowpass filter for the position sensor;
- $K(p)$ is the transfer function of the position sensor;
- $K_1(p)$ is the transfer function of the vibration sensor;
- $H_1(p)$ is the filter transfer function of the vibration sensor corresponding to a bandpass filter between 30 and 800 Hz;
- $F_1(p)$ is the transfer function of the vibration corrector.

Several comments can be made from these measurements of the closed loop transfer function of the block diagram (Fig. 6.30):

- at 150 Hz, a resonance frequency exists and increases the response of the capacitive sensor;
- in low and high frequencies, a phase shift exists between the accelerometer response and its filtered response;
- it is confirmed that the capacitive position sensor, linked to the payload is able to measure the payloads position, despite the microvibration.

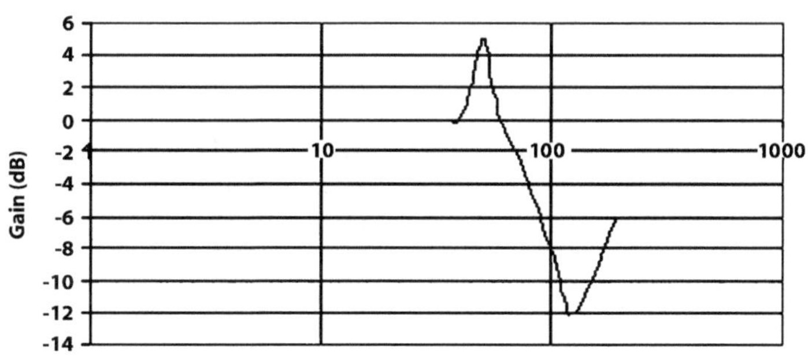

Fig. 6.30. Transfer function of the attenuation profile of the isolation closed loop

One also checks that the amplified piezo actuator is able to counteract acceleration perturbations in the order of 20 mg to 40 mg: under these levels which are largely representative of spacecraft microvibrations, the studied system remains linear.

In a second step, the system is tested in a closed loop. One must consider the proportional integral corrector that is used to isolate the payload from the microvibrations. The surtension noticed at 40 Hz could be improved by a better controller with an integral gain. A compromise exists between the capability of the acceleration loop to counteract the microvibrations and its stability.

The position closed loop is effective below 5 Hz; the isolation vibration closed loop is effective up to 60 Hz. The achieved performances are the following:

- −40 dB/decade roll off;
- cut off frequency close to 50 Hz;
- over shoot 5 dB @ 50 Hz;
- maximum attenuation: 10 dB.

6.2.8 Energy Harvesting Application Using Piezoelectric Actuators

Energy harvesting may be useful to energize low consumption sensors or radio emitters, without using batteries. For instance, health monitoring sensors embedded in a aircraft may benefit from this approach, since the vibration of the aircraft would be used to supply the sensor. As a result, the cables routing is no longer necessary.

A demonstrator has been built [20] to show the interest of the piezoelectric actuator in this technique (Fig. 6.31). When the vibration is in the range 100 ... 400 Hz, the required piezoelectric actuators are much more compact than any electromagnetic actuators. Secondly, an efficiency in the range of 50% has been demonstrated.

6.2.9 Outlook

Piezoelectric actuators are more and more often used for their accuracy and fast response. They are used in industrial applications together with a dedicated driver and a control loop. Optical applications were the first to use piezoelectric multilayer actuators. The past years have seen the development of adaptronic applications in machine tools and large scale application in automotive (gazole injectors) systems.

When choosing a piezo actuator for an adaptronic application, it is essential to correctly tailor the no-load displacement and the blocked force of the piezo actuator.

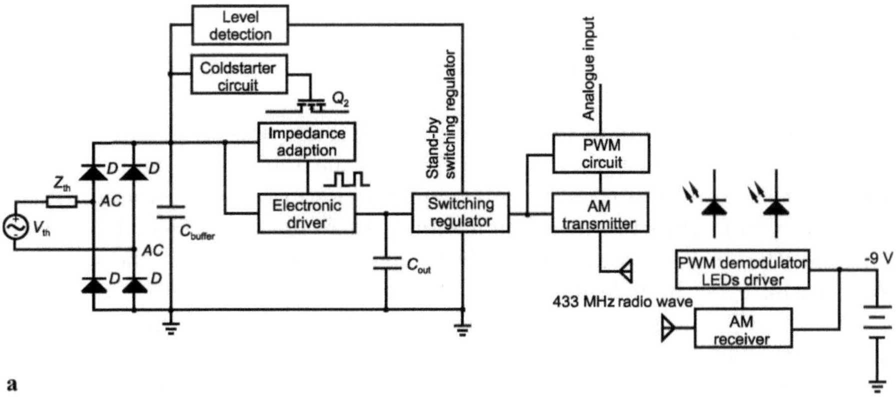

a

b

Fig. 6.31. Piezoelectric energy harvesting. **a** Synoptic of the demonstrator, **b** view of the demonstrator

Piezoelectric actuators are still the subject of much research such as:

- looking for the use of single crystal material in the multilayer technique: this will allow the taking of benefits from the high piezoelectric effect in single crystal materials at low voltage;
- increasing the reliability of piezoelectric material under aggressive environmental conditions and establishing reliability figures remains important to increase the number of industrial applications.

6.3 Magnetostrictive Actuators

F. Claeyssen, G. Engdahl

Magnetostriction occurs in most ferromagnetic materials and leads to many effects [21,22]. The most useful is referred to as the Joule effect, and is responsible for the expansion (positive magnetostriction) or the contraction (negative) of a rod subjected to a longitudinal static magnetic field. In a given material, this magnetostrain is quadratic and occurs always in the same direction whatever the field direction.

Rare-earth-iron giant magnetostrictive alloys (GMAs), discovered by A.E. Clark [23], feature magnetostrains that are two orders of magnitude larger than nickel. Among them, $Tb_{0.3}Dy_{0.7}Fe_{1.9}$, often called Terfenol-D, presents at room temperature the best compromise between a large magnetostrain and a low magnetic field. Positive magnetostrains of 1000...2000 ppm obtained with fields of 50... 200 kA/m are reported for bulk materials [23,24]. New composite materials of Feredyn offer an interesting possibility for high frequency ultrasonic applications [25]. More recently, high magnetostrains (in the range of 500... 1000 ppm) have also been obtained in rare-earth-iron thin films [26]. However, these expansion strains are rarely used directly because most applications require a linear behavior. The linearity is obtained by applying a magnetic bias and a mechanical prestress in the active material. Moreover, in the case of applications based on a mechanical resonance, it is a condition of producing huge dynamic strains that their peak-to-peak amplitude is greater than that for the static magnetostrain [27].

The static magnetostrain of the GMAs permits the building of linear actuators offering small displacements (20... 200 μm) and large forces (500... 5000 N) at low voltage. These linear actuators are constructed to be used directly, for instance for micropositioning tools or for damping structures. They can also be used as components of a more complex actuator, such as inchworm motors. Such motors present holding forces/torques that are often much higher than piezoelectric inchworm motors; they also provide good positioning accuracy. Their main disadvantage is a low efficiency, which is due to their static operating conditions. Huge dynamic strains (up to 4000 ppm) can be produced in Terfenol-D linear actuators using the device at mechanical resonance, even when working against a high load; in such conditions, large power and rather good efficiency can be achieved. Using these properties, some magnetostrictive underwater transducers already outperform PZT transducers in the low-frequency domain and receive a great deal of attention. Some research works are being pursued in order to also use mechanical resonance in magnetostrictive motors, aiming at greater mechanical power and a better efficiency than in inchworm motors.

Although there is no large-volume application for magnetostrictive actuators at the moment, some are already used for specific applications in domains such as pumps, micropositioners, and transducers, and research into other applications is growing. It is likely that we shall also see magnetostriction finding applications in the microactuators domain in the future.

6.3.1 Theory of Magnetostriction in Magnetostrictive Devices

Constitutive Equations

In the most general way, the behavior of magnetostrictive materials is nonlinear [21, 22] and has to be described with nonlinear relations:

$$\boldsymbol{S} = f(\boldsymbol{T}, \boldsymbol{H}) \tag{6.5}$$

$$\boldsymbol{B} = g(\boldsymbol{T}, \boldsymbol{H}) , \tag{6.6}$$

relating $\boldsymbol{S}$ and $\boldsymbol{T}$, the tensors of strain and stress, to $\boldsymbol{B}$ and $\boldsymbol{H}$, the vectors of induction and magnetic field. The functions f and g may be obtained by measuring the magnetostriction and the magnetization against the applied field and the external stress [28]. Then functions f and g can be described numerically by an interpolation method [29, 30]. This technique, feasible for the finite difference method, is used in lumped element models [31], where nonlinearity and hysteresis effects can be treated. Another method could consist of developing f and g as a Fourier series, taking some first-order terms, and such an approach is being applied in the ATILA software, based on a finite element method [32], for modeling the nonlinear behavior of three-dimensional (3D) structures, including electrostrictive materials [33].

However, although magnetostrictive materials are nonlinear, the behavior of most magnetostrictive devices may be rather well described using a linear theory, because the active materials are biased. Experimental results obtained on a high power transducer (see Sect. 6.3.2) show that linearity can be rather good even with large excitation fields and large dynamic strains.

The bias conditions are defined by the magnetic bias H_0 and the mechanical prestress T_0, applied along the magnetostrictive rod axis, which is referred to as the third axis. Then, considering only the variations around this initial bias state, the material behaves in a quasi-linear manner and follows piezomagnetic laws [34]:

$$\boldsymbol{S}_{\mathrm{i}} = \boldsymbol{s}^{\mathrm{H}}_{\mathrm{ij}} \boldsymbol{T}_{\mathrm{j}} + \boldsymbol{d}_{\mathrm{ni}} \boldsymbol{H}_{\mathrm{n}} \qquad (i, j = 1, \ldots, 6) \tag{6.7}$$

$$\boldsymbol{B}_{\mathrm{m}} = \boldsymbol{d}_{\mathrm{mj}} \boldsymbol{T}_{\mathrm{j}} + \boldsymbol{\mu}^{\mathrm{T}}_{\mathrm{mn}} \boldsymbol{H}_{\mathrm{n}} \quad (m, n = 1, \ldots, 3) \tag{6.8}$$

where $\boldsymbol{s}^{\mathrm{H}}$, $\boldsymbol{d}$ and $\boldsymbol{\mu}^{\mathrm{T}}$ are the tensors of constant-H compliance, piezomagnetic constants and constant-T permeabilities, respectively. They are called the magneto-elastic coefficients. $\boldsymbol{S}$ and $\boldsymbol{T}$ are the tensors of varying strain and stress, $\boldsymbol{B}$ and $\boldsymbol{H}$ are the vectors of varying induction and magnetic field. In the actuators, $\boldsymbol{H}$ is called the excitation field.

The real situation in the material can be reconstructed by adding the bias static situation to the variations. For instance, the real field in the material is the vector sum of static magnetic bias H_0 and the varying magnetic field $\boldsymbol{H}$. Note also that the values of the coefficients of the materials tensors depend strongly on the bias and the prestress [28, 34]. Complete sets of values for the tensors $\boldsymbol{s}^{\mathrm{H}}$, $\boldsymbol{d}$ and $\boldsymbol{\mu}^{\mathrm{T}}$ and other equivalent tensors of Terfenol-D have already been established [34–36]. Longitudinal coefficients ('33') and shear coefficients ('15') may be determined using length expansion and shear resonators such as the MB [35] and DCC [37] types (as described in Sect. 6.3.2 and shown in Fig. 6.37). Other coefficients may be found using some special assumptions [34].

Terfenol-D is often used in long rods, subjected to an excitation field parallel to the rod axis. In this case, the simple theory of the longitudinal mode can be applied. Such theory can be used to obtain a preliminary system design, before the use of numerical models to refine it. In such a situation, it is presumed that the transverse excitation fields are negligible ($H_1 = H_2 = 0$). In theory, a pure longitudinal mode (33-mode) is then obtained starting from the assumption that radial stresses are equal to zero ($T_1 = T_2 = 0$) and that there is no shear effect ($T_4 = T_5 = T_6 = 0$), leading to the following equations:

$$S_1 = S_2 = s_{13}^{\mathrm{H}} T_3 + d_{31} H_3 \tag{6.9}$$

$$S_3 = s_{33}^{\mathrm{H}} T_3 + d_{33} H_3 \tag{6.10}$$

$$B_3 = d_{33} T_3 + \mu_{33}^{\mathrm{T}} H_3 \; . \tag{6.11}$$

The 33-mode coupling coefficient associated with this mode is given by

$$k_{33}^2 = \frac{d_{33}^2}{s_{33}^{\mathrm{H}} \mu_{33}^{\mathrm{T}}} \; . \tag{6.12}$$

This coefficient represents the capability of the material to convert electric energy into elastic energy. Its value is high in Terfenol-D even with high prestress and bias [28] (see Table 6.1). As will be shown later, the combination of a high coupling, a high prestress and a high bias is required to obtain giant dynamic strains and very high output powers [27].

Simplified Theory of Magnetostrictive Linear Actuators

It is interesting to analyse the behavior of linear actuators because most applications are based on such actuators. To simplify the presentation, we can consider an actuator with one end working either free (no load) or against a purely resistive load R_{load} (in kg/s); the other end of the actuator is clamped. The vibration against this load produces an output power (either mechanical or acoustic), and its behavior is representative of any magnetostrictive device. Most of them can be analysed as whole systems, including a compliance k^{H} (at constant field), an effective mass M and a mechanical

Table 6.1. Magneto-elastic longitudinal coefficients of Terfenol-D at about 90 kA/m bias versus prestress T_0

T_0	(MPa)	30	35	40	50
Y^{H}	(GPa)	29	21	23	40
s_{33}^{H}	(1/GPa)	0.034	0.048	0.043	0.025
Q^{H}		4.6	3.5	4.3	8.3
$\mu_{33}^{\mathrm{T}}/\mu_0$		3.7	4.2	3.8	3.0
Q^{T}		2.0	1.9	2.2	2.8
d_{33}	(nm/A)	8.0	11.0	9.7	5.0
k_{33}	(%)	63.1	69.3	67.4	52.0

resistance R_{m} (in kg/s) due to internal mechanical losses. The magnetostrictive part is activated by a longitudinal field H_3 produced by a coil driven by an excitation current I. In such a system, all the strain is converted to displacement of the free mass.

Under quasi-static conditions, according to (6.9) and neglecting prestress spring stiffnesses for a first approximation (which gives $T_3 = 0$), the strain S_3 of Terfenol-D in an unloaded actuator is:

$$S_3 = d_{33} H_3 \,. \tag{6.13}$$

A maximum excitation field H_3 equal to the bias H_0 can be applied. Higher values lead to a frequency-doubling effect. In this situation, the actuator is field-limited. The heating of the coil is another limitation often encountered in static conditions. A high excitation field needs a high current density in the coil wires, typically in the range of 10 A/mm^2. As it is a rather high value, a significant heating may occur and it is therefore necessary either to use the actuator during short pulses or to cool the coil.

When the unloaded actuator is excited with a constant field amplitude against frequency, a sharp peak is obtained for the induced vibration. A typical example of strain curves (Fig. 6.32) without load or with a load is given by a linear Terfenol-D actuator based on a driver such as MAP (described in Sect. 6.3.2) (A load value of $R_{\mathrm{load}} = 10^4$ kg/s is used in Figs. 6.32 to 6.34,

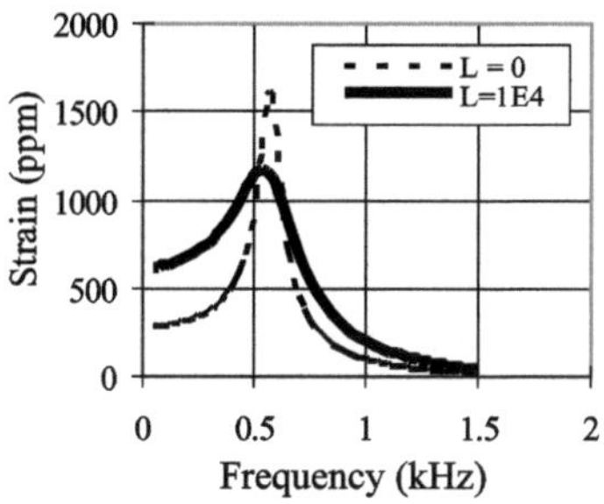

Fig. 6.32. Strain S_3 versus frequency at constant currents, without load and with a load equivalent to $Q_{\mathrm{m}} = 2$

and is there denoted by $L = 10^4$). The natural longitudinal vibration mode occurs, and because of the coupling, this mode is magnetically excited. Compared with static strains, the strains at resonance are magnified by a factor called the 'mechanical quality factor' Q_m (the load value being equivalent to $Q_m = 2$) where

$$S_3 = Q_m d_{33} H_3 \,. \tag{6.14}$$

This mechanical quality factor defines the damping of the resonance. When the vibrating end is unloaded, the damping is only due to internal mechanical losses and Q_m is equal to the material mechanical quality factor Q^H. When a load is applied, the resistive part of the load provides an additional damping that reduces the devices mechanical quality factor. Typical values for Q^H in Terfenol-D are in the range of 3 to 20. Consequently, for a very first approximation, the maximum strain S_3 at resonance under such conditions is determined from (6.9) and (6.14), by the stress T_3, since $d_{33}H_3$ is necessarily small compared with $s_{33}^H T_3$. We thus have

$$S_3 = s_{33}^H T_3 \,. \tag{6.15}$$

Without load (or also with a small load), the actuator is limited at resonance by the stress: the dynamic stress level T_3 reaches the prestress value T_0. With use of a high prestress, the maximum dynamic peak-to-peak amplitude of strain may be much larger than the maximum static strain (1600 ppm for this material) [27]. For instance, with $T_0 = 40\,\text{MPa}$, the peak-to-peak strain is about $S_{pp} = 2S_3 = 3500\,\text{ppm}$ according to (6.15) and Table 6.1. This high strain is also permitted by the good coupling factor of Terfenol-D at such high prestress, and can be obtained under low load with a low field amplitude $H_3 = 40\,\text{kA/m}$ according to (6.14) and Table 6.1. Intensive research on giant strains is being conducted and has allowed experimental work with peak-to-peak strains of 3500 ppm and more (see Sect. 6.3.2).

Due to the strong coupling, the mechanical resonance obtained at constant current is associated with the electrical antiresonance f_a, the maximum impedance (Fig. 6.33). Using a constant voltage, the mechanical resonance would occur at the electrical resonance f_r, the minimum impedance.

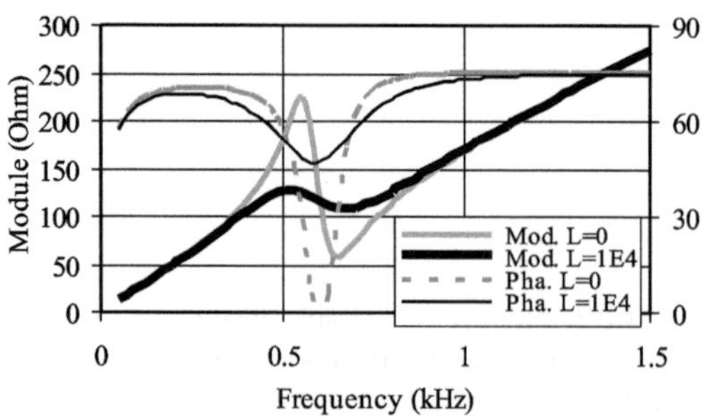

Fig. 6.33. Module and phase impedance versus frequency, without load and with a load equivalent to $Q_m = 2$

These resonances determine the effective coupling factor k_{eff} of the device,

$$k_{\text{eff}} = \sqrt{1 - (f_a/f_r)^2} \,, \tag{6.16}$$

and this factor represents the capability of the device to convert electrical energy to elastic energy. As shown below, the output power of a device depends strongly on this factor. In the best theoretical case, it is equal to the material coupling factor; in the best actuators, the measured k_{eff} may reach 55... 60%.

The high power handling capability of Terfenol-D can be observed by applying a high load. A high load condition is achieved when the mechanical quality factor Q_m of the vibration mode of the system is low (load higher than the optimal load). In this case, the actuator is field-limited, even at resonance. Then the maximum excitation field that can be applied is equal to the bias. It is important to notice that even against such high loads – and unlike PZT actuators under the same condition – the maximum strain of Terfenol-D actuators remains very high (Fig. 6.32).

A special case is obtained with an optimal load. Both stress and field limits are reached. This permits production of the absolute maximum power. The optimal load of an actuator can be determined theoretically. Typically (see Table 6.2, Sect. 6.3.2), it leads to a mechanical quality factor in the range of 2 to 3, which also shows the ability of Terfenol-D to work against high loads.

The output power can be compared with the electric power through the efficiency (Fig. 6.34). The curve of efficiency against frequency shows that the best way to produce a significant output power with an actuator or a transdu cer is to work at resonance. A good efficiency ($\geq 50\%$) may be obtained with a high load ($Q_m \leq 2$).

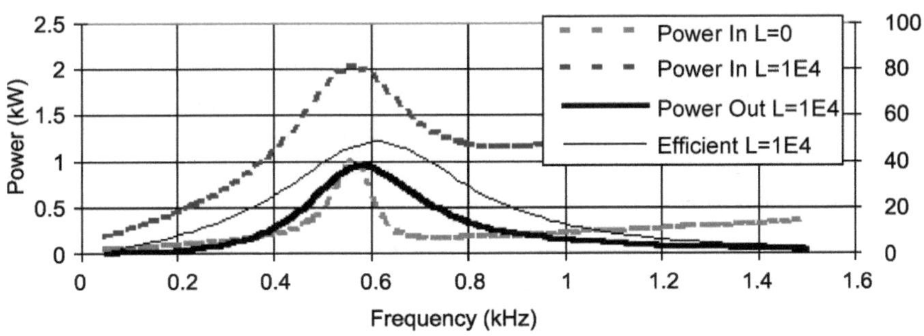

Fig. 6.34. Powers and efficiency versus frequency, without load and with a load L equivalent to $Q_m = 2$

The expression of the output power at resonance [34] permits examination of the role of some parameters:

$$P_{\text{out}} = \omega e_{\text{m}} k_{\text{eff}}^2 Q_{\text{m}} (L_{\text{LF}} I^2/2) \,, \tag{6.17}$$

where $L_{\text{LF}} I^2/2$ is the electric energy stored in the low-frequency inductance L_{LF} of the device, $e_{\text{m}} \equiv 1/(1 + R_{\text{m}}/R_{\text{load}})$ is a mechanical efficiency and ω is the resonance pulsation. In general, the pulsation and the load are often prescribed by the application. When it is possible to select the load, a high load is preferred to obtain a high efficiency. The stored electric energy can be increased using higher prestress, bias, and current. However, bias values much greater than 100 kA/m are difficult to produce with permanent magnets. The effective coupling factor can be optimised by improvement on the basic design.

The maximum force that can be produced by the actuator is the clamped force. This force F is given by G, the force factor (also called the electromechanical conversion factor)

$$F = GI \tag{6.18}$$

with

$$G = k_{\text{eff}} \sqrt{L_{\text{LF}} k^{\text{H}}} \,. \tag{6.19}$$

This is also the blocked force of the main mode of the actuator at resonance. So, it is an important parameter for several applications: for example, in both quasi-static and resonant motors it strongly influences the maximum force/torque of the motors.

This simplified theory provides an understanding of some important features of linear magnetostrictive drivers of actuators, transducers, etc. It shows, for instance, that a driver may be limited either by the stress or by the field, and that the strain at resonance may be much larger than that of a static system and yet may require much less field. However, because of the assumptions on the field shape, the strain uniformity and so on, it is not possible to accurately predict the behavior of the device, especially its exact limits. So, without a good knowledge of these limits, it is difficult to use the full potential of the device. That is why a more accurate model is required and has been developed.

Nonlinear Modeling Approach

One characteristic of linear models is that they only are valid for small signal excitations, where account for bias magnetisation level and prestress are taken by adjusting the magnetostrictive linear tensor parameters $\boldsymbol{d}$, $\boldsymbol{s}^{\text{H}}$, $\boldsymbol{s}^{\text{B}}$, $\boldsymbol{\mu}^{\text{T}}$, and $\boldsymbol{\mu}^{\text{S}}$ in an appropriate way. Besides, linear models cannot give an appropriate

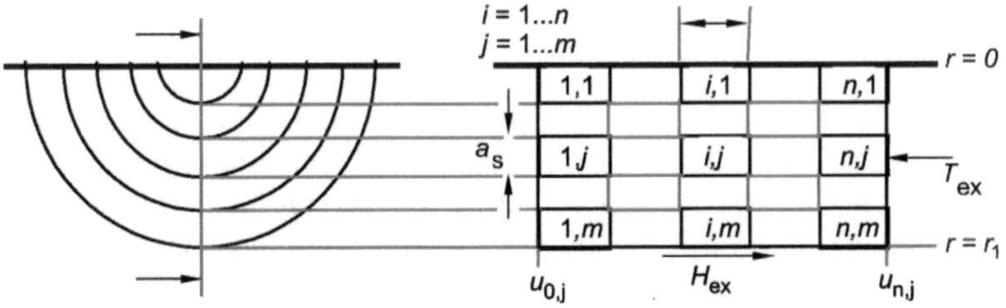

Fig. 6.35. A magnetostrictive rod with radius r_1 sectionized in axial and radial directions with an applied field and stress H_{ex} and T_{ex} respectively, and axial section boundaries u_{ij}

description of hysteresis effects that can be of significant importance even at low excitations.

A feasible way to include nonlinear effects in an actuator structure is to use the lumped element approach, which means that parts of the geometry that have similar potential, current, mass, magnetic flux, mechanical stress, strain or other relevant properties are lumped together to be represented by one discrete component.

The governing idea in this approach is to delimit the state variables to stress and strain in a finite number of sections of a rod of the magnetostrictive material. In a radial-axial model [69] an example of such sections is shown in Fig. 6.35. The constitutive equation for the field distribution inside the rod when it exposed to a longitudinal field is $\frac{\partial^2 H}{\partial r^2} + \frac{1}{r}\frac{\partial H}{\partial r} = \sigma \frac{\partial B}{\partial t}$, where r is the radial coordinate in a cylindrical coordinate system with its symmetry axis coinciding with the rod symmetry axis. By discretizing and combining this equation with Newtons second law, and magnetostriction and magnetizing experimental data for each section, a differential algebraic equation system can be set up.

Such systems can be solved by program packages such as SANDYS, SABER, DYMOLA etc.. In that approach the mechanical boundary conditions are defined by the mechanical load conditions. The rod ends can be clamped, free or attached to some load defined by a network of passive mechanical components. In a more general case delivered forces and/or displacements can be prescribed explicitly.

For an applied H field H_{ex} a boundary condition according to $H_{\text{i,m+1}} = H_{\text{ex}}$ can be defined, or for a continuous flux a condition according to $\frac{1}{m}\sum_{\text{j=1}}^{\text{m}} B_{\text{i,j}} = B_{\text{ex}}$, or for the total flux and the derivative of the applied field a conditions according to $\left.\frac{\partial H}{\partial r}\right|_{r=r_1} = \frac{\sigma}{2\pi r_1}\frac{\text{d}\phi_{\text{r}}}{\text{d}t}$, where σ is the electric conductivity of the active material, r_1 the rod radius and ϕ_{r} the total flux through the rod. The boundary condition at $r = 0$ is $\left.\frac{\partial H}{\partial r}\right|_{r=0} = 0$ or $\left.\frac{\partial B}{\partial r}\right|_{r=0} = 0$.

As long as one only considers H_{ex} and B_{ex} as the driving quantities one does not need to specify the magnetic circuit because if H_{ex} is specified it is always possible to obtain a corresponding B_{ex} from the material data base or vice versa.

In real applications one is, however, rather interested in driving quantities as imposed current I_{ex} and/or imposed voltage V_{ex}. To make it possible to excite the rod with input currents and/or voltages in the model it is necessary to specify the magnetizing system or in other words the magnetic circuit.

Magnetic Circuit

In principle such a magnetizing system involves a reluctance and a coil flux leakage, see Fig. 6.36. In this 1D model it is sufficient to estimate the reluctances R_{p} and R_{l} (in 1/Ωs) in order to take the magnetic circuit into account. Assuming equivalent cross-sectional areas and lengths of the flux return and leakage paths one can obtain a rough estimate of $R_{\mathrm{p}} = l_{\mathrm{p}}/(\mu_{\mathrm{p}} A_{\mathrm{p}})$ and $R_{\mathrm{l}} = l_{\mathrm{l}}/(\mu_{\mathrm{l}} A_{\mathrm{l}})$, where l, μ and A are appropriate effective lengths, permeabilities and areas of the magnetic return flux path and of the leakage flux, respectively.

The relation between imposed current I_{ex} and imposed magnetic field H_{ex} then can be described by the equations [70]:

$$\begin{aligned} N I_{\mathrm{ex}} &= R_{\mathrm{p}} \phi + \frac{R_{\mathrm{l}} R_{\mathrm{r}}}{R_{\mathrm{l}} + R_{\mathrm{r}}} \phi \\ \phi &= \phi_{\mathrm{l}} + \phi_{\mathrm{r}} \\ H_{\mathrm{ex}} &= \frac{\phi_{\mathrm{r}} R_{\mathrm{r}}}{l_{\mathrm{r}}} = \frac{\phi_{\mathrm{l}} R_{\mathrm{l}}}{l_{\mathrm{r}}} \ . \end{aligned} \tag{6.20}$$

Similarly the relation between imposed voltage V_{ex} and imposed magnetic field B_{ex} can be described by the equations:

$$\begin{aligned} V_{\mathrm{ex}} &= R_{\mathrm{coil}} I + L_{\mathrm{leak}} \frac{\mathrm{d}}{\mathrm{d}t} (I - I_{\mathrm{p}}) + L_{\mathrm{rod}} \frac{\mathrm{d}}{\mathrm{d}t} (I - I_{\mathrm{p}}) \\ V_{\mathrm{ex}} &= R_{\mathrm{coil}} I + L_{\mathrm{path}} \frac{\mathrm{d}}{\mathrm{d}t} I_{\mathrm{p}} \\ A_{\mathrm{r}} B_{\mathrm{ex}} &= L_{\mathrm{rod}} (I - I_{\mathrm{p}}) \ , \end{aligned} \tag{6.21}$$

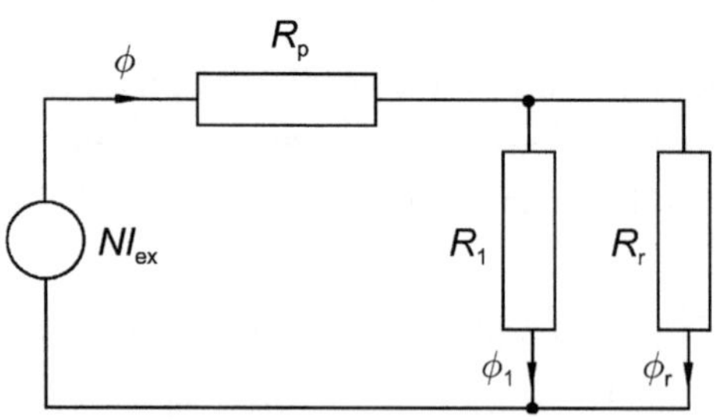

Fig. 6.36. Reluctance description of the magnetic circuit of a magnetostrictive actuator with N coil turns and a magnetostrictive rod reluctance R_{r}

with the coil resistance R_{coil} (in Ω), $L_{\mathrm{leak}} = \frac{\mu_{\mathrm{l}} N^2 A_{\mathrm{l}}}{l_{\mathrm{l}}}$, $L_{\mathrm{path}} = \frac{\mu_{\mathrm{p}} N^2 A_{\mathrm{p}}}{l_{\mathrm{p}}}$, and $L_{\mathrm{rod}} = \frac{\mu_{\mathrm{r}} N^2 A_{\mathrm{r}}}{l_{\mathrm{r}}}$. I is the resulting current through the actuator when the imposed voltage V_{ex} is applied – I_{p} is the fraction of this current that corresponds to the leakage flux in Fig. 6.36.

Magnetostrictive Losses

It is assumed that the magnetic field quantities are directed along the rod axis, which implies that the eddy current density $J_{\mathrm{i,j}}$ in section (i, j) is equal to $-\frac{\partial H_{\mathrm{i,j}}}{\partial r}$. Thus the total dissipated eddy current losses P_{eddy} in the active material will be $P_{\mathrm{eddy}} = \rho \sum_{\mathrm{j=1}}^{\mathrm{m}} \left(\sum_{\mathrm{i=1}}^{\mathrm{n}} \left(\frac{\partial H_{\mathrm{i,j}}}{\partial r} \right)^2 V_{\mathrm{i,j}} \right)$, where $V_{\mathrm{i,j}}$ is the volume of segment (i, j). The hysteresis losses can be estimated by a model based on thermodynamics [71]. A basic assumption in this model approach is that magnetic and magnetostrictive hysteresis are analogous to dry mechanical friction (so-called Coulomb friction). When using this model the strain S and magnetic flux density B values will be taken from the above hysteresis model instead of from the data base comprising de-hysterised data, i. e. the numerical values are given by

$$S_{\mathrm{i,j}} = S_{\mathrm{hyst\ model}}(H_{\mathrm{i,j}}, T_{\mathrm{i,j}}) \tag{6.22}$$

$$B_{\mathrm{i,j}} = \mu_0 H_{\mathrm{i,j}} + M_{\mathrm{hyst\ model}}(H_{\mathrm{i,j}}, T_{\mathrm{i,j}})\ . \tag{6.23}$$

Magnetic and Mechanical Operation Ranges

To minimize the required active material one should magnetize it as high as possible. However, there is a trade off between the amount of required material and efficiency i. e. the required cooling capability. A rule of thumb is that an optimal mechanical operation point for high mechanical loads corresponds to 30 MPa maximal mechanical output. The mechanical prestress then should be slightly higher than 35 MPa. There also is an approximate general relation [70] between the magnetic bias field level H_{bias} (in A/m) and applied prestress T_{bias} for optimal operation according to

$$T_{\mathrm{bias}} = 480 \cdot H_{\mathrm{bias}} + 10^6\ [\mathrm{N/m^2}]\ . \tag{6.24}$$

6.3.2 Principles and Properties of Various Applications

Linear Actuators and Drivers

Many linear actuators have been built [43–50]. For example, Etrema [43] has a wide range of products of different sizes, all of which are adapted for quasi-static use. The 50/6MP, for instance, [43] is based on a 50 mm-long by 6 mm-diameter Terfenol-D rod. It is biased with a field H_0 of about

40 kA/m. Low prestress and bias have the advantage of yielding to the highest d_{33} values; consequently, via (6.14), a high static strain S_3 of the unloaded actuator is obtained with a small field H_3. The maximum static strain of 500 ppm, leading to a displacement of 25 µm, can be achieved with a field of about 35 kA/m. It gives a strain of 14 ppm per kA/m, better than that of the MAP actuator of Cedrat Recherche, which offers only 7 ppm per kA/m. However, the prestress T_0 of the 50/6MP model is lower than 20 MPa, much smaller than that of MAP. So the maximum dynamic strain of the 50/6MP is limited to about 1000 ppm and is also much smaller than that of MAP, which reaches more than 3000 ppm. This example shows that each actuator should be designed for its specific application.

The design problems of magnetostrictive linear drivers have been addressed at Cedrat through several actuators [36, 42, 51]. These actuators (see Fig. 6.37) are identical except in their bias system. They are all based on one driver and two symmetrical head-masses. Their driver contains a total length of Terfenol-D of 100 mm. The rod diameter is 20 mm. The first actuator, called MB, is biased with a DC current in a coil giving a bias field from 0 to 160 kA/m. The second actuator, MAP, is biased with permanent magnets placed outside the dynamic flux circuit and produces a field of about 90 kA/m bias. A 10 mm thick coil permits using it against high loads, although because of the magnets and the coil, the diameter (excluding the masses) is about 70 mm. The third actuator, MAS, is biased with cylindrical permanent magnets placed in series between slices of Terfenol-D. The magnets shape has been optimized [36] with FLUX2D [52], and produces a 90 kA/m bias field. MAS also has a 10 mm thick coil, slightly longer than for the other types, but its diameter is only 50 mm. Some experimental properties of the MB, MAP, MAS drivers are compared in Table 6.2.

The MAS type of driver is an interesting example, both from the results obtained and the modeling point of view. It has the smallest coupling factor, due to the series magnets that introduce magnetic reluctances, uncoupled

Table 6.2. Experimental properties of the MB, MAP, MAS drivers

			MB	MAP	MAS
Bias	H_0	(kA/m)	100	90	90
Prestress	T_0	(MPa)	30	40	35
Coupling coefficient	k_{eff}	(%)	52	55	35
Max. magnetic energy density	$\varepsilon'_{\mathrm{m}}$	(kJ/m^3)	6.3	3.0	4.0
Max. elastic energy density	$\varepsilon'_{\mathrm{e}}$	(kJ/m^3)	15.3	30	48.5
Max. dynamic strain	S_{pp}	(ppm)	2020	3000	3500
Max. dynamic stroke	Δl	µm	202	300	350
Optimal mechanical quality factor	Q_{mopt}		≈1.5	≈2.5	≈3.5
Max. dissipated energy density	$\varepsilon''_{\mathrm{opt}}$	(kJ/m^3)	≈10	≈10	≈12

Fig. 6.37. MB, MAS and DCC actuators (respectively, from *left* to *right*)

longitudinal compliances and radial stiffnesses [36]. These last mechanical effects cannot be correctly explained by simplified theory (see Sect. 6.3.1) but they are clearly predicted by ATILA software. In spite of these effects, the MAS presents high dynamic strains. The research of the absolute strain limits of linear drivers shows that the highest strains are obtained below resonance.

The curve of the absolute maximum strain against the frequency of the unloaded MAS (Fig. 6.38) has been calculated and tested taking into account both the field limit and the stress limit at each frequency. It defines a law of current that depends on frequency. This new strain curve is above the classical curve of strain at constant current, based on the maximum current acceptable at resonance. It possesses a large pass band, which might be used in several applications such as active damping, low frequency projectors, etc.

The maximal dissipated energy density is the maximum energy per volume of Terfenol-D that can be dissipated in the load, which is achieved in the case of the optimal load. All the experimental values converge to 10 ... 12 kJ/m^3. This value is between five and ten times higher than that of PZT. It indicates that all these actuators can dissipate 0.4 J providing, for instance, at 1 kHz, an output power of 2.5 kW on optimal load. Linear actuators are studied for building micropositioners [44,45], fuel injectors [46], fast hydraulic drives [29], high pressure pumps [47], active damping applications [48,49], helicopter blade control [50], etc. In all of these applications, the expected advantages over piezoelectric or conventional solutions are the

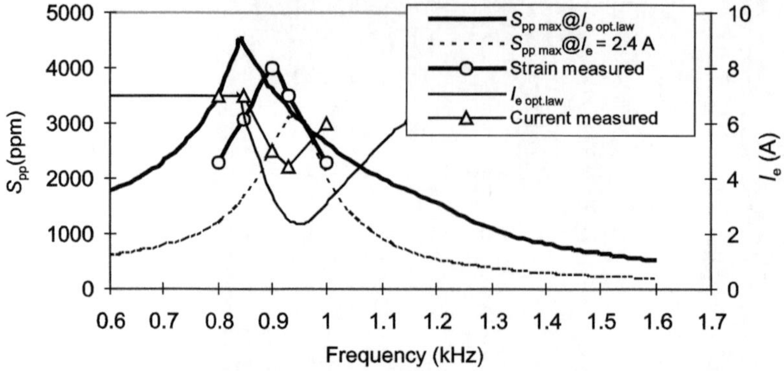

Fig. 6.38. Calculated curves of MAS: peak-to-peak strain S_{pp} at constant current $I_{\mathrm{e}} = 2.4\,\mathrm{A}$, strain with optimised current $I_{\mathrm{e\,law}}$, and corresponding current $I_{\mathrm{e\,law}}$, and measured values of strains of MAS and corresponding currents

large displacements and large forces associated with low voltages. Their main drawback is the rather high electric power requirement.

Transducers

The significance of the giant dynamic strains of Terfenol-D has been grasped rather quickly by transducer designers. Such strain levels, as well as high field limits, high coupling and high compliance, are well suited for high power transducers both for acoustics (loudspeakers, sonars) [53–55] and for mechanics (welding, sealing, cleaning, machining, cutting, etc.) [25, 56].

The Tripode Tonpilz-type sonar transducer [40] (Fig. 6.39) is a good example for showing the high power capability of Terfenol-D. It is 31 cm long and 30 cm in diameter. It is based on three drivers, each of them including a 100 mm long by 20 mm diameter Terfenol-D rod. The maximum theoretical expectation was a head mass displacement of 110 µm, a Terfenol-D strain of 3250 ppm, an output power of 3.8 kW and a source level of 208.6 dB ref. 1 µPa at 1 m. Experimentation was performed to achieve about 90% of the theoretical performance. The head mass displacement was measured with an accelerometer, giving 98 µm at 1.2 kHz (Fig. 6.41). It corresponds to a 2900 ppm peak-to-peak strain in Terfenol-D, an output power of 3 kW and a sound level of 208 dB (Fig. 6.40). This performance is achieved with an acceptable linearity. High power densities achieved now in Terfenol-D are ten times higher than those of PZT transducers. These results are interesting in the knowledge that the bias problem is now solved in different ways thanks to specific permanent magnet configurations (see earlier in this section) and are being applied [39, 57]. Such applications are seen to be promising candidates for development.

Fig. 6.39. Tripode sonar transducer

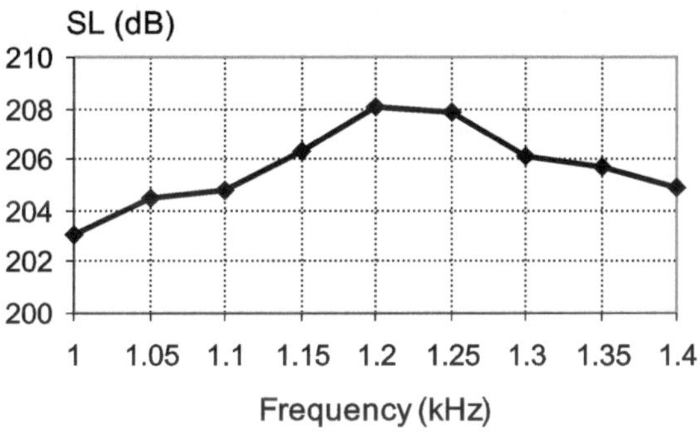

Fig. 6.40. Tripode sonar transducer: measured sound level versus frequency

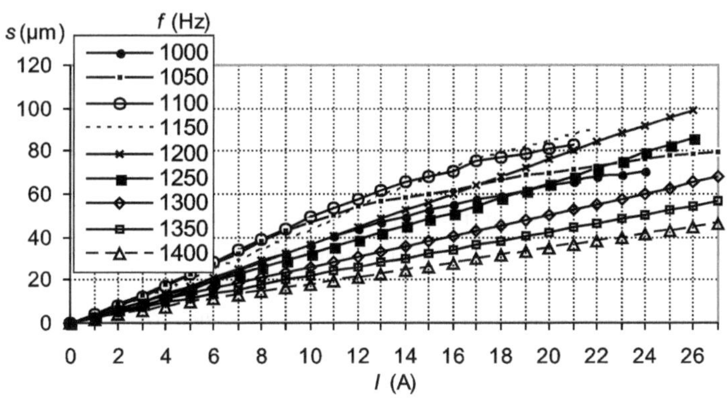

Fig. 6.41. Displacement s of the head mass versus excitation current I at different frequencies

Motors

Magnetostrictive linear actuators are able to produce static displacements in the range 20 to 200 µm. These displacements being larger than mechanical tolerances, they render possible the successful building of inchworm motors [58, 59] offering high forces/torques and good resolution, at low speed. Such properties are very difficult to obtain with conventional electromagnetic motors. Inchworm motors need a gearbox to obtain high torques, which then introduce angular play: poor efficiency and wear are thus the weaknesses of inchworms, and these factors limit the number of applications.

J.M. Vranish [59] has constructed a rotating stepping motor with the highest torque (12.2 Nm) ever reported among all the piezoactive motors; its holding torque is also very high. Its speed limit (0.5 rpm) is small. Its angular resolution is better than 800 µrad. As typically with inchworms, its output power is low (<1 W) compared with the electric power required (600 W).

L. Kiesewetter [41, 58] has built a linear inchworm motor which has been commercialised by Dynamotive in the paper industry. It uses both longitudinal and radial strains in the moving part of a Terfenol-D rod. According to Dynamotive, typical results for a motor based on a 120 mm long by 10 mm diameter rod are a maximum speed of 20 mm/s, a maximum force of 1000 N and a resolution of 2 µm. Friction motors (also called ultrasonic motors) offer a new field of applications for magnetostriction. These motors use the vibrations of a stator to transmit a motion to a rotor or a driven member. Such motors based on piezoelectric ceramics already exist on a large commercial scale; they offer large dynamic and holding torques, along with low speeds and good efficiencies through resonance.

Following a principle used in piezoelectric ultrasonic motors [60], T. Akuta [61] has built the first magnetostrictive friction motor. This stator is made of pairs of orthogonal actuators excited with sinusoidal 90° phase-shift currents, which produce an elliptical vibration. The modeling of such magnetostrictive stators [42] has shown that in quasi-static operation a good elliptical motion is produced. It has also been shown that there are many coupled modes, but none of them provides a satisfactory elliptical motion. Therefore, unlike piezoelectric motors, this motor cannot operate at resonance. As a consequence and in relation to the previous analysis of power (Fig. 6.34), the efficiency is comparatively weak. Its other characteristics are a speed of 40°/s and a torque of 1.8 Nm [62].

It is difficult to convert existing piezo-motors to magnetostrictive versions; new designs have to be found. A first magnetostrictive motor using the mechanical resonance of two vibration modes has been built and tested by Cedrat Recherche [63] (Fig. 6.42). Its stator modules are made of a ring and two Terfenol-D linear actuators. The translation mode of the stator produces a vibration that is tangential to the contact zone (Fig. 6.45a). The flexure mode produces a vibration that is normal to the contact zone (Fig. 6.45b). These modes are coupled using a 90° phase shift, in order to produce ellipti-

Fig. 6.42. Multi-mode magnetostrictive FLEX-M1 motor

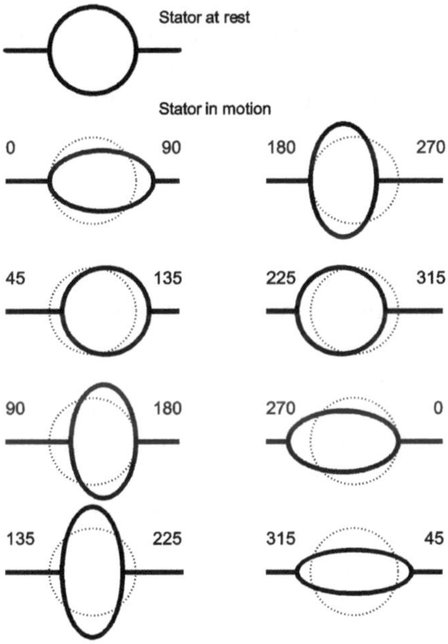

Fig. 6.43. Principle of FLEX-M1 stator Stator at rest and in motion versus the actuators phases

cal vibrations (Fig. 6.43) that are used to transmit a motion to two rotors by friction. A low rotating speed of 100°/s, and a torque of 2.1 Nm are achieved (Fig. 6.44). The goal was to show that Terfenol-D can be used for making high torque motors and that resonance can be beneficial for that purpose and for improving efficiency.

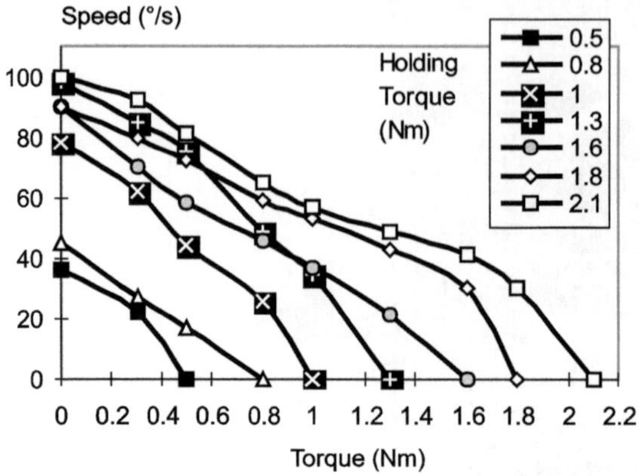

Fig. 6.44. Measured torque-speed characteristics of FLEX-M1 motor with different holding torques

The same principle can be applied to building various kinds of motors according to the number of modules and design choices: linear or rotating, stepping or ultrasonic, etc. It has been used recently to build an interesting ultrasonic piezomotor [64].

Micromotors and Actuators

T. Fukuda [65] has opened the field of miniature magnetostrictive actuators and motors taking advantage of wireless magnetic excitation. He has experimented with two small self-moving linear motors (some of cubic centimeter dimensions) based on a conversion-mode principle.

The first linear micromotor, based on magnetostrictive thin films deposited on a 7 µm polyamide film, was built in Japan in 1994 [66]. The 13 mm long prototype used a 200 Hz vibration induced by magnetostriction to obtain one-way motion at 5 mm/s. This is a mode conversion ultrasonic motor (MCUM) according to the Japanese classification of piezoelectric motors.

The torsion-based, drift-free microactuator [67], invented by CNRS Grenoble, is basically a unimorph structure composed of a single magnetostrictive film deposited on a passive substrate. The new feature is a square shape maintained by hinges at three corners (Fig. 6.46).

The useful displacement due to magnetostriction is obtained at the fourth (free) corner and without thermal displacement. The different deformed shapes are due to the anisotropy of magnetostrictive strains and the isotropy of thermal strains. Modeling with ATILA (Fig. 6.47) has permitted the design of appropriate microhinges. Prototypes have been realised by micromachining a Silicon substrate and by depositing a magnetostrictive film by sputtering. Measurements using laser interferometry have confirmed the modeling expectations.

a

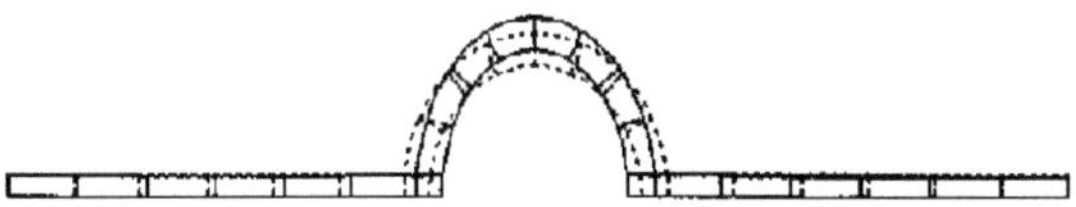

b

Fig. 6.45. Vibration modes of a stator module, computed with ATILA. **a** Translation mode, **b** flexure mode

Fig. 6.46. Microactuator

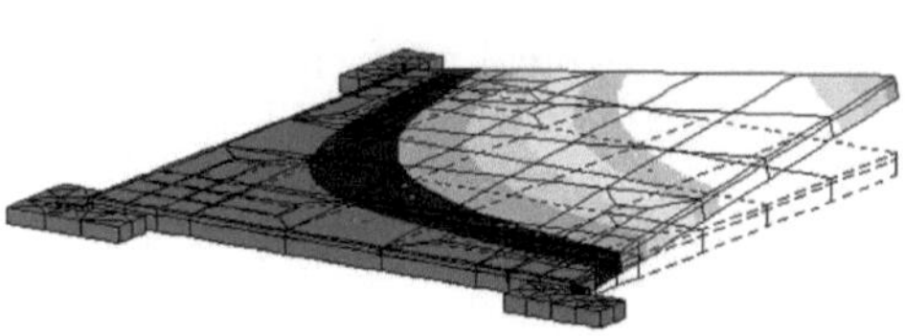

a

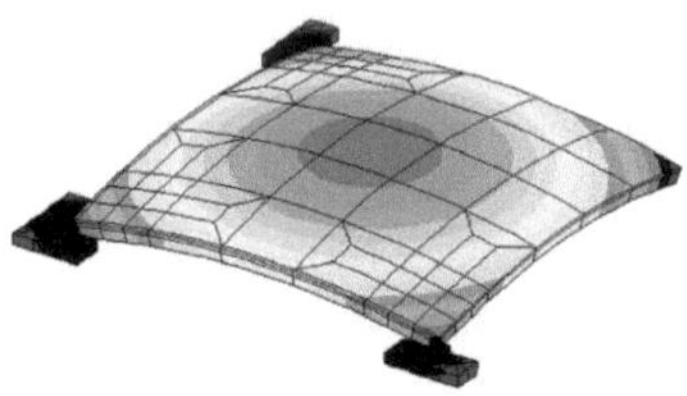

b

Fig. 6.47. Modeling of the microactuator shown in Fig. 6.46. **a** Magnetostrictive deformation, **b** thermal deformation

Several standing-wave ultrasonic motors (SWUMs), have been designed at Cedrat [68] (Figs. 6.48 and 6.49). A linear motor is a self-moving silicon plate including magnetostrictive film. It is submitted to a 10 mT dynamic field produced by an external coil, which may be placed at some centimeters distance

a

b

Fig. 6.48. Two wireless micromotors. **a** Standard form, **b** rotating form

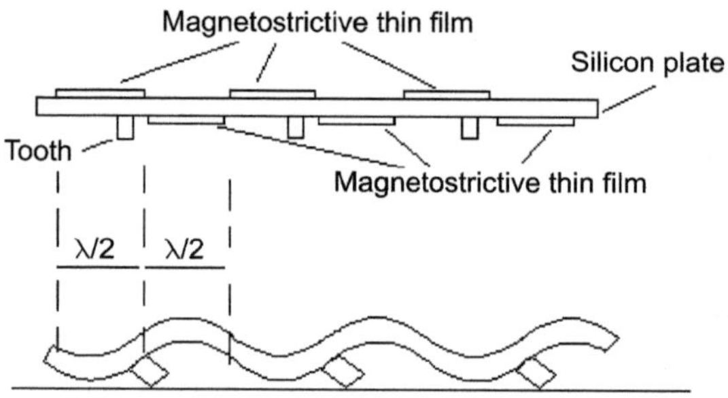

Fig. 6.49. Principle of the wireless linear micromotor

from the motor. At resonance, this field excites a flexure mode, producing vibrations in the plate, which in turn induces by friction a motor motion at 10 ... 20 mm/s. A rotating version has been also created (Fig. 6.48b) that uses a slightly different principle [68]: the vibrating rotor is based on a 100 μm thick by 20 mm diameter plate with 10 μm deposited magnetostrictive films, which are wireless and excited by a small coil. Typical performance is a rotating speed of 30 rpm and a torque of 1.6 μNm, with a 20 mT excitation field.

These examples demonstrate some of the special advantages of magnetostriction, especially the fact that the moving parts are wireless. The disadvantage is the coil, which is difficult to miniaturise because of the field requirements. These considerations are driving the development of films with magnetostriction at low fields. Note that, as these devices are very small, the price of the material is not a problem, and so such actuators could find large scale applications – for instance in optics, in medicine, or in the automobile industry.

6.3.3 Summary

Several new giant magnetostrictive materials have been developed in recent years for actuation applications. Bulk rare-earth-iron alloys Terfenol-D present typical static magnetostrains of 1000 ... 2000 ppm, which permit the building of low-frequency actuators and transducers. New composite materials offer an interesting possibility for high frequency ultrasonic applications. More recently, rare-earth-iron thin films have been also explored for actuation in microsystems.

Among all applications based on magnetostrictive materials, devices based on mechanical resonance, such as underwater transducers, ultrasonic transducers, resonant motors and micromotors, are of special interest. Using resonance, giant dynamic strains (up to 4000 ppm) can be produced, which can lead to very large power/force densities and rather good efficiency. Although there is no large-volume application at the moment, some devices are already used for specific applications in domains such as pumps, micropositioners, transducers, and research into other applications is growing. It is likely that we shall see magnetostriction finding application in the microactuator domain in the near future.

6.3.4 Acknowledgement

The authors would like to thank: D. Boucher (DCN) and A. Colin (DRET) for the financial support of Cedrat works on acoustic applications; C. Sol (French Ministry of Research) for the support on electrical engineering applications; the European Commission for the financial support on microsystems applications (BRE2–0536 MAGNIFIT); the partners of MAGNIFIT, especially Laboratory Louis Néel CNRS Grenoble, Kassel Universität and Forschungszentrum Karlsruhe, for their efforts in producing microsystems; R. Bossut (ISEN); and the team at ISEN acoustic laboratory for their continuous efforts in developing ATILA.

6.4 Shape Memory Actuators

J. Hesselbach

The shape memory effect was first discovered at the end of the 1940s in a gold-cadmium alloy. Since this extraordinary effect was recognized in the early 1950s as being caused by a martensitic transformation, new and improved shape memory alloys have been found. As prices for shape memory alloys are dropping, more and more commercial applications – ranging from aviation to medicine – make use of the functional properties of those materials. In this contribution we will focus on the new and innovative field of actuator applications.

6.4.1 Properties of Shape Memory Alloys

Shape Memory Effect

The term shape memory (SM) refers to the ability of certain materials to annihilate a deformation and to recover a predefined or imprinted shape. Even though the shape memory behavior is also attributed to some plastic materials, in this text only shape memory alloys are considered. The SM effect is based on a solid–solid phase transition of the shape memory alloy that takes place within a specific temperature interval.

The properties of the shape memory alloy vary with its temperature. Above the transition temperature, the alloys crystallic structure takes on the austenitic state. Its structure is symmetric and the alloy shows a high elastic modulus. The martensitic crystalline structure will be more stable for thermodynamical reasons if the materials temperature drops below the transformation temperature. Martensite can evolve from austenitic crystals in various crystallographic directions and will form a twinned structure. Boundaries of twinned martensite can easily be moved; for that reason SM elements can be deformed with quite low forces in the martensitic state.

When heated up, the austenitic structure will be established again. At the same time the SM element will return to its original shape because neither the phase transformation nor the de-twinning of the martensitic structure involves changes to the atomic lattice. The SM element may exert high forces when recovering its predefined shape; therefore the SM effect can be employed as a new actuator principle.

Forward and reverse transformation occur at different temperatures, resulting in a hysteresis as can be seen in Fig. 6.50. The start and end of the transformation from martensite to austenite are given by A_s (austenite start temperature) and A_f (austenite finish temperature). The reverse transformation takes place in the temperature interval from M_s to M_f (martensite start and finish temperatures). The shape of the hysteresis curve in Fig. 6.50 strongly depends on the thermomechanical treatment of the shape memory alloy (see also Sect. 6.4.1).

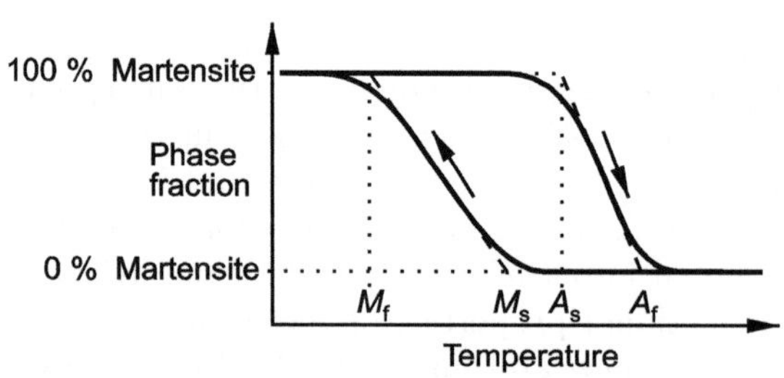

Fig. 6.50. Transformation temperatures and their hysteresis

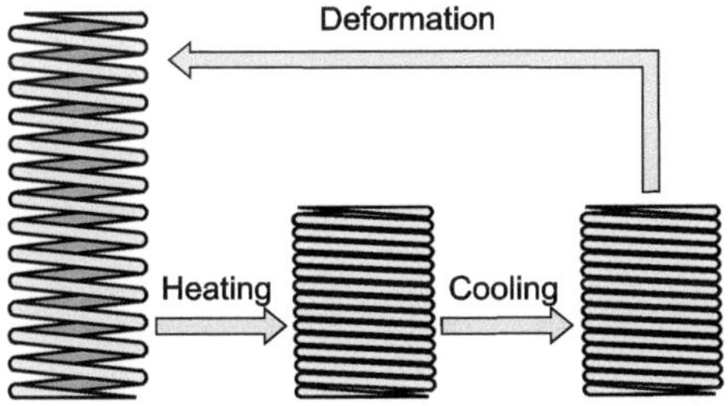

Fig. 6.51. One-way effect of a shape memory alloy spring

The shape memory effect may be divided into three categories, each showing different functional behavior. They will be described briefly in the following subsections. More detailed discussions can be found in [72] or [73].

One-Way Effect

Stretching a shape memory coil spring when it is in a martensitic (i. e. cold) state will de-twin the differently oriented martensite. The result is an almost homogenously oriented martensitic structure. Similar to a common plastic deformation the SM coil spring will stay in the stretched shape when unloaded (Fig. 6.51).

If the shape memory alloy spring is heated and the temperature surpasses the A_s temperature, the shape memory material starts transforming to austenite and the coil spring returns to the unstretched form. On reaching the A_f temperature, the transformation is completed. It is characteristic for the one-way effect that a shape recovery occurs only when the SM element is heated. There is no shape change when the element is cooled. The cold SM element must be deformed by an external force in order to achieve a movement when heated again.

The one-way effect is mainly utilized for fastening and clamping devices. Among the commercially most successful products are coupling sleeves made of shape memory alloy with the one-way effect, which guarantees very reliable connections of hydraulic pipes in airplanes.

Two-Way Effect

Shape memory elements with a two-way effect will remember a high-temperature shape as well as a low-temperature shape. The element flips between both shapes depending on the temperature.

If a SM coil spring with the two-way effect is heated, it will return to its predefined high-temperature shape – in Fig. 6.52 this is the compressed form. Upon cooling, the spring stretches to reach its low-temperature

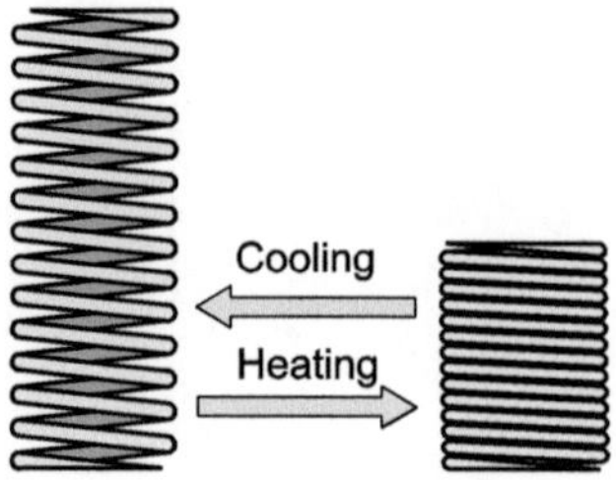

Fig. 6.52. Intrinsic two-way effect of a shape memory alloy spring

shape, ideally without the need of a supporting external force. SM elements with a two-way effect therefore represent thermally activated actuators.

A shape memory element must be specially trained to display the two-way effect. Training typically consists of several cycles of deformation to the desired low-temperature shape and subsequent shape recovery by heating [73]. Only very low forces can be exerted when the SM element changes from its high-temperature shape to its low-temperature shape; for this reason, steel springs or other elastic elements are added to the system to guarantee a full return to the low-temperature shape.

Pseudo-Elasticity

Above the transition temperature an extraordinary elasticity can be observed in the shape memory alloys. Figure 6.53 shows a uniaxial stress-strain diagram of a pseudo-elastic (or super-elastic) SMA.

Initially the pseudo-elastic material is in its austenitic phase at room temperature. Initially the material in the austenitic phase deforms like a conventional material linear elastic under load. With increasing loads a stress-induced transformation of the austenitic to the martensitic phase is initiated at the pseudo-yield stress R_{pe}. This transformation is accompanied with large reversible strains at nearly constant stresses, resulting in a stress plateau shown in Fig. 6.53. At the end of the stress plateau the sample is completely transformed into martensite. Additional loading passing the upper stress plateau causes a conventional elastic and subsequently plastic deformation of the martensitic material. If the load is decreased within the plateau and the stress reaches the lower stress level $R_{\mathrm{pe}}^{\mathrm{A}}$ a reverse transformation from martensite to austenite occurs. Since the strains are fully reversible the material and the sample respectively is completely recovered to its underformed shape. These strains are often called pseudo-elastic because the reversible deformation is caused by a reversible phase transformation and is not only due to a translation of atoms out of their former equilibrium position [74].

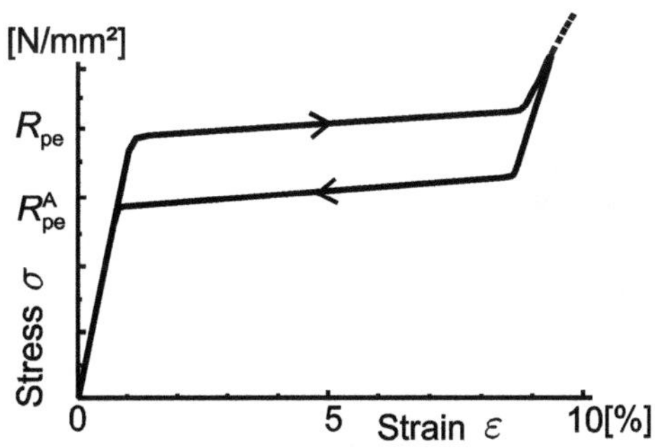

Fig. 6.53. Stress-strain diagram of pseudo-elastic shape memory alloys

The large strains at nearly constant stresses can be used in many applications in which great displacement must be set up at constant forces, e. g. minimal surgery instruments, orthodontic wires and glass frames.

Actuator Activation

Shape memory alloys with the two-way effect can be utilized as actuators to generate repeated movements. To activate the shape memory effect and the corresponding movement, the SM element must be heated above the transition temperature A_f. Heating may be accomplished in different ways:

Thermal Activation. Most of the application examples of SM actuators presented so far rely on a thermal activation of the shape memory effect, i. e. the actuator element reacts according to the ambient temperature. Here is a short list of various application areas:

- automatic transmission (to shift points adjustment for cold start);
- ventilation flaps of greenhouses (temperature-dependent opening angle);
- fan clutches (control of engine temperature);
- headlamp concealment devices (open when light is switched on);
- fire protection (closes windows or opens sprinkler valves); and
- anti-scald safety valves (to cut off hot water).

As an example, Fig. 6.54 shows the principle of a control valve with a shape memory coil spring in the automatic transmission of a limousine [75]. Depending on the oil temperature, the SM spring lowers the pressure of the transmission oil, resulting in a smoother shifting of gears.

A major advantage of thermally activated SM actuators is the conversion of thermal energy of the surrounding medium to an actuator movement. These actuators do not need any additional (for instance electrical) power supplies, a property highly appreciated for all sorts of safety devices. Such an SM element makes up a complete system, consisting of temperature sensor and actuator.

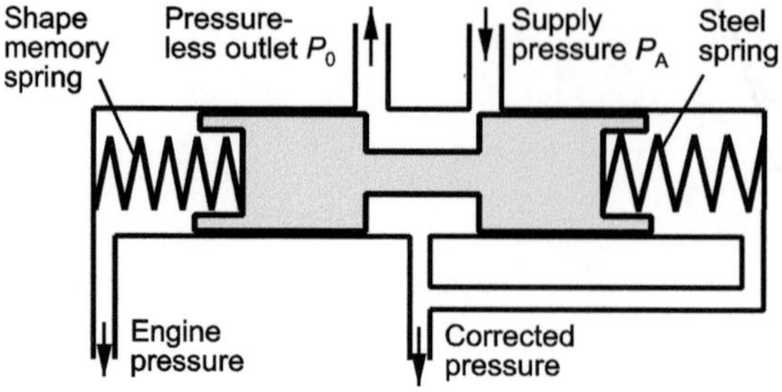

Fig. 6.54. Control valve with SM spring in an automatic transmission [75]

Electrical activation. The required heat energy may also be generated directly within the SM element by electrical current. Joule heating allows the construction of very small and compact electrically controlled SM actuators. Experimental applications for robotic devices and grippers (such as silicon wafer grippers) have proven the feasibility of such actuators [72]. The remaining part of this article will mostly cover electrically heated SM actuators because they can be utilized in automation devices.

Available Shape Memory Alloys

There are many alloys with a shape memory effect. They strongly differ amongst themselves with respect to transformation temperatures, effect amplitude and other material properties.

Shape memory alloys designated for actuator purposes (for example in automation systems) should meet the following requirements:

- large shape memory effect resulting in a long actuator stroke;
- high transformation temperatures: phase transformation should occur at high temperatures like 150 ... 200 °C to avoid unwanted activation by warm ambient air, and high transition temperatures also guarantee a complete phase transformation to the martensitic state;
- a high number of activations and a stable SM effect; and
- a small hysteresis width between forward and reverse transformations.

Table 6.3 summarizes the most important properties of some commonly used shape memory alloys.

A comparison of the required properties and the data of the shape memory alloys available (or under development) leads to the following conclusions:

- nickel-titanium (NiTi) and nickel-titanium-copper (NiTiCu) have the best properties for actuator purposes. For that reason, industrial applications almost exclusively rely on nickel-titanium-based SM alloys. there is one single drawback to NiTi: its transformation temperature is limited so far to approximately 100 °C.

Table 6.3. Comparison of shape memory alloy systems

	NiTi	**CuZnAl**	**CuAlNi**	**FeNiCoTi**	Unit
Range of transformation temperature	−100 to +90	−200 to +100	−150 to 200	−150 to +550	°C
Hysteresis width	30	15	20		K
Max. one-way effect	8	4	6	1	%
Max. two-way effect	4	0.8	1	0.5	%
Fatigue strength	800 ... 1000	400 ... 700	700 ... 800	600 ... 900	N/mm^2
Admissible stress for actuator cycling	150	75	100	250	N/mm^2
Typ. number of cycles	>100 000	10 000	5 000	50	
Density	6450	7900	7150	8000	kg/m^3
El. resistivity	80 ... 100	7 ... 12	10 ... 14		$10^{-8}\,\Omega\,m$
Young's modulus	50	70 ... 100	80 ... 100	170 ... 190	GPa
Corrosion resistance	very good	fair	good	bad	

- copper-based shape memory alloys (CuZnAl, CuAlNi) can be designed for higher transformation temperatures and are less expensive than NiTi. Due to a lower lifespan and lower work output, they are not feasible for electrical actuator applications. Elements of CuZnAl are successfully implemented as thermal actuators in fire safety devices.
- other shape memory alloys such as FeNiCoTi or NiTiHf, NiTiPd or NiAl have not yet been perfected for commercial use. The properties being sought are high transition temperatures combined with good SM effects [76–78].

Due to the superior actuator properties and the commercial impact of NiTi alloys, the following discussions will focus on these alloys. NiTi- and NiTiCu-based wires are commercially available with a range of transformation temperatures and in all diameters down to 25 µm. The same alloys are also supplied in the shape of flat-rolled wire or stripes in various sizes. Nickel-titanium is usually vacuum melted and then drawn or sheet-rolled. To reduce this time-consuming and expensive manufacturing process, new procedures suited especially for small-sized SM actuators are being investigated:

- rapid quenching: by pouring the melted alloy on fast-spinning cylinders, the alloy is cooled within milliseconds and forms thin (100 µm or thinner) films with the desired width [79]; and
- sputter-deposition: different sputtering techniques are available to deposit NiTi or NiTiCu on a substrate [80, 81]. The thin films have a thickness of up to 10 µm. This technique opens up the possibility of using SM actuators in micro-mechanical systems.

6.4.2 Electrical Shape Memory Actuators

Actuator Shape and Stroke

The shape that the SM actuator recovers to when heated is imprinted into the alloy by an annealing process. For instance, to fabricate a coil spring a SM wire is wound around a mandrel and annealed for 1... 2 hours at 350... 500 °C. Annealing temperature and duration have a strong influence on the actuators properties, such as the trainable two-way effect, the effect stability, and the hysteresis behavior.

The shape change between high-temperature and low-temperature shape defines the actuator stroke. Table 6.4 lists some commonly used actuator shapes and actuator strokes.

The two-way effect will be stabilized after 20... 100 thermal and mechanical cycles. Due to the ability of the martensite (low-temperature phase) to form a twinned crystalline structure, different areas of the actuator element may be strained in different ways: extension, compression, or shear are deformations that will be reverted to by heating. This variety offers the interesting opportunity to adapt the actuators shape change to the special needs of the actuating task. By this means, transmission links or gears may be eliminated, which helps reduce the size and price of a system.

The actuator stroke is limited only by the reversible strain that the martensitic structure can accommodate by de-twinning – otherwise irreversible strain will occur. The admissible strain is determined by the type of shape memory alloy as well as the desired number of activation cycles. If the effect is to be employed only once (for example, for tube connectors), NiTi-based alloys may be strained up to 8 %. For actuator use with more than 100 000 activations, only smaller strains are permitted, namely extensions $\varepsilon_{adm} < 3\,\%$, shear $\gamma_{adm} < 4\,\%$, and stresses up to $\sigma_{adm} < 150\,\mathrm{N/mm^2}$ or $\tau_{adm} < 100\,\mathrm{N/mm^2}$. Table 6.5 gives an overview of the design data of the most commonly utilized SM actuator geometries.

Table 6.4. Examples of actuator shapes

Actuator stroke	Material deformation	Actuator shape
Translation	Contraction	Tensile wire, bar, or tube 
Translation	Extension	Compression bar or tube
Translation	Shear	Coil spring 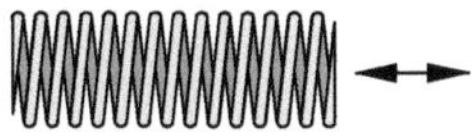
Rotation	Bending	Leaf spring 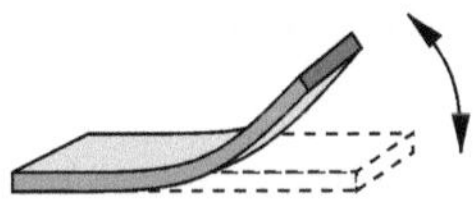
Rotation	Bending	Torsion helical spring 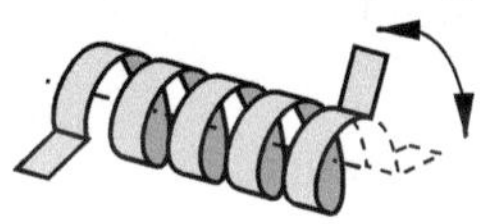
Rotation	Shear	Torsion wire, bar, or tube 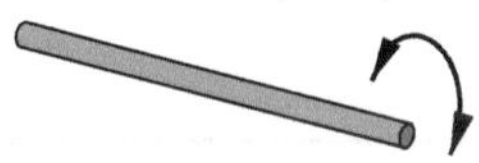

Dynamic Response

A central point of consideration when using shape memory alloys as electrically activated actuators is their response time between commanding signal and actuator movement. In theory, the phase transformation propagates with the speed of sound, but only if the necessary heat energy is supplied or dissipated fast enough.

Heating. Heating up the SM element is relatively simple. When conducting an electrical current, heat is generated Due to Joule losses directly within the SM actuator. By controlling the current appropriately, very quick heating is possible. As an example, the response of a SM wire (diameter 0.22 mm) to different heating currents is shown in Fig. 6.55. With an additional short-time current pulse (line 'b') the actuator reacts much faster than with a constant heating current (line 'a'). The positioning time is faster than 0.5 s.

Cooling. The process of cooling down is strongly influenced by the medium surrounding the actuator. Therefore only external or constructional mea-

Table 6.5. Data of SM actuators

Symbols: F_{max}, M_{max}, ΔL_{max}, $\Delta\varphi_{\mathrm{max}}$: maximum actuator force, torque, stroke, and angle respectively; σ_{adm}, τ_{adm}, $\varepsilon_{\mathrm{adm}}$, γ_{adm}: admissible tensile stress, shear stress, extension, and shear respectively; D: SM wire diameter; L: SM wire length; D_{m}: coil diameter; i_{f}: number of turns; b, h: width and thickness of SM flat wires or bars.

Actuator shape	Max. force/torque	Max. stroke/angle
Tension wire or bar, compression bar (round cross section)	$F_{\mathrm{max}} = \frac{\pi}{4} D^2 \sigma_{\mathrm{adm}}$	$\Delta L_{\mathrm{max}} = \varepsilon_{\mathrm{adm}} L$
Tension wire or bar, compression bar (rectangular cross section)	$F_{\mathrm{max}} = bh\sigma_{\mathrm{adm}}$	$\Delta L_{\mathrm{max}} = \varepsilon_{\mathrm{adm}} L$
Torsion wire or bar (round cross section)	$M_{\mathrm{max}} = \frac{\pi}{16} D^3 \tau_{\mathrm{adm}}$	$\Delta\varphi_{\mathrm{max}} = \frac{2L}{D} \gamma_{\mathrm{adm}}$
Torsion helical spring (made of flat wire)	$M_{\mathrm{max}} = \frac{1}{6} bh^2 \sigma_{\mathrm{adm}}$	$\Delta\varphi_{\mathrm{max}} = 2\pi i_{\mathrm{f}} \frac{D_{\mathrm{m}}}{h} \varepsilon_{\mathrm{adm}}$
Coil spring (tension or compression)	$F_{\mathrm{max}} = \frac{\pi D^3}{8kD_{\mathrm{m}}} \tau_{\mathrm{adm}}$ $k = \frac{2D_{\mathrm{m}}+D}{2D_{\mathrm{m}}-D}$	$\Delta L_{\mathrm{max}} = \pi i_{\mathrm{f}} \frac{D_{\mathrm{m}}^2}{D} \gamma_{\mathrm{adm}}$

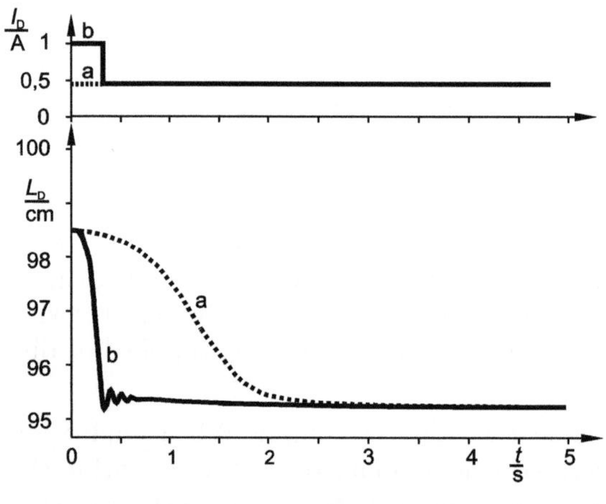

Fig. 6.55. Response under heating (SM wire, length L_{D}, diameter 0.22 mm) [82]

sures determine the actuator behavior at cooling time. Cooling can be greatly sped up by choosing a different surrounding medium, as shown by the plots in Fig. 6.56. It shows the cooling behavior of a SM wire in calm air, turbulent air, and water. As can be seen, a SM actuator in water will cool

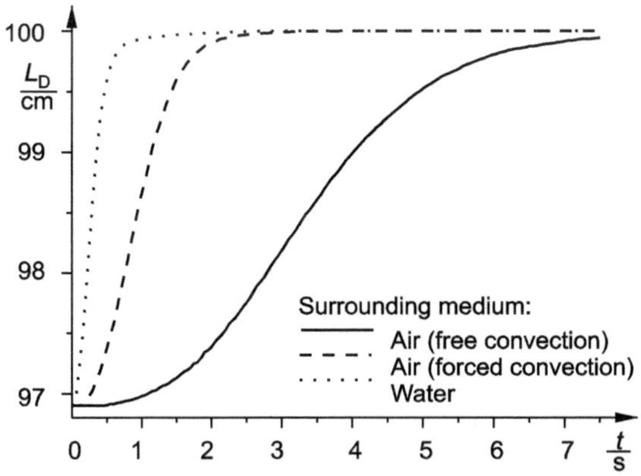

Fig. 6.56. Response under cooling (SM wire, length L_D, diameter 0.22 mm)

more than ten times faster than the same actuator in air at room temperature.

Further possibilities to accelerate the cooling process are:

- enlargement of the ratio between actuator surface and volume: one way to accomplish this is to make use of flat-rolled wire instead of round wire;
- increasing the difference between the actuator temperature and the temperature of the surrounding fluid: for that reason SM alloys with high transformation temperatures should be preferred for actuators; and
- active cooling by forced convection.

Position Control and Internal Sensoric Effect

If a SM actuator is employed only to switch between two different positions, a simple on/off-control of the heating current will be adequate. However, most SM actuator applications require fine positioning, which will be dealt with in this subsection.

To reach and hold a defined position cannot be accomplished by a feedforward control because the relationship between heating current and actuator stroke displays a hysteresis and is therefore ambiguous. Considering the

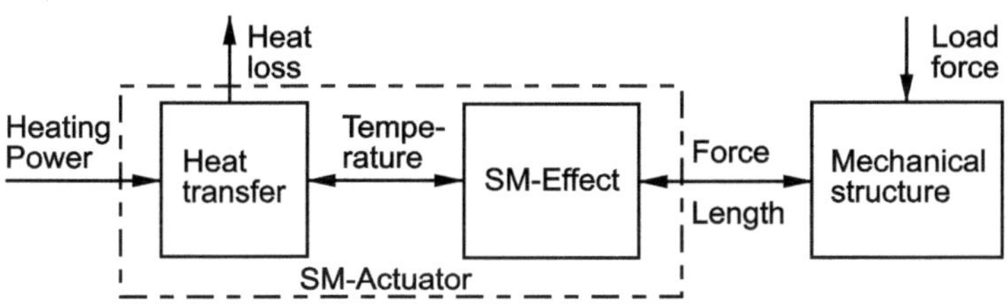

Fig. 6.57. Model of SM actuator system [83]

physical effects involved, a system with an electrically heated SM actuator may be described by a mathematical model consisting of three parts [83] (see Fig. 6.57):

- The heat transfer model describes the heating of the actuator alloy by Joule energy as well as the heat losses to the surrounding air. The heat transferred from the actuator to the environment is a strongly nonlinear function of actuator temperature, ambient temperature and type of convection.
- The model of the shape memory effect is based on thermodynamic laws of phase transitions in solids. Due to inner friction and losses of the phase transformation, the simulation of the hysteresis by means of the Preisach-model [83] must be modified and linked with the thermodynamic equations.
- A kinematic model of the mechanical structure into which the SM actuator is integrated.

Based on this concept, the shape memory actuator system can be simulated by a nonlinear dynamical model.

Not only ambiguity due to hysteresis but also the influence of disturbances such as load force and heat loss on the actuator position make it clear that steady positioning of a SM actuator can only be achieved with a position sensor and feedback control.

The installation of an additional position sensor is not always possible. Reasons may be costs and/or unavailable space. In this case, the internal sensoric effect displayed by some NiTiCu-alloys may be employed for indirect position sensing [84]. This leads to the use of a self-sensing actuator (cf. Sect. 6.9). In Fig. 6.58 the actuator length L_{D} of a NiTiCu shape memory wire is plotted against its electrical resistance R_{D}. The relation is free of hysteresis and is only slightly shifted by the actuators load.

The almost-linear behavior between wire length and resistance can be explained by the fact that the actuator stroke is approximately proportional to the fraction of austenite and martensite in the alloy. Since the resistivity of martensite and austenite is different, the resistance of the SM actuator will vary according to the phase fractions: only in the fully martensitic or austenitic state does the relationship between actuator stroke and resistance become non-linear. The stroke-resistance relation is independent of the ambient temperature (respectively, the type of surrounding medium or the type of convection) because it is affected only by the martensite fraction in the actuator material. However, a load force will induce a small amount of elastic strain, resulting in a shift of the stroke-resistance relation.

These explanations establish that the actuators resistance may be used as an indirect positional feedback signal. A block diagram of such a feedback control circuit is displayed in Fig. 6.59. A PI-algorithm is implemented in order to calculate the electrical heating power P_{el}.

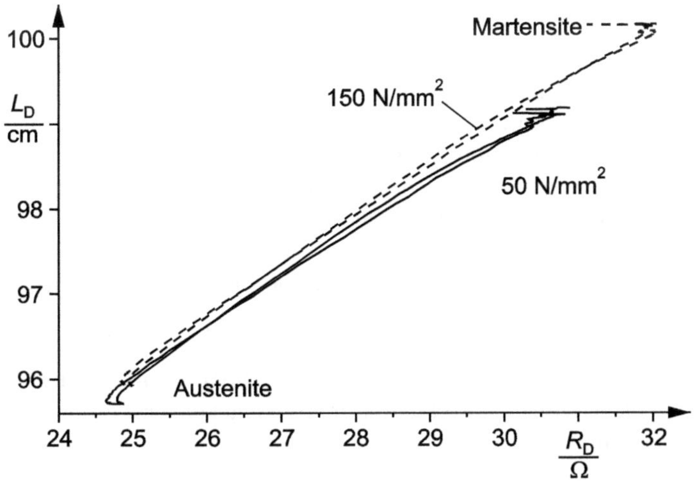

Fig. 6.58. Length of SM wire (diameter 0.22 mm) with respect to electrical resistance

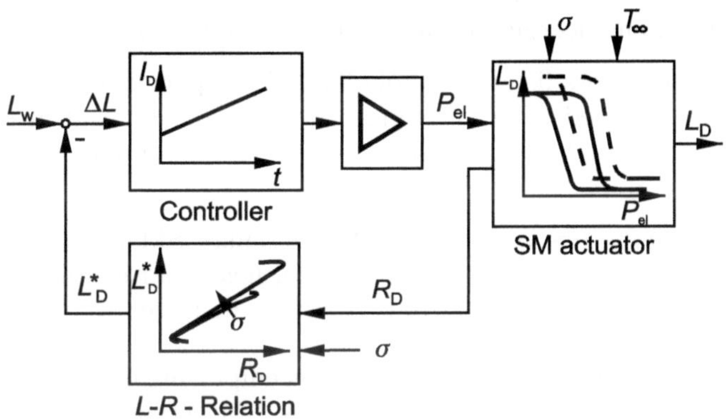

Fig. 6.59. Control circuit with resistance feedback

6.4.3 Perspectives for Shape Memory Actuators

The properties of electrically activated shape memory actuators described so far have indicated that these actuators are well-suited to drive mechanical mechanisms. The advantages and disadvantages of this kind of new actuator principle are summarized in Table 6.6.

Shape memory actuators offer a lot of advantages, but there are also some quite serious drawbacks. When comparing SM actuators with other actuator principles (such as piezoelectric stacks or solenoids), it should be taken into consideration that research for improved shape memory materials is relatively young. With shape memory alloys and actuators slowly gaining commercial importance it is expected that in the next few years new SM alloys will emerge that have higher transition temperatures and good effect stability [85, 86].

The disadvantage of low efficiency (below 2 %) is determined by thermodynamics: most of the input energy is transformed to heat within the SM element. Furthermore, during the cooling process the heat is lost to the surrounding region and cannot be converted back to electrical or another reusable form of energy. Due to the low efficiency, limited effect duration, and low speed, it must be understood that shape memory actuators are not intended for applications where electrical motors or pneumatic cylinders are well established. Instead, electrical shape memory actuators offer a good choice for very special or new applications where conventional motors would either require expensive modifications or are not available.

The analysis of the advantages and disadvantages reveals good feasibility and opportunities for electrically heated shape memory actuators, especially in two fields of application:

- compact and light auxiliary actuating devices – as an example they may be used to increase the flexibility of automation devices, such as adjusting the range of grippers [82, 87]; and
- actuators for precision engineering and micromechanical systems.

The advantages of SM actuators listed in Table 6.6 gain importance where small mechanisms are concerned. The following properties recommend utilization of shape memory actuators in millimeter- or micrometer-sized mechanical mechanisms:

- Compared with SM actuators with large volumes, small SM actuators offer a much higher surface-to-volume ratio. Hence, heat transfer to the surrounding medium is strongly improved, resulting in faster response times of the actuators.
- Small NiTi-strips or thin-films may be fabricated by employing new methods, such as rapid quenching or sputter techniques. SM actuators fabri-

Table 6.6. Advantages and disadvantages of SM actuators

Advantages	Disadvantages
+ Large energy density	– Uncertainty about duration and stability of the effect
+ Small and compact	
+ Simple mechanisms	– Relatively low velocity
+ Variable shapes	– Very low efficiency
+ Linear or rotational motion	– Limited range of transformation temperatures
+ Miniaturisable	
+ Usable in clean room environment	
+ Good corrosion resistance	
+ Low voltage (<40 V)	
+ Silent	
+ Intrinsic sensoric effect	

cated in this way are less expensive than SM wires because less material is needed and the element is produced in the necessary size.

- Sputtering is a typical fabrication process also employed for micro parts. Therefore, sputtering of NiTi can be integrated in the manufacturing process more easily.
- The very high work-per-volume ratio of approx. $4\,\mathrm{J/cm^3}$ is highly valued if space is limited. SM actuators offer high forces and strokes. For example, a piezoelectric actuator with the same force and stroke would have to use up to ten times the space necessary for a SM actuator.
- Low efficiency is less important because overall energy consumption is low.
- As self-sensing actuators, the internal sensoric effect can be used for position sensing (see Sects. 6.1.4 and 6.9). Again, this property is useful in small-sized applications where additional sensors cannot be accommodated for space and weight reasons.

There is a strong demand for miniature devices and sophisticated designs in many areas of technology. Examples for such fields of application are as follows:

- **Micro assembly.** While most of the technology used to fabricate parts of millimeter or micrometer size could be copied and modified from microelectronic fabrication processes, this is not true for micro assembly. There are scarcely any devices suitable for a small- or medium-scale automated assembly of micromechanical systems. Micro assembly opens a broad field of potential applications for new actuator principles. An example is the handling of millimeter-sized, lightweight parts under clean-room conditions and with small operational space available.
- **Inspection tasks.** It is often necessary to inspect inner cavities of machines or pipe systems without disassembly or destruction. Endoscopes are available for this task, but these (albeit flexible devices) are insufficient if the object under inspection has a very complex geometry or exact positioning is required. Here, SM actuators promise more degrees of freedom and more controllability, and are formed into smaller devices.
- **Medical devices.** Similar to the inspection of technical devices, more flexibility and controllability are desired for medical devices such as surgery instruments for minimally-invasive operations. Other medical applications may be drug-release systems implanted in the body.

6.4.4 Innovative Application Examples

In this section some examples of precision engineering prototypes are presented that apply electrically heated shape memory actuators as driving elements. Further on flexure hinges of pseudo-elastic SM alloys will be presented.

Mechanical Grippers for Micro Assembly

General Aspects. Mechanical grippers have a variety of applications and for that reason will very likely be the most often used grippers also in micro assembly. There are some differences to common assembly procedures to be considered when assembling very small parts. For grippers, the required properties are briefly summarized:

- compact size;
- good controllability of gripping forces (in the range 1 to 100 mN);
- suited for a clean-room environment;
- control of adhesive forces;
- centering of gripping object.

Miniaturized copies of conventional grippers designed for macro handling of systems cannot meet the special requirements posed by the small dimensions of micro parts. Small gripper size and clean-room suitability are achieved, for example, by observing the following rules:

- Conventional slide or roller bearings should be replaced by flexure hinges. Flexure hinges are created by specially formed notches in the material, causing a much lower flexural strength at that point. Miniature flexure hinges can be produced with little effort and are suited for clean-room usage.
- Use of solid state actuators. Small shape memory actuators can apply relatively high forces and strokes, can be well integrated into the grippers mechanical structure, and do not emit particles into the clean-room environment.

Only small SM elements are required to actuate miniature grippers. Hence, fast opening and closing times of the gripping jaws can be expected. Two examples of micro grippers built according to these design principles are described in the next paragraphs.

Parallel Gripper. The parallel gripper mechanism in Fig. 6.60 consists of a single piece of plastic with interchangeable jaws. For this prototype, flexure hinges were cut out of the material by micro milling, but for higher production quantities injection moulding is possible. The jaws are closed by heating the SM wires (length 12 mm, diameter 0.15 mm). The gripping mechanism translates the actuators stroke into a movement of the jaws of 1.5 mm with a gripping force of 0.15 N. The gripping-/loosening time averages by 0.3 s.

SMA Actuated Miniature Silicon Gripper. By using flexure hinges the micro gripper is designed in a compliant mechanism. To provide the gripper with a centering capability a four-bar-linkage mechanism with a transmission of −1 between input and output crank is used, where the ends of the cranks represent the gripping jaws [89]. The SMA actuators are connected

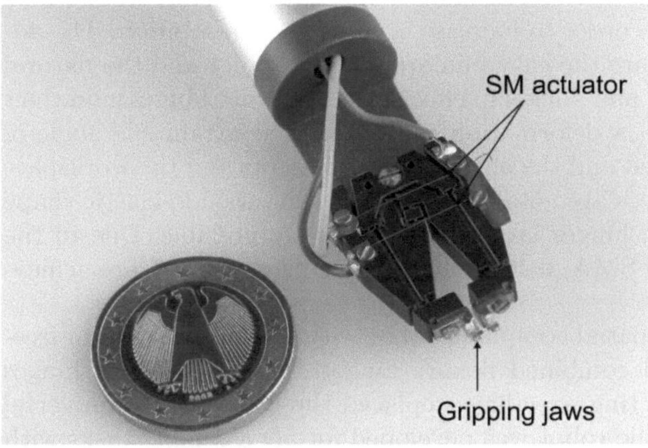

Fig. 6.60. Parallel gripper with interchangeable jaws [88]

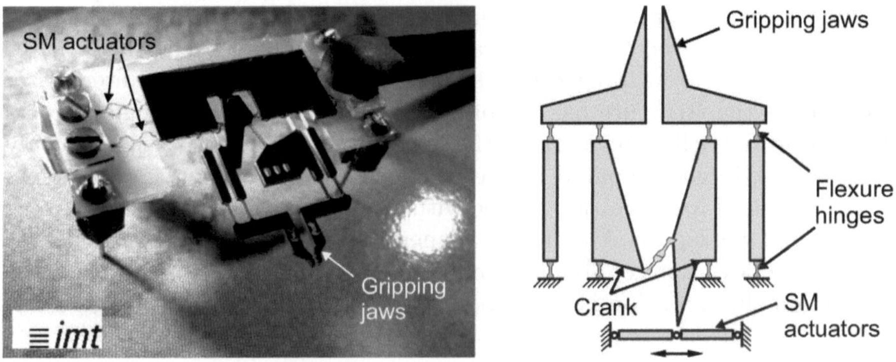

Fig. 6.61. Microstructured NiTi actuator mounted to the silicon surface of the gripper [90]

to one crank, forming a serial differential type actuator. A parallel movement of the gripping jaws can be achieved with two additional linkages (see Fig. 6.61).

The micro gripper in Fig. 6.61 consists of a silicon structure with a dimension of approximately $7 \times 4\,\text{mm}^2$. In the open position the gripping jaws are 0.5 mm apart. The flexure hinges have a minimum thickness of 30 µm. By machining a sputtered NiTi foil the SMA actuator has been realized with a minimum thickness of 30 µm. The gripping force averages by circa 11 mN.

Pseudo-Elastic Flexure Hinges

High-precision positioning devices are often required for micro system production. Therefore miniaturized robots or fine positioning systems are built up using flexure hinges in order to increase accuracy and resolution. The advantages of these hinges are the easy miniaturization ability and the natural lack of backlash, friction, and stick-slip effects. Since flexure hinges gain their mobility exclusively from a deformation of matter their attainable angle of rotation is strongly limited and the achievable movements and the workspace of these positioning devices are notably small. By using pseudo-elastic shape memory alloys as flexure hinges larger movements are possible. Due to the large reversible strains of SMA, deflections of the hinges of $\pm 30\,^\circ$ are achievable.

Figure 6.62 shows a spatial compliant robot with 3 DOF (degrees of freedom) and six integrated combined flexure hinges. These combined hinges with 2 DOF and intersecting axes have replaced the conventional universal joints. The structure of the robot was developed for 3D assembly tasks with movements in x-, y- and z-directions. The robot is driven by three linear direct drives. Each drive is connected with the working platform by two links forming a parallelogram, allowing only translational movements of the platform and keeping the platform parallel to the base plane. The three drives of the structure are arranged star-shaped in the base plane at intervals of $120\,^\circ$. Thus the structure has a workspace which is nearly triangle-shaped. Restricting the deflection angle of the hinges to $\pm 30\,^\circ$ the workspace is only minimally reduced compared to the workspace with no angular restriction. With the actual configuration a cube with dimensions of $112 \times 112 \times 112\,\mathrm{mm}^3$ fits into the workspace. A planar movement in an area of $60 \times 60\,\mathrm{mm}^2$ leads to maximal deviations of about 1.5 mm, and if the movement is 60 mm in the z-direction the deviation is only about 0.05 mm. Using flexure hinges made

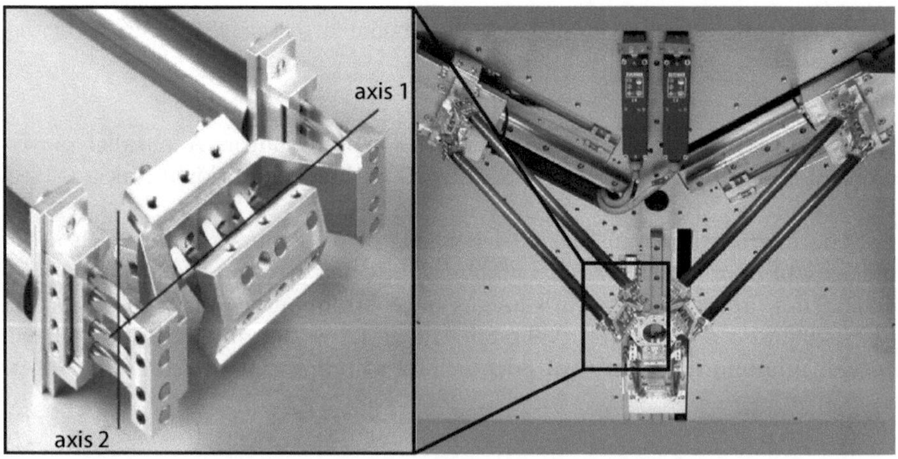

Fig. 6.62. Compliant spatial robot with 3 DOF and optimized design of a flexure hinge with 2 DOF [91]

of spring steel the resulting workspace is a hundred times smaller compared to the workspace when using pseudo-elastic flexure hinges [92].

6.4.5 Conclusion

The advantages offered by shape memory actuators become most obvious where small-sized devices are concerned. Due to the very high work-per-volume ratio, SM actuators of millimeter or micrometer dimensions have large actuator strokes and forces. The response time strongly decreases with shrinking actuator size.

SM actuators may have very many different shapes and offer a variety of shape changes (i. e. actuator strokes). This property can be exploited so as to adapt the SM elements shape to the actuating task. As an application example, a miniature parallel gripper with electrically heated SM wires integrated into its mechanical structure was presented. Further on the performance of pseudo-elastic shape memory flexure hinges in parallel robots for micro-assembly tasks was shown. The future opportunity for thin-film SM actuators to drive micromechanical systems and devices was demonstrated by a miniature silicon gripper.

6.5 Electrorheological Fluid Actuators

W.A. Bullough

6.5.1 Particulate Fluids

An electrorheological fluid (dispersion type), as defined for this purpose, is a mixture of micron-sized, high dielectric constant particles carried in an insulating base oil. When an electric field is applied transverse to the direction of any motion of the fluid, it causes an interaction between the particles, the field and the dispersant, and this results in an increase in the resistance to the flow of the mixture.

The first, and perhaps the most important thing to understand is that there is more than one so-called electroviscous effect [93]. Many researchers have reported these phenomena [94]. The fluids have not all been of the slurrified electrorheological (ER) type that were first investigated in depth and developed by Winslow [95] – see later. Indeed, high-speed recordings of the response of a promising particulate ER fluid show at least two separate responses to the applied voltage excitation. However, so far as the engineer is concerned, this event may be considered to be part of a single response. The overall response, in this case the resistance to an existing flow, which manifests itself as an increase in driving pressure or shear stress due to an applied voltage change, must not be confused with (say) the time response of a hydraulic servo valve which demonstrates basically a change of flow rate or piston displacement brought about by an electric input signal.

The mechanism(s) of a particulate fluid electroviscous effect is still not fully resolved and quantified. It is not strictly relevant to this work and is therefore not dealt with in detail. At this stage it can only be said that it is a very multi-parameter and multidisciplinary event and, secondly, it should be understood that there is little change in the viscosity μ of the fluid as it is normally defined in its continuum context save for a derived effective or non Newtonian viscosity sense. The term electroviscous, which has often been used to describe the present class of fluids, is misleading in this case. Rather, the field imposes a yield stress type of property on the fluid which is similar to, but not the same as, that which is a feature of the ideal Bingham plastic. This can readily be seen by referring to Figs. 6.63 to 6.66 inclusive. It is alternatively possible to claim that either the plastic viscosity changes with shear rate or the electrode surface yield stress does.

Testing Particulate ER Fluids

The shear mode of operation is the term generally given to the simple shearing of the fluid, as in a Couette rotational or parallel plate type of viscometer but with an electric field applied between the moving and the stationary electrodes of gap size h (Fig. 6.63). With zero voltage ($V = 0$) applied, most ER fluids exhibit near-Newtonian properties. When an electric field ($E = V/h$) is applied to the fluid, there is an increased resistance to its movement which must be overcome before motion can take place (see Fig. 6.64 which is an idealised representation). Conventional constant temperature Θ and speed ω Couette laboratory techniques can normally only encompass shear rates ($\dot{\gamma} = \omega R/h$) up to several hundred s^{-1} although cooled purpose made industrial clutch-type devices of similar geometry may reach $6000\,s^{-1}$.

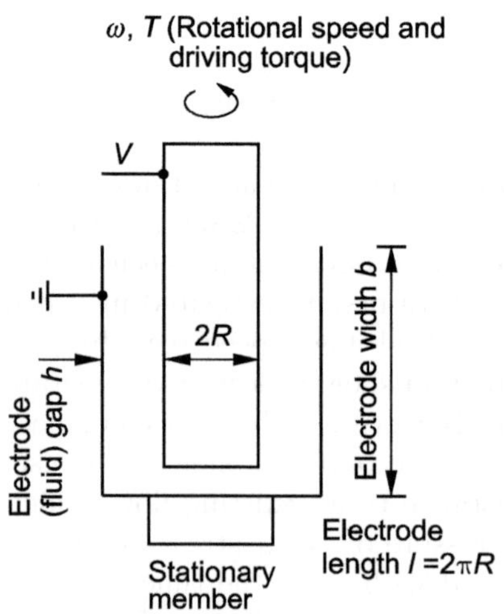

Fig. 6.63. Shear-mode viscometer

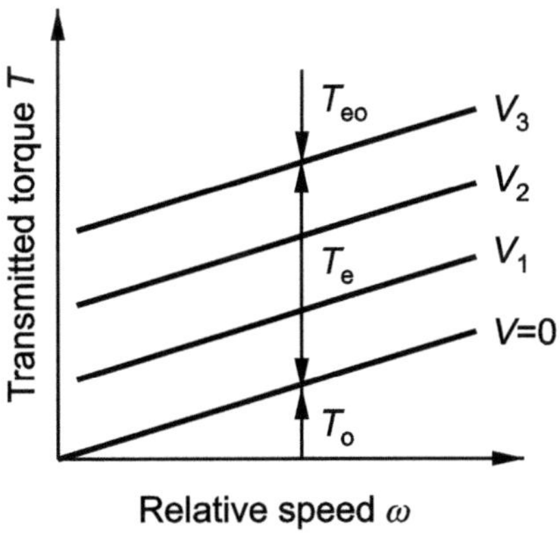

Fig. 6.64. Shear mode – diagrammatic test results

Because of the problems associated with the manufacture of sample slurries of the ER type and on account of difficulties brought about by high shear stress (τ) heating the small amount of fluid in a viscometer (at least, so far as scientific readings are concerned), much of the developmental testing of the fluids has been done in the static yield situation, namely the point at which yield shear stress τ_e is overcome, at a given voltage, so that motion can commence. The shear-stress/shear-rate characteristic beyond this static yield point has often been measured by rotating a viscometer with zero volts applied, at a much lower shear stress than the static yield point level. It will be noted that whilst this test procedure and configuration is obviously very useful for the small batch development of the fluids by chemists, physicists and rheologists, the data produced in this way needs to be treated with some reserve. It is, after all, a system, comprising the same identifiable fluid matter throughout the tests, and in all a situation not often encountered in, say, a hydraulic power mechanism where, usually, a throughflow is required to procure the displacement of a linear or rotary piston, and a high level of compressive stress (pressure) is needed in order to keep the size of that piston down. Also, no allowance is made for the effect, of high $\dot{\gamma}$ on μ or τ_e or, due to changes in the structure of the particle matrix.

The flow mode is thus perhaps of more interest to the engineer who is seeking to control and design high-force/torque, low-weight/volume, power transmission systems. Here it is necessary to have the facility to remove the working fluid from the source of heat generation to a convenient location, where it may be cooled and then recirculated, thus preserving the fluids lubrication properties. In this configuration the fluid is normally tested by pumping it between fixed parallel plates across which a voltage is applied (Fig. 6.65). Though this method clearly requires larger amounts of fluid for the tests, it is nearer to the more familiar control volume situation – the study of well defined geometrical configurations through which fluid is continually flowing. The high $C_v\rho$ (specific heat × density) product of the liquid serves

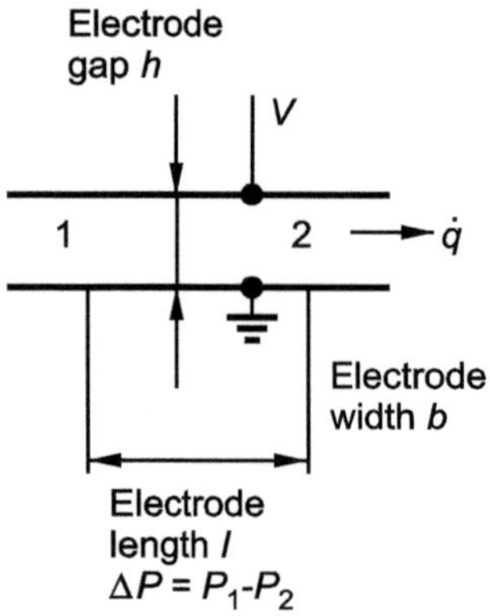

Fig. 6.65. Flow mode test electrodes

to keep temperature excursions in check – a situation not always encountered in a closed system, where the negative temperature coefficient of resistance of what is effectively a hydraulic semiconductor can cause a conductance based thermal runaway. Industrial power rotational clutches will probably need throughflow and/or on-off operation to supplement casing heat convection; hence a combination of shear and flow mode operation.

Most of the tests performed in this (flow) mode have concentrated on keeping the flow rate $\dot{q}$ constant and measuring the pressure response ΔP to an applied voltage step. The use of some kind of pump over a period of time, with associated flow meters, pressure transducers, strainers, tanks, coolers, connectors etc. being in contact with the fluid, is a more convincing test of the serviceability and durability of the fluids, in a fluid power sense, than a closed system shear mode or other low fluid volume test. However, the simultaneous high-speed recording of pressure drop, flow rate, voltage, temperature and electric current I, given even the advanced instrumentation available nowadays, involves some problems. For example, it is very easy to erroneously measure some wave action in the hydraulic or electrical part of the test circuit or equipment, or indeed the drive motor regulation in taking the extra pressure load, unless great care is taken [96]. Driver noise can be a problem.

The separation of the true response time of the electroviscous, Winslow or electrorheological effect from unsteady pressure recordings of the step input type is a complex and tedious affair. Experimentation carried out in this domain is expensive and time-consuming, not least on account of the many variations of electrode separation and length, the number of variables and the different input frequencies involved. Again, because of these problems any data presented for appraisal should be treated with caution. It is necessary to ensure exactly what information has been put forward and, from what kind of test and how it was derived. Also, a model of the constitutive form

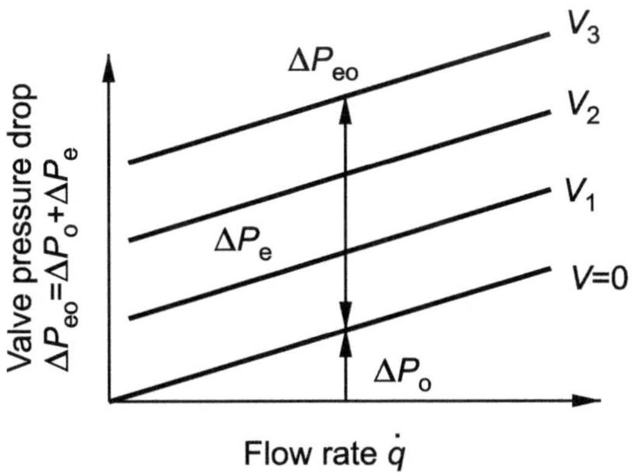

Fig. 6.66. Flow mode – diagrammatic test results

and velocity profile is often assumed, with no subsequent iteration towards a precise solution.

Flow visualisation tests are easier to carry out in the flow mode than the shear. A simple two-dimensional test-cell, using a very dilute suspension of solid, shows clearly that the particles adopt a semi-regular matrix pattern in stationary fluid in a valve and/or a columnar structure as the voltage is applied, and will then be distorted but held there until the forcing pressure reaches the yield value dictated by the voltage. At this juncture, particles appear to flow with the base liquid, though possibly not always at the same speed. Small electrode gap sizes prohibit true velocity-distribution studies across them; the usual technique of building oversized devices to facilitate such studies is prevented by the increased voltage demand. However, a Bingham-plastic-type velocity-profile/core-flow analysis on the problem [97] ties in quantitatively with each set of experimental flow-rate/pressure-drop results. Some photographic evidence [98] of plug flow is available. One interesting feature of an unyielded situation (when pressure is applied but the solid particles are still held by the field) is that small droplets of pure base oil are periodically observed leaving the outlet of the valve as if the matrix was behaving like a filter. The formation, distortion and breaking of the matrix is presumed to lead to complex rheological situations that have been observed as a hysteresis type of effect. CFD (computational fluid dynamics) can be very useful in indicating velocity profiles – in the continuum sense, albeit in a 2d case with non separating electrodes.

A further advantage of flow-mode testing is that the shear-rate magnitudes that would be encountered in a practical hydraulic device, often in excess of $40\,000\,\mathrm{s}^{-1}$, can be achieved [99]. However, the definition of shear rate needs to be subjected to scrutiny: it is often derived from the Newtonian/Poiseuille formula, albeit when plug flow is present [100]. In a Couette viscometer care must be taken to avoid plug flow: the plug formed by radial effects at low speeds distorts the $\tau, \dot{\gamma}$-plot [101], and electrostatic breakdown can occur at relatively low shear rates.

Characterizing the Fluid for Design Considerations

In many practical applications of hydraulic machine engineering it is not necessary to design to a high degree of precision. This is due to a number of factors such as, for example, the difficulty in predicting the load accurately over the cycle of operation; hence the working temperature of the fluid is not entirely assessable. Further, the advantages in capital-cost economy brought about by the mass production of pumps and motors etc. limits the number of truly purpose made devices and, lack of precise fluid data.

Within limits a particulate electrorheological fluid (ERF) can have its properties fixed to suit a particular task [102]. This could be done by, say, adjusting the water content within the solid phase of a 'wet' fluid or its volume/mass fraction in relation to the base liquid or by adjusting the particle size distribution; thus the rheological behaviour of the mixture will be affected. The same applies to 'dry' (no added water) fluids, which may be polymeric by nature. Surfactants can change performance beneficially.

There are many operating variables in an ER power system, not all of which can be controlled easily or simultaneously, and for this and for all of the above reasons it is probably not too productive at this stage of the development of ERF to spend an inordinate amount of time in perfecting precise steady-state and time-dependent analytical rheological models. These will no doubt be called for in due time when more standard fluids are produced or as applications demand computational fluid dynamic (CFD) prototyping. Existing CFD practices can accommodate elastic shear moduli and non ideal $\tau_e \vee \dot{\gamma} \vee V/h$ models, and thermal effects [103].

The diagrammatic raw steady-state test results of Figs. 6.64 and 6.66 are typical approximations of those that would be achieved in shear- and flow-mode tests, respectively. Notably, in this idealization the slope of the lines in Fig. 6.64 has the same constant value whatever the magnitude of the applied voltage or relative shear condition. In Fig. 6.66 much the same can be claimed but the slopes may not have the same value as in Fig. 6.64 – see later in this section. This infers that the field/yield effect can, for analytical reasons, be sometimes assumed to be independent of the fluid motion/shear rate/flow rate and that the field does not affect the flow/forcing term – slope parameter – a very important approximation and often permissible simplification so far as a design procedure is concerned. Much the same applies for current flow. The overriding considerations outlined in this section have led to the choice of a simple Bingham plastic fluid model (Fig. 6.67) on which to base steady state performance estimations for a conceptual electrorheological-shear-stress/electric-field device. (In unsteady operation a viscoelastic Bingham CFD model is required).

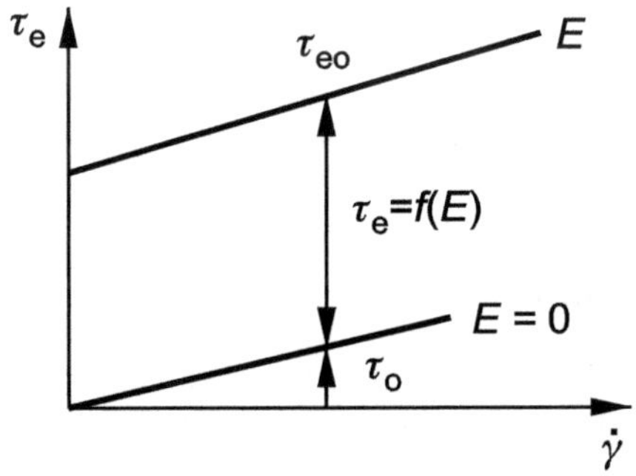

Fig. 6.67. Bingham-plastic-type model

In order to do this, the viscous pressure drop ΔP_o of Fig. 6.66 is deducted from the total pressure drop ΔP_{eo} to give the electro or yield pressure ΔP_e for a particular voltage. Given the valve dimensions and using the dynamic viscosity as calculated from the zero-volts line and the valve dimensions, the yield stress at the wall may be isolated and $\dot{\gamma}$ calculated. Likewise, shear stress τ and $\dot{\gamma}(= \omega R/h)$ can be calculated from the shear mode test data – by neglecting radial effects. The well known relevant Poiseuille and Couette flow analysis are often used in these procedures. In both modes μ is derived from the zero-volts test. On this basis, flow- and shear- mode data will not necessarily correspond in the $\tau, \dot{\gamma}$-plane.

Having produced results in the form of Fig. 6.67 for a given fluid, the yield stress dependence on voltage is derived at a nominal shear rate and is shown typically in Fig. 6.68. This property may be modeled in two ways: either as a deadband or barrier of field strength E_o of zero yield magnitude, followed by a linear relationship between τ_e and E (for $E > E_o$); or as τ_e proportional to E^2. In both cases the approximation will produce reasonable enough but rarely precise designs over the full voltage range of operation. Only when the mechanism of the effect is fully understood will the

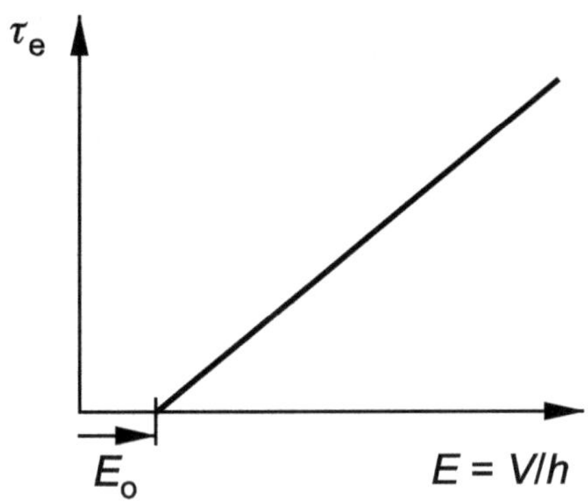

Fig. 6.68. Electro yield property

correct presentation be clear. CFD can be made to provide more precise results forms from real test data. Also, the Bingham plastic/Buckingham parallel plate relationships should give better values of τ_e and $\dot{\gamma}$ – see Sect. 6.6.

Design Formulas for Estimation Purposes

By adopting the approach explained above, calculations for a device may be approached in the following manner, the flow normally being laminar in fashion:

Clutch Type Controller. Torque T on a radial rotor element at a general radius r is given by

$$T_\mathrm{eo} = 2\pi r^2 \tau_\mathrm{eo} \delta r \,, \tag{6.25}$$

where

$$\tau_\mathrm{eo} = \tau_\mathrm{e} + \tau_\mathrm{o} \,, \tag{6.26}$$

$$\tau_\mathrm{e} = f(E) \,, \tag{6.27}$$

and

$$\tau_\mathrm{o} = \mu\omega r / h \,. \tag{6.28}$$

The inner core of the plates has little effect save to consume electricity.

For a cylindrical clutch this becomes

$$T = 2\pi R^2 l \tau \,, \tag{6.29}$$

where R is the mean radius and l the length of the cylinder.

In some applications there will be obvious limits to the use of the simple solutions on account of heating, radial and centrifugal effects, and flow stability. The behavior of a device like this in pick up and drop load situations will depend to a major extent on the driver and load characteristics; only rarely will the speed of the ER effect per se come into question. There is little difference in performance between well designed radial and cylindrical clutch types. Likewise the mass of the ERF can be neglected in most inertial calculations [104, 105].

Valve Controller. In this case

$$\dot{q} = bh^3 \Delta P_\mathrm{o} / 12\mu l \tag{6.30}$$

and the wall shear-yield stress derived from a control volume placed around the electrode gap is

$$\Delta P_\mathrm{e} = 2\tau_\mathrm{e} bl / bh \,. \tag{6.31}$$

To facilitate this simple design procedure, and for other reasons, the shear stress and shear rate in a valve should generally be quoted in flow-mode-derived characteristics. Again, the pump actuator and load characteristics and the elasticity of the fluid may have a significant effect on the time response of the system to a change of input voltage signal [105], especially under extreme conditions of operation.

Quasi-Steady Calculations

Generally speaking, a particulate ER fluid being operated at the correct temperature will respond rapidly to a voltage signal [106]. Fortunately, the implication is that so long as the delay between say a step voltage and the short-term steady-state shear-stress response to it (t^*_{m}) is not great, then the ER fluid design can be treated the same as for a normal hydraulic fluid for quasi-steady design purposes. Typical values of $t^*_{\mathrm{m}} \leq 1\,\mathrm{ms}$ at the best operating condition (for E, $\dot{\gamma}$, Θ) and normally can be neglected for all except electrical supply-circuit and electronic-control purposes. For example, usually in the run up time for a clutch the load torque is essentially equal to the inertia of solid parts times its angular acceleration, all at the correct operating temperature.

A much more detailed appraisal of the ER machine/device controller and typical treatments by CFD can be seen in [103,105]. All applications of CFD to MR and ER systems are likely to be reasonably comprehensive and precise once better performance data for the fluids becomes available.

Electrical Quantifications – Particulate ER Fluids

The resistance R and capacitance C of an ER device of electrode area A follow (approximately) the classic forms of $C \propto A/h$ and $R \propto h/A$, respectively, with a fixed time constant RC. Alas, both parameters depend on temperature, shear rate and voltage level/rate of slew. If the electrodes have too large a surface area, then peak current values are large, as well as the magnitude of the conductance. Rapid switching of electrostatic catches in the high tension (volts)/direct current (HT/DC) circuits can then be a limiting factor on ERF application. Special drivedown facilities are required for these step-voltage-producing devices, and even then the controller may not discharge fully before charging is required again. Nevertheless, control of hysteresis in locking devices and proneness of the fluid to electrophoresis may require binary digital control. In general, modeling in an electrical sense is highly nonlinear and its use is restricted more often than not to the design of control and excitation circuitry – see the next section.

Typical ER Particulate Fluid Properties

There are many different types of particulate ER Fluids: different base oils, solid material and solid fraction, different size and size distribution of particles, surfactants, and, if wet, different levels of water content therein can all be used. Dependent on the application, one characteristic is more important than another.

The comparison of ERF performance in fluids is made difficult through the lack of unsteady-state test data and an account of the differing effects of shear rate, temperature and form of field dependance from fluid to fluid. Table 6.7 shows a nominal comparison of shear stress levels and corresponding conductances, drawn from typical commercial fluid data available in the public domain at the time of writing.

Table 6.7. Typical ERF characteristics taken from a modified shock absorber with valve control 100 mm long by 13 mm outside diameter and 0.75 mm electrode gap. Date are derived from steady direct current excitation, valve pressure drop (assuming Poisieulle flow), speed of piston and valve geometry. Volume fraction 50 ... 60%, density $\approx 1.04\,\mathrm{g/cm^3}$

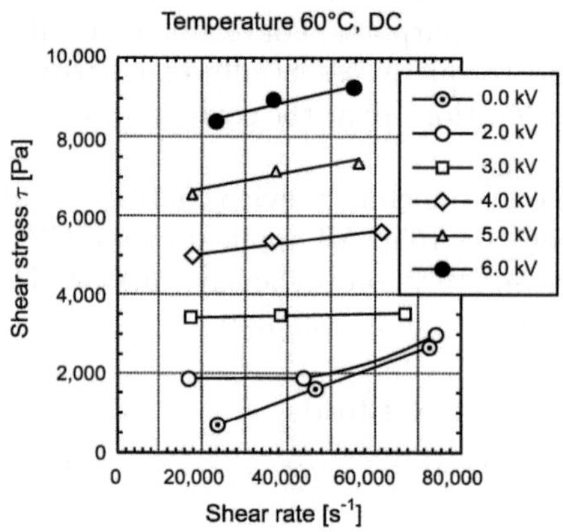

Shear stress as a function of shear rate and field strength

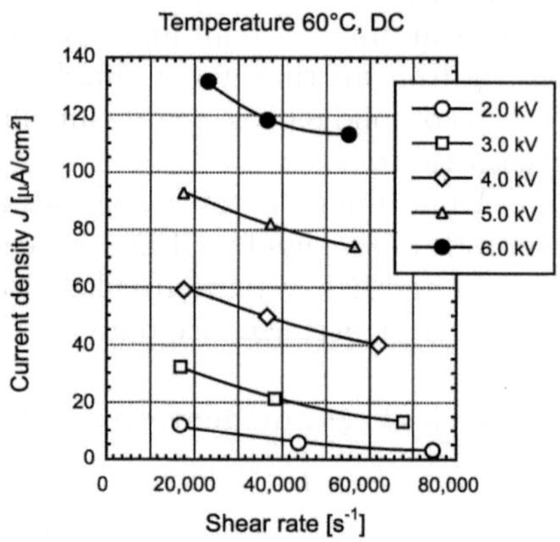

Current density as a function of temperature and field strength

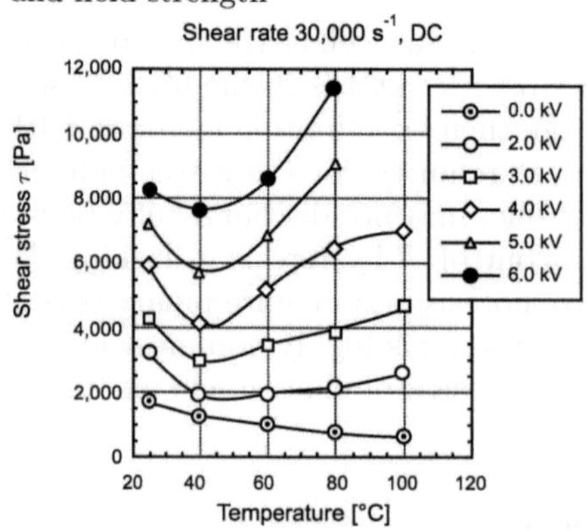

Shear stress as a function of temperature and field strength

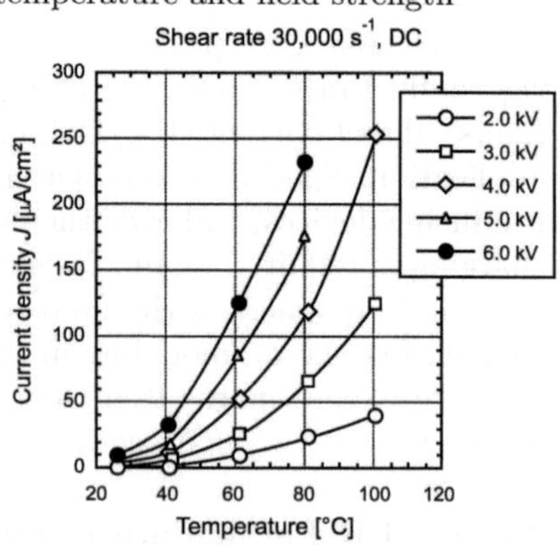

Current density as function of temperature and field strength

Rheobay Electrorheological Fluid. (Provisional product information TP AI 3566. Miles Industrial Chemical Division, USA.) Although the effect of a change of temperature will be evidenced in the above properties (and may affect the settlement rate of the solid, which can only be matched perfectly with the base fluid at one temperature), none is more significant than its effect on current density J. Here, a small increase in temperature can raise the current consumption considerably, and vice versa. This is a major problem area of ERF. Certainly this factor alone is sufficient justification for the use of a simple approach to the initial characterisation, fluid comparability problem and design procedures. Different temperatures can change slopes of the $\tau, \dot{\gamma}$-characteristic: increased temperature can often increase current to that for the optimum t^*_{m} (however small it is) and τ_{e} performance and/or give a level $\tau_{\mathrm{e}} \neq f(\dot{\gamma})$ characteristic or thereabouts.

Work is still going on as to what the time domain response of the ER effect really depends on [105] or what is its meaning in terms of the fluid design. The initial and true ER effect in a valve, for example, is not the main concern: it is also not the true time of the full pressure rise, the valve geometry, flow rate and other factors being involved. In the shear mode the position is similar yet the torque/input voltage response limit in response to small sine waves has been claimed to be as high as 1000 Hz.

Design Variables and Controller Shape

From an inspection of Figs. 6.64 and 6.66 and (6.29), (6.30) and (6.31), it can be seen that at any given relative speed or flow rate the ratio between the excited and unexcited torque or pressure drop, in the shear and flow modes, respectively, can be influenced by the choice of b and/or h for a particular fluid. The ratios $\tau_{\mathrm{e}}/\tau_{\mathrm{eo}}$ and $\Delta P_{\mathrm{e}}/\Delta P_{\mathrm{oe}}$ are often very important considerations in a practical controller mechanism, see Sect. 6.6 for empirical coefficients.

Equation (6.31) gives some indication of how to amplify the yield stress, i. e. by fixing the value of l/h in the valve. For example, if $l = 100\,\mathrm{mm}$ and $h = 0.5\,\mathrm{mm}$, then $\Delta P_{\mathrm{e}} = 400\tau_{\mathrm{e}}$. This type of manipulation is, of course, familiar to a hydraulics engineer, who often uses a small shear stress to effect a high pressure drop to drive a piston. It is, after all, why shock absorbers and actuators are usually of the piston type rather than of the shear plate variety, and why fluid-based damper devices are preferred to purely electrical types. However the choice of shear plates can significantly reduce shear rate and hence parasitic drag.

For the flow-mode situation, (6.30) and (6.31) show that the control ratio $\Delta P_{\mathrm{e}}/\Delta P_{\mathrm{o}}$ is independent of the valve width b and that the volumetric flow rate, for a given pressure ratio, is proportional to b. Similarly this pressure ratio for a given flow rate does not depend on l but the pressure drop does. If the surfaces of the electrodes are increased in area, the current demand will rise in proportion. If the gap h is increased, then ΔP_{o} falls dramatically – roughly as the cube of h since the flow is usually laminar and near-Newtonian.

However, if (say) h were to be doubled, the voltage needs to be increased in line to maintain the yield stress, and this would in turn only give one half of the previous ΔP_{e}. This can be seen from the shear/pressure force balance around the electrode gap, as per (6.31). The designer has some choice – but at cost. Also, consider the effect of these manipulations on, say, the capacitance and conductance of a controller, and the unsteady state performance [105].

Although this design situation poses many questions, a particular control duty that (say) a valve is required to perform will depend on the application and will not be further discussed here, since the flow rate may not always be constant as it is for the above example and, in addition, $\tau_{\mathrm{eo}} = \tau_{\mathrm{e}}/\tau_{\mathrm{o}}$ and is therefore not a simple function of voltage (or shear rate). The true Bingham-plastic solution involves cubic terms (see Sect. 6.6) and is much more tedious to handle than the simple procedure.

There exist a few specialist combinations of the flow and shear modes. These comprise the squeeze mode, which is basically the flow mode achieved by flow between plates that are approaching one another, but with additional dominant in-line forces and the Rayleigh step mode in which pressure is generated hydro-dynamically by a change in a flow section. The former is usually associated with low-frequency operations of small displacement, e. g. in engine mounts. The latter has so far shown little promise in respect of rapidly centered bearings: the time constant t^*_{m} has proved too large for the intended operation for the nano particle fluid required in the small journal/shell interface [107]. However, variable stiffness operation remains a possibility. Two dimensional flows are under investigation as a means of cooling slipping clutch drives [103].

6.5.2 Limitations to the Concept of Particulate Electrorheological Fluids

Particulate electrorheological fluids are now considered in greater detail and with respect to their application in devices that are aimed at featuring electronically designated motion and flexible operation via adaptronics, or in third-wave machines. In effect, this also sets down the state-of-the-art position of research in the field and outlines the salient factors that determine research trends for fluid developers, whilst at the same time giving some idea of ERF machine performance implications. The type of artifact under the spotlight is one wherein its function can be rapidly regulated without a change of geometry of the solid parts and at the behest of an electric signal alone i. e. speed of response is all important.

ER Models and Characterization

Perhaps the most problematic area that obstructs the control of these changes by the use of electrorheological fluids is the particle mechanics or continuum conundrum for characterizing the flow of the dense slurry (unexcited) or yielding plastic (excited). Whilst many years have been spent in computer analysis based on dielectric polarisation models of the multidielectric excited particle fluid structure, the kinematic yield-stress level remains underpredicted from the constituents by an order of magnitude. Only recently have models begun to realistically account for the yield-stress levels achieved in practical fluids at the static yield point [107]; however, it is not difficult to find texts that claim that the performance of the same fluid in Couette-shear flow and Poiseuille valve flow cannot be related i. e. the fluid cannot be treated as a continuum albeit in plane shear flows [100].

It is now evident that polarisation is not the only mechanism at work and that hydrodynamic effects [107] plus conductivity [96, 108, 109] at least need to be included in multi body effect models designed to illustrate the modus operandi of the effect and to link it quantitatively to solid particle/fluid properties, flow conditions and excitation levels (see Fig. 6.69).

Much insight has recently been gleaned into the relative importance of disparate fluid/particle conductivities and dielectric properties, where and when they are important, how they relate to the physical charge processes, and the dependence of the yield stress upon them. This has mainly been confirmed in steady-state-based investigations of the attractive forces between single spheres. There is, however, some way to go before any optimization procedure (for a given application) can be quantified in terms of materials make-up, especially in the time domain and where clustering of particle chains and particles are important. Meanwhile engineers need to use what empirical characterization data is available for commercial fluids, and this explains the layout of previous subsections. A notable extension of particle aggregation studies has produced [107] fluids which demonstrate static yield stresses greater than 100 kPa. The further investigation of such fluids is proceeding.

The possibilities of characterising ER fluids in flow as a continuum (otherwise the properties of viscosity and density have little significance and design

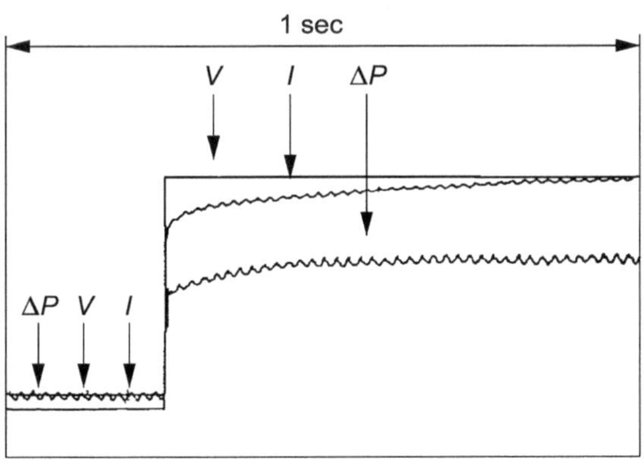

Fig. 6.69. Conductivity is seen to be important in the secondary response (at least). Steady flow in a valve with step voltage V applied; ΔP is valve pressure drop; and I is the electrical current. Similar behaviour is seen in a clutch

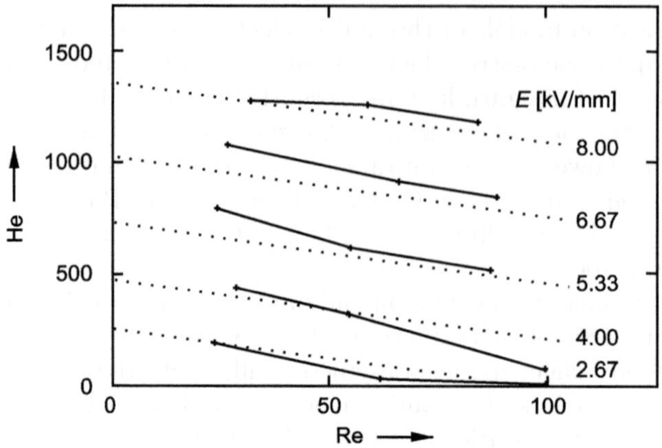

Fig. 6.70. Hedström number versus Reynolds number for a valve (*full line* experimental result) as predicted from clutch (Couette mode) experimental data (*dotted*) for different values of $V/h = E$

techniques become entirely empirical) have been given a boost by work that indicates a link between such a fluids performance in the shear and flow modes of employment and others (e. g. static shear) for different fluids. This is done by the use of nondimensional Hedstrom and Reynolds numbers via use of Buckinghams relationships for a Bingham plastic [110] for steady flow, (Fig. 6.70). This is welcome yet perhaps surprising, since the thickness of the shearing fluid layer near a boundary can approximate to a particle (often of variable geometry) size. More generally the fluid is non isotropic i. e. with respect to motion transverse to the electrodes and, no effect of the electrode surface was included.

The extended concentration by fluid developers on the details of slow steady flow belies the necessity to confront the intensely unsteady (and indeed steady) high shear-rate motions that will be required in practical machine work cycles. Very often misleading appraisals of situations arise from the lack of fluid/machine performance details.

Third Wave Machines

In a flexible adaptronic machine capable of a high resolution of (smooth) force, velocity or displacement variation, there is little scope for the rapid generation of, say, large-scale motion by an inductive or relatively heavy rotor electromagnetic drive or the by generation of a shaped control current. In both cases, respectively, latching onto a high inertia source of steady motion (and the braking of it) and a high-capacity, high-tension supply line (and discharge by earth shorting) produces a digital event via engagement of the ERF [111, 112]. Both AC and DC excitation components are present in step switching and dwell periods respectively, and yet there is a tendency to separate fluids research into AC or DC types. The realisation is growing that potent AC fluids depend on particle polarisation in a poorly conducting fluid, whilst apparently good DC fluids need appreciable current flow – both conclusions being based on only steady-state strength and current flow appraisal.

Referring to the ultra high acceleration/low-inertia flexible machine regime, it is predicted that the limiting change of (99%) speed response time t_0, in the digital mode of operation is heavily influenced by the inter-electrode gap size, and the fluid density and viscosity [103]. The solid part of the power transmission mode dominates the mechanics.

For a change-speed response time of less than, say, 20 ms a $4 \cdot 10^5$ V/s signal-rise/fall time rate is required. This has implications for the fluid capacitance, which is difficult to model as a function of shear rate [113]. When the voltage is rapidly applied, the yield shear stress follows at a time constant of approximately RC, the resistance and capacitance product of the inter-electrode space. There is little point in accelerating the load rapidly if the torque initiation lags much behind the step rates of change of excitation, although this lag can be difficult to measure [114]. Fortunately the lag seems to decrease the harder the fluid is being punished in terms of E, $\dot{\gamma}$ and Θ (electric field, shear rate, and temperature respectively) (Fig. 6.71).

This factor becomes important if the generation of a motion profile in a third-wave machine is considered. Without getting involved with digital technology: if the x direction speed provided is constant, then the y penetration (driven by a bang-bang application of voltage and a yield stress of sufficient magnitude to give the relevant part high and instant acceleration) must be maintained over a very small time interval (fixed by the switching

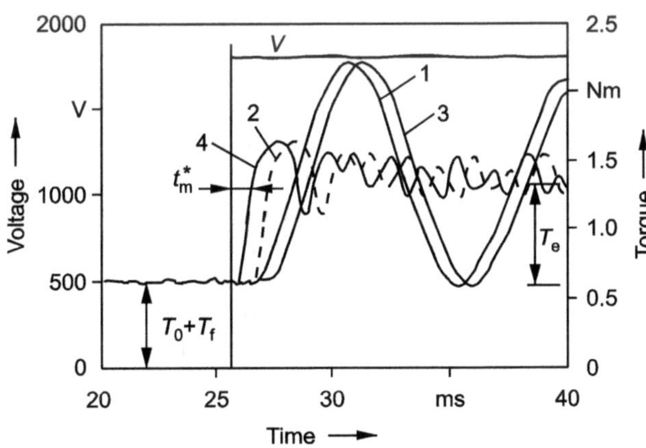

Fig. 6.71. Numerical transformation of step torque: *1.* measured torque-transducer signal; *2.* first estimate of ER clutch torque response; *3.* predicted torque-transducer response for first estimate; *4.* final estimate of ER clutch torque response. $T_o + T_f$ are viscous and real friction torques, with T_e due to application of step voltage V

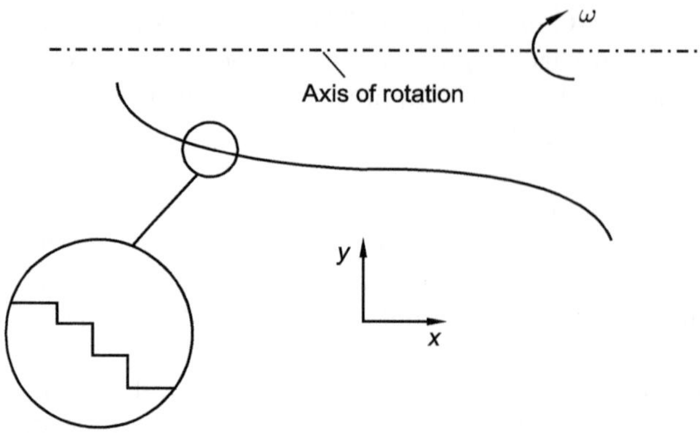

Fig. 6.72. Digital ER motion synthesizer (concept): y direction is shown ER controlled in switched steps of equal time elements, with steady speed traverse in (say) a lattice in the x direction

speed) if the resolution is not to be too crude. DC operation seems virtually mandatory, with any hysteretic and electrophoretic tendencies being arrested by a conjunction of binary switching and high $\dot{\gamma}$ (see Fig. 6.72).

Mechatronics and Testing

It does not seem possible to provide a figure of merit for a fluid that possesses these sundry needs, but (see Sect. 6.6) the linear traverse mechanism will demonstrably test total capability in that respect [115]. In this device, two contra-rotating, high-inertia, constant-velocity rotors provide motion sources with HT (high tension) and earth 'busbars', the excitation being controlled via switches.

Two driven clutches, spaced from their drivers co axially by the ERF, are each connected to a pulley, both of which are connected by a belt. The ERF, in opposing clutch drives (Fig. 6.73), is excited alternatively to make the belt reciprocate, with typical steady speed of up to $\pm 5\,\mathrm{m/s}$ separated by turnround times determined by the fluid properties: τ_{e}, μ, Θ and t^*_{m}; high μ can distort the traverse profile, if excessive. A good-quality fluid should turn round in 20 ms. Thermal runaway should be avoided by as large a margin as possible; the heat transfer rate from the outer driving rotor is about the maximum per unit area that is achievable into the atmosphere. The full-speed centrifugal field on the particles is up to $100\,g$ and the belt acceleration around $50\,g$. Fluid degradation has been only generally described [116].

With an analysis of such performance data, the fundamental compatibility arises in relation to a low ERF time constant, the heating effect of the viscous shearing and conduction loads, and the level of voltage. Alas, a failure of fluid on this machine implies its separate analysis on each of several simple

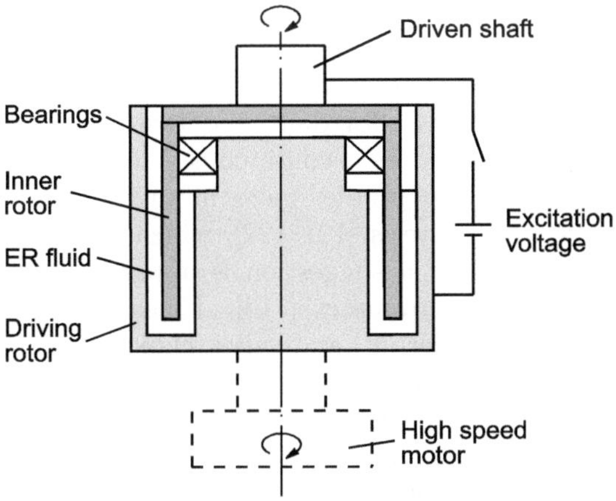

Fig. 6.73. Cylindrical clutch for ER traverse gear: on driven shafts (contra-rotating), pulleys are connected by belt; alternate excitation of clutches causes reciprocity with the belts, which carry the product to be wound on a bobbin (not shown)

characteristics tests – in order to isolate the problem area. The test does, however, give a good example of the machine side of the overall electrical-chemical-rheological-thermal fluid/machine optimization.

It will be appreciated that the inertia, geometry and stress and strain in mechanical parts are linked to operating conditions, and particularly to acceleration. For example, the uniform speed of the traverse could be obtained (presumably) by having a long, small-radius clutch and having a large pulley and a low rotational speed. Likewise, fluid performance τ_e, t^*_m, and μ depends on the solids content, the materials properties, and the size and shape of particles, Θ and $\dot{\gamma}$ [117]. With present lowish kinematic fluid yield-strength properties, the optimization process is made more hit-and-miss if full fluid data is not available [118]. Dynamic analysis is required.

In connection with (for example) the traverse mechanism, the need for a yield stress and a rising τ characteristic with $\dot{\gamma}$ is noted – Figs. 6.64 and 6.66. This is necessary e. g. for any clutch drive where an overload may cause slippage if τ_e was to fall. However, the Bingham-plastic characteristic per se is not obligatory: in other types of flexible machines such as the vibration isolator, a rising linear force/velocity characteristic seems preferable [119] (non dynamic operation). In the flow mode of operation, much the same factors come into play as in the shear mode. The benefits of a high yield-stress magnitude is to reduce the amount of fluid volume for a given force/stroke requirement but, high pressures may mean high compressibility effects.

Hysteresis and Control

The valve control application in general exemplifies an interesting control problem. Whilst good reasons have been given for digital control of a particulate ERF, the damper could be envisaged as a continuous ride member under analogue excitation/control [120]. Past studies have, however, shown a pseudo, or perhaps time-dependent, hysteresis [121, 122], which is better treated by bang-bang operation; more than a suggestion is apparent that voltage alone is insufficient as a control parameter [114]. These and the sometimes-experienced violent clutch (shuddering) and valve (choking) [123] may yet prove to be not separate phenomena but related characteristics linked to structure formation and destruction. These effects plus electrophoresis are to be avoided, save for their further investigation (Fig. 6.74). Shear modulus G', specific heat capacity C_v and bulk modulus β under field need to be known, since they can also determine the precision of any controlled positioning device.

All of the foreseen effects put a limit on the performance of an adaptronic type machine and set the requirements for τ_e and t^*_m $f(\dot{\gamma}, \Theta, E)$ in the ER fluid. There may be competing limiting factors: fluid elasticity and volume, heating and cooling etc. Further limitations arise from lubrication – for example particles will only move through an elastohydrodynamic region at low speeds and anti-wear boundary lubricity is hence very important [124]. ER fluids are generally poorer boundary lubricants than MR-fluids.

The self-weight/inertial loading problem [125] can easily be avoided so far as solid material at its critical breaking length is concerned, but strain will limit the overall acceleration (on grounds of precision) – only a few materials will exhibit less than 0.01% inherent strain at an acceleration of $100\,g$. Accelerations above $100\,g$ are regularly attained in conventional machines and cause one to wonder at the rate of separation of particles and any possible cavitation effects in the fluid.

6.5.3 Future Aims and Present Problems

The whole aim behind present ERF developments is to provide a means of control/adaptronics (that is easy to apply and economical) to a mechanism that has to be by its nature flexible in force, displacement or speed and hydraulically operated. Electronic solid-state semiconductor devices and computers are powerful, inexpensive and adequate for many applications in control, and yet their interface with hydraulic machines usually involves a bulky solenoid or an expensive servo-valve, often with a pilot stage and a power supply. The aim of ER research is to be able to influence a hydraulic mechanism directly with a current low enough to allow the integration of a system made up of electronic transducing and signal-conditioning equipment, computer processors and feedback monitoring, solid-state controllable field-excitation

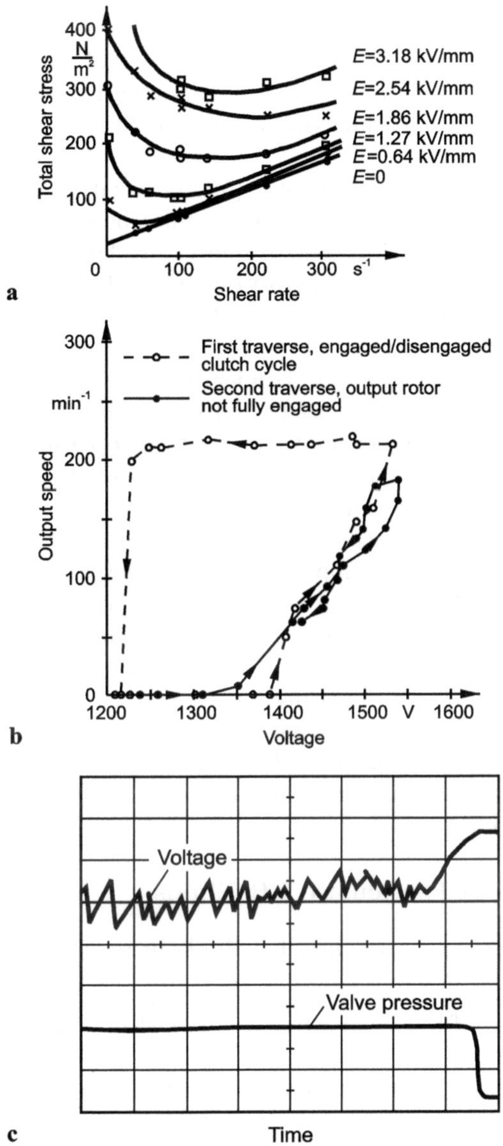

Fig. 6.74. Hysteresis/structure-related effects in **a** a Couette viscometer, **b** a clutch in on-off DC operation, and **c** a valve experiencing choking phenomena at nominally constant flow rate, where the uppermost trace is ongoing DC voltage and the lower is valve pressure drop versus time

supply and the hydraulic device itself. If this can be achieved without moving parts, then so much the better. This is the implementation problem of ERF.

At present, the problems of high current density and particularly its sensitivity to temperature, and low yield strength in commercially available fluids restrict the range of practical application of ER. An ER valve network is only competitive with (say) a servo-valve in certain passive control situations. Nevertheless, ER fluid is potentially preferable to a magnetic (i. e. long time-constant and large-excitation system) fluid on account of its speed of operation.

The matching of the nonlinear yield characteristic (or linearization) to a device has so far not proved to be a problem. The same cannot be said for high pressure operation or heavy duty position control. Here, very large and leak proof valves would be required to give a locked position and load stiffness where large disturbing forces are involved: magnetic fluids or conventional electromagnetic devices are better on heavy duty.

The subject of high-speed, flexibly operated electronically reconfigurable (adaptronic) machines based on ER fluids is intensely multidisciplinary and highly nonlinear in terms of analysis, furthermore the limits of operation cannot be graphically represented, such is the degree of interaction between fluid design, motion and machine. This section cannot give comprehensive cover to the interface problems that exist; rather, it lists the more apparent and important factors. Having done this, it is hoped that the targets to suit both fluid developers and applications engineers are set more effectively than hitherto. Specific CFD studies are required for pre prototype comparisons.

Finally a word of caution: a $\tau, \dot{\gamma}$-characteristic for an ER fluid will not give exactly the same shape of torque, speed, pressure, or flowrate-curves for an ER device, and viscometers should be designed and operated so that the fluid rather than the device characteristic is measured [101,117]. Other areas of ER fluid development requiring specialist attention include lubrication; hysteresis, stabilization and the exclusion of impurities.

6.5.4 Summary of Advantages of Particulate ER Fluids

The general comparative advantages of particulate ER fluids in relation to other electrostructured fluids are:

- speed of action: the achievement of full yield stress occurs virtually when voltage is applied;
- heavy ferrous components are not required. Acceleration can be rapid;
- the powerpack that drives the electrodes can be remote from the controller, and its size is not usually a problem;
- steady-state currents can be low, albeit that they are provided at high voltages.

6.5.5 Homogenous ERF

Despite the attention paid to the particulate or dispersion type ERF in the preceding sections (mainly on account of its fast response time and potential for industrial adaptronics), it is the liquid crystal (LC) polymer type that has provided the first commercial application of ERF.

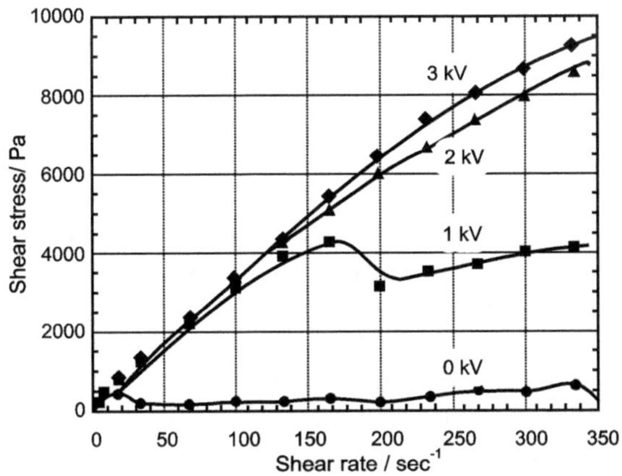

Fig. 6.75. Shear stress versus shear rate for ERF at different levels of applied excitation – Asahi homogenous fluid

The Asahi-Castor walker [126] has been devised for patients with walking problems. Clutches similar to those in Fig. 6.73 are filled with a grease like LC polysiloxzine plus dilutant, the characteristics of which are typified in Fig. 6.75. Since no particles are present and the zero volt viscosity is large ($\approx$10 Pa s) there is little sedimentation and good stability provided the mixture does not crystalise at low temperatures ($\approx$10 °C).

The clutches are fitted to the rear wheels of a zimmer type of frame and, because of the low current demand (1 μA/cm^2 of electrode area) can adapt to patient stumbling or run away down a slope or, enable the assembly to act as a trainer – self adaptable to the weight of the patient. Also, the sensors can pick up any irregular movement in gait to procure a safe situation.

About 2 kV/mm is required to produce 8 kPa of shear stress at a few hundred s^{-1} shear rate and a two wheel braking torque of 16 Nm. The draw back to further adaptronic application is the 20 to 80 ms time constant of shear stress to voltage and possibly the Wiesenberg effect arising due to rotation of the polymer at high shear rate.

The brakes are about 10 cm diameter ×1.5 cm long with a control current at 200 μA via a CPU, supplied from a 6 V battery. The fluid is not abrasive.

Calculations for homogenous fluids follow the same pattern as for particulate fluids but the $\tau = f(E, \dot{\gamma})$ and $\tau_e = 0$.

6.5.6 Other ER Fluids

Several further methods of achieving ER effects have come to light but have not yet been comprehensively investigated.

Immiscible liquid-liquid suspensions, like liquid crystals, do not exhibit a yield effect but change the limit and slope of viscosity with electric field strength. In this case the suspended droplets extend in the field direction like the rod molecules in an LC. It is possible to achieve both positive and negative ER effects in both.

In fibre type ER situations high shear stress modulation under field derive from cellulose fibres trailing from a woven material which covers one electrode. Field application causes the fibres to attach to a plain material which covers the surface of the other electrode.

Some particulate fluids are affected by both electrical and magnetic fields with a high degree of synergy arising. Dependant on the relative direction of the fields a range of characteristics can be produced.

A summary of these fluids can be found in [127] with occasional attempts at application appearing in the regular international conference(s) on ER Fluids and MR Suspension proceedings, from which more details of the specific ER fluid engineering experiences of relevance to adaptronics may be found, see also [128].

6.6 Magnetorheological Fluid Actuators

J.D. Carlson

Magnetorheological or MR fluids are materials that respond to an applied magnetic field with a dramatic change in their rheological behavior [129]. They are magnetic analogues to electrorheological fluids (see Sect. 6.5). The essential characteristic of MR fluids is their ability to reversibly change from a free-flowing liquid to a semi-solid having controllable yield strength in milliseconds when exposed to a magnetic field. In the absence of an applied magnetic field, MR fluids are generally well modeled as Newtonian liquids characterized by their viscosity. When a magnetic field is applied, a simple Bingham-plastic model is effective at describing their essential field-dependent fluid characteristic [130]. In this model, the total yield stress τ_{total} is given by

$$\tau_{\text{total}} = \tau_{\text{MR}}(H)\text{sgn}\dot{\gamma} + \eta_{\text{p}}\dot{\gamma} \,. \tag{6.32}$$

Here, $\tau_{\text{MR}}(H)$ is the yield stress caused by the applied magnetic field H, $\dot{\gamma}$ is the shear rate and η_{p} is the field-independent plastic viscosity defined as the slope of the measured shear stress against the shear strain rate.

Magnetorheological fluids are non-colloidal suspensions of micron-sized, paramagnetic or soft ferromagnetic particles. Virtually all practical MR fluids consist of elemental iron particles that are a few microns in diameter and suspended in a carrier liquid. Magnetorheological fluids should not be confused with colloidal ferrofluids in which the particles are about one thousand times smaller than those found in typical MR fluids. Like ER fluids, MR fluids have an early history that dates from the late 1940s. Beginning in the early 1990s a resurgence of interest in MR fluids and applications emerged in response to many of the practical limitations encountered with ER fluids [131, 132].

Magnetorheological fluids offered substantially higher yield strength plus the ability to operate at higher and lower temperatures. Most importantly, high voltages were not required to provide the necessary magnetic fields required to activate MR fluids. Common, low-voltage power supplies, e. g. 12 volt systems, could directly power the electromagnets in MR fluid devices. Scientists and engineers at several organizations, including TRW, QED and Lord, demonstrated that practical MR fluids and devices could be made which actually could achieve many of the unrealized hopes for ER fluids [133–139].

The initial discovery and development of MR fluids and devices can be credited to Jacob Rabinow at the U.S. National Bureau of Standards in the 1940s [140–142]. This work was almost concurrent with Willis Winslows pioneering work on ER fluids. Today, MR fluid technology has progressed to the point where it is routinely used on a commercial scale to provide semi-active control in a variety of automotive and industrial applications. A number of these applications are described later in this section. The long sought goal of mass-produced, controllable fluid automotive shock absorber systems was finally realized in early 2002 with the introduction of the Magne-Ride suspension system as standard equipment on the Cadillac Seville with MR fluid made by Lord Corporation and shock absorbers and struts made by Delphi [139, 143].

Magnetorheological fluid production levels in 2005 are of the order of hundreds of metric tons per year (or tens of thousands of liters) such that commercial applications on several automotive platforms are supported. A factor of ten or more increase in volume over the next decade is anticipated. It is estimated that there are presently more than one hundred thousand MR dampers, shock absorbers, brakes and clutches in use worldwide. This number is expected to rise into the millions as more automotive platforms adopt smart MR fluid suspensions and clutch systems.

6.6.1 Description of MR Fluids

A typical magnetorheological fluid consists of 20 . . . 40% by volume of relatively pure, elemental iron particles suspended in a carrier liquid such as mineral oil, synthetic oil, water and/or glycol. A variety of proprietary additives, similar to those found in commercial lubricants, that inhibit gravitational settling and promote particle suspension, enhance lubricity, modify viscosity, and inhibit wear are commonly added.

The ultimate strength of MR fluid depends on the square of the saturation magnetization of the suspended particle [144–146]. The key to a strong MR fluid is to choose a particle with a large saturation magnetization. Ideally, the best available particles are alloys of iron and cobalt known as permendur, which have saturation magnetizations of about 2.4 Tesla [144, 147]. Unfortunately, due to their high cobalt content such alloys are prohibitively expensive for all but the most exotic applications. The best practical particles are pure elemental iron with a saturation magnetization of 2.15 Tesla. Virtually all

Table 6.8. Typical magnetorheological fluid properties [courtesy of Lord Corporation]

Property	Normal Range
Particle volume fraction, Φ	0.20 to 0.45
Particle weight fraction	0.70 to 0.90
Density	2 to 4 g/cm^3
Yield strength, τ_{MR} @ 100 kA/m	10 to 55 kPa
Yield strength, τ_{MR} @ saturation	25 to 100 kPa
Plastic viscosity, η_{P} @ 40°C, $\dot{\gamma} > 500\,\mathrm{s}^{-1}$	50 to 200 mPa·s
Temperature range	−40 to +150°C
Magnetic permeability, relative @ low field	3.5 to 10
Fig. of merit, $\tau_{\mathrm{MR}}^2/\eta_{\mathrm{P}}$	10^{10} to 10^{11} Pa/s
Response time	<0.001 s

other ferromagnetic metals, alloys and oxides have saturation magnetizations significantly lower than that of iron, resulting in substantially weaker MR fluids. The most widely used form of iron particles for MR fluids is a material called carbonyl iron. This is the common name given to iron particles that are formed from the thermal decomposition of iron pentacarbonyl. The resulting particles are highly spherical in shape with sizes in the 1 to 10 micron range with an elemental iron content >98%.

Depending on the volume fraction of iron particles, MR fluids can have maximum yield strengths ranging from 30 to 80 kPa for an applied magnetic field of 150... 250 kA/m. Magnetorheological fluids are not highly sensitive to contaminants or impurities such as are commonly encountered during manufacture and usage. As the magnetic particle polarization mechanism is not affected by surfactants and additives, it is relatively straightforward to stabilize MR fluids against gravitational separation of the particles in spite of the large density mismatch. Antiwear and lubricity additives can also be included in the formulation without affecting strength and power requirements. A listing of typical MR fluid properties is given in Table 6.8.

6.6.2 Advantages and Concerns

Interest in MR fluids stems from the benefits they enable in mechatronic systems. Much of the current interest in MR fluids can be traced directly to the need for a simple, robust, fast-acting valve necessary to enable semi-active vibration control systems [148–150]. Such a valve was the holy grail of semi-active vibration-control technology for nearly two decades beginning in the mid 1970s. Magnetorheological fluid technology has proven to be enabling technology for such semi-active systems.

The primary advantage of MR fluids stems from the large, controlled yield stress they can achieve. Typically, the maximum yield stress of an MR fluid is an order of magnitude or more greater than the best ER fluids, while their viscosities are comparable. This has a very important ramification for ultimate device size. As discussed in Sect. 6.6.4, the minimum amount of active fluid in a controllable fluid device is proportional to the plastic viscosity and inversely proportional to the square of the maximum field-induced yield stress. This means that for comparable mechanical performance the amount of active fluid needed in an MR fluid device will be about two orders of magnitude smaller than that for an ER device. From a more fundamental physics perspective, the large strength of MR fluid is related to the very high magnetic-energy density that can be established in the fluid before complete magnetic saturation of the particles occurs. For a typical iron-based MR fluid, this is of the order of $0.1\,\mathrm{J/cm^3}$. Electrorheological fluids, however, are limited not by polarization saturation but by dielectric breakdown. This limits the maximum field strength and consequently the maximum energy density that can be established in an ER fluid to about $0.001\,\mathrm{J/cm^3}$. For comparable device performance, MR and ER devices need to control the same magnitude of total field energy. Hence, the smaller amount of active fluid needed for MR.

From a more practical perspective, a key advantage of MR fluids is the form of electric power needed to create the magnetic field. While the total electric power for comparable performing MR and ER devices are approximately equal [131,132], the advantage of MR lies in the fact that they can be powered directly from common, low-voltage sources such as batteries, 12 volt automotive supplies, or inexpensive AC to DC converters. High-voltages are not required. Standard low-cost electrical connectors, wires, and feedthroughs can be reliably used, even in mechanically aggressive and dirty environments, without fear of dielectric breakdown. This is particularly important in cost-sensitive applications such as automobile suspension systems and domestic appliances such as washing machines.

Another important advantage of MR fluids is their relative insensitivity to temperature changes and contamination. This arises from the fact that the magnetic polarization of the particles is not influenced by the presence or movement of ions or electric charges near or on the surface of the particles. Surfactants and additives that affect the electrochemistry of the fluid do not play a role in the magnetic polarizability of the particles. Further, bubbles or voids in the fluid can never cause a catastrophic dielectric breakdown in an MR fluid.

A concern that is often expressed about MR fluids is the possibility of gravitational settling of the dense iron particles. While particulate settling is indeed a phenomenon that can occur, it can be controlled and has not been a barrier to the successful commercial application of MR fluids. As early as 1950 Jacob Rabinow pointed out that complete suspension stability was not necessary for most MR fluid devices [151]. Most MR fluid devices such as dampers and shock absorbers are highly efficient mixing devices. As long as the MR fluid does not settle into a hard sediment, normal motion of the device is adequate to cause sufficient flow to remix any stratified MR fluid back to a homogeneous state. For a small MR fluid damper such as the Lord Motion Master RD-1005-3 [152], two or three strokes of a damper that has sat motionless for several months are sufficient to return it to a completely remixed condition. Testing of automotive MR fluid shock absorbers made by Delphi Corporation has shown that with as little as one stroke these devices will return to their original condition even after one year of settling [153]. For special cases, such as dampers designed for seismic damage mitigation in civil engineering structures and devices used to absorb energy during a crash incident of an automobile, MR fluids can be formulated to remain homogeneous indefinitely. In these instances, additives are included in the fluid formulation that convert them into shear-thinning, thixotropic gels.

MR fluids have the potential to be abrasive. In fact, one application of a special class of MR fluids is as a polishing media for optical components. These MR fluids are actually formulated with abrasive additives such as cerium oxide powder that allows them to efficiently remove surface material from glass optics under the control of a magnetic field. For most MR fluid devices, wear or abrasion of components is not desirable and the MR fluids are formulated to minimize such. The choice of the specific iron particles is important in this regard. High-purity, soft-iron particles are less aggressive than non-reduced, hard varieties. Proprietary additives similar to those used in lubricating oils are also effective at mitigating wear. Of particular importance is wear of the dynamic elastomeric shaft seals that are necessary in all shock absorbers and dampers. It is important to insure that the surface finish the shafts in these devices is fine enough to ensure that no particles become stuck in surface imperfections where they cannot be scraped off by the seal. If the particles are subsequently carried through the seal line they can act like a rasp and rapidly degrade the effectiveness of the seal. The surface finish of shaft used in MR fluid dampers is typically specified to be much finer that the minimum particle size of the MR fluid [154]. If care is taken in this regard it is possible to have dynamic devices that will sustain tens of millions of cycles or more and many hundreds of kilometers of cumulative seal travel.

Centrifugal effects are a concern in high-speed rotary applications. For brakes in which the housing is stationary, centrifugation is generally less of a concern because of the continual shear induced remixing. Centrifugation is much more of a concern for high-speed clutches. In general, drum geometries in which the entire MR fluid gap is at the same diameter as opposed to disk geometries, are preferred for mitigating centrifugal effects. Well-designed MR fluid clutches can be operated at speeds of 5000 rpm or more [155].

Depending on the conditions of the specific application, all MR fluids will eventually show some degree of deterioration. Such deterioration is usually manifested as a thickening of the fluid often referred to as 'in-use-thickening' or IUT. In general, IUT manifests itself as a progressive increase in the off-state viscosity of the fluid. While the amount and rate of thickening will depend on shear rate and temperature, the most important factor seems to be the specific amount of mechanical energy that is converted to heat in the MR fluid. An ad hoc measure that has proven useful in estimating the expected life of a MR fluid in a particular application is the lifetime dissipated energy or LDE [156] defined in (6.33):

$$\mathrm{LDE} = \frac{1}{V}\int_0^{\mathrm{life}} P\mathrm{d}t\ , \tag{6.33}$$

where P is the instantaneous mechanical power converted to heat in the MR device and V is the total volume of MR fluid in the device. The lifetime dissipated energy is simply the total mechanical energy converted to heat per unit volume of MR fluid over the life of a device. The best MR fluids today can sustain a LDE on the order of $10^7\,\mathrm{J/cm^3}$ before they thicken to the point where device performance is compromised. Poor MR fluids, on the other hand, may become unusable with LDEs as low as $10^5\,\mathrm{J/cm^3}$. Today, good MR fluids are capable of lasting hundreds of thousands of kilometers in automotive shock absorbers.

6.6.3 MR Fluid Devices

Virtually all devices that use controllable MR fluids operate in a valve-mode, direct-shear mode, or a combination of these two modes. Diagrams of the basic valve and direct-shear modes are shown in Fig. 6.76. Examples of valve-mode devices include dampers, and shock absorbers. Examples of direct shear-mode devices include clutches, brakes, chucking and locking devices, and some dampers.

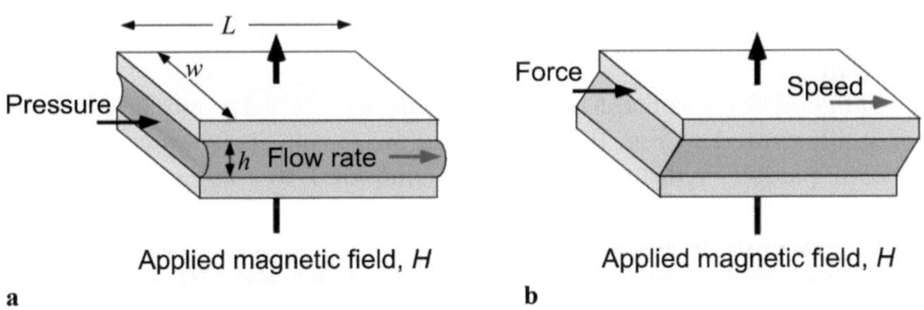

Fig. 6.76. Two modes of MR fluid operation: **a** valve-mode, **b** direct-shear mode

Valve-Mode

The pressure drop developed by a valve-mode device can be divided into two components, the pressure ΔP_η due to the fluid viscosity and ΔP_{MR} due to the magnetic field-induced yield. These pressures may be approximated by [131,157,158]:

$$\Delta P_\eta = \frac{12\eta_\mathrm{p} QL}{h^3 w} \tag{6.34}$$

$$\Delta P_{\mathrm{MR}} = \frac{c\tau_{\mathrm{MR}}(H)L}{h}, \tag{6.35}$$

where Q is the volumetric flow rate. The parameter c has a value that ranges from a minimum value of 2 for $\Delta P_{\mathrm{MR}}/\Delta P_\eta$ less than $\approx$1 to a maximum value of 3 for $\Delta P_{\mathrm{MR}}/\Delta P_\eta$ greater than $\approx$100. The total pressure drop in a valve-mode device is approximately equal to the sum of ΔP_η and ΔP_{MR}. The force developed by a valve-mode damper will thus be the total pressure multiplied by the effective piston area.

An example of a simple valve-mode device is the RD-1005-3 linear damper by Lord Corporation shown in Fig. 6.77 [152]. As in the vast majority of all commercial MR fluid dampers, these dampers have an internal, axi-symmetric valve with an annular flow path. In this case the damper is a single-ended, mono-tube style having an internal rod volume accumulator pressurized with nitrogen. As indicated in the Fig. 6.77, downward motion of the piston causes MR fluid to flow up through the annular flow channel. Application of current to the coil creates a magnetic field that interacts with the MR fluid in two regions where the magnetic flux crosses the flow channel.

The damper body is 41.4 mm in diameter and 144 mm long. Maximum allowable travel of the piston is 53 mm. The MR fluid valve and associated magnetic circuit is fully contained within the piston. Current is fed to the electromagnetic coil via the leads through the hollow shaft. Input power of 5 W is required to operate the damper at its nominal design current of 1 A. Although the damper contains about 70 cm^3 of MR fluid, the actual amount of fluid that is activated in the magnetic field at any given instant is only about 0.4 cm^3.

The range of force control that is possible with a valve-mode MR fluid damper is illustrated in Fig. 6.78. Here the force/velocity character that is typical of a passive hydraulic damper is compared to the range of forces possible with a MR damper. With appropriate control based on displacement, velocity or acceleration, any force profile between the upper and lower bounds can be realized. Unlike passive viscous dampers, with the MR damper it is easy to achieve large force at very low speed.

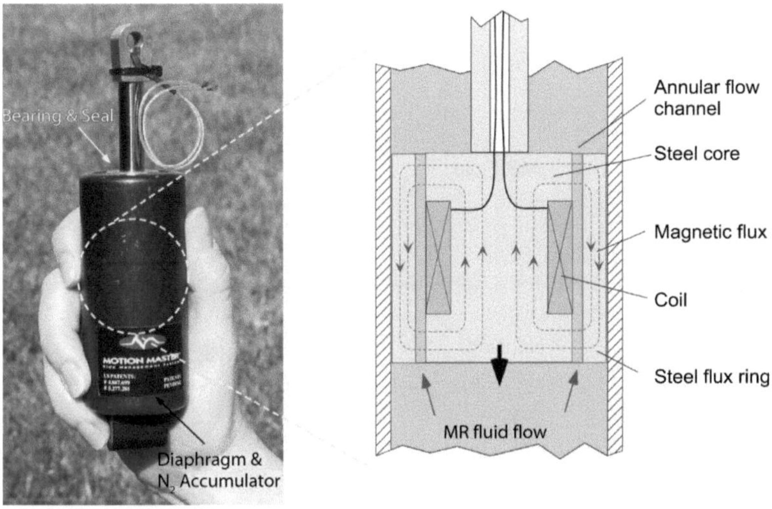

Fig. 6.77. Basic MR fluid damper with axi-symmetric valve geometry

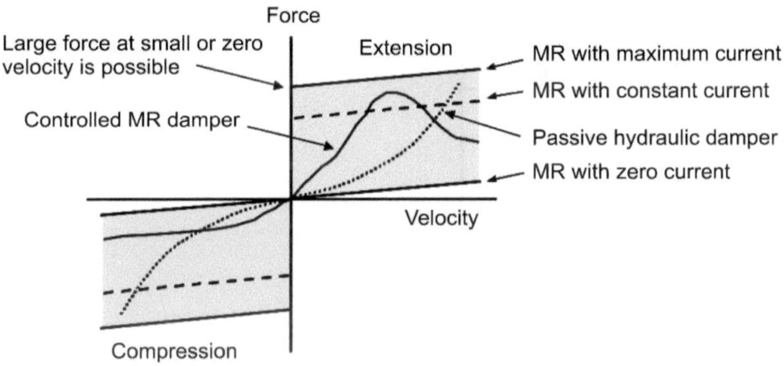

Fig. 6.78. Controllable force range possible with MR fluid damper

Direct-Shear Mode

In a similar fashion, the force developed by a direct-shear device can be divided into F_η the force due to the viscous drag of the fluid and F_{MR} the force due to magnetic field induced shear stress:

$$F_\eta = \frac{\eta_{\mathrm{p}} v_{\mathrm{S}} L w}{h} \tag{6.36}$$

$$F_{\mathrm{MR}} = \tau_{\mathrm{MR}}(H) L w \,, \tag{6.37}$$

where v_{S} is the relative velocity. The total force developed by the direct-shear device is the sum of F_η and F_{MR}.

An example of a simple, direct-shear device is shown in Fig. 6.79. In this brake MR fluid is located between the faces of the disc-shaped rotor and the

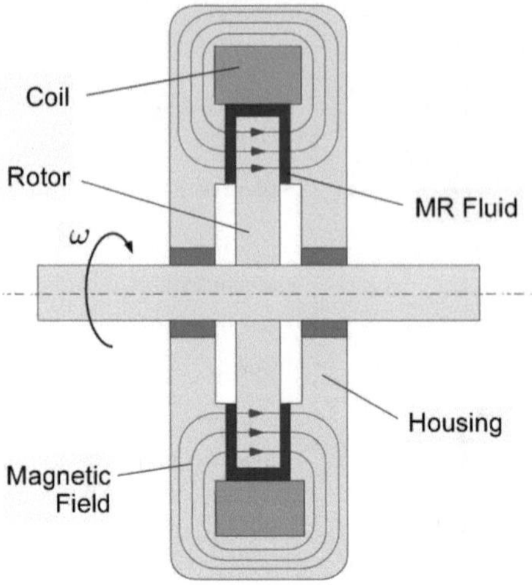

Fig. 6.79. Simple MR fluid direct-shear rotary brake with disc geometry

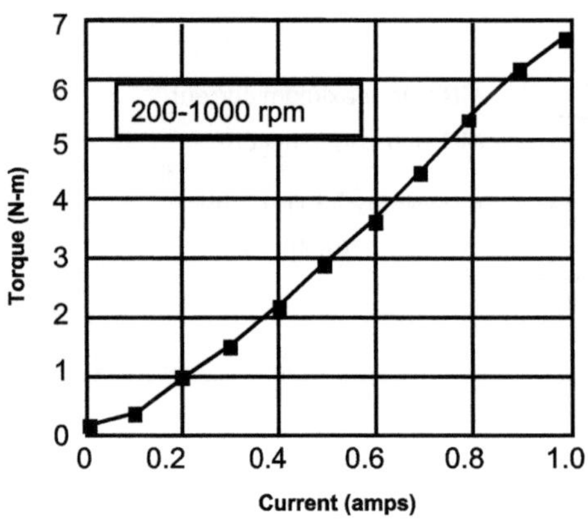

Fig. 6.80. Typical braking torque versus current for direct-shear brake

stationary housing. Rotation of the shaft causes the MR fluid to be directly sheared as the rotor moves relative to the housing. A coil fixed in the housing produces a toroidal shaped magnetic field that interacts with the MR fluid in the fluid gaps on each side of the rotor. Torque versus current for the small MRB-2107 brake by Lord Corporation is shown in Fig. 6.80.

6.6.4 Basic MR Device Design Considerations

Measured on-state yield strength τ_{MR} and flux density B versus magnetic field intensity H for several standard MR fluids from Lord Corporation are given in Fig. 6.81 and 6.82. Also shown in these Figures are a series of predicted

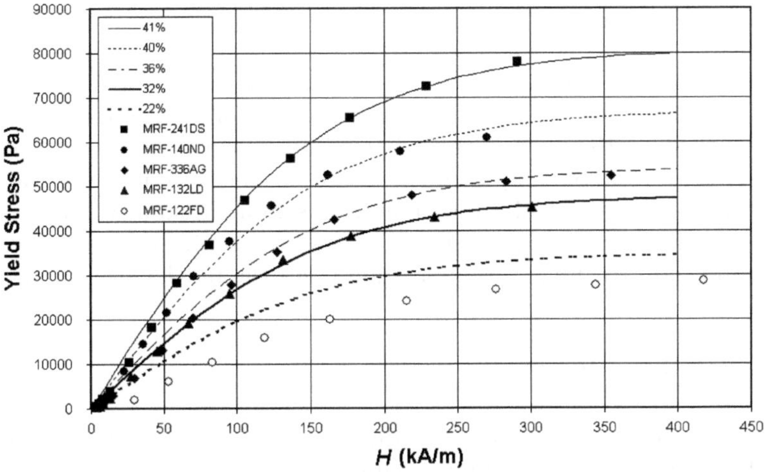

Fig. 6.81. Measured and predicted yield strength versus H for typical MR fluids

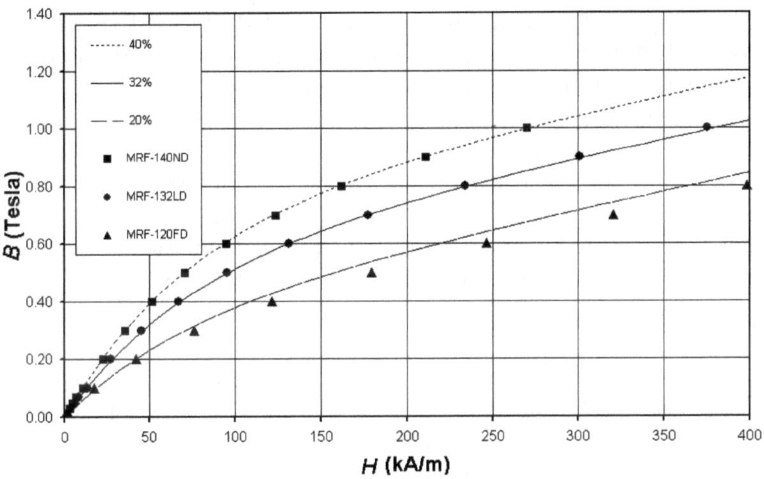

Fig. 6.82. Measured and predicted B versus H for several MR fluids

curves based on empirical equations [159]:

$$\tau_{\mathrm{MR}} = C\,271700\,\Phi^{1.5239}\tanh(6.33\cdot 10^{-6}H) \tag{6.38}$$

$$B = 1.91\Phi^{1.133}[1-\exp(-10.97(\mathrm{m}^2/\mathrm{Vs})\mu_0 H)] + \mu_0 H\ , \tag{6.39}$$

where Φ is the volume fraction of iron particles, τ_{MR} is in Pa, H is in A/m, μ_0 is the magnetic constant equal to $4\pi\cdot 10^{-7}\,\mathrm{Vs/Am}$ and the constant C equals 1.0, 1.16 or 0.95 depending on whether the carrier fluid is hydrocarbon oil, water or silicone oil. These equations have been developed to provide a practical and convenient description of any MR fluid.

MR Device Size and Feasibility

The equations describing the on-state and off-state pressures or forces in MR fluid devices can be combined into a simple expression for the minimum active fluid volume, i. e. the volume of fluid acted upon by the magnetic field in a MR fluid valve [131]. Such an expression is useful because it allows one to estimate the necessary size of a device and determine feasibility prior to developing a detailed engineering design. For many of the most widely used standard commercial MR fluids this expression takes the particularly simple form [159]:

$$V_{\text{min}} = \alpha \left(\frac{F_{\text{on}}}{F_{\text{off}}} \right) F_{\text{on}} v 10^{-10} \,. \tag{6.40}$$

In this expression, forces are in N (or torques in Nm), speed v in m/s (or rad/s) and V_{min} in m^3. The constant α equals 1 for direct-shear devices, while for valve-mode devices it has a value of approximately 2. This approximation is valid for any MR fluid having $\tau^2(H)/\eta$ that is on the order of 10^{10} Pa/s. Examples of such fluids are Lord MRF-122ES, MRF-132AD and MRF-336AG [160–162]. The minimum active fluid volume estimated by (6.40) is generally accurate to within about a factor of two.

For valve-mode devices the estimated minimum active fluid volume of (6.40) can be used to make a further estimate of the overall size of the MR fluid valve. Based on experience with a wide spectrum of MR fluid devices ranging from tiny laboratory dampers to very large dampers for seismic damage mitigation, the overall size of a well-designed and magnetically efficient MR fluid valve is 25 to 50 times the minimum active fluid volume [159]. Thus:

$$V_{\text{valve}} \approx (25 \ldots 50) V_{\text{min}} \tag{6.41}$$

where V_{valve} comprises all the materials that make up the valve and magnetic circuit including active MR fluid, copper coil windings and steel poles and magnetic flux conduits. For a well-designed MR fluid damper having a valve in the piston, V_{valve} is essentially the total volume of the damper piston. Thus, without having an a priori detailed knowledge of the device geometry one is still able to estimate the overall size of the MR valve and make an initial determination of feasibly.

Based on the above minimum active fluid volume, it is also possible to estimate the electric power required to power the electromagnet. For MR fluid that is operating near its maximum yield strength, the magnetic field energy density that needs to be established in the fluid is approximately 0.1 J/cm^3. Thus, in order to establish the required magnetic field H within a desired time interval Δt, the power source must be capable of supplying a minimum electric power P_{el} (in watts) that equals 0.1 J/cm^3 times the fluid volume V_{min} (in cm^3) divided by the time interval Δt (in seconds):

$$P_{\text{el}} = \frac{0.1 \, V_{\text{min}}}{\Delta t} \,. \tag{6.42}$$

Thus, for any application, the minimum information needed to estimate active fluid volume, minimum electric power and overall valve size is:

- F_{on}: minimum on-state force or torque needed (N or Nm).
- F_{off}: maximum off-state force or torque that may be tolerated (N or Nm).
- v: maximum speed or angular velocity for F_{off} (m/s or rad/s).
- Δt: desired switching speed (seconds).

Response Time

The speed of an MR fluid device is largely determined by factors extrinsic to the MR fluid, particularly the inductance of the MR device and the characteristics of the current source (amplifier). Recently, some experimental time-response data on practical MR fluids has become available. Goncalves has made measurements of the response of MR fluids as a function of fluid dwell-time in a well-defined MR fluid valve [163]. Based on the observed roll-off in MR response as dwell-time in the magnetic field becomes very small, one can conclude that the response time of an MR fluid is much less than one millisecond. Experimental, transient response-time measurements on the RD-1005-3 damper have shown that the damper can reach rheological equilibrium within approximately 6 ms after a step voltage input to the current driver [164]. This same damper driven by a current amplifier having an even higher voltage compliance can be switched from off to on in less than 2 ms. The response time for most practical MR devices is controlled by the time it takes for the current source to establish the magnetic field in the fluid, i. e. how fast the power supply can deliver the necessary energy into the magnetic field. The key factors will thus be the resistance and inductance of the electromagnet, eddy currents in the surrounding ferrous materials and the output characteristics of the current amplifier, particularly its ability to over-voltage the inductor in order to raise the current more quickly.

Complete MR Device Design

Creation of an efficient, high-performance MR fluid device requires simultaneous consideration of many inter-related and highly coupled factors. These include: specific MR fluid properties; size, weight and shape constraints; required forces or torques (on-state and off-state); nonlinear magnetic properties and magnetic saturation; an efficient electromagnet including fringing, and boundary loss considerations; fluid dynamics including dynamic pressures and Reynolds number; electrical constraints such as voltage, current and inductance limits; durability of seals, fluid and bearings; thermal expansion; and, ultimately, manufacturability and cost. Optimization of device parameters to achieve high on-state and low off-state with a compact, low-power, fast-response electromagnet can be challenging.

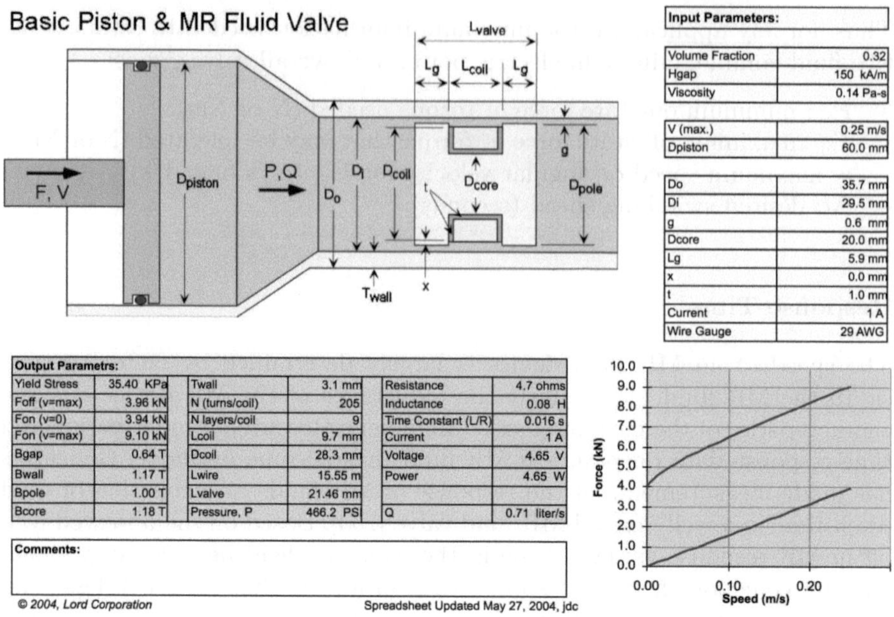

Fig. 6.83. Simple design spreadsheet for axi-symmetric MR fluid valve

One approach is to solve the inverse problem wherein the optimum MR fluid valve geometry, magnetic field and MR fluid properties that will result in desired on- and off-state forces are determined. Inverse problems are, however, extremely difficult to solve. In contrast, the direct problem wherein resultant on-state and off-state forces for a given valve geometry, magnetic field and specific MR fluid are calculated is straightforward. It is relatively simple to explicitly and simultaneously take into account the nonlinear magnetic properties of MR fluid and associated ferrous elements, the nonlinear dependence of MR yield strength on magnetic field, and the vastly different functional dependence of on-state and off-state pressure on MR valve geometry. Solutions to the direct problem are readily amenable to spreadsheet calculation such as Microsofts EXCEL. Beginning with a set of starting parameters, one calculates resultant device performance and then, with a modicum of experience, adjusts the input parameters to achieve desired performance while meeting all of the necessary geometric and electrical constraints. Such MR fluid design spreadsheets can be quite accurate in their predictions. An example of such a simple spreadsheet tool for a basic axi-symmetric MR fluid valve is shown in Fig. 6.83 [159].

6.6.5 Examples of MR Devices and Systems

MR fluids have been used commercially since the mid-1990s. The first application was a small controllable MR fluid brake in aerobic exercise equipment manufactured by Nautilus [165]. In retrospect, this was not a particularly good application for MR fluid owing to the inherent fickleness of the exercise equipment market and the extreme use to which some exercise equipment can be subjected. However, it did demonstrate the efficacy of MR fluids for providing real-time control in mechanical systems. In 1998, a small, real-time controlled MR fluid damper system (the RD-1005-3 described above) was introduced commercially into the heavy-duty truck and off-highway vehicle market for suspended seat applications [152]. That same year, a controllable MR fluid based primary suspension shock absorber for NASCAR race-vehicles was introduced by Carrera [166].

Auto Primary Suspensions

Today, the greatest driving force behind MR fluid technology is automotive, particularly real-time controlled primary suspensions systems. In January 2002, the Cadillac Seville automobile, shown in Fig. 6.84, was introduced by General Motors with a MagneRide™ suspension system having real-time controllable MR fluid shock absorbers and struts as standard equipment [167,168]. The Magneride™ shock absorbers are made by Delphi Corporation with the MR fluid being made by Lord Corporation. Similar, controllable MR fluid-based suspension systems have since become available on numerous other vehicle models including: Corvette sports car [169], Cadillac SRX roadster, Cadillac XLR sport utility vehicle [170,171], Cadillac STS sedan, Cadillac DTS [172] and Buick Lucerne [173]. All of these systems are

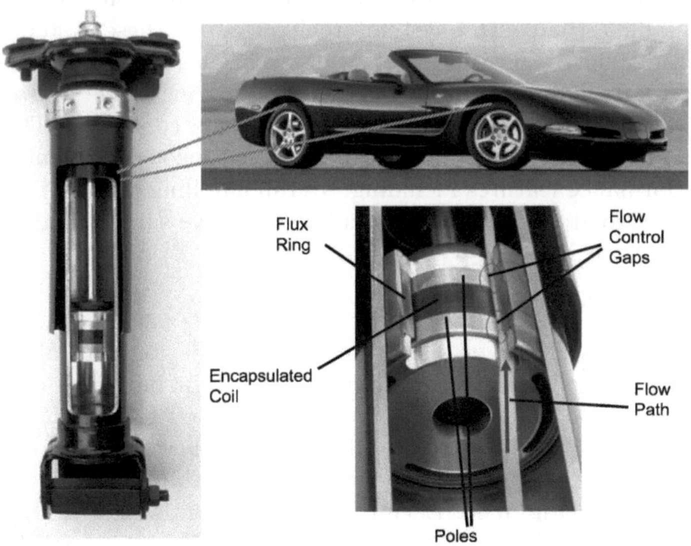

Fig. 6.84. Detail of MR fluid shock absorbers on Corvette sports car

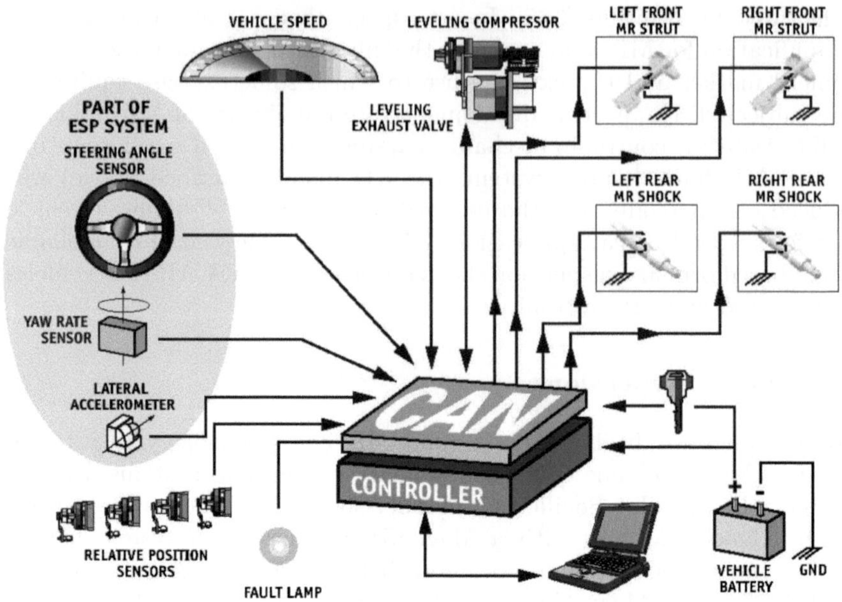

Fig. 6.85. Control system architecture for automotive MR fluid shock absorbers

based on monotube shock absorbers that have a single-stage, axi-symmetric MR valve contained within the piston.

The MR fluid-based suspension systems implemented on these various vehicles enable simultaneous ride comfort control and body motion control. As indicated in Fig. 6.85, the control system architecture for these systems processes inputs from relative position sensors at each wheel. In addition, inputs from a lateral accelerometer, yaw rate sensor, steering angle sensor and speed sensor all feed by way of a CAN BUS into the controller. The control algorithms are quite complex and seek to simultaneously optimize a wide range of performance features including: overall handling, overall ride comfort, body control, road noise, head toss and a subjective safe feeling.

Civil Engineering Structures

Magnetorheological fluid technology offers unique solutions for control of vibration and motion caused by wind or seismic activity in buildings and bridges. Magnetorheological fluid dampers are readily scaled to very large sizes that can provide controllable forces appropriate for large civil structures. Several large MR fluid dampers capable of controllable forces up to about 200 kN are shown in Fig. 6.86. Each of these dampers weighs 280 kg and contains 15 liters of MR. While the electromagnetic coils are located in the pistons, the heavy-walled, steel damper housing provides the magnetic flux return path as indicated in Fig. 6.87.

Fig. 6.86. 200 kN MR dampers for controlling seismic motions in buildings

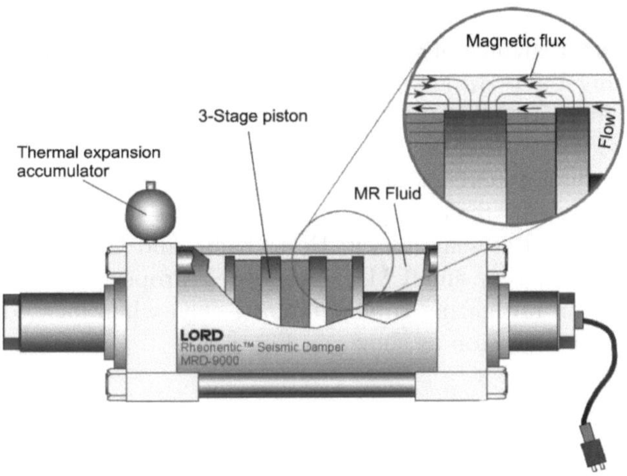

Fig. 6.87. Schematic of MR fluid seismic damper

Japans new National Museum of Emerging Science and Innovation in Tokyo (Nihon-Kagaku-Miraikan) has been constructed with an earthquake control system that includes 300 kN MR fluid dampers located within the structural framework as shown in Fig. 6.88. In this instance the MR dampers have an external bypass valve outside of the main damper body [174]. The MR dampers also form part of the museums exhibits. In the event of an earthquake, the dampers, drawing power from batteries, would sense the amount of energy affecting the building and then respond to dissipate energy before it reaches destructive levels.

In another civil engineering application, MR fluid dampers have been used to mitigate potentially damaging wind-induced cable vibrations in a cable-

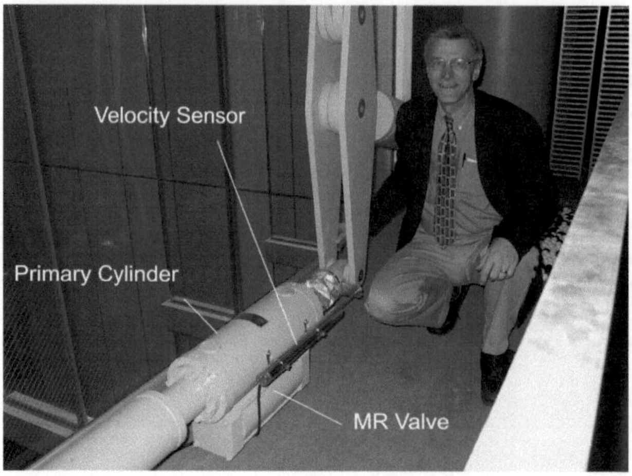

Fig. 6.88. MR dampers for seismic damage mitigation in National Museum for Emerging Science and Innovation (Nihon-Kagaku-Miraikan) in Tokyo

stayed bridge in the Hunan Province of China [175, 176]. The MR dampers installed on the Dong Ting Lake Bridge are basically the Lord RD-1005-3 damper as described earlier. To preserve the graceful architecture of the bridge, dampers must be located near the bottom end of the cable, typically at a distance of no more than 1 percent or 2 percent of the cables overall length from the anchor points. At this location normal passive dampers have limited effectiveness. In contrast, researchers in Hong Kong and Changsha, China have demonstrated that very small MR dampers, if properly tuned, can have a profound effect on mitigating cable galloping even when located very close to the cable anchor location as shown in Fig. 6.89.

Fig. 6.89. MR fluid dampers used control wind-induced cable vibrations on Dong ting Lake cable-stayed bridge in central China

Fig. 6.90. Haptic feedback for steer-by-wire systems

Steer-By-Wire

The trend in vehicle industries toward control-by-wire (steer-by-wire, shift-by-wire, throttle-by-wire, brake-by-wire, etc.) has created a need for highly controllable, rugged, cost-effective haptic devices to provide realistic force-feedback sensations to the operator, whether the manual device is a wheel, a joystick, a pedal, or a lever. British forklift manufacturer Linde uses MR brakes to control over-steer in their R14 industrial forklift [177]. The R14 vehicle, shown in Fig. 6.90, is an all-electric forklift intended for close maneuvering and manipulation in confined, clean-spaces such as food handling warehouses with large drive-in freezers. There is no mechanical connection between the steering wheel and the ground wheels. Steering is accomplished entirely by electrical control. Rotation of the steering wheel turns an optical encoder, which supplies an electrical signal that is transmitted to the drive ground wheel and causes a motor to orient them in the desired direction. The steering wheel and the optical encoder are both mounted to the shaft of a MR brake. The brake provides a variable amount of rotational resistance depending on the instantaneous vehicular motion and orientation of the ground wheels. Such tactile feedback to the operator is necessary to insure stable operation. The MR brake and magnetic rotary encoder are packaged into a common package as shown in Fig. 6.90 and mount directly to the dashboard of the forklift.

Smart Prosthetic Knee

As a final example of a MR fluid controlled adaptronic system, the smart prosthesis knee developed by Biedermann Motech GmbH [178–181] is presented. This system shown in Fig. 6.91 is a complete artificial knee that automatically adapts and responds in real-time to changing conditions to provide the most natural gait possible for above-knee amputees. The heart of this system is a small magnetorheological fluid damper that is used to semi-actively control the motion of the knee based on inputs from a group

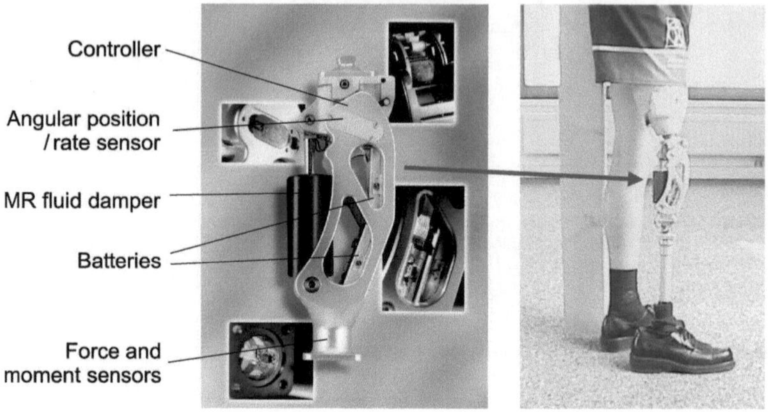

Fig. 6.91. Above-knee prosthesis with real-time control provided MR fluid damper

of sensors located in the prosthesis. The damper is a modification of the RD-1005-3 damper described above. An embedded microprocessor controller interprets input signals (axial force, bending moment, knee-angle and speed) to determine what the person is attempting, e. g. walk fast, walk slow, navigate a slope or navigate stairs. The controller then adjusts the current to the MR damper to provide more or less damping such that the actual gait profile matches an ideal profile stored in memory. The benefit of such an artificial knee is a more natural gait that automatically adapts to changing gait conditions, i. e. walking speed, inclination of the terrain, presence of stairs, weight of footwear, etc. The basic arrangement of the control unit is shown in Fig. 6.92. Details of the damper control algorithm are shown in Fig. 6.93.

In operation, the control of the leg prosthesis works as follows. Measured data from the sensors for knee angle and force are transferred to the control unit. The control unit produces a time varying current to the MR fluid damper as a function of the instantaneous gait condition. Typically, the overall response time of the system is about 30 milliseconds. This is similar to the muscle-neural response time in a living leg. In special circumstances it is possible that the damper can be activated in a time span that is short enough to act as a relapse brake. For instance, if the person wearing the prosthesis stumbles, the folding of the lower-leg can be avoided by a very fast increase in the damper force. A critical aspect in the development of the MR fluid controlled artificial knee has been the availability of compact, lightweight, high-power density Li-ion batteries. The MR fluid prosthetic knee system is capable of providing about two days of use before the battery requires recharging.

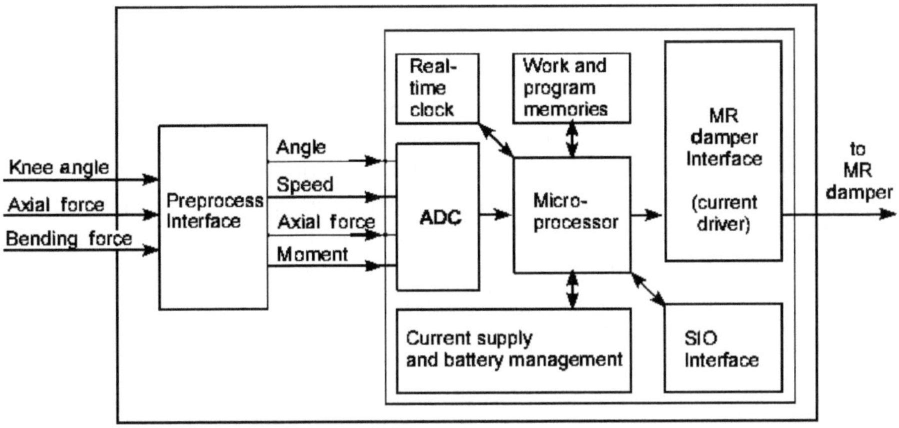

Fig. 6.92. Basic elements of control electronics for MR controlled knee prosthesis

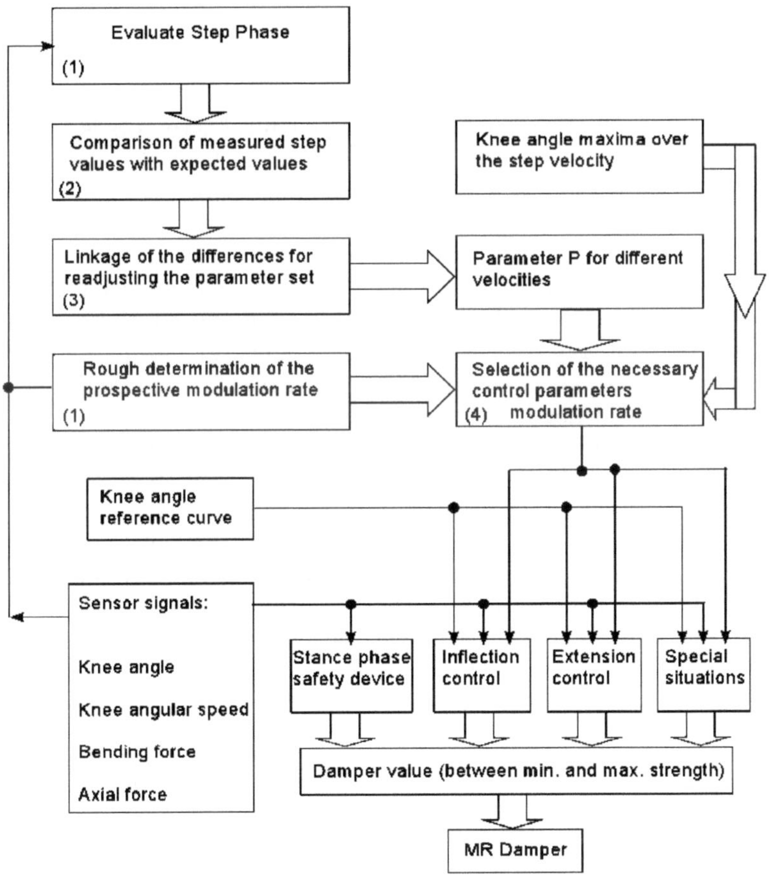

Fig. 6.93. Algorithm for controlling the MR fluid damper in the artificial knee

6.6.6 Conclusion

Magnetorheological fluid actuators provide technology that enables effective semi-active control in a number of real-world applications. Automotive applications of MR fluid are significant and growing rapidly. Annual production of MR fluid is now on the order of hundreds of tons. It is estimated that more than one hundred thousand MR fluid devices are presently in use. This number is expected to rise into the millions as more automotive platforms adopt MR fluid based real-time controlled motion control systems. Due to their simplicity, low power, and inherent robustness, MR fluid devices have proven themselves in a wide variety of commercial applications.

6.7 Electroactive Polymer Actuators

A. Mazzoldi†, F. Carpi, D. De Rossi

6.7.1 Introduction

The construction of small but powerful electromechanical actuators is one of the most important aims for several applications in the field of drive technologies. The miniaturization of traditional components, as for instance in the case of microelectronics, may not always be a successful approach. Dimensioning problems and material issues prevent conventional drives from being excessively scaled down. Therefore, new drive principles, technologies and materials are required to achieve innovative solutions for these problems. Materials that can transduce a certain form of energy into mechanical energy, withstanding high loads and having large strokes, are needed. Ideally, these materials should not be driven by high electric or magnetic fields, nor large temperature gradients. Polymer actuators are a promising alternative to conventional drives. They can convert electrical power (but also other sources of energy, such as heat, light, chemicals, etc.) into mechanical power, so as to transfer motion to loads. Polymer based materials which are able to transduce electrical into mechanical energy are called electroactive polymers (EAP) [182, 183]. They are classified principally in two main categories, as summarised in Fig. 6.94: ionic EAP whose actuation is based on diffusions of ions and solvents and electronic EAP whose actuation is based on electronic charging of the material. Each of these two classes presents the following sub-division in specific groups (Fig. 6.94):

- ionic EAP: polyelectrolyte gels, such as modified poly(acrylonitrile); ionic polymer metal composites (IPMC), such as Nafion/Pt; conducting polymers, such as polypyrrole (PPy) and polyaniline (PANi); carbon nanotubes, currently classified as EAP even though they are non-polymeric macromolecular materials;

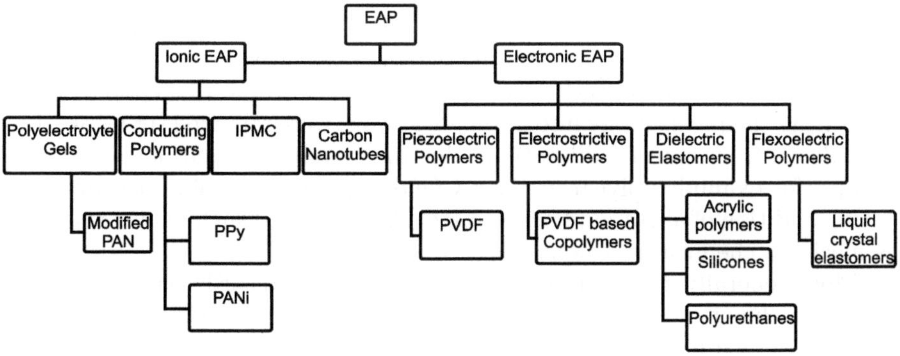

Fig. 6.94. EAP classification and examples of materials

– electronic EAP: piezoelectric polymers, such as PVDF; electrostrictive polymers, such as copolymers based on PVDF; dielectric elastomers, such as silicone; flexoelectric polymers, such as liquid crystal elastomers.

These polymers are studied as candidate materials for pseudo-muscular actuators. Such devices are conceived to promote a functional biomimesis of natural muscles [182, 183]. The following sections provide a brief description of the basic features of the less diffused EAP materials and devices: ionic EAP and dielectric elastomers.

6.7.2 Polyelectrolyte Gels (PG)

Working Principle of PG Actuators

A polymer gel consists of an elastic cross-linked polymer network and a fluid filling its interstitial space. Gels are wet and soft and look like a solid polymer material, but are capable of undergoing large deformations through swelling and de-swelling. Polymer gels can be easily deformed by external stimuli and generate force or execute work externally. If such responses can be translated

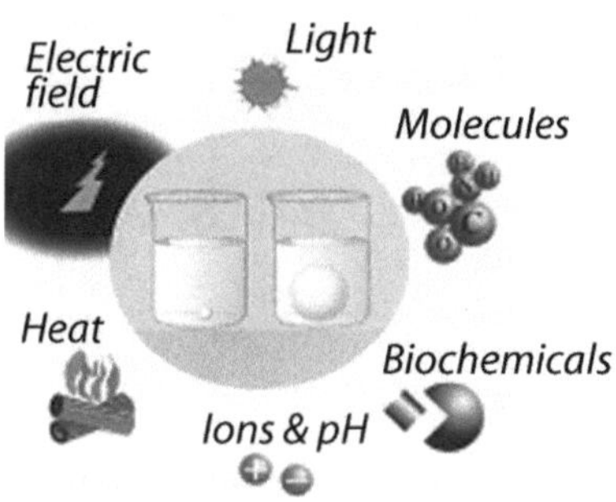

Fig. 6.95. Stimuli enabling mechanical responses of polyelectrolyte gels

from the microscopic level to a macroscopic scale, a conversion of chemical free energy into exploitable mechanical work is achieved.

As early as the end of the forties, studies about water-swollen polymer gels converting chemical energy into mechanical work were reported [184–186]. Reversible contractions and dilatations, due to reversible ionizations of suitable groups (for example polycarboxilic (–COOH) groups), are obtained by alternating addition of alkalis and acids. Katchalsky denoted such transformations as mechano-chemical reactions.

More generally, gels can undergo reversible order-disorder transitions, induced by changes either in temperature, irradiation, electric fields, pH (by chemical or electrochemical activation) or solvent properties. Figure 6.95 lists such different types of stimuli enabling a mechanical response of a polyelectrolyte gel.

PG Actuators

Water swollen hydrogels are generally amorphous without any particularly ordered structure at molecular level. For many years, polymer gels have been studied for the development of low-voltage soft actuators [187–193]. As an example, they can be used to construct thermo-responsive diaphragms capable of automatically opening and closing a valve [194]. They can also show shape memory effects. For instance, a thermal activation of a shape memory gel is shown in Fig. 6.96.

Concerning solvent-controlled activations, the structure of a gel can shift to a disordered state by means of an immersion in ethanol or tetrahydrofuran, to produce swelling. More generally, gels swell in organic solvents and undergo spontaneous motion when they are placed in water [195, 196]. The driving force of the gel motion originates from the spreading of the inner organic solvent out of the material when it is placed in water (Fig. 6.97); this is in a certain sense similar to what happens in jet motors [196].

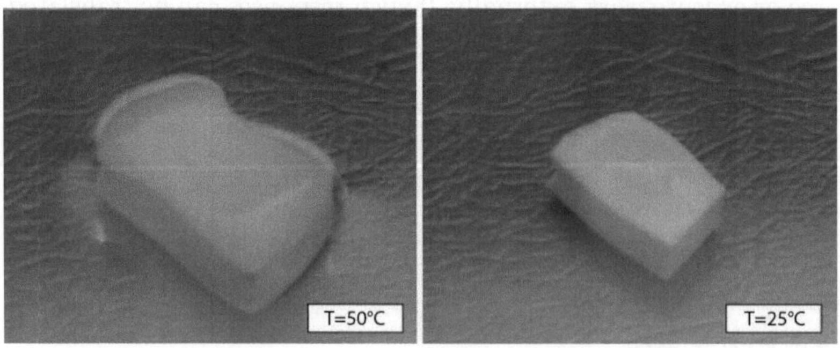

Fig. 6.96. Activation of a shape memory gel due to a temperature variation from 50 to 25 °C

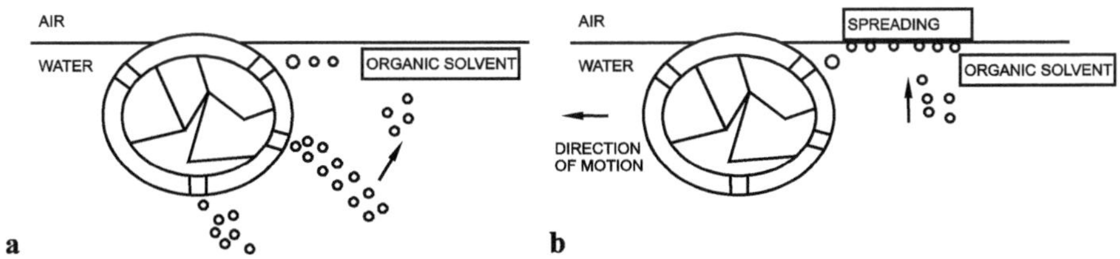

Fig. 6.97. Gel motion due to a spreading process of organic solvent

If a water-swollen cross-linked polyelectrolyte gel is inserted between a pair of planar electrodes and a voltage difference is applied, the material can undergo anisotropic contractions and concomitant fluid exudations [197, 198]. Electrically induced contractions of the gel are caused by transport of hydrated ions and water in the network (electrokinetic phenomena). In fact, when an outer electric field is applied across a gel, both macro- and micro-ions are subjected to electrical forces in opposite directions. However, macro-ions are typically in a stationary phase, being chemically fixed to the polymer network, while counter ions are mobile and are capable of migrating along the electric field, dragging water molecules with them.

Several active devices have been realized by using these phenomena with different actuating configurations, such as films, strips, membranes and fibers. An example consists of an electrically activated chemical valve membrane, which reversibly expands and contract its pore size in response to an electrical stimulus [199]. When the electro-chemo-mechanical contraction is developed isometrically, i. e. keeping the membrane dimensions constant, the contractile stress generated in the membrane expands its pore channels, through which solute and solvent permeate. By applying on/off constant potential cycles, the chemical valve membrane increases and decreases the water permeability, according to the applied electrical stimulus. It may be possible to use such a system as a permeation-selective membrane continuously separating solute mixture with different molecular sizes.

As another example, a gel-looper was proposed. A piece of gel was suspended from a long plastic ratchet bar, following its immersion in a solution. When a voltage was applied through a pair of long plate carbon electrodes placed at upper an lower positions of the ratchet bar, and the polarity varied at regular intervals, the gel moved forward in the solution like a 'looper', by repeating bending and stretching movements [200, 201].

Actuators with fiber configuration have also been demonstrated. They can be particularly interesting because a small thickness permits reduction of the response time. Modifications of the pH of aqueous media around the fibers (e. g. by electrolysis) are frequently used to induce their dimensional changes [202].

Active PG fibers can be obtained from PAN fibers by means of a heating at 220 °C in atmospheric pressure for 5 hours, and then saponification with boiling in 1 M (molar) NaOH for 30 min, following the process reported in [203]. The procedure transforms the original PAN fibers into swollen fibers of amphoteric amino-carboxylic polyelectrolyte gels. An example of preparation of PG samples is also reported on the following web site: http://ndeaa.jpl.nasa.gov/nasa-nde/lommas/eap/EAP-recipe-UA.htm.

6.7.3 Ion-Polymer Metal Composites (IPMC)

Working Principle of IPMC Actuators

Most ionic polymeric membranes swell in solvents and are hydrophilic. This gives rise to the ability of the membrane to swell in water, which can be controlled in an electric field, due to the ionic nature of the membrane. By placing two electrodes in close proximity of the membrane and applying a low voltage (below the threshold for electrolysis), the forced transport of ions within a solution through the membrane becomes possible at microscopic level. The occurring local swelling and de-swelling of the membrane can be controlled, depending on the polarity of the nearby electrode.

Such a basic principle is exploited in the so-called ion-polymer metal composites (IPMC) actuators. They are used to realize actuators showing large deformations in response to low applied voltages and offering low electrical and mechanical impedance [204, 205]. In more detail, materials used for IPMC actuators (such as Nafion by Du Pont) have many ionizable groups in their molecular chain. These groups can be dissociated in various solvents, showing a resulting net charge, which is compensated by the presence of mobile counterions. The net charges of the network macromolecules are called polyions. Electrophoretic migrations (due to an imposed electric field) of the mobile ions within the macromolecular network can cause the network to be deformed accordingly [204–218]. In fact, the shifting of ions of the same

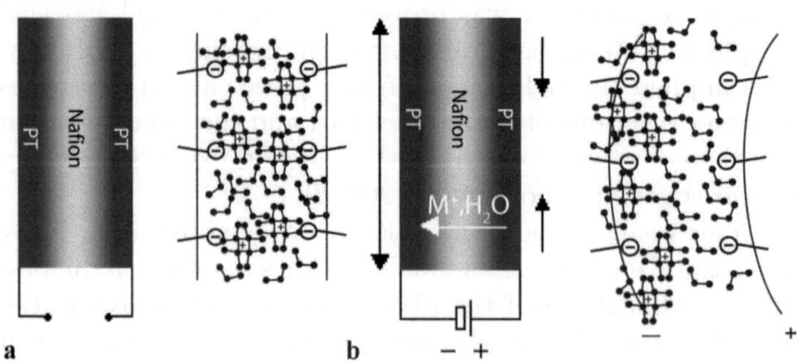

Fig. 6.98. Schematic drawing of the working principle of an IPMC actuator: **a** device at rest, **b** device under activation

polarity within the network results in both electrostatic interactions with the fixed charges of opposite polarity (contained in the side groups of the polymer chains) and transport of solvent molecules. Both these factors concur to produce a stress gradient between the opposite sides of the membrane, where local collapse and expansion occur, causing a macroscopic bending of the structure. A schematic drawing of the resulting electro-chemo-mechanical activation is shown in Fig. 6.98.

IPMC Actuators

A typical material used to assemble IPMC actuators consists of a film of Nafion-117 (Du Pont), an ion exchange membrane (IEM). Platinum electrodes are deposited on both sides of the film. The thickness of the actuator is typically of the order of 0.20 mm. To maintain the actuation capability, the film usually needs to be kept continuously moist. The structure and properties of Nafion membranes have been subjected to numerous investigations. One of the interesting properties of this material is its ability to absorb large amounts of polar solvents, i. e. water. Platinum ions, which are dispersed throughout the hydrophilic regions of the polymer, are subsequently reduced to the corresponding metal atoms.

When equilibrated with aqueous solutions, the membranes are swollen and they contain a certain amount of water. Swelling equilibrium results from a balance between the elastic recovery force of the polymeric matrix and the water affinity to the fixed ion exchanging sites and the moving counterions. The water content depends not only on the hydrophilic properties of the ionic species inside the membranes, but also on the electrolyte concentration of the external solution. When an external voltage (usually of the

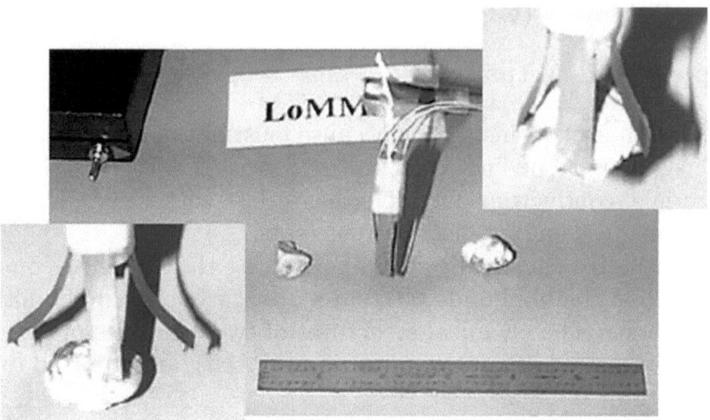

Fig. 6.99. Gripper with IPMC end-effectors (adapted from [217])

Fig. 6.100. Swimming robotic system (adapted from [217])

order of 1 V) is applied to an IPMC composite film, it bends toward the anode. An increase of the voltage level causes a larger bending. When an alternate voltage is applied, the film undergoes movements like a swing. The displacement depends not only on the voltage magnitude, but also on the frequency (lower frequencies lead to higher displacements, according to the device bandwidth) [204–218].

IPMC actuators usually operate best in a humid environment, even though they can be made as encapsulated devices to operate in dry conditions. Several applications have been investigated. These include fingers of an end-effector for a miniature low-mass robotic arm (Fig. 6.99), cilia systems and swimming robotic structures (Fig. 6.100) [217].

An example of the preparation of IPMC samples is reported on the following web site: http://ndeaa.jpl.nasa.gov/nasa-nde/lommas/eap/IPMC_PrepProcedure.htm.

6.7.4 Conducting Polymers (CP)

Working Principle of CP Actuators

Polymers have been often used as insulators because most of them are unable to conduct electricity. This trend has been changed in the last years since a new class of materials, conducting polymers, has been synthesized. These polymers are in fact able to conduct electrical currents.

Conducting polymers are chemically characterized by the so-called conjugation, in which carbon double bonds alternate with carbon single bonds along a polymer backbone. The chemical structures of two examples of conducting polymers, polypyrrole (PPy) and polyaniline (PANi), are reported in Fig. 6.101.

Conducting polymers can be characterized by a high conductivity when doped with ions (Fig. 6.102). Their conductivity can be reversibly changed by orders of magnitude, by changing the doping level. Unlike silicon, dopants

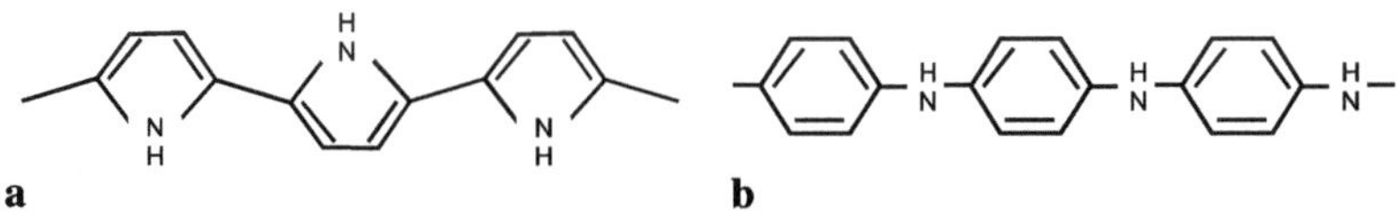

Fig. 6.101. Chemical structure of **a** PPy and **b** PANi

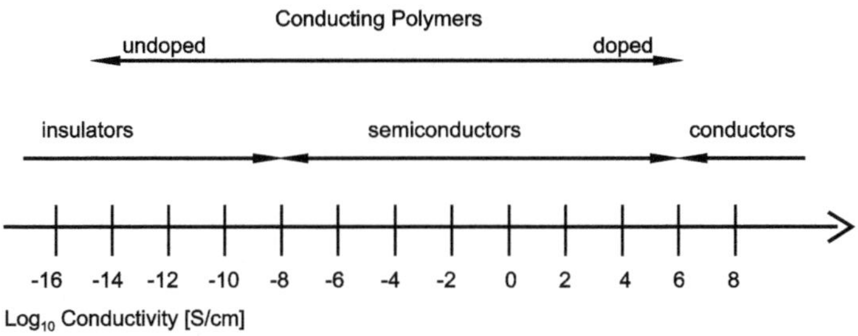

Fig. 6.102. Conductive properties of CP

can be easily inserted and removed from the spaces they occupy between the polymer chains. Moreover, in comparison with other semiconducting materials, the doping level can be very high: approximately one dopant counterion per three or four monomers.

Conducting polymers are being studied for several fields of application. Since these materials are able to store a large amount of charge, they are of interest for use in batteries. Another interesting property is their bandgap that allows electron-hole recombination, which has made these materials appealing for light-emitting diodes. Their optical properties (especially light absorption) can be voltage controlled, so that conducting polymers have also been investigated for electrochromic devices.

They are studied for actuation tasks too. For this purpose, they are used as components of an electrochemical cell, whose basic structure includes two electrodes immersed in an electrolyte. The conducting polymer material constitutes of one or both of the electrodes of the cell. By applying a potential difference between them, red-ox reactions cause strongly anisotropic and reversible volume variations of the material [219], which can be used for actuation [220–242].

It has been found that the following three effects are responsible for dimensional and volume changes in conducting polymers: interactions between polymer chains, variation of the chain conformation and insertion of counterions. The third effect is generally considered to be the most dominant. In fact, the commonly accepted explanation of the observed deformations attributes the dimensional changes to the input/output of ions (exchanged with the surrounding media) into/from the polymer sample, driven by an applied voltage. In particular, the voltage produces a variation of the polymer oxidation state, causing the necessary modification of the number of ions associated to each chain, in order to maintain the global electro-neutrality.

CP Actuators

The most diffused actuating configuration, in which these materials are used, is represented by the so-called unimorph bilayer bender. This kind of actuator consists of a film of active material coupled to a passive supporting layer. The bilayer structure is operated within an electrochemical cell, having a liquid electrolyte in which the device is immersed. The active polymeric layer of the actuator works as one electrode of the cell, while a counter electrode and a third reference electrode are separately immersed in the electrolyte. One end of the bilayer is constrained, while the other is free. The potential difference applied between the electrodes causes red-ox reactions of the conducting polymer. Since the CP and the passive layers are mechanically interlocked, when the polymer swells/shrinks the passive layer, which can not modify its dimensions, transforms the CP linear displacement into a bending movement of the structure [238–242]. Very similar is the bimorph structure. In this case the passive layer is substituted by a second CP film and they work in opposition of phase.

Both unimorph and bimorph benders can be used to realize useful active systems, such as small clamps to move small objects, manipulators conceived for minimally invasive surgery and devices to control the bending of catheters or endoscopes [243].

Fiber actuators made of conducting polymers have been also proposed, consisting of an extruded fiber, covered by a thin layer of solid polymer electrolyte (SPE) and a counter-electrode of polypyrrole [230]. Conducting polymer fibers have today become available. For instance, Santa Fe Science and Technology produces polyaniline (PAni) fibers under the trademark of Panion™. They have been used to fabricate linear actuators (Fig. 6.103): a bundle of Panion™ fibers (operating as an actuating electrode) is inserted into a Panion™ hollow fiber (counter electrode) with a separator/electrolyte medium. This kind of actuator, tested with a [BMIM][BF4] ionic liquid electrolyte, has reported strains of about 0.3%, stresses of about 1.8 MPa and red-ox cycle lifetimes in excess of 104 cycles [225].

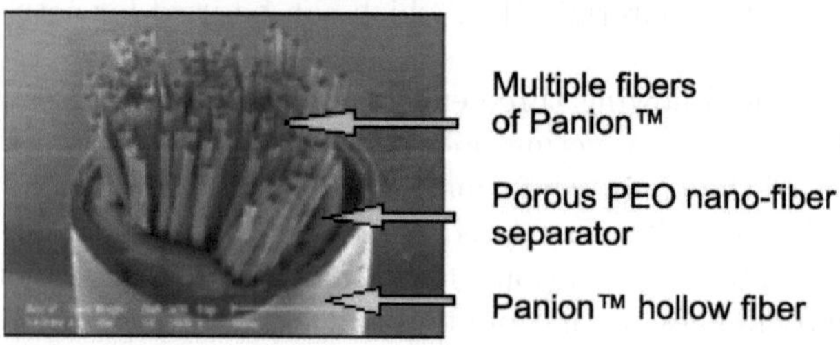

Fig. 6.103. Pani fiber based actuator developed by Santa Fe Science and Technology (adapted from [244])

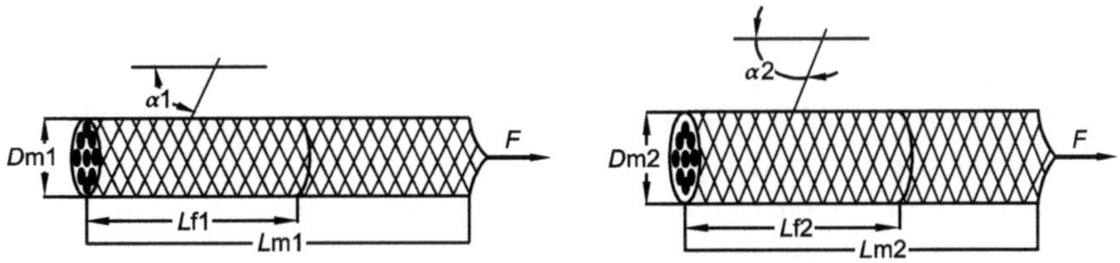

Fig. 6.104. Schematic drawing of a modified Mc Kibben actuator

State-of-the-art CP devices need very low driving voltages (order of 1 V), producing strains of the order of 1 ... 10% for linear actuators and rotations up to ±90° for benders, with large active stresses (up to tens of MPa). Nevertheless, such interesting performances correspond to several drawbacks, such as high response times and short lifetimes, whose relevance has to be evaluated in relation to the specific application of interest.

Approaches to enlarge achievable displacements are needed. As a first method, because CP are typically poor ion conductors, it is useful to make thin polymer layers and to add water filled pores or tunnels in order to allow fast diffusions of ions inside the polymer. As a second point, it can be useful to store the ions instead of transporting them. This can be done by using a solid polymer electrolyte, SPE, (electrolyte storage configuration) or switching the ions between two different polymer layers through a SPE (electrode storage configuration).

A method to transfer and amplify the radial strain of a CP fiber into an axial strain has been proposed, inspired to a Mc Kibben actuator [245]. In the classical version of this latter device, a cylindrical rubber bladder is covered by a braid mesh, made of flexible, but not extensible, threads. Both ends of the bladder are connected to the mesh. By changing the force applied to the free end of the mesh and the pressure inside the bladder, the mesh shape change dimensions: its diameter increases and its length decreases. In the CP version of the Mc Kibben actuator (sketched in Fig. 6.104), the bladder is substituted with a bundle of conducting polymer hollow fibers. In the center of each hollow fiber a rigid metal wire works as a counterelectrode. A filling liquid electrolyte completes the system.

The actuation mechanics of such a device has been studied, by performing an electro-chemo-mechanical analysis [246]. According to results of that study, this type of structure might enable axial strains of different magnitude, ranging from 25% up to 80%, depending on the inclination angle of the mesh. Unfortunately, practical reasons related to the complexity of the manufacturing process of the mesh limit the feasibility of fabrication of appropriate inclination angles of the threads. Moreover, such theoretical predictions have to deal with the inevitable losses due to internal friction.

The actuation technology based on conducting polymers has opened interesting perspectives, so that the first commercial applications in the biomedical field, such as blood vessel connectors, Braille displays and cochlear implants, are being developed today [247].

Fabrication and, in particular, microfabrication of conducting polymer based structures is usually performed by using a large number of technologies, implementing either pre-, post- or direct- microstructuring of the material. Concerning the fabrication of macroactuators (main dimensions of the order of centimeters or tens of centimeters), different techniques have been proposed so far. They consist of classical procedures borrowed from many industrial sectors, where they are employed for different uses. Electrochemical deposition, casting, deep- and spin-coating are the most notable examples.

Electrochemical deposition (or *electropolymerisation*) is performed by using an electrochemical cell, whose liquid electrolyte contains the monomer under polymerisation. The procedure consists of a growth of polymer layers typically via monomer oxidation. In particular, the polymer is deposited on the electrode where oxidation takes place (anode) [248,249]. This method can be used for direct fabrication of electrode/polymer bilayers. Alternatively, the active polymeric layer can be successively peeled from the deposition electrode, so that to be coupled to another type of passive substrate.

Casting, *deep-* and *spin-coating* and *extrusion* can be used for film and fiber fabrication if the material is available in solution phase. Following the material processing and shaping, the polymer solution is dried in an oven or by exposure to an infrared lamp. These techniques have been largely used for polyaniline [250,251] and certain forms of polypyrrole [252–255]. Bender actuators fabricated with such techniques can present, when fatigued, a separation of the film from the support (delamination), due to shear stresses generated at the layer interface by the bending movement during operation. Interface roughening, enriching the mechanical interlock between the two layers, has been demonstrated as being useful in order to reduce such a problem [256]. Owing to the insolubility of several conducting polymers, these fabrication procedures are not widely applicable to many representative materials of the conducting polymer family.

In order to microfabricate small-scale (down to micron size) conducting polymer based actuators, the most used microtechnologies consist of conventional procedures of surface and bulk micromachining derived from *photolithography*. They are implemented as sequential steps of layer depositions and etching removals [257, 258]. With such methods, several examples of bending actuators have been reported, mainly related to Au/PPy bilayers fabricated onto silicon wafers with polymer thickness even down to 1 μm [257–261]. Many interesting applications of this kind of actuators have been described, including microgrippers [257, 261], gates for 'cell clinics' [257,261], self-assembling boxes [262,263], microrobots [257,261,264] and positioning microhinges [260].

Recently, more innovative methods such as ink-jet printing, soft lithography and deposition via controlled-volume or pneumatic microsyringes have been proposed.

Ink-jet printing is a simple and fairly economical technique consisting of a drop-by-drop deposition of a polymer, previously dissolved in a volatile solvent, by using a printing head [265–267].

Soft lithography is a methodology derived from photolithography, which has been pioneered by the group of Whitesides at Harvard University [268]. This technique includes microinjection moulding in capillaries and microcontact printing [268–270]. Microinjection moulding uses microfabricated stamps made of poly(dimethylsiloxane) (PDMS). The elastomeric stamps are filled up with a polymer solution and the excess of solvent is evaporated, so that the polymer filling the microchannels assumes a specified geometry. The realised microstructure is then removed from the mould via lift-off [271].

The use of *microsyringes* as extruders mounted on micropositioning systems enables the deposition of polymers in two- and three-dimensional structures [272]. According to the principle of extrusion, and, in particular, of the method used in order to apply and modulate the pressure gradient expelling the reservoir solution, two types of systems can be recognised: 1) those with pneumatic microsyringes, where the solution flow is enabled and regulated by compressed air; 2) those with volumetric microsyringes, driven by the controlled movement of a piston. All these systems have been used to fabricate benders, as shown for instance in [273].

An example of preparation of a CP based bender actuator is reported here. The considered structure presents two CP layers that enclose a solid polymer electrolyte (SPE) film. Polyaniline can be selected as a suitable conducting polymer for the fabrication of the actuators. For the realisation of the active layers of the bender, a polyaniline suspension in 1-methyl-2-pyrrolidone can be used. It is mixed with a gelification inhibitor, e. g. heptamethyleneimine; this compound limits the formation of gelatinous lumps, which, however, can be eliminated by heating the suspension in an oven. A solid polymer electrolyte can be obtained by dissolving polyacrylonitrile in a solution of ethylencarbonate/ propylenecarbonate/ sodiumperchlorate. An aluminium coated Mylar® film can be used as both the deposition substrate and a conductive layer, working as a current collector. On the top of it, CP and SPE layers have to be sequentially deposited.

An example of preparation of CP samples is also reported on the following web site: http://ndeaa.jpl.nasa.gov/nasa-nde/lommas/eap/Polypirrole-PrepProcedure.htm.

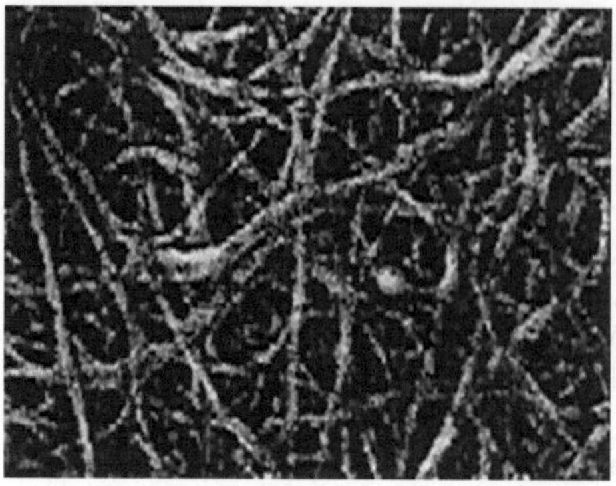

Fig. 6.105. Micrograph of a nanotube sheet

6.7.5 Carbon Nanotubes (CNT)

Working Principle of CNT Actuators

Carbon nanotubes are a recent addition to the class of electroactive organic materials. They can be described as a graphite monoatomic sheet rolled to form a tube [274]. Carbon nanotubes have lengths about 1000 times that of their width (typical diameters are of the order of 1 nm, while typical lengths are about 1 μm). Moreover, they are typically combined in bundles with diameters of 10 nm. Carbon nanotubes can be divided in two classes: single-walled and multi-walled. A single-walled CNT consists of a single film rolled to make a tube, while a multi-walled CNT is made of several films rolled together. Mechanical performances of multi-walled tubes are predicted to be lower, with respect to those predicted for the single-walled ones, according to the lower forces between the layers. Figure 6.105 presents an image of several bundles combined to form a sheet.

Unlike conducting polymers, which can act as batteries, CNT can be used as electrochemical supercapacitors [275]. CNT actuators can be realized by using sheets of single-walled nanotubes. Their actuation properties have been demonstrated by employing an electrochemical cell with at least one CNT electrode (characterised by a very high surface area). A change of the applied cell voltage results in a double-layer charge injection for this electrode, with a related deformation [276]. The actuating principle is represented by this charge-injection, which is able to produce dimensional changes in the CNT structure. These originate from quantum chemical and double-layer electrostatic effects [276].

CNT Actuators

Early investigations on CNT bending actuators showed active strains of the order of 0.2%, depending on the experimental conditions, when an applied voltage was limited to the electrochemical stability of the electrolytes

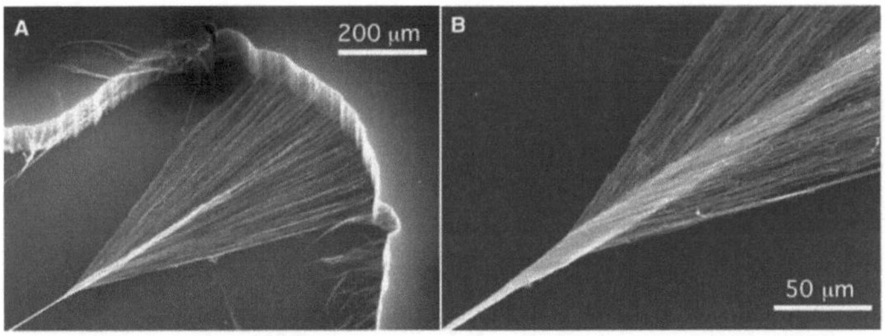

Fig. 6.106. Carbon nanotube yarns (adapted from [279])

(−1 V to +1 V, versus saturated calomel electrode (SCE), for aqueous electrolytes) [276]. Higher strains were reported when larger voltages were applied to the CNT in an aqueous NaCl electrolyte. In particular, reversible contractions up to 2% were achieved, by applying pulses between −0.5 and +1.5 V in 5 M NaCl. Additional superimposed phenomena responsible for increased strains were also described [277].

CNT were used to realize unimorph micro-benders for clamps. CNT were embedded in a gel matrix (obtained by adding CNT in DMA in a mixture of PVA-PAA), which constitutes only a supporting scaffold without substantially altering the typical electrical characteristics of CNT [278].

Recently, high-quality nanotube thin fibers and yarns were realized by the University of Texas at Dallas (Fig. 6.106) [279]. These results may open new investigations towards CNT fiber actuators.

An example of the preparation of CNT samples is reported on the following web site: http://ndeaa.jpl.nasa.gov/nasa-nde/lommas/eap/Nanotube-PrepProcedure.htm.

6.7.6 Dielectric Elastomers (DE)

Working Principle of DE Actuators

Macromolecular actuators made of dielectric elastomers are stimulating a growing interest, due to their excellent electromechanical properties. These materials consist of dielectric polymers with a low elastic modulus, which can present significant electrically-induced strains. In particular, a dielectric elastomer actuator consists of a thin layer of an insulating rubber-like material sandwiched between two compliant electrodes (e. g. made of carbon conductive grease), which are electrically charged by a high voltage difference. Following the electrical activation, the material undergoes an electric field-sustained deformation at constant volume, consisting of a thickness squeezing and a related surface expansion (Fig. 6.107) [280–283].

This deformation is mainly due to a Coulombic effect, arising from the electrostatic interactions among the electrode free charges. The stress of the Coulomb force acting between the electrode free charges is responsible for

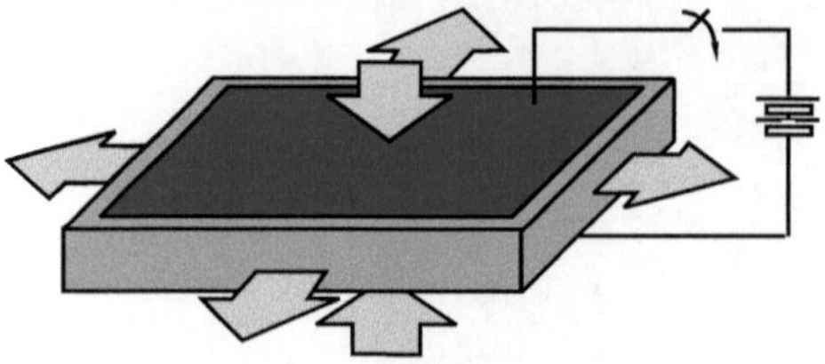

Fig. 6.107. Working principle of a dielectric elastomer planar actuator

the so-called Maxwell stress for this electromechanical phenomenon. This kind of stress acts in any kind of dielectric material subjected to an applied electric field. However, the corresponding deformations are emphasized by the eventual compliance of the electrodes, as well as by the polymer softness. These key-features basically distinguish actuating devices made of dielectric elastomers from those based on different electric-field-driven dielectrics, such as piezoelectric or electrostrictive materials.

Thickness strains S can be analytically described, by assuming that the dielectric elastomer is a linearly elastic body, with a Youngs modulus Y and a relative dielectric constant ϵ_r, as follows ($\epsilon_0 = 8.85 \cdot 10^{12}\,\mathrm{F/m}$ is the free-space dielectric permittivity) [280–283]:

$$S = -\frac{1}{Y}\epsilon_0\epsilon_r E^2 \,. \tag{6.43}$$

This equation shows that such materials exhibit a quadratic dependence of the strain on the applied field, as it happens for electrostrictive polymers. However, in comparison with these polymers, dielectric elastomers are capable of significantly larger deformations, even though at reduced forces, as reported in the following subsection.

DE Actuators

Acrylic and silicone rubbers are the most significant types of the dielectric elastomers used for actuation. Such kinds of polymers comprehend representative materials which can be very compliant, being able of showing the highest actuating deformations among all EAP [281].

High-level actuation capabilities have been reported for certain types of acrylic polymers (or acrylates): thickness strains up to 60 ... 70% at 400 V/µm, area strains up to 200% at 200 V/µm and corresponding stresses of some MPa [281]. Such performances are enabled by low elastic moduli and high dielectric strengths (dielectric breakdown can occur at electric fields up to about 500 V/µm). The highest active performances were achieved by prestretching the material: this operation was demonstrated to increase the dielectric strength, permitting the application of higher electric fields [281].

Beyond acrylates, silicones (mainly poly-dimethylsiloxanes) offer attracting characteristics: they are easily processable (by spin coating, casting, etc.) and permit the realization of rubber-like dielectrics with suitable elastic properties, arising from the flexibility of the material molecular chains. Certain silicone elastomers have been actuated with electric fields up to 100 ... 350 V/µm, enabling thickness strains up to 40 ... 50% and area strains up to 100%, with related stresses of 0.3 ... 0.4 MPa [281].

Owing to the excellent figures of merit shown by several dielectric elastomers (very high actuation strains, considerable stresses, very fast response speeds, high efficiency, stability, reliability and durability), this class of EAP is considered today as one of the most outstanding for polymer actuation. Nevertheless, some drawbacks still affect this technology. The most significant is certainly represented by the high driving electric fields needed (order of 100 V/µm). For a definite polymer thickness, such field levels can be reached by applying high voltages, which may be disadvantageous in several applications. In order to reduce such driving fields, polymers with unusually high dielectric constants would be advantageous (6.43). Accordingly, some research efforts are today devoted to the development of new elastomers with enriched dielectric permittivity.

One of the simplest approaches relies on the realisation of composite materials: by filling an ordinary elastomer with a highly dielectric component (e. g. ceramics), it is possible to obtain a resulting material showing the combination of the advantageous matrix elasticity and filler permittivity. As an example, promising results have been obtained with a silicone elastomer mixed with a titanium dioxide powder [284].

Several configurations for dielectric elastomer actuators have been proposed and demonstrated so far: planar, tube, roll, extender, diaphragm, bimorph and unimorph bender represent the most significant.

Linear (i. e. working along a line) actuators can be obtained by adopting the tube-like and the roll-like configurations, depicted in Fig. 6.108. The first one consists of an elastomeric tube having compliant electrodes on the inner and outer surfaces; by applying a high voltage difference between them, the wall of the tube is squeezed and the structure elongates [289]. The roll-type actuator is made of thin electroded layers of elastomers rolled so as to obtain the compact structure sketched in Fig. 6.108; a high voltage input causes an axial elongation of the device [285, 286].

As mentioned, both these devices elongate under electrical activation. This property has been exploited to provide excellent actuating functions to biomimetic robots [285, 286]. However, certain applications may specifically require devices capable of active contractions, instead of elongations. Accordingly, different configurations are necessary. The simplest one, from a conceptual point of view, consists of a stack of elementary actuating units, made of planar actuators connected in electrical parallel and mechanical series [280, 290]. The thickness contraction of each element causes the axial contraction of the entire structure. This configuration can enable very interesting per-

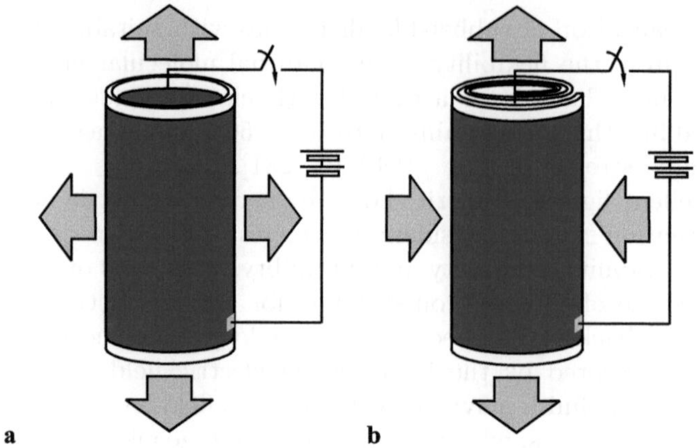

Fig. 6.108. Linear elongating dielectric elastomer actuators: **a** tube and **b** roll configuration

formances [290]. Nevertheless, its discontinuous structure can complicate its fabrication. Therefore, new solutions for contractile actuators may be of help.

As an example, two types of new configurations have been recently presented, as shown in Fig. 6.109. The first is termed an helical dielectric elastomer actuator (Fig. 6.109a) [291]. It consists of a hollow cylinder of dielectric elastomer, having two helical compliant electrodes integrated within its wall. The second is termed a folded dielectric elastomer actuator (Fig. 6.109b) [292]. It is made of a monolithic strip of electroded elastomer which is folded up. For both these configurations, a high voltage difference applied between the electrodes induces attractions among opposite charges of the two electrodes, as well as repulsions among the same type of charges of each electrode: accordingly, these effects determine the compression of the dielectric included between the electrodes, causing an axial contraction and a radial expansion of the structure.

Such devices might be useful for applications requiring spring-like contractions of an elastomeric device activated and modulated by an electrical signal.

6.7.7 Electroactive Polymers as Sensors

In this paragraph a short description of basic sensing properties of electroactive polymers is reported. In fact, they can also be used for different types of physical and chemical sensing, according to different effects, as described in the following.

Sensing devices can be divided into active sensors and passive sensors. We classify here as active sensors those that intrinsically convert the input energy into a useful electrical potential difference. Differently, those sensors that require an external power source to convert the input into a usable output are

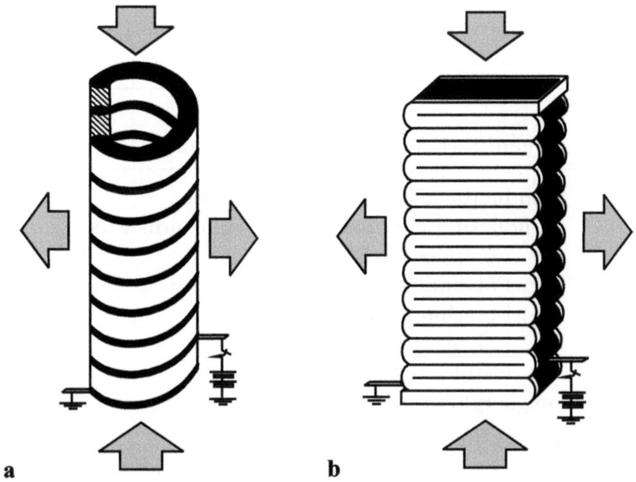

Fig. 6.109. Linear contractile dielectric elastomer actuators: **a** helical and **b** folded configuration

defined as passive. Table 6.9 presents a non-exhaustive list of electroactive polymers and conventional inorganic counterparts currently used for the indicated types of passive sensing. Likewise, physical effects and related devices for fundamental active sensing are listed in Table 6.10.

Among the possible different types of sensing, the most advantageous for adaptronics are mentioned here.

Electroactive polymers can be used for piezoresistive strain sensing, i. e. as polymer strain gages. These sensors work according to the piezoresistive effect: their electrical resistance is modified by an imposed strain of the material. Most performing piezoresistive EAP are listed in Table 6.9. A large number of applications are possible. As an example, EAP based sensors have been used to confer strain sensing properties to garments, in order to monitor body-kinematics, such as position and movement of articulation segments. In this respect, both conducting polymers [293] and carbon-loaded elastomers [294–297] have been studied.

A second type of relevant sensing exploits piezoelectricity. According to the well-known direct piezoelectric effect, the application of a stress along one of the main axes of a piezoelectric material causes its polarisation, generating net opposite charges on opposite surfaces. The electric potential difference produced by the opposite charge distribution can be detected by embedding the material between two electrodes. The most exploited piezoelectric inorganic materials for several commercial applications are titanate ceramics, such as lead zirconate titanate (PZT). Polyvinylidene fluoride (PVDF) is the most commonly used piezoelectric polymer. It has a typical piezoelectric coefficient (d_{31}) of 24... 27 pC/N [298]. A list of common piezoelectric organic materials is presented in Table 6.10.

Table 6.9. Most used EAP for passive sensing and conventional inorganic counterparts

Physical effect	Sensing devices	Materials Organics (EAPs)	Inorganics
Piezoresistivity	Strain gages	Conductor-loaded rubbers Conducting polymers e. g. Polypyrrole (PPy) e. g. Polyaniline (PAni) e. g. Polythiophene (PT) e. g. Polyacetylene (PA) e. g. Pyrolized polyacrylonitrile (PAN)	Metals Semiconductors
Thermoresistivity	Bolometers	Poly(p-phenylene vinylene) (PPV)	Metals Metal oxides Titanate ceramics Semiconductors
Magnetoresistivity	Magneto-resistive sensors	Polyacetylene (PA) Pyrolized polyvinylacetate (PVAc)	Nickel-iron alloys Nickel-cobalt alloys
Chemioresistivity	Chemio-resistive sensors	Polypyrrole (PPy) Polythiophene (PT) Ionic conducting polymers Charge transfer complexes	Palladium Metal oxides Titanates Zirconia
Photoresistivity	Photo-resistive sensors	Copper phthalocyanines Polythiophene complexes	Intrinsic and extrinsic (doped) semi-conductors

Beyond piezoresistivity and piezoelectricity, of course several other sensing effects exploitable with electroactive polymers for adaptronic systems could be mentioned too. For instance, the piezocapacitive effect is largely used for electrostatic devices, such as for dielectric elastomer actuators with intrinsic strain sensing properties. As another relevant effect, we also mention that conducting polymer actuators have been recently demonstrated to be capable of sensing a load. For this purpose, the correlation between the variation of the charging current and the applied load is particularly useful [224]. Finally, a couple of further effects deserve to be reported for IPMC and polyelectrolyte gels. In fact, the passive bending of an IPMC actuator can originate from an electric potential difference between its electrodes, as a result of internal ion migrations driven by the applied stress [217]. Concerning gels, the compression of a piece of these materials can induce a pH change, associated with a changing ionization of carboxyl groups under deformation. This can cause a resulting change in the electric potential between opposite electrodes placed in contact with the material. Hence, similarly to the touch-sensing

Table 6.10. Most used EAP for active sensing and conventional inorganic counterparts

Physical effect	Sensing devices	Materials	
		Organics (EAPs)	Inorganics
Piezo-electricity	Piezoelectric transducers	Polyvinylidene fluoride (PVDF)	Lead zirconate titanate ($Pb(Zr,Ti)O_3$) (PZT)
		Polyvinylfluoride (PVF)	Lead based lanthanum-doped zirconate
		Poly(vinylidene fluoride – trifluoroethylene) (P(VDF-TrFE))	titanate ($(Pb,La)(Zr,Ti)O_3$) (PLZT) Quartz (SiO_2)
		Poly(vinylidene fluoride – hexafluoropropylene) (P(VDF-HFP))	Zinc oxide (ZnO) Barium titanate ($BaTiO_3$)
		Poly(vinylidene fluoride – tetraflouoroethylene (P(VDF-TFE))	Potassium niobate ($KNbO_3$) Lithium niobate ($LiNbO_3$)
		Polyamides e. g. Nylon-11	Lithium tantalate ($LiTaO_3$) Bismuth ferrite ($BiFeO_3$)
		Liquid crystalline polymers (flexoelectricity)	Triglycine sulfate (TGS) $Ba_2NaNb_5O_5$ $Pb_2KNb_5O_{15}$
Thermo-electricity	Thermocouples	Polyacetylene (PA)	Silicon, Bismuth, Nickel, Cobalt, Palladium,
		Polyaniline (PAni)	Platinum Uranium, Copper, Manganese,
		Polypyrrole (PPy)	Titanium, Mercury, Lead, Tin, Chromium,
		Polythiophene (PT)	Molybdenum, Rhodinium, Iridium, Gold,
		Polyphthalocyanines	Silver, Aluminum, Zinc, Tungsten, Cadmium,
		Nitrile based polymers	Iron, Arsenic, Tellurium, Germanium. Lead telluride (PbTe) Lead selenide (PbSe) Cadmium selenide (CdSe) Cadmium telluride (CdTe) Bismuth selenide (Bi_2Se_3) Bismuth telluride (Bi_2Te_3) Antimony Telluride (Sb_2Te_3) $Cu_{100}/Cu_{57}Ni_{43}$
Pyro-electricity	Pyroelectric transducers	PVDF P(VDF-TrFE) P(VDF-HFP) PVF	PZT PLZT $BaTiO_3$ $LiTaO_3$ TGS

Table 6.10. (continued)

Physical effect	Sensing devices	Materials Organics (EAPs)	Inorganics
Photo-electricity	Photovoltaic cells	Polythiophene (PT) Polyaniline (PAni) Polypyrrole (PPy) Poly(N-vinyl carbazole) (PVCZ) Polyacetylene/n-zinc sulfide (PAS) Poly(p-phenylenevinylene) (PPV) Poly(2-vinylpyridine) (P2VP) Oligothiophenes Phthalocyanines	Silicon (Si) Germanium (Ge) Gallium arsenide (GaAs) Gallium aluminium arsenide (GaAlAs) Gallium indium phosphide (GaInP) Gallium indium arsenide (GaInAs) Gallium indium arsenide phosphide (GaInAsP) Copper indium diselenide ($CuInSe_2$) Indium antimonide (InSb) indium phosphide (InP) Indium gallium nitride (InGaN) Cadmium telluride (CdTe)

system of the human skin, the gel is able to convert mechanical energy into electrical energy, behaving like a type of soft and wet piezoelectric material, useful for developing tactile-sensing devices [299, 300].

6.7.8 Final Remarks and Conclusions

This section has briefly highlighted key issues related to the development of electroactive polymer actuators. According to their structure, polyelectrolyte gels and ionic polymer metal composite typically offer high strains but low stresses, while an opposite behaviour is shown by conducting polymers. Although these materials can be advantageously driven by low voltages, they are limited by a low response speed, due to a diffusion control, and a poor efficiency and durability, due to the electrochemical activation. Similarly, carbon nanotubes present low strains while interesting potential forces, even though their technological development is not mature.

On the contrary, dielectric elastomer actuators are characterized by the necessity of high driving voltages, while offering interesting electromechanical performances, consisting of large, fast and stable deformations at moderate stresses.

The usability of such actuators for practical applications still requires the solution of several problems in the case of ionic EAP, while possible uses are expected in the near future for dielectric elastomer devices.

6.8 Microactuators

H. Seidel

6.8.1 Introduction

Microactuators are key elements in adaptronic systems. Due to their small size they can often be combined with sensory functions to provide a self-sensing actuator, which can readily be integrated into a smart structure. Microactuators are not only characterized by their smaller size in comparison to classical actuators, but define themselves much more prominently by their way of production, which is derived from microsystem technology and is based on batch-processing steps. This means that rather than fabricating individual devices one by one in a serial approach, a large number of devices, usually on the order of hundreds to thousands, are being produced simultaneously in a parallel way. Typical steps of fabrication include lithography, thin film deposition techniques, and thermal processes (e. g. thermal oxidation) as well as steps for etching and for doping. Deposition techniques include chemical vapour deposition (CVD) at atmospheric or reduced ambient pressure, plasma enhanced methods for reducing the deposition temperature, and physical methods, such as sputtering and evaporation. These techniques can be applied for metals, dielectrics and functional layers, such as piezoelectric ceramics.

The most widely used substrate material is silicon, which is extremely well known from its use in microelectronic industries. Other crystalline substrates including quartz (SiO_2), gallium arsenide (GaAs), or lithium niobate ($LiNbO_3$) that exhibit piezoelectric properties, and are typically used for resonator applications. Quartz is still dominating frequency reference applications for various oscillators, be it the clock of a microprocessor or of watches. In recent years, polymer based materials are rapidly gaining importance especially for microfluidic or life science oriented applications. This is mainly due to their lower cost per unit area and to their mechanical flexibility, which can be a desired property in some applications. Ceramic materials that are traditionally well represented in packaging technologies and in hybrid integration start finding their specific microactuator niches in harsh environment applications too.

Silicon based microtechnologies are classified into bulk and surface micromachining, where the first one exploits the full depth of the substrate as the structural material, whereas the latter is based on deposited layers, which are typically polysilicon, for forming the structures and silicon dioxides of various compositions as so called sacrificial layers, defining gaps between the substrate and the structure.

Polymer microstructures are typically fabricated by embossing and by moulding techniques, which by now have attained a high level of sophistication. Lithographic methods based on LIGA (lithography and galvanoforming, derived from the German expression) or on SU-8 resist technology (a special photo resist that can be employed in thick layers, up to several hundred microns) can be exploited for making highly precise mould inserts for embossing and injection moulding on the µm or even sub-µm scale.

The principles used for generating forces in microactuators are the same as those encountered in classical actuators. However, due to the different scaling behaviour and the different compatibility with microsystem technologies, other forces dominate the scene.

By their very nature, microactuators allow only small displacements and forces, leading to a natural limitation of their application potential. They have the largest potential where only small forces are needed and where miniaturization is an advantage per se, e. g. because an array setup is required. Thus, applications that are aimed at the switching of small electrical currents, at the manipulation of light or of small volumes of fluids have a high potential. The ink-jet printer head, which controls the ejection of tiny droplets of ink onto the print medium, is amongst the most successful high volume device in all microsystem technology. Similarly, analytical devices in life science applications, including control valves and micropumps, are rapidly gaining importance. Another highly successful microactuator is the digital mirror device from Texas Instruments. It consists of an array of electrically movable micromirrors, that can be addressed individually to project a pixel defined picture on a screen and forms the key element of most modern digital projectors. Switches and high frequency micromechanical oscillators for microwave applications in the GHz domain are rapidly gaining importance, even defining a new subclass of microsystem technology, called RF-MEMS (radio frequency micro-electrical-mechanical systems). A further important application is the implementation of self-test capabilities in sensor-systems aimed for safety-relevant applications. A typical example for this is the airbag accelerometer, which requires an actuation of the seismic mass to prove its functionality during its lifetime. Some sensors, especially micromechanical gyroscopes for measuring an inertial angular rate, even require a means of actuation from their very principle of operation, putting them into a constantly vibrating mode.

6.8.2 Driving Mechanisms, Scaling Laws, and Materials

The most dominant driving mechanism in conventional actuators is the electromagnetic force. With the exception of combustion engines, almost all other motors, aimed for a large variety of applications, are based on this principle. This goes from very small motors on the centimeter scale up to large motors generating in excess of 1 MW, e. g. for driving high-speed trains. The success of this principle is mainly due to the ease of generation of strong magnetic fields by electromagnetic coils and to the relatively long range of the magnetic force on the scale of several centimeters, or even more. Electrostatic forces, however, only play a side roll in conventional actuators. Although they can also be generated quite easily in a parallel capacitor plate configuration, their strength decreases inversely proportional to the distance of the plates. This limits their practical applicability usually down to the µm range.

When we now look at the laws of scaling of these forces down to smaller dimensions, the situation observed in conventional actuators gets to be reversed. By applying a linear geometric scaling factor related to the length dimension l to a mechanical structure, electrostatic forces are scaled down by a factor proportional to l^2, when the field strength is assumed to remain constant at its maximal value. Since volumes and inertial masses are scaled down by l^3, electrostatic forces are actually gaining in relative strength by reducing the size of a structure. This effect becomes even more favourable, when taking into account that the breakdown field strength in an air gap condenser increases when the gap shrinks to the dimensions of the mean free path of the molecules filling the gap. This is called the Paschen effect and leads to an approximately linear scaling of electrostatic forces with the geometric dimension l.

The scaling laws for electromagnetic forces are somewhat more complicated, depending on the assumptions made. The limiting factor in shrinking a magnetic coil is the current density that the electrical conductor can carry. This limit, in return, is linked to the conditions of heat transfer in the structure, because excessive heat resulting from the inevitable power loss inside the coil would lead to its self-destruction. When a constant current density is assumed, the electromagnetic force scales with l^4, leading to a very unfavourable situation. This can be improved to approximately l^3, when a more efficient heat transfer is implemented in the structure, taking advantage of the improved surface-to-mass ratio. Surface area is linked to heat transfer, whereas mass or volume is linked to heat generation. The interaction of a permanent magnet with a coil also scales with l^3. In any case, electromagnetic forces lose upon miniaturization in comparison to electrostatic forces.

Besides these purely mathematical considerations of scaling, there are further restrictions deriving from process compatibility issues. Coils turn out to behave rather problematic from a planar process integration point of view, as can be observed from their nearly negligible role in integrated circuit technology. Only recently, some groups working on RF-MEMS structures have successfully tried to implement truly three-dimensional coils in planar technology. An overview on electromagnetic microactuators can be found in [301].

Air gap separated parallel plate condensers, however, can readily be integrated in microstructures. In the most common configuration, one of the plates is flexible or flexibly suspended (Fig. 6.110), the other is firmly attached to the substrate. Application of a voltage causes the flexible electrode to be drawn toward the rigid electrode. Reasonable forces with realistic driving voltages, however, can only be implemented when the gap comes down to the µm range. In many applications it is required to limit the voltage to the

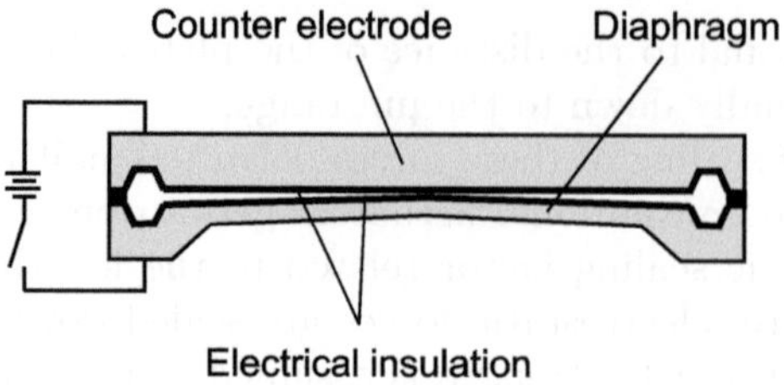

Fig. 6.110. Electrostatic diaphragm actuator

typical range encountered in IC technology, i. e. 5 V or up to 15 V. This poses a severe limitation on the achievable forces and, thus, on the applications. For generating higher forces, it may become necessary to raise this operating voltage to the range of 100 ... 200 V, which means that a special high voltage electronic circuitry needs to be implemented. A positive point of electrostatic actuation is its inherently low temperature drift.

Another phenomenon that needs to be considered is the so-called pull-in effect: the force between the plates is inversely proportional to the gap and thus highly non-linear. For this reason, the plates can only be displaced in a controlled manner by a maximum distance of one third of the original air gap. When the driving voltage is further increased beyond this point, the plates are suddenly attracted to each other until they reach direct contact. To avoid electrical shorting, an insulator is required between the two plates. After having reached contact, the voltage must be reduced substantially to get the plates back into their original position. Thus, a hysteresis effect can be observed. Due to the relative strength of adhesional forces in microstructures, there is a real danger of irreversible sticking of the capacitor plates, which must be prevented by appropriate geometrical means in the layout to reduce the contact surface.

A new type of structure was invented to overcome the limited deflection capability of the parallel plate capacitor: the so-called comb drive [302]. An example of this is shown in Fig. 6.111. This structure can easily be im-

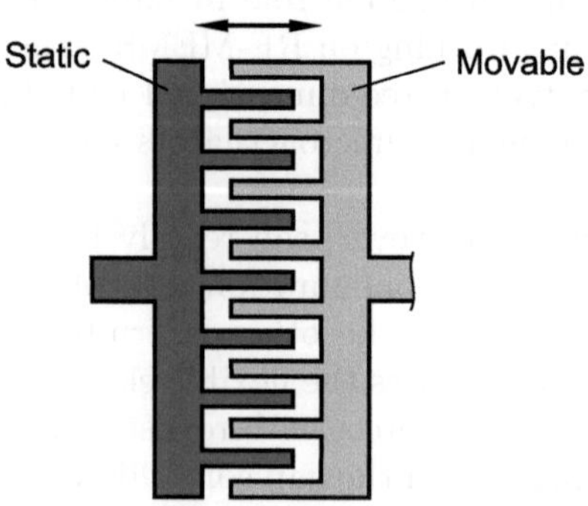

Fig. 6.111. Electrostatic comb drive element

plemented, especially when applying surface micromachining technologies. In contrast to the parallel plate configuration, a comb drive exhibits a completely linear behaviour over its range of operation. Its strength can be scaled by the number of combs and by minimizing the air gap between the movable and the rigid structure. These structures have now become very popular in both sensing and actuating applications, e. g. in gyroscopes or in optical switches for telecommunication.

Despite their unfavourable scaling behaviour, electromagnetic forces find application in microstructures when large deflections are required or when the possibility of reversing the direction of force is important. The magnetic force can be nearly kept constant over a large geometric range and can also be reversed in direction by an opposing driving current. Electrostatic forces, in contrast, are always attractive in practical applications. An example for electromagnetic actuation that found its way into production is the micromechanical gyroscope by Bosch, where the excitation of the resonator is achieved via Lorentz force by a permanent magnet in combination with a driving current passing over the resonant structure [303].

The piezoelectric driving mechanism is also well known from macroscopic applications. Materials exhibiting this effect change their mechanical shape upon application of an external voltage. The dimensional change is usually very small, on the order of a few µm. However, this force is strong and can attain extremely high speeds, up to the GHz range. The classical approach is to use piezoceramic plates and attach (glue) them to the structures (e. g. diaphragms) that need to be deflected. This has to be done for every device individually, and is therefore a limitation to the cost reduction potential. Application of a voltage generates a transverse contraction of the piezoceramic material, leading to a vertical deformation of the layered composite. The displacements, however, are limited to a few micrometers at voltages on the order of 100 ... 200 V.

For larger displacements of 10 ... 30 µm (about 0.15% of the thickness), forces of several hundred Newton and surface pressures of 30 MPa can be achieved with piezo stacks. However, such structures cannot easily be integrated in microdevices. Cantilever type piezoelectric bimorph and unimorph transducers are capable of generating displacements of several hundred micrometers – depending on the transducer dimensions – although at considerably lower forces. The highest compatibility with planar batch technology can be achieved by depositing thin film piezoelectric materials [304]. Today, the best choices of materials, considering piezoelectric coefficients and process compatibility issues, are reactively sputtered aluminium nitride (AlN), directly sputtered lead zirconium titanate (PZT) and, to a lesser extent, zinc oxide (ZnO). As commonly observed in thin film technology, these layers do not quite attain the coefficients known from their bulk material equivalents, but are still very reasonable values.

Another very promising approach is the use of piezoelectric polymer materials such as PVDF and its copolymers that usually are extruded into foils with typical thicknesses of several tens of micrometers and are subsequently polarized. Combining such a foil with another elastic material (usually a metal for electric activation) by deposition or gluing, a unimorph structure can be created, whereas two active layers packed together form a bimorph structure. The copolymer PVDF-TrFE is of particular interest because, in contrast to standard PVDF, it can be deposited by spin coating with a subsequent polarization step. This allows the direct integration of an active polymer layer into a microdevice.

The operating range of piezoelectric materials is limited to well below their Curie temperature, which is typically 150... 300°C for ceramics and only 70... 90°C for polymers. Piezoelectric drives generally exhibit hysteresis which can be compensated by sophisticated electronic circuitry. For applications where a purely resonant mode of operation is desired, hysteresis poses no problem.

The magnetostrictive effect is a change of length of materials in the presence of a strong magnetic field (cf. Sect. 6.3 and references there). It is proportional to the square of the field strength ensuing a frequency doubling effect in oscillatory conditions. Both a positive and a negative effect can be observed, leading to a lengthening or shortening of the original structure. For most common materials this effect is rather small, with typical relative strains on the order of 10 ppm. However, an exotic class of materials based on rare earth elements exhibits a so called giant magnetostrictive effect, achieving strains up to 2000 ppm or 0.2%. These materials are known under the brand name Terfenol-D [305] with a composition of $Tb_xDy_{1-x}Fe_y$. It has been shown that these materials can be sputter deposited as thin films which makes them attractive for microactuator applications [306]. However, the large power requirement for generating strong magnetic fields limits the practical applicability of this method.

Thermal actuators commonly exploit differential thermomechanical expansion of materials, known as the thermomechanical effect for solids or as thermopneumatic effect when gases are involved. In some cases the liquid-vapour phase change is exploited, generating a substantially larger increase in volume or pressure, as can be achieved otherwise. The thermomechanical principle commonly exploits different thermal expansion coefficients of two materials, such as two metals, or a metal and a semiconducting or dielectric material. This is called the bimetal effect. Upon electrical heating by thermal resistors, two materials joined together bend due to their differential expansion, leading to a displacement of the actuator [307]. An alternative thermomechanical approach is the differential heating of neighbouring sections of the same material. This method is usually restricted to small displacements, due to the difficulty of maintaining large temperature gradients in a small structure. However, the latter principle is not sensitive to changes in ambient temperature, which is a major limitation for the practical application of bimetal actuators.

When a sealed gas filled chamber is heated, an increase of pressure is induced that can displace an elastic diaphragm. This thermopneumatic effect is used for actuating valves and pumps. Through thermal losses, the electrical power consumption of actuators based on this effect is relatively high – typically 0.1 ... 2 W. Response times for heating are on the order of a few milliseconds, for cooling on the order of 100 ms, and up to 100 µm displacements are typical. The achievable forces can be increased substantially through the vaporization of a liquid in a partly filled chamber. The liquid phase can be converted to the gaseous phase without effecting a change in temperature.

All thermal actuators have a reputation of being rather slow, due to thermal time constants typically in the upper millisecond range, particularly for the cooling phase. Large forces can usually be achieved at the expense of considerable power consumption. In small structures, however, it has been shown that substantially higher speeds can be attained, due to reduced thermal time constants. Thus, an accelerometer based on a thermally actuated resonant read-out principle was shown to operate at a frequency of 400 kHz [308].

A relatively novel principal of actuation is based on the use of shape memory alloys (SMAs) [309] which can produce large dimensional changes upon heating, due to a phase transition between martensitic crystalline state at lower temperatures and austenitic state at higher temperatures (cf. Sect. 6.4). The achievable relative strains are on the order of several percent (1 ... 8%). However, these materials usually require the application of an initial mechanical strain, the so-called training phase. The best known materials to show this effect are NiTi-based alloys (Nitinol) with the possible addition of Cu, Pt or Fe. CuZn and CuFeZn are also known to show this effect. These materials can also be deposited as thin films by sputtering techniques which makes this effect applicable for microactuators. A microvalve operating on this principle has been demonstrated (cf. Sect. 6.4) but only few practical microdevices have found their way into production.

In electrochemical actuators, the flow of an electrical current chemically converts a liquid to a gas in an electrochemical cell. Hydrogen is generated, for example, in a chamber filled with water, causing an increase of pressure in the cell. A reverse of the current decreases the pressure by oxidation of the hydrogen to form water. A diaphragm bounding the chamber can be brought into periodic motion as a result of the current (and thus pressure) changes. Typical response times of these actuator types are on the order of several seconds at displacements of several millimeters [310].

For some applications the simultaneous incorporation of two actuation principles in a hybrid way can be of interest. An example for this is the combined use of electromagnetic and electrostatic forces in a microvalve [311]. The electromagnetic force is applied for generating large displacements in opening or closing the valve, whereas the electrostatic force can keep the valve in its closed position with very little power consumption. Similarly, the combination of electromagnetic and piezoelectric forces was demonstrated for operating a microswitch [312], leading to an advantage both in maximum achievable deflection and in reducing the power consumption in its closed position.

6.8.3 Microfluidic Systems and Components

The best known and most widespread microfluidic system today is the ink-jet printer. It is described in detail in the next section. Initial pioneering work was started in the 1950s, when attempts were made to build an analogue hardcopy device. In the mid 1980s several discrete microactuators for flow control were introduced, including microvalves and micropumps (see subsequent subsections). Then starting from the 1990s, the advance in biotechnology stimulated intensive research in new microfluidic systems for life science applications, including lab-on-a-chip analytical devices and drug-delivery systems.

Ink-Jet Printer Heads

Ink-jet technology can be characterized as a contact free dot matrix printing procedure. Ink is ejected from a small aperture nozzle directly onto a specific position of the print medium [313]. The generation of small droplets with well defined volume is based on the Raleigh-instability of free liquid jets. The enabling component for this system with a multi-billion Euro annual turnover is the ink-jet printer head. This is an example of a true microactuator that made it to a readily available product with extreme commercial success, being produced in ever increasing numbers. Due to their small fabrication costs, these printer heads nowadays are typically marketed as disposables, avoiding the tedious change of ink cartouches. This strategy increases the production numbers to the largest heights of any commercial microsystem product that is currently available.

The principle setup of this device can be described as follows: an ink reservoir feeds a pressure chamber which is in direct contact with a linear arrangement of microscopic nozzles, shooting out droplets of ink on demand towards the print medium (usually paper). Two principles of actuation for the pressure chamber have been successfully implemented. In the more traditional setup a piezoelectric element is employed to contract a wall of the chamber, thus increasing the pressure which leads to the ejection of an ink droplet. This principle is employed by companies such as Epson, Sharp and Tektronix, to mention a few. The limit of this technology is the size reduction of the piezoelectric actuators. In most cases, piezoceramic elements are used in hybrid integration. More recently, however, PZT deposited in thick film technology has been employed by Epson.

As an alternative, the application of a phase-change thermopneumatic principle has become very popular. A short heating pulse induced by an electric resistor ($<10\,\mu s$, $10\,mW$) vaporizes the ink in the chamber, generating a gas bubble which leads to a substantial increase in pressure. A pressure wave is created in the liquid, causing a small droplet of ink to be ejected from the nozzle. After turning off the heating pulse, the gas bubble collapses, returning the printing head to its original condition. This technology is employed by Hewlett-Packard, Canon, Olivetti and Lexmark, among others. Its main advantage is the ease of integration of a thermal resistance heater with all driving electronic circuitry into a standard CMOS-chip. The nozzles are implemented in a separate silicon wafer and finally bonded together.

Typical droplets have a diameter of about 20 µm and volumes on the order of 50 pl. Actuation frequencies up to the kilohertz range can readily be achieved.

More recently, the ink-jet principle has been extended to new applications for dispensing other liquid substances at defined positions. One of the fastest growing is the dispensing of bioliquids for the fabrication of microarrays, also known as biochips, where samples of DNA, individual nucleotides or proteins are brought onto a carrier substrate in a dot matrix arrangement. Another target application is rapid prototyping of three-dimensional structures by printing successive layers of curable polymers on a substrate.

Microvalves

A pneumatic valve, produced using conventional fine mechanical methods but which can be regarded as a microvalve with respect to its power consumption, is being produced and sold by the German company Hoerbiger [314]. The core of the 3/2-way diverter valve is a piezoelectric bending transducer that closes either the compressed air inlet or the air-bleed nozzle depending on the end position, thereby controlling the pressure in the chamber connected to the outlet (Fig. 6.112). The piezoactuator enables the valve to be switched within 2 ms while drawing negligible power. The valve is available with working pressures up to 200 kPa and has a nominal throughput of 1.5 l/min. It can be implemented either as an on-off valve or as a proportional valve. Additional companies (Bürkert, Joucomatic) are following the trend of powerless pneumatics and offering similar valves based on piezoelectric bending actuators.

The American company IC-Sensors offers a thermomechanically driven microvalve (with 2/2-way functionality) for gases (Fig. 6.113) [315]. The valve consists of an elastically suspended valve reed and a rigid valve base with a valve opening. The valve reed can be put into motion using the thermomechanical (bimetallic) effect. The temperature of the bimetallic structure of aluminium and silicon is used to control the actuator force. The normally closed valves are designed for a maximum pressure of 10 ... 200 kPa. Gas flows of up to 0.1 l/min are achieved with a driving power of 300 mW.

The American company Redwood MicroSystems offers thermopneumatically operated microvalves for gases. Both normally open and normally

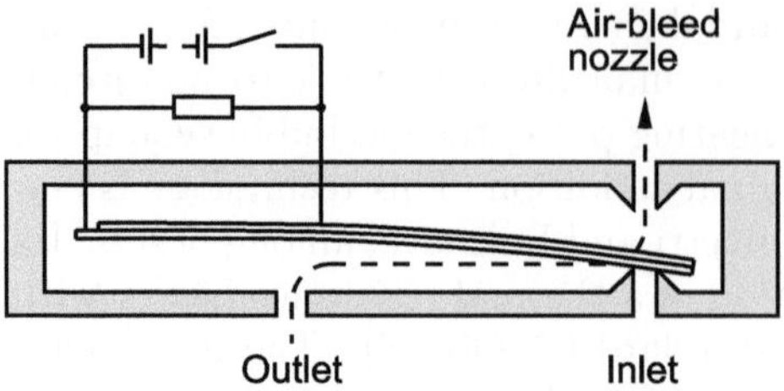

Fig. 6.112. Piezovalve according to [314] (dimensions: $30 \times 19 \times 8\,\text{mm}^3$)

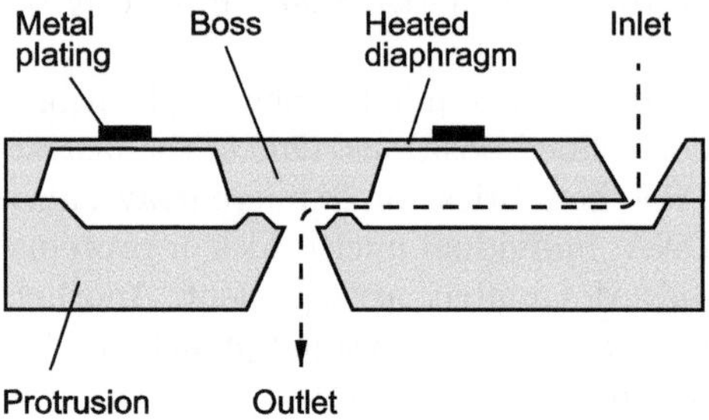

Fig. 6.113. Thermomechanical valve [315]

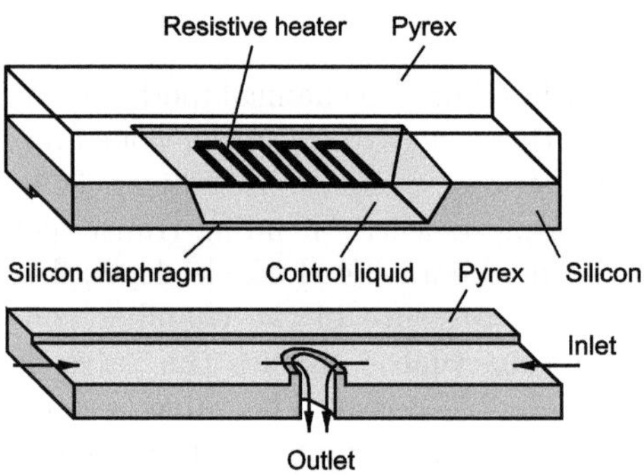

Fig. 6.114. Thermopneumatic valve [316]

closed models are available. The basic structure consists of three components (Fig. 6.114). A drive chamber etched into silicon and filled with a liquid is bordered on one side by a thin silicon diaphragm and on the other side by a rigid glass cover plate (Pyrex). Heating of the chamber causes the liquid to vaporize and displace the diaphragm [316]. A valve opening located in a third component is opened or closed depending on the position of the diaphragm. The normally open valve variation can be converted to a normally closed variation with appropriate adaptation of the mechanics. The valves are designed for a maximum pressure of 700 kPa. A driving power of 1.5 W can control gas flows up to 1.5 l/min. A particle filter with a pore size of 10 µm must be used. The operating temperatures are limited to the

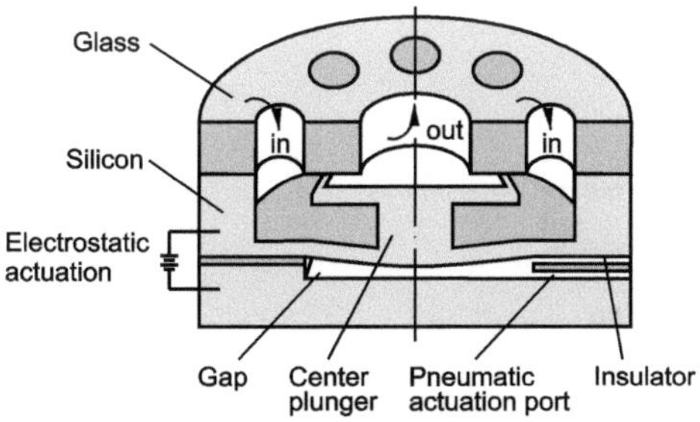

Fig. 6.115. Electrostatic valve according to [317]

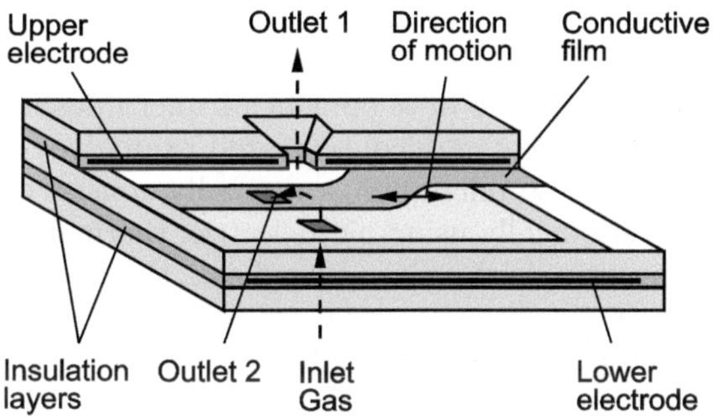

Fig. 6.116. Electrostatic valve according to [318]

range 0 ... 55°C, and the switching times are typically of the order of one second.

In an approach pursued by Bosch, a valve reed in a normally closed design is opened by electrostatic forces (Fig. 6.115). Pressures of the order of 10 kPa and a throughput of 0.2 l/min are controlled with an operating voltage of 200 V [317]. In another approach realized by Hitachi, a thin conductive film is electrostatically set into motion over an opening (Fig. 6.116). The valve with 3/2-way functionality can be activated to control a maximum pressure of 30 kPa, and it requires nearly zero power at an operating voltage of 200 V [318].

A thermally driven bimetallic valve was developed at the Institute for Microtechnology and Information Engineering (IMIT) of the Hahn-Schickard Society, Villingen-Schwenningen, Germany [307]. The valve is suitable for use with both liquids and gases (Fig. 6.117). The valve consists of two components, a flexible valve reed and a valve seat. Placement of the valve seat on a flexible compensating diaphragm decouples the required actuation force from the input pressure. The valve was designed for a maximum pressure of 100 kPa. The maximum throughput is 0.5 l/min for gases and 1 ml/min for liquids. The electrical power consumption is 1 W.

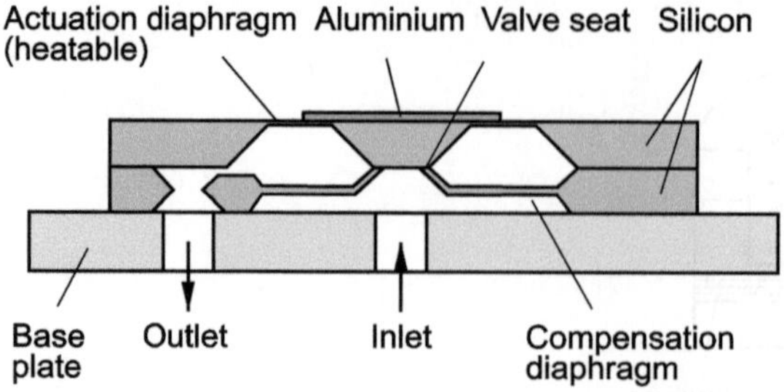

Fig. 6.117. Thermomechanical valve according to [307]

Micropumps

A second topic paralleling microvalves is the development of micro dosing elements and micropumps. The developments are concentrated primarily on the miniaturized diaphragm pump. These micropumps normally consist of a displacement diaphragm driven periodically using piezoelectric, thermal or electrostatic principles, and two passive check valves that direct the flow of liquid from the inlet to the outlet.

The electrostatically driven micropump displayed in Fig. 6.118 was developed at the Fraunhofer Institute for Solid-State Technology (IFT) in Munich. Maximum pumping rates of 1 ml/min and a maximum hydrostatic counter pressure of 30 kPa can be achieved with this device [319]. The external dimensions of the pump are $7 \times 7 \times 2\,\mathrm{mm}^3$. The electrical drive signal is composed of a pulsed DC voltage with an amplitude of 200 V. The electrical power consumption of the pump unit depends upon the operating frequency, which lies typically in the range 1 ... 20 mW. The pump is normally operated at frequencies between 1 and 1000 Hz. A volumetric displacement of approximately $0.01 \ldots 0.05\,\mathrm{mm}^3$ is achieved in each cycle. Filters with a pore width of 5 µm are implemented to prevent contamination. An increase of the operating frequency above the mechanical resonance frequency of the valve (2000 ... 6000 Hz) causes a reversal of the pumping direction; thus the pump can be implemented as a bi-directional unit. This effect results from a phase shift between the motion of the valve and that of the fluid [320].

Micropumps with flow nozzles fulfilling a rectifying function do not require a check valve [321].

A micropump driven by piezoelectrics was developed at the Technical University of Ilmenau. The privileged direction of flow is determined by two pyramid-shaped diffusers etched into silicon (Fig. 6.119). Due to their geometry, these diffusers exhibit different flow resistances for each direction at higher flow speeds (Reynolds number > 100). This characteristic enables alternating flows to be rectified through a two paces forward, one pace back principle. With valve channel widths between 80 µm and 300 µm depending

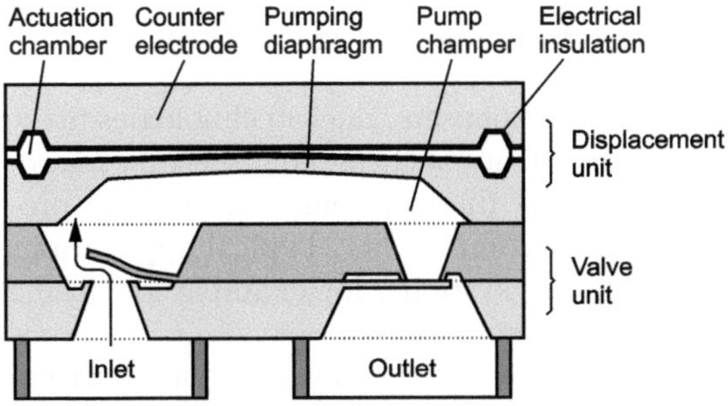

Fig. 6.118. Electrostatic micropump

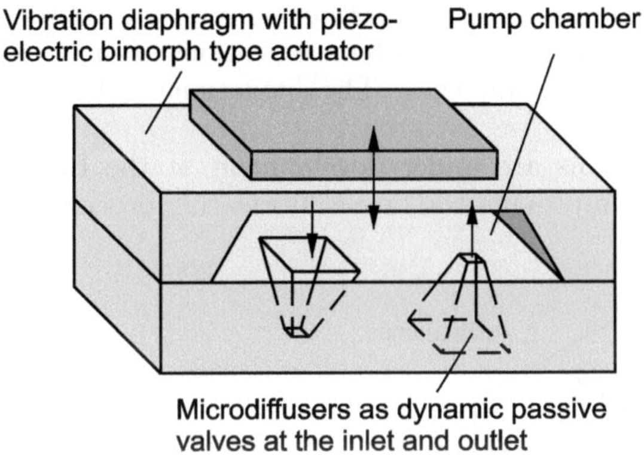

Fig. 6.119. Valveless micropump. After [321]

upon the type, these pumps are also less susceptible to contamination than those with check valves. Maximum pump rates of 400 µl/min and a maximum hydrostatic counter pressure of 7 kPa were achieved with a watery solution for a unit measuring $7 \times 7 \times 1\,\mathrm{mm}^3$. Pump rates of 1 ... 10 µl/min are achievable with gases.

At Chalmers University in Stockholm, the nozzles were etched laterally into the silicon. A maximum hydrostatic counter pressure of 25 kPa (for a pump with larger external dimensions) was measured following an optimization of the flare angle [322]. The pumps are suitable for direct feed of fluids and gases. Care is to be taken when shutting off valveless micropumps because the transported medium will flow back in the presence of a hydrostatic counter pressure.

An infusion pump for painkillers was developed at Trinity College in Dublin [323]. The pump will be worn on a wristwatch and is based on an electrochemical form of actuation. The flow through an electrochemical cell generates a gas and a corresponding pneumatic pressure. The pressure displaces the medication stored in a compressible reservoir (10 ml). The pump is currently undergoing clinical testing on patients receiving painkilling medication.

Microfluidic Systems for Analysis, Lubrication and Dosing

In addition to valves, pumps, nozzles and dispensers, microtechnologies make available other modular fluid-flow components such as flow sensors, micromixers and reaction chambers. Customized fluid systems can now be produced solely on the basis of these modular components. Typical applications are microanalysis systems and microdosing systems, for example for dosing medications, chemical reagents, lubricants and adhesives.

A microsystem for analysing water (Fig. 6.120) was developed within the scope of a joint project (VIMAS) funded by the German Ministry of Education and Research (BMBF) under the leadership of the Fraunhofer Institute for Solid-State Technology. Using appropriate sensors, this system determines environmentally relevant parameters (concentrations of nitrates, oxygen and carbolic acid; pH values; opaqueness). The dimensions of the base plate are $31 \times 32\,\mathrm{mm}^2$.

Miniaturized lubricating systems are under development at the IMIT. The first application will be for improving the 'wick' lubricating process. In

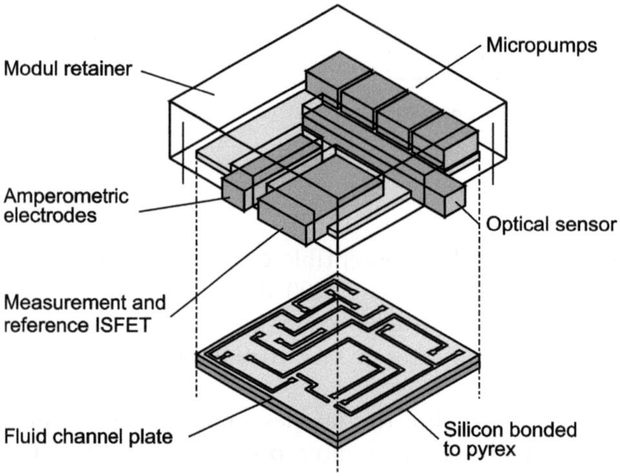

Fig. 6.120. Microanalysis system [source: IFT]

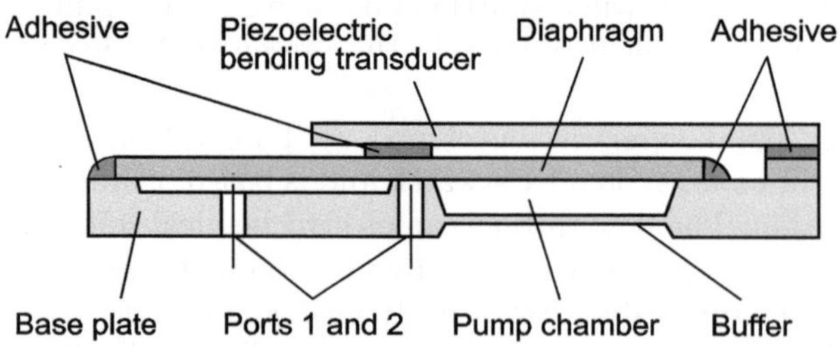

Fig. 6.121. Micropump. After [324]

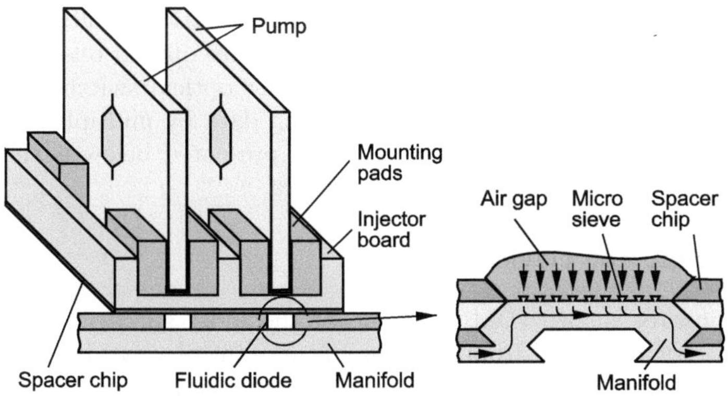

Fig. 6.122. Microdrop injector. After [325]

this process, a film of oil is carried by capillary action from a container to the part to be lubricated, typically a rotating part. Problems can sometimes arise when undesired excess lubrication and strongly varying oil consumption result from varying rotating speeds. A microsystem consisting of the dosing pump (as presented in Fig. 6.121), a microbuffer (volume $< 5\,\mathrm{mm}^3$) and an oil sensor offers a viable solution [324]. The oil sensor measures the level in the buffer and the pump provides the lubrication as needed.

Dosing of the smallest quantities of liquid on the order of nanoliters and microliters was the goal in the cooperation between the Research Centre of Rossendorf and the GeSiM company of Dresden in the development of a microdrop injector [325]. The unit consists of a micro-injection pump (MEP) and a microsieve functioning as a diode for liquids (Fig. 6.122). The piezoelectrically driven injection pump functions similar to an ink-jet printer head, applying microdrops to the sieve. These droplets mix themselves with the liquid located below through surface tension. The microsieve makes use of surface tension effects to prevent the carrier liquid from soaking through into the injection chamber containing air. The disadvantage of half-opened systems is offset by the advantageous ideal liquid separation between the injection and carrier liquids by the air/sieve interface. Applications for this unit can be found in the fields of chemical sensing, pharmacy, medicine and biotechnology.

6.8.4 Actuators in Microoptical Systems

Microactuators have a large potential in optical applications, since no large forces are required for the manipulation of light. One of the largest and most rapidly growing markets in this field is for projection displays with an annual turnover in the multi-billion Euro range. Presently, such displays are mainly focussed on business and educational applications, but they can be expected to gain a major share in future consumer TV markets with larger screens.

At the heart of this application stands an array of electrostatically actuated micromirrors that will be described below. In modern fibre optical telecommunication networks there is an increasing demand for optical switches at hubs that redirect and distribute streams of incoming data by multiplexing into the right optical channels. Another application presented below is the manipulation of optical waveguides by microactuators.

The Digital Micromirror Device

The digital micromirror device (DMD) is one of the most successful microsystem devices ever produced from an economical point of view. Its idea goes back to an invention made by L.J. Hornbeck at Texas Instruments in 1987 [326]. At its heart stands a pixelated array of deflectable micromirrors that can be addressed and actuated individually to display an image on a projector screen, when combined with the illumination and optics required for this purpose. The mirror structures are fabricated after the completion of the CMOS process flow that creates all the underlying circuit elements required for driving and ultimately displacing the mirrors by electrostatic forces. The micromirrors are 16 µm squares of a highly reflective aluminium alloy.

The hinges are hidden underneath the mirror, so that they cannot defract light, thus achieving a high contrast ratio of the image. The micromirrors are arranged in an x–y array, and the chip also contains row drivers, column drivers and timing circuitry. The addressing circuitry under each mirror pixel is a memory cell (a CMOS SRAM) that drives two electrodes under the mirror with complementary voltages. The electrodes are arrayed on opposite sides of the rotational axis that turns through the torsion bar attachments. The mirror is held at ground potential through an electrical connection provided by the support pillars and the torsion bar attachments. A micrograph of a group of micromirrors can be seen in Fig. 6.123. One element has been removed to provide visual access to the underlying hinge-support structure.

Depending on the state of the SRAM cell (a ‘1’ or ‘0’ in the memory) the mirror is electrostatically attracted by a combination of the bias and address voltage to one or the other of the address electrodes. The mirror rotates until its tip touches on a landing electrode fabricated from the same level of metal as the electrode. The electrode is held to the same potential as the mirror. The mirror can rotate $+/-\ 10°$. A ‘1’ in the memory causes the mirror to rotate $+10°$, while a ‘0’ in the memory causes the mirror to rotate $-10°$. A mirror rotated to $+10°$ reflects incoming light into the pupil of the projection lens and the mirror appears bright (on) at the projection screen, whereas in its opposite state the reflected light misses the pupil of the projection lens, and appears dark (off). The input data rates and data bus widths are designed and specified so that the entire memory/mirror array

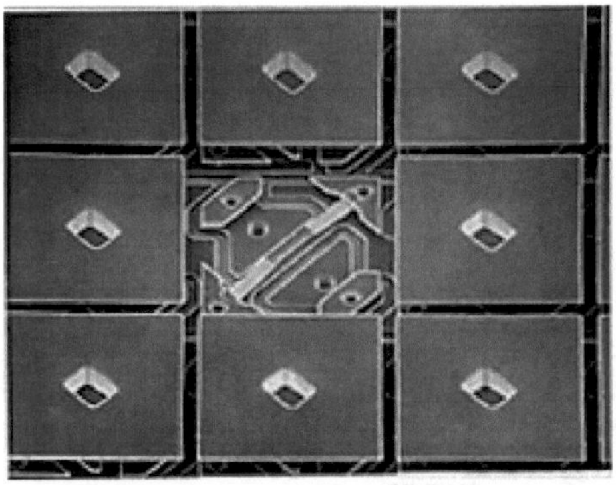

Fig. 6.123. Digital Mirror Device from Texas Instruments

can be refreshed 48 ... 60 times during a single video frame and to provide a display with 16 million possible colours.

The DMD chip has become the key element of modern digital display projector technology. Up to now there is almost a monopoly situation for this device, which is reflected by the pricing policy.

Adjustment of Optical Waveguides

An integrated microsystem, designed at the Technical University of Ilmenau in Germany, is able to handle the tight mechanical tolerances of mono-mode waveguide couplings by a controlled adjustment. It basically contains a two-axis microactuator for moving a fibre or a microlens, an optical sensor for position detection and a control circuit. The piezoelectric drive has a bimorph cantilever movable normal to the wafer surface [327]. Its second direction, the in-plane movement, employs a compliant mechanism in order to enlarge the very small strains of a piezoelectric monomorph. It contains a set of elastic hinges arranged as two-stage gear. Figure 6.124 shows the structure and the kinematic principle of the compliant gear.

6.8.5 Microdrives

Micromotors

Electrostatic micromotors built in silicon based surface micromachining technology were first presented by Berkeley University in 1989 [328]. A typical example is shown in Fig. 6.125.

The rotor built out of polycrystalline silicone has a diameter of about 100 µm and includes a number of radial teeth. It is surrounded on its periphery by electrodes that can be addressed individually. Their number is larger than

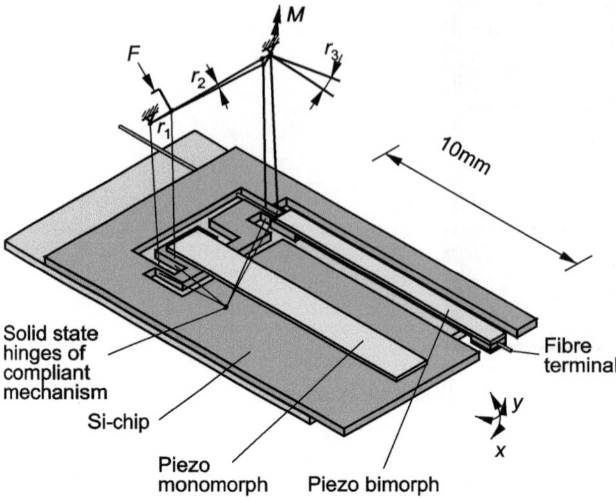

Fig. 6.124. Piezoactuator with compliant gear

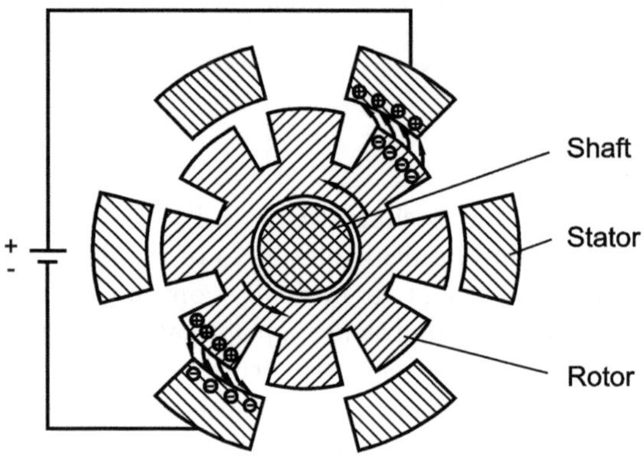

Fig. 6.125. Electrostatic micromotor fabricated in polysilicon surface micromachining technology [328]

the number of teeth (e. g. in a ration of 4:3) so that an attractive force arises between the activated electrodes and nearby rotor teeth, due to an induced electric charging on the electrically insulated rotor. The motors have been brought to spin at rates higher than 10 000 min^{-1}. The most severe problem is the occurrence of friction and high wear in the bushing, severely limiting the practical lifetime of such a motor. Up to now, these motors have mainly been used to demonstrate the capability of the technology but are still waiting for real world applications.

The Institut für Mikrotechnik, Mainz in Germany, has developed a highly reliable micromotor [329]. A synchronous motor scheme with a rotating permanent magnet has been employed (Fig. 6.126a). The micromotor featur-

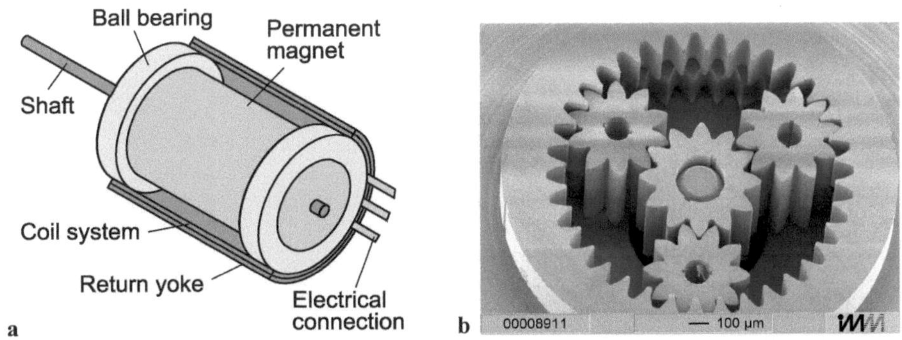

Fig. 6.126. Micromotor with integrated gear box: **a** schematic of the micromotor, **b** assembled planetary gear system [329]

ing an outer diameter of 1.9 mm exhibits a maximum measured torque of 5 µNm in continuous operation. The lifetime is considerably longer than 6 months at 10 000 min^{-1}. A micro gear box of the Wolfrom type has been developed using individually modified involute tooth profiles, whose components are fabricated in metal and polymer materials by means of the LIGA process (Fig. 6.126b). The motor with integrated gear box increases the available torque and helps to open new fields of application – for example, communication and information technology as well as consumer electronics.

Hybrid concepts make use of the most suitable material and the most appropriate process in the fabrication of each component. Such a heteromorphic construction is typical of many microsystems and also demonstrates a broad need for efficient construction, connection and microassembly techniques, and standardized electrical and mechanical interfaces.

Electrostatic Linear Actuators

In the linear actuator depicted in Fig. 6.127, the slide moves over the stator supported by air. Electrostatic forces are generated between comb-like or striped electrodes located on the opposing surfaces of the stator and slide, causing motion of the slide along the x-axis, binding in the y direction and attraction in the z direction. Additional electrodes act as sensors for determining the position in the x direction and the distance of separation in the z direction. All structures are sputtered onto a glass substrate using conventional methods.

The actuator is a 3-phase stepping motor represented by an open control loop. This actuator has a range of displacement in the x direction of 25 mm with a positioning uncertainty of 5 µm. It can also perform small rotations φ_x, φ_y. The holding force can reach 50 mN and the maximum displacement speed 50 mm/s. The electrostatic actuator offers a graphic representation

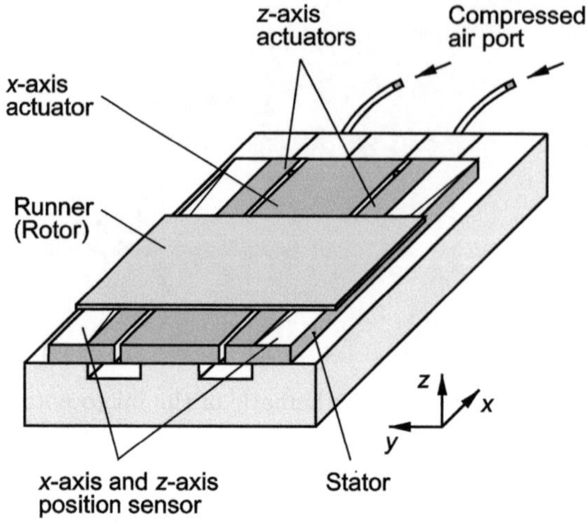

Fig. 6.127. Construction of an electrostatic linear actuator based on microtechnologies. (Source: PASIM Mikrosystemtechnik, Suhl in Germany)

of the trend toward milliactuators: this species of actuator is constructed using microtechnologies but generates forces and displacements on a more macroscopic scale.

6.8.6 Conclusion and Outlook

Ink-jet printer heads and digital mirror devices can presently be regarded as the most successful microactuators on the market. From an economic point of view the ink-jet printer head can even be said to be the most successful device of all currently available microsystems. This demonstrates the extraordinary commercial potential of microactuators in an impressive way.

Microfluidic actuators for controlling fluids have a very high potential of penetrating into new applications outside the printer field. This includes infusion pumps for use in medicine; industrial and micromechanical valves for various applications, microdosing of fluids in microarray technology and other applications.

The market outlook appears to be particularly favourable for microvalves, driven by piezoelectric bending transducers or thermal principles. Such devices are already being produced and sold by several companies. The possible applications of these valves will increase when they can be mass-produced inexpensively and operated with very low power consumption. Stimulus can be expected particularly from valves with 3/2-way functionality, to be introduced as pilot valves in many areas of automation.

Micropumps for transporting and dosing small liquid quantities represent another main group of microfluid actuators. The fact that these pumps are not capable of producing suction currently impedes their use in industry, but should be solvable in the near future.

Strongly miniaturized micromotors based on electromagnetic forces on the threshold of commercial application, whereas electrostatic micromotors are still in a demonstrator stadium.

For the future it can be expected that high frequency RF-MEMS switches and oscillators will rapidly penetrate into commercial applications, opening up new miniaturization and cost reduction potentials in telecommunication applications.

6.9 Self-Sensing Solid-State Actuators

H. Janocha, K. Kuhnen

6.9.1 Introduction

A typical feature of adaptronics is the integration of sensory, actuator and control functions in structures and systems. The degree of function density is particularly high if one and the same component exhibits sensory and actuator properties. Such multifunctionality is enabled for example through the application of piezoelectric, electrostrictive or magnetostrictive materials as well as shape memory alloys. Actuators based on such materials hold – independently of auxiliary sensors – information about the mechanical output quantities force and displacement as well as about the electrical input quantities. The concept of a so-called self-sensing actuator [330, 331] encompasses certain signal processing techniques and will be explained in detail for piezoelectric and magnetostrictive solide-state actuators in this section.

Since the 1990s, great effort has been put into researching the application of self-sensing actuators. Figure 6.128 displays one of the obvious application fields of self-sensing actuators. The drawing on the left, Fig. 6.128a, shows a customary closed control loop. A key function consists in measuring the characteristic system or process quantities which are then pre-processed in the measurement electronics and fed into the controller. The controller compares the measured quantities with the given set values and, depending on the difference between the two, determines the control signal for the power electronics by means of algorithms in accordance with a control strategy which has been installed in the computer.

In Fig. 6.128b, the actuator has been replaced by a self-sensing actuator. Due to the self-sensing effect, it is possible to reconstruct the current process quantities force F and displacement s by means of measured electrical quantities, and a special force sensor or an additional displacement sensor is not required. A more detailed description of the procedure of generating the reconstructed process quantities F_r und s_r will be given in Sect. 6.9.4. Additionally, Fig. 6.128b shows that it is possible to implement closed-loop control without any explicit sensor technology. If knowing the values of force F,

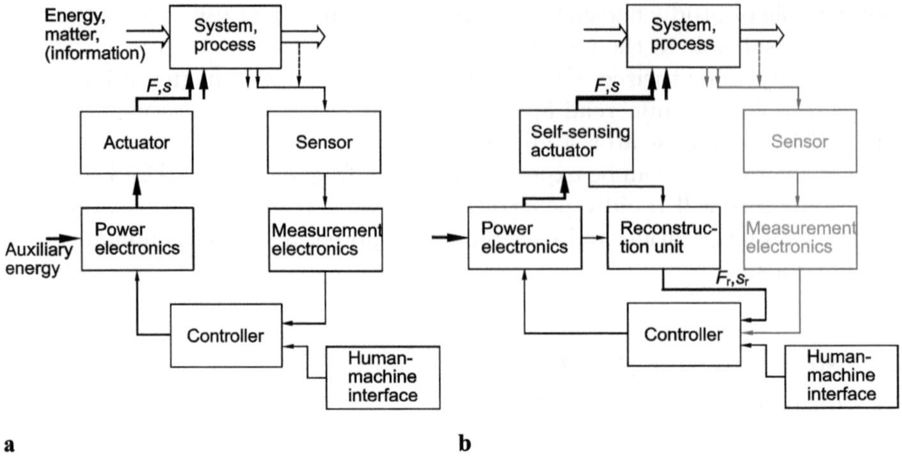

Fig. 6.128. Control of systems and processes. **a** Customary closed-loop control, **b** closed-loop control with self-sensing actuator

displacement s, and their time-dependent derivatives and integrals suffices, the system or process can even manage without the entire right-hand sensor branch (sensor and measurement electronics).

For technical applications it is a great advantage that the self-sensing effect has a wide range of other positive properties. Modeling the output-input characteristic of the self-sensing actuator is a prerequisite for reconstructing F_r and s_r. It can be used for the software-based linearisation of the hysteretic transmission behaviour, which is characteristic of piezoelectric and magnetostrictive actuators (Sects. 6.2 and 6.3). With an extended error model it is additionally possible to compensate creep effects, whose consequences for static operation have often been underestimated particularly in piezoelectric actuators and which often result in position errors. When modern methods of signal processing are applied, the above mentioned types of compensation can also be implemented in real time [332].

Further advantages of the self-sensing effect can be seen in the example of piezoelectric laminar transducers (transversal mode, d_{31} mode), applied to a plate-shaped or bowl-shaped structure (see Fig. 6.129). The self-sensing actuators exchange sensory information with each other and with the host processor, for instance regarding the structures eigenmode. The controller implements a structural model which uses this information to generate control signals for the actuator operation. These signals are fed directly into the self-sensing actuators allowing the user to control the surface form. The fact that actuators and sensors are collocated has the positive effect of making it easier to maintain stability in the control loop, which in turn proves advantageous for the design and operation of the controller [333].

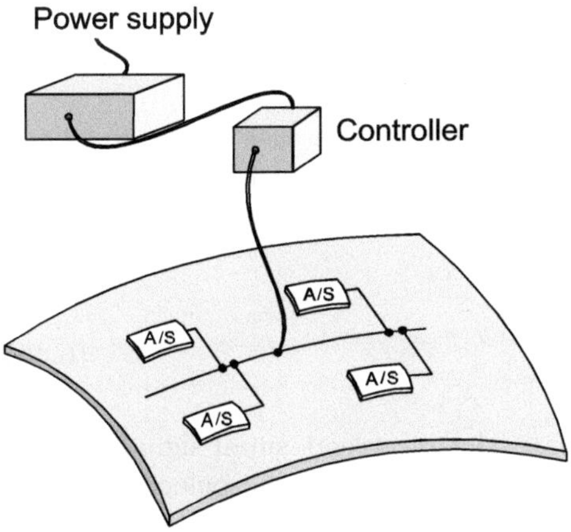

Fig. 6.129. Controlling of freeform structures with networked self-sensing actuators

Finally, the natural combination of sensor and actuator properties in self-sensing actuators constitutes a good basis for the implementation of mature health monitoring systems. A number of piezoelectric self-sensing actuators, for instance, which are applied to or built into a certain structure can be operated such as to transmit test signals into the structure, while some of the transducers collect the response signals and send them to an assessment computer for analysis. The sender and receiver transducers can be cycled according to specific strategies in order to infer faults in the material or structural integrity based on deviations of the transmission behaviour with respect to the reference behaviour. Furthermore, the self-sensing effect offers the possibility of straight-forward and reliable in-process self diagnoses of the piezotransducers in order to check their function.

The following sections focus on the self-sensing effect in piezoelectric and magnetostrictive actuators. Therefore, the most important and basic principles of this group of solid-state actuators will be given below from the standpoint of system theory.

6.9.2 Solid-State Actuators

Piezoelectric, electrostrictive and magnetostrictive transducers are able to transform electrical energy into mechanical energy and vice versa, almost without any delay. This property is the base for self-sensing solid-state actuators.

Piezoelectric Actuator

Small-Signal Equivalent Circuit Diagram. In stack transducers the vectors of the electric field strength $\boldsymbol{E}$ and the dielectric displacement $\boldsymbol{D}$ as well as the tensors of the stress $\boldsymbol{T}$ and the strain $\boldsymbol{S}$ can be replaced by their scalar components voltage V, electrical charge q, force F and displacement s [330].

The linear (6.1) and (6.2) in Sect. 6.2 lead in this case to the following system equations for the integral electrical and mechanical quantities of the piezoelectric transducer:

$$q(t) = CV(t) + d_{\mathrm{p}}F(t) \, , \tag{6.44}$$

$$s(t) = d_{\mathrm{p}}V(t) + \frac{1}{c_{\mathrm{p}}}F(t) \, . \tag{6.45}$$

The parameters in these equations are the electrical small-signal capacitance C, the small-signal stiffness c_{P} and the effective piezoelectric charge constant d_{P}, compare Fig. 6.130.

The total current I_{g} on the electrical side of the transducer is the sum of the polarisation current component $\mathrm{d}q/\mathrm{d}t$ and a component corresponding to the conductance G, which results from the ceramics non-ideal insulation properties:

$$I_{\mathrm{g}}(t) = \frac{\mathrm{d}}{\mathrm{d}t}q(t) + GV(t) \, . \tag{6.46}$$

The resulting force F_{g} on the mechanical side can be approximated by summing the force F inside the piezoelectric transducer with a force component resulting from the inertia of the transducers effective mass m:

$$F_{\mathrm{g}}(t) = F(t) + m\frac{\mathrm{d}^2}{\mathrm{d}t^2}s(t) \, . \tag{6.47}$$

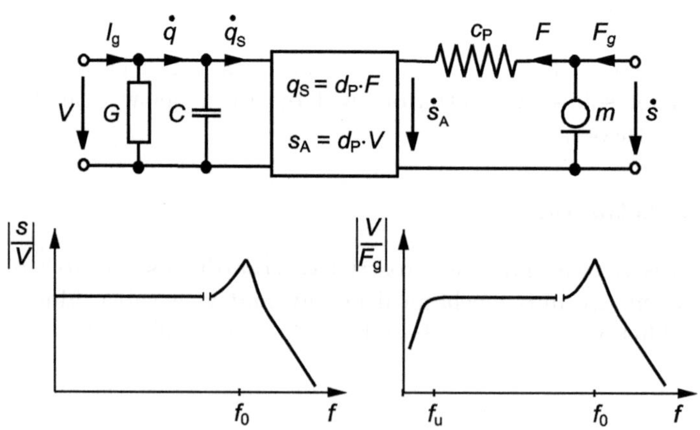

Fig. 6.130. Electromechanical equivalent circuit diagram and amplitude responses of actuator and sensor transfer characteristic in small signal operation

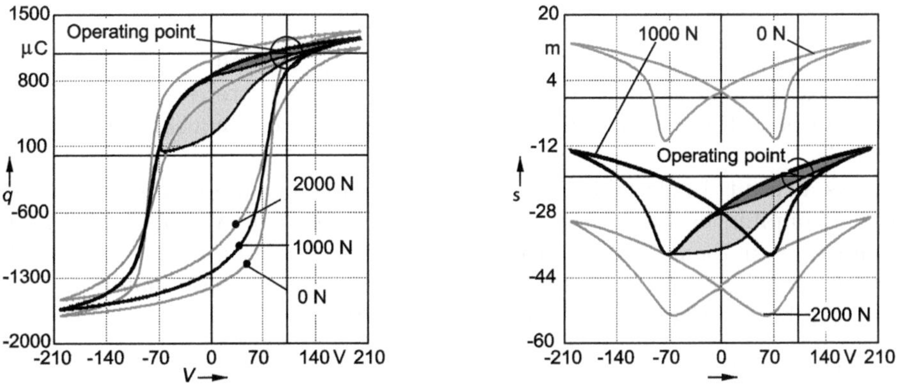

Fig. 6.131. Typical hysteretic characteristics of a piezoelectric stack transducer for different mechanical loads with marked operating regions and operating points

The interpretation of (6.44), (6.45), (6.46) and (6.47) is illustrated in Fig. 6.130, showing an electromechanical equivalent circuit diagram. Accordingly, the input of a piezoelectric transducer can be considered as an electrical capacitor with the capacitance C and its output as a mechanical spring with the stiffness c_{P}. As in reality C is always lossy and c_{P} has always a mass and a structural damping behaviour, the amplitude response $|V/F_{\mathrm{g}}|$ of the piezoelectric transducer has a definite lower cut-off frequency f_{u} and a mechanically determined natural frequency f_0 for an open electrical port ($I_{\mathrm{g}} = 0$), and the amplitude response $|s/V|$ has a mechanically determined natural frequency f_0 for an open mechanical port ($F_{\mathrm{g}} = 0$).

Operation Range and Operating Point. The maximum achievable displacement of piezoelectric ceramics is limited by saturation and repolarization. In practice, usually only the operating region of the displacement-voltage characteristic which is dark grey-shaded in Fig. 6.131 is employed. For special applications, it is also possible to expand the operating region to the light grey-shaded area. However, the negative operating voltage may not exceed about 30% of the maximum voltage, as otherwise an electrical repolarization will occur. In order to achieve bipolar operation of the piezoelectric transducer, the transducer is electrically biased by a constant voltage at about half of its operating range. The mechanical operating point is given by the mechanical pre-stress in the transducer casing.

Large-Signal Characteristic. In order to produce noteworthy displacements during actuator operation, a piezoelectric transducer is driven by an electrical control voltage V, which excite unwanted domain switching processes in the active material. These domain switching processes cause more or less strong macroscopically observable hysteresis and creep effects in the electrical q–V characteristic and the actuator s–V characteristic. The consequences are the complex branching characteristic shown in Fig. 6.131. Moreover, during actuator operation, the solid-state transducer is loaded with mechanical forces F, which lead to mechanically excited domain switching processes if the force amplitudes are large enough. These mechanically excited domain switching processes also cause hysteresis and creep effects in both the sensor q–F characteristic and the mechanical s–F characteristic.

As a result, in electrical and mechanical large-signal operation there exists a coupling between the voltage V and the force F which, in principle, requires a mathematical description by means of vectorial operators [332] which considers the hysteresis and creep of the piezoelectric material. This fact can be described by the operator notation

$$q(t) = \Gamma_{\mathrm{S}}[V, F](t) \tag{6.48}$$

$$s(t) = \Gamma_{\mathrm{A}}[V, F](t) \ , \tag{6.49}$$

instead of (6.44) and (6.45) which are only a good approximation of the material behaviour in the small-signal range[1]. Equation (6.48) is called the sensor equation and (6.49) the actuator equation of the piezoelectric transducer for large-signal operation.

Magnetostrictive Actuator

Small-Signal Equivalent Circuit Diagram. This type of actuator is based on highly magnetostrictive materials, which are typically implemented in a rod shape. In this case the vectorial quantities of the magnetic field strength $\boldsymbol{H}$ and the flux density $\boldsymbol{B}$ as well as the tensors of the stress $\boldsymbol{T}$ and the strain $\boldsymbol{S}$ can be replaced by their scalar components current I, magnetic flux ψ, force F and displacement s. Subsequently, instead of (6.5) in Sect. 6.3 the following system of equations applies to the integral electromagnetic and mechanical quantities of the magnetostrictive transducer:

$$\psi(t) = LI(t) + d_{\mathrm{M}}F(t) \tag{6.50}$$

$$s(t) = d_{\mathrm{M}}I(t) + \frac{1}{c_{\mathrm{M}}}F(t) \ . \tag{6.51}$$

The parameters in this equation are the small-signal inductance L, the small-signal stiffness c_{M} and the effective magnetostrictive constant d_{M}, compare Fig. 6.132. On the electrical side, the voltage V_{g} is the sum of the voltage evoked by induction and the voltage drop across the copper resistance R of the coil:

$$V_{\mathrm{g}}(t) = \frac{\mathrm{d}}{\mathrm{d}t}\psi(t) + RI(t) \ . \tag{6.52}$$

The resulting force F_{g} on the mechanical side, analogous to (6.47), is approximated by the sum of the force F of the magnetostrictive transducer

[1] Operators are here used to mathematically describe the mapping between the input and output time functions of dynamical systems.

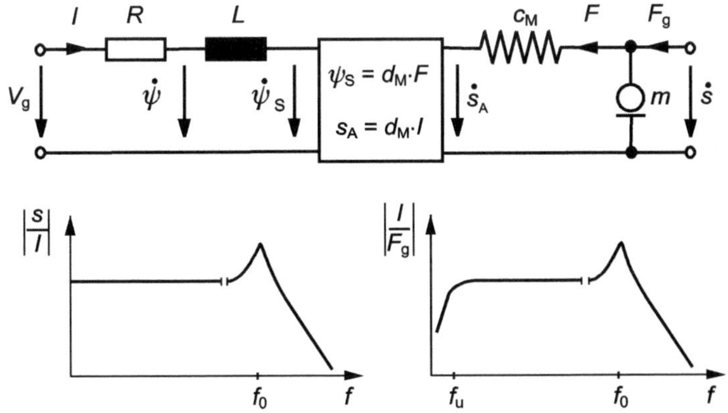

Fig. 6.132. Electromechanical equivalent circuit diagram and amplitude responses of the actuator and sensor transfer characteristic in small-signal operation

and a force component resulting from the inertia of the transducer mass m. In a small-signal equivalent circuit diagram the electrical behaviour of the magnetostrictive transducer can be considered as a lossy inductance L, see Fig. 6.132.

Analog to the piezoelectric transducer, the mechanical behaviour can be described by a spring with the mass m and the stiffness c_M. The amplitude response $|I/F_g|$ of the magnetostrictive transducer has a definite lower cut-off frequency f_u and a mechanically determined natural frequency f_0 for a short-circuit electrical port ($V_g = 0$), and the amplitude response $|s/I|$ has a mechanically determined natural frequency f_0 for an open mechanical port ($F_g = 0$).

Operating Range and Operating Point. In magnetostrictive transducers the positive branch of the relationship between the displacement s and the current I is normally used. The magnetic operating point is usually placed in the middle of the operating range. It is set by a bias current via a magnetic coil or by permanent magnets. The relationship between ψ and I displays a highly sensitive inherent sensory effect in the magnetic operating point shown in Fig. 6.133. Starting with the choice of the magnetic operating point, the operating range of the transducer maximally extends to the reversal point of the s–I characteristic on the left hand side, and on the right hand side to the amplitude range in which ferromagnetic saturation effects restrict the further displacement of the transducer (see grey-shaded area).

Large-Signal Characteristic. For large-signal amplitudes the interaction between the driving current I, the magnetic flux ψ and the displacement s shows the complex branching characteristics displayed in Fig. 6.133. The changes in ψ and s which are produced by the mechanical load F can also be observed in Fig. 6.133.

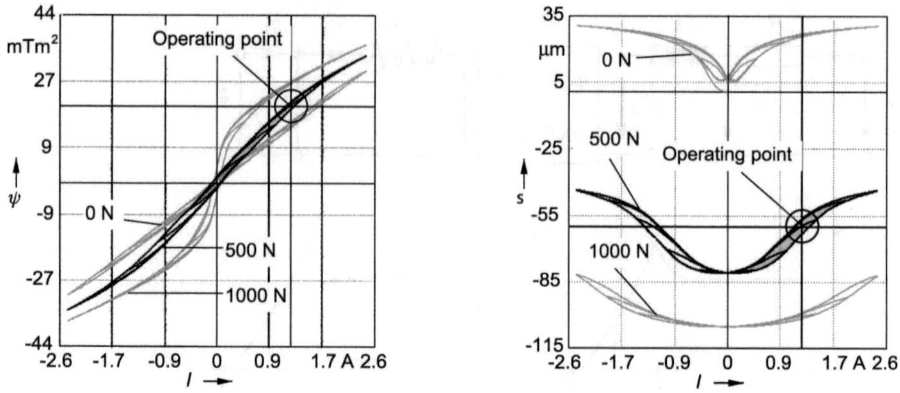

Fig. 6.133. Typical hysteretic characteristics of a magnetostrictive transducer for different mechanical loads with marked operating region and operating point

Similar to the characteristics of piezoelectric transducers, the complex branching in the ψ–I and s–I relationships, and the ψ–F and s–F relationships, which are not pictured here, result from unwanted domain switching processes within the material. In contrast to the piezoelectric material, the domain switching processes inside the magnetostrictive material occur nearly undelayed over a wide range of frequency, meaning that the branching in Fig. 6.133 is purely hysteretic in this range of frequency. Creep is negligible here. However, in contrast to piezoelectric transducers, magnetostrictive transducers exhibit eddy current losses at higher frequencies.

As a result, in large-signal operation there exists a hysteretic coupling between the current I and the force F which, in principle, requires also a mathematical description by means of vectorial operators [334] which considers the hysteresis of the magnetostrictive material. For higher frequencies this description has to be extended to consider eddy current effects. With the general operator notation this fact can also be described by

$$\psi(t) = \Gamma_{\mathrm{S}}[I, F](t) \tag{6.53}$$

$$s(t) = \Gamma_{\mathrm{A}}[I, F](t) \,, \tag{6.54}$$

instead of (6.50) and (6.51). Equation (6.53) is called the sensor equation and (6.54) the actuator equation of the magnetostrictive transducer for large-signal operation.

6.9.3 Self-Sensing Model for Solid-State Actuators

The description of the transfer characteristic of solid-state actuators can be generalised if the system equations which have been introduced in Sect. 6.9.2 are not interpreted as electromechanical equivalent circuit diagrams but as signal flow charts [335]. The result is shown in Fig. 6.134.

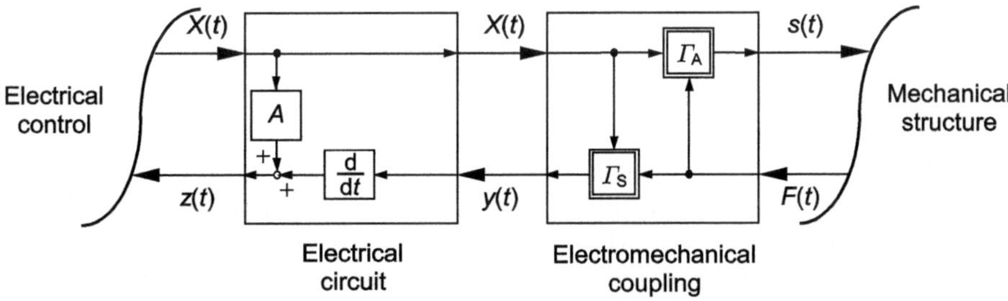

Fig. 6.134. Signal flow chart of the generalized solid-state transducer model for large-signal operation

The intrinsic sensing property of the solid-state actuators result in a dependence of the electrical quantity on the mechanical load, which is thermodynamically dual to the electrical control quantity. In piezoelectric actuators, the dual quantity which contains the sensory information is given by the electrical charge q if voltage V is controlled and by the voltage V if charge q is controlled. Accordingly, the sensory information in magnetostrictive actuators with current control is provided by the magnetic flux ψ through the magnetostrictive material, and in those with flux control it is provided in the coil current I. In order to give a uniform notation, the electrical control quantity V or I will be represented by the electrical input parameter X, whereas the dual electrical quantity q or ψ which contains the sensory information will be represented by the electrical output parameter y. Additionally if we consider the coupled nonlinear memory behaviour of the materials in large-signal operation by means of the general operator notation for the sensor equation

$$y(t) = \Gamma_{\mathrm{S}}[X, F](t) \; , \tag{6.55}$$

and the actuator equation

$$s(t) = \Gamma_{\mathrm{A}}[X, F](t) \; , \tag{6.56}$$

we obtain the self-sensing model for solid-state actuators shown in Fig. 6.134. In this case the electrical input circuit is described by

$$z(t) = \frac{\mathrm{d}}{\mathrm{d}t} y(t) + AX(t) \tag{6.57}$$

with A as the conductance G in the piezoelectric or electrostrictive case and the resistance R in the magnetostrictive case. The abstract variable z describes the electrical port variable current I_{g} in the former case and the voltage V_{g} in the latter case and contains the sensor information due to the inherent sensor property of the material. Based on this generalized transducer model general concepts for the use of the inherent sensor effect in solid-state actuators are discussed in the next section.

6.9.4 Concept of Self-Sensing Solid-State Actuators

The inherent sensor effect in active materials allows, in combination with proper measurement and signal processing methods, the simultaneous use of piezoelectric or magnetostrictive transducers as both sensors and actuators. At present there exists two different methods, a state quantity-related and a parameter-related for using these inherent sensor effects [336]. In both cases the mechanical values of F and s must be reconstructed from the measured electrical quantities.

State Quantity-Related Approach

The state quantity-related sensing method makes use of the dependence of the thermodynamical dual electrical quantity $y(q,\psi)$ on the electrical control quantity $X(V,I)$ and the mechanical load F according to (6.55). The reconstructed mechanical load F_r is gained by means of measurements X_m and y_m of the quantities X and y according to

$$F_\mathrm{r}(t) = \Gamma_\mathrm{S}^{-1}[X_\mathrm{m}, y_\mathrm{m}](t) \; . \tag{6.58}$$

For this purpose the inverse of the y–F mapping with X as a parameter must be calculated. The reconstructed transducer displacement s_r is then obtained in a second step by fitting in the reconstructed force F_r into the actuator equation (6.56). The corresponding reconstruction filter equation is

$$s_\mathrm{r}(t) = \Gamma_\mathrm{A}[X_\mathrm{m}, F_\mathrm{r}](t) \; . \tag{6.59}$$

This is done in a so-called reconstruction filter unit, compare Fig. 6.135. The measured values X_m and y_m are determined from the port quantities X and z by means of special electrical measurement circuits. Together with the driving electronics for the transducer, they form part of the measurement circuit and power electronics unit illustrated in Fig. 6.135.

At their outputs they generate the two measuring voltages V_X and V_y, whereas V_X is proportional to X over the entire frequency range, and V_y is only proportional to y for frequencies well above the cut-off frequency f_z of

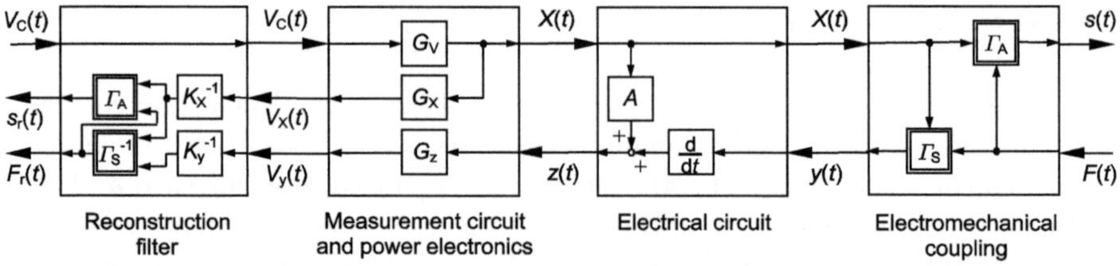

Fig. 6.135. Self-sensing solid-state actuator with state quantity-related use of the inherent sensor effect

the filter transfer function G_z, see e.g. [337] for further details. Scaling the measurement voltages results in the measuring values X_m and y_m which will serve for further processing in the reconstruction filter unit. The scaling as well as the reconstruction of F and s take place in the reconstruction filter unit. The control value X is generated by the power electronics unit. The control value X contains the position control information from the superior electrical control in the form of the control voltage V_C.

Parameter-Related Approach

The parameter-related sensing method uses the dependence of the small-signal electrical parameter

$$\gamma_E(X(t), F(t)) := \frac{\partial \Gamma_S(X(t), F(t))}{\partial X(t)} \tag{6.60}$$

which is defined as the partial derivative of the sensor characteristic Γ_S with respect to the electrical driving quantity X on the mechanical load F. γ_E replaces the small-signal capacitance C of the piezoelectric and the small-signal inductance L of the magnetostrictive transducer, respectively [336, 338, 339].

For the experimental determination of the small-signal electrical parameter the driving voltage V_{CA} which contains the control information will be superimposed by a sinusoidal high-frequency test voltage V_{CT} with small amplitude. This is described by the signal flow chart in Fig. 6.136.

From the control voltage V_C the power electronic unit generates the control quantity

$$X(t) = X_A(t) + X_T(t) . \tag{6.61}$$

With (6.57) this leads to

$$z(t) = \frac{d}{dt}\Gamma_S(X(t), F(t)) + AX(t) . \tag{6.62}$$

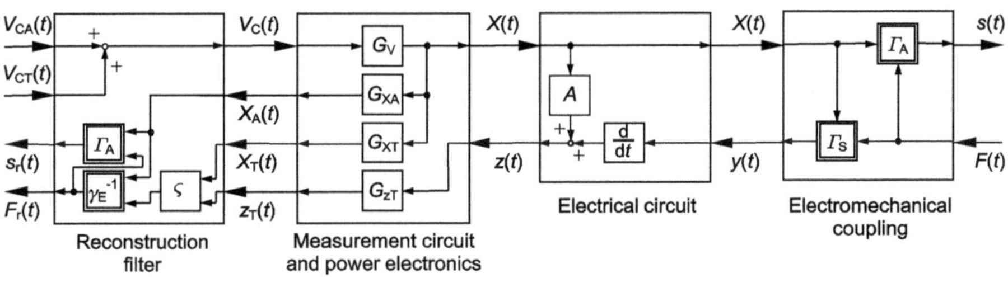

Fig. 6.136. Self-sensing solid-state actuator with parameter-related use of the inherent sensor effect

According to (6.60) the small-signal parameter γ_E can be interpreted as the effective slope of the y–X mapping in the operating point defined by the driving quantity X and the mechanical load F. If the amplitude of the test signal is sufficiently small the small-signal high-frequency variation of γ_E produced by X_T is neglectably small against the large-signal low-frequency variation produced by X_A. In this case the influence of the test signal can be neglected in the argument of γ_E. Thus the electrical quantity z consists of a high-frequency part z_T which can be separated from the low-frequency part z_A by means of a bandpass filter with a transfer function G_zT. X_T is also determined from X by means of a bandpass filter with a transfer function G_XT. An experimental determination of the small-signal parameter from the measurements of X_T and z_T, i. e. a measurement value γ_Em, follows from a phase-selective demodulation or a parameter identification or a signal analysis based on a discrete Fourier transformation (DFT) [340].

These procedures realize a mapping ζ which maps the measured high-frequency components of X and z to the measurement values γ_Em of the small-signal parameter γ_E and is described here by the notation

$$\gamma_\mathrm{Em}(t) = \zeta(X_\mathrm{T}(t), z_\mathrm{T}(t)) \;, \tag{6.63}$$

see Fig. 6.136. Finally the force reconstruction requires an inversion

$$F_\mathrm{r}(t) = \gamma_\mathrm{E}^{-1}(X_\mathrm{A}(t), \gamma_\mathrm{Em}(t)) \tag{6.64}$$

of the parameter model $\gamma_\mathrm{Em}(X_\mathrm{A}, F)$ with respect to the mechanical load F and with the low-frequency driving quantity X_A as a parameter. In this case X_A is determined from X by means of a lowpass filter with a transfer function G_XA. As in the state quantity-related approach the reconstructed transducer displacement s_r is obtained according to (6.59).

Prerequisite for Reconstruction and System Inversion

As just shown, the use of the self-sensing effect requires an inversion of the y–F mapping according to (6.58) in the case of the state quantity-related approach and an inversion of the γ_E–F mapping according to (6.64) in the case of the parameter-related approach. Therefore, above all, the precondition for a successful inversion and thus a successful reconstruction of the mechanical load has to be specified. This object should now be discussed representatively by means of the y–F mapping.

At time t the force reconstruction unit determines the reconstructed force value $F_\mathrm{r}(t)$ which generates the measured value $y_\mathrm{m}(t)$ for the measured driving value $X_\mathrm{m}(t)$. This can be done with the sensor model (6.55) by means of solving the implicit equation

$$y_\mathrm{m}(t) - \Gamma_\mathrm{S}[X_\mathrm{m}, F_\mathrm{r}](t) = 0 \,. \tag{6.65}$$

This equation possesses a unique solution for time t if and only if the continuous y–F mapping is strongly monotonous for all X. In this case the different branches of hysteretic nonlinearities differ with respect to the actual hysteretic state of the operator Γ_S. An additional consideration of this information allows us to solve (6.65) uniquely, and thus we can calculate the inverse mapping (6.58).

In the case of a non strongly monotone y_m–F_r mapping, we have amplitude ranges for y_m with a multivalued solution for the same history of the system and thus an unique inverse mapping Γ_S^{-1} does not exist. From this it follows that the continuity and strong monotony of the y–F and the γ_E–F relationships suffice for the feasibility of self-sensing solid-state actuators. This constitutes a restriction for the parameter-related approach, as the sensory relation of y and F may contain inflection points in large-signal operation. These inflection points result in a maxima in the functional relationship between the electrical parameter γ_E and the force F and therefore lead to a non-monotonous behaviour.

6.9.5 Modeling Hierarchy of Self-Sensing Actuators

As the amplitude of the control signal grows, the domain processes within the solid-state actuators experience an increasing excitation resulting in a stronger non-linear transfer characteristic of the solid-state transducer. In order to keep the mathematical models and the reconstruction equations derived from them as simple as possible, they must be adapted to the amplitude ranges. As a consequence, there are varyingly complex models for different amplitude ranges. These will be described in more detail below.

Linear System Model

As shown in Sect. 6.9.2 in small-signal range the operators Γ_S and Γ_A can be approximated by the linear system equations

$$y(t) = \gamma_E X(t) + \gamma_S F(t) \tag{6.66}$$

$$s(t) = \gamma_A X(t) + \gamma_M F(t)\ . \tag{6.67}$$

Here the coefficients γ_E, $\gamma_S = \gamma_A$, and γ_M correspond to the small-signal capacitance C, the effective piezoelectric charge constant d_P and the inverse of the small-signal stiffness c_P for a piezoelectric transducer and to the small-signal inductance L, the effective magnetostrictive constant d_M and the inverse of the small-signal stiffness c_M for a magnetostrictive transducer, respectively.

In this linear case the inverse operator (6.58) can be derived analytically by an evaluation of (6.66) and (6.67). Then the linear reconstruction model

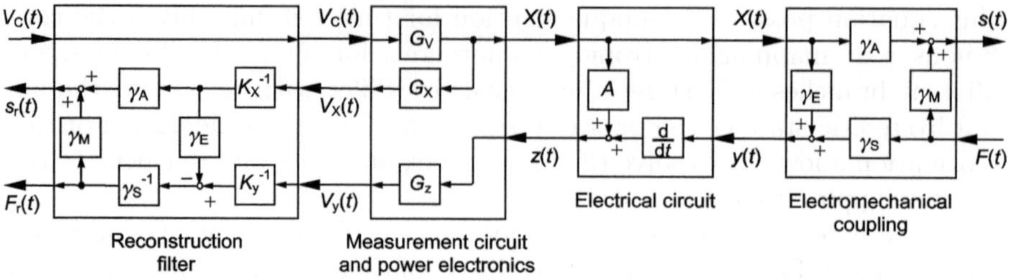

Fig. 6.137. Self-sensing solid-state actuator with linear reconstruction filter

of the self-sensing actuator shown in Fig. 6.137 results in

$$F_r(t) = \gamma_S^{-1}(y_m(t) - \gamma_E X_m(t)) \tag{6.68}$$

and

$$s_r(t) = \gamma_A X_m(t) + \gamma_M F_r(t) \; . \tag{6.69}$$

At first (6.68) was used to evaluate the inherent sensor effect of piezoelectric transducers. The substraction of $\gamma_E X_m$ from y_m can be realized in an analog signal processing with a capacitive bridge circuit in which a piezoelectric transducer is one bridge element [342]. Then the voltage across the bridge is proportional to the force F which now can be determined without additional force sensors. With such a self-sensing actuator different mechanical systems were equipped [343–348].

All these applications have confirmed the principle of a self-sensing actuator, they have also shown that the linear reconstruction model (6.68) and (6.69) is restricted for small amplitudes of the voltage and force. Furthermore the bridge circuit is strongly affected by external disturbances e. g. from temperature leading to a wrong evaluation of the sensory information.

Nonlinear Hysteresis-Free System Model

A further step to extend the validity of the model beyond the small-signal range is a description of the operators Γ_S and Γ_A in (6.55) and (6.56) with smooth nonlinear multidimensional characteristics

$$y(t) = \Gamma_S(X(t), F(t)) \tag{6.70}$$

and

$$s(t) = \Gamma_A(X(t), F(t)) \tag{6.71}$$

without memory. Due to the memory-free[2] character of these mappings the sensor model (6.70) and the actuator model (6.71) are able to model nonlinear

[2] Memory-free mapping means that the present output value in time depends only on the present input value in time and not on the past history of the input signal. Thus memory-free mappings are function mappings.

phenomena like saturation or electrostrictive effects but can not consider hysteretic or creep effects in the characteristics of solid-state transducers.

The formulation (6.70) and (6.71) permits the application of the state quantity-related approach described in Sect. 6.9.4 for the reconstruction of the mechanical quantities s and F. In this case the reconstruction model of the self-sensing actuator results in

$$F_{\mathrm{r}}(t) = \mathit{\Gamma}_{\mathrm{S}}^{-1}(X_{\mathrm{m}}(t), y_{\mathrm{m}}(t)) \tag{6.72}$$

and

$$s_{\mathrm{r}}(t) = \mathit{\Gamma}_{\mathrm{A}}(X_{\mathrm{m}}(t), F_{\mathrm{r}}(t)) \tag{6.73}$$

and requires an invertible memory-free sensor model and an actuator model for a successful implementation.

From the description model (6.70) and (6.71) the linear system equations can be derived as a special case by a linearization in a fixed operating point. In this case the functions γ_{E}, $\gamma_{\mathrm{S}} = \gamma_{\mathrm{A}}$, and γ_{M} correspond to the small-signal capacitance $C(V, F)$, the effective piezoelectric charge constant $d_{\mathrm{P}}(V, F)$ and the small-signal elasticity $1/c_{\mathrm{P}}(V, F)$ for a piezoelectric transducer and to the small-signal inductance $L(I, F)$, the effective magnetostrictive constant $d_{\mathrm{M}}(I, F)$ and the small-signal elasticity $1/c_{\mathrm{M}}(I, F)$ for a magnetostrictive transducer, respectively. The partial derivatives γ_{E}, $\gamma_{\mathrm{S}} = \gamma_{\mathrm{A}}$, and γ_{M} represent the local slopes of the functions $\mathit{\Gamma}_{\mathrm{S}}$ and $\mathit{\Gamma}_{\mathrm{A}}$ in (6.70) and (6.71) in the present operating point (X, F). As discussed in Sect. 6.9.4 the dependence of the electrical small-signal parameter γ_{E} from the operating point (X, F) is used to determine the force.

The method used in [349] is based on a transducer model in which the dependence of the small-signal capacitance C on the voltage V is noticeable already at small amplitudes of the voltage. This dependence was measured at a bending transducer and stored as a characteristic. The dependence of C on the mechanical quantity F was not considered. The bending transducer was used in vibration damping of a cantilever beam. The polarisation charge q was measured by a so-called Sawyer-Tower circuit and the substraction with $C(V)$ was realized by a digital signal processor considering the stored characteristic for $C(V)$. This processor also calculates the phase-inverted driving signal for the transducer according to the reconstruction equation

$$\mathrm{d}F(t) = \gamma_{\mathrm{S}}^{-1}(\mathrm{d}y(t) - \gamma_{\mathrm{E}}(X(t))\mathrm{d}X(t)) \tag{6.74}$$

for the differential quantities. Utilizing this self-sensing actuator the time needed to bring the beam to rest was shortened by a factor of 60, while the assumption of a constant C leads only to a reduction by a factor of 20.

Nonlinear Hysteretic System Model

The next step to extend the validity of the system model is carried out by the modeling of complex hysteretic nonlinearities by the so called complex hysteresis operators. These complex hysteretic nonlinearities are present in varying degrees in virtually all solid-state actuators provided that they are driven with sufficiently high amplitudes [349].

The best-known examples of these so-called complex hysteretic nonlinearities are the Preisach- or Krasnosel'skii-Pokrovskii operator R, the Prandtl-Ishlinskii operator H and the modified Prandtl-Ishlinskii operator $M := S(H)$ which is constructed as a concatenation of a Prandtl-Ishlinskii operator H and an asymmetrical scalar function S of Prandtl-Ishlinskii type which models the deviation of the real hysteretic nonlinearity from the class of Prandtl-Ishlinskii operators [332, 353, 355]. All these operators belong to the class of operators with a Preisach memory P [341].

If the electrical excitation and the mechanical load are limited to amplitude ranges where the dependence of the characteristic of the electrical transfer path and the actuator transfer path on the mechanical load as well as the dependence of the characteristic of the sensor transfer path and the mechanical transfer path on the electrical excitation can be neglected, then the vectorial operators in sensor equation (6.55) and in actuator equation (6.56) can be simplified to a linear superposition of scalar operators:

$$y(t) = \Gamma_{\mathrm{E}}[X](t) + \Gamma_{\mathrm{S}}[F](t) \tag{6.75}$$

$$s(t) = \Gamma_{\mathrm{A}}[X](t) + \Gamma_{\mathrm{M}}[F](t) \,. \tag{6.76}$$

If the mappings Γ in the sensor equation (6.75) and the actuator equation (6.76) are purely hysteretic they can be modeled by a Prandtl-Ishlinskii operator H, a modified Prandtl-Ishlinskii operator M or a Preisach hysteresis operator R depending on the degree of symmetry of the branching behaviour. The calculation of these hysteresis operators and the corresponding compensators from the measured output-input characteristic requires special computer-aided synthesis procedures which is based on system identification methods. Due to a lack of space, this article cannot further comment on these synthesis methods. However, a detailed description of both the synthesis method and the mathematical basics can be found in the literature [332, 341, 350–352, 356].

With the decoupled system model (6.75) and (6.76) the reconstruction model corresponding to (6.58) and (6.59) is given by

$$F_{\mathrm{r}}(t) = \Gamma_{\mathrm{S}}^{-1}[y_{\mathrm{m}} - \Gamma_{\mathrm{E}}[X_{\mathrm{m}}]](t) \tag{6.77}$$

$$s_{\mathrm{r}}(t) = \Gamma_{\mathrm{A}}[X_{\mathrm{m}}](t) + \Gamma_{\mathrm{M}}[F_{\mathrm{r}}](t) \tag{6.78}$$

and requires the compensator Γ_{S}^{-1} of the sensor mapping Γ_{S} in (6.75). It is shown in Fig. 6.138. In contrast to the Preisach hysteresis modeling approach,

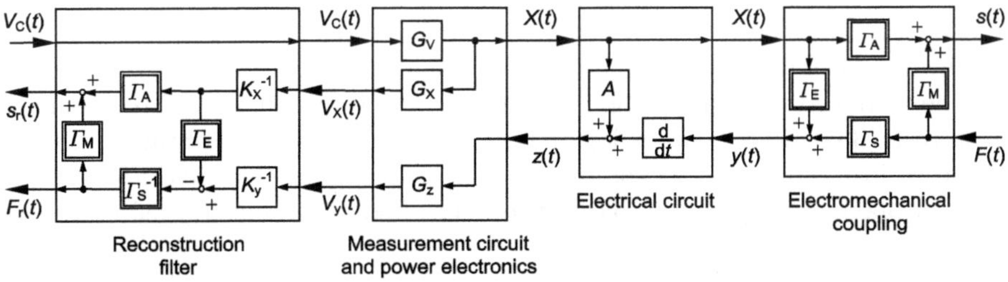

Fig. 6.138. Self-sensing solid-state actuator with hysteretic reconstruction filter

an invertible Prandtl-Ishlinskii operator and a modified Prandtl-Ishlinskii operator permits an analytical design of the corresponding compensator, see e. g. [354, 357]. According to (6.77) this is an important feature for the realisation of the reconstruction filter in real-time.

A reconstruction model, which is able to consider the hysteretic nonlinearities in the characteristic of a piezoelectric transducer, was first introduced by Jones and Garcia in 1997 [358]. In their application they use a charge amplifier instead of a voltage amplifier to drive the piezoelectric self-sensing actuator. Therefore, the polarisation charge q must be regarded as the independent quantity X and the voltage V as the dependent quantity y. In this model only the scalar hysteretic relation between the voltage V and the polarisation charge q will be considered by a scalar Prandtl-Ishlinskii hysteresis operator $\Gamma_{\mathrm{E}} := H_{\mathrm{E}}$. The relation between the force F and the voltage V is assumed to be linear. Therefore, the influence of the large-signal amplitudes on the force, which leads to vectorial hysteresis effects, will not be considered, and this model is only valid for small amplitudes of the force. The strongly nonlinear creep phenomena, which have an influence on the transfer characteristic worth to be mentioned, will not be considered either. It is an advantage of this piecewise linear model that the reconstruction model can be developed analytically from the system model.

Nonlinear Hysteretic and Creeping System Model

Unfortunately, in addition to complex hysteretic nonlinearities, actuators and sensors based on the technologically important piezoelectric ceramics contain also $\log(t)$-type creep dynamics to a degree which is not neglectable in wideband applications like positioning systems.

The term creep is used in the literature primarily in connection with the delayed deformation behaviour of solid materials due to sudden mechanical loading [359]. Very similar behaviour can be observed to different degrees in the relationship between the respective physical parameters in ferromagnetic and ferroelectric materials as well as in magnetostrictive and – even more pronounced – in piezoelectric actuators. And so the term creep came to stand for more than just the delayed response between mechanical input and output parameters. It is not a far step then to go beyond the bounds of physics to obtain a purely phenomenological description of creep, which will be achieved using the elegant operator based approach used to describe hysteresis [332]. The complex log(t)-type creep effect is representative of that observed in many of the technologically important piezoelectric ceramics and as such plays an important role in the field of solid-state self-sensing actuation.

A scalar operator which considers simultaneously complex hysteresis effects, log(t)-type creep effects as well as saturation effects can be constructed by the parallel connection of a Prandtl-Ishlinskii hysteresis operator H and a Prandtl-Ishlinskii log(t)-type creep operator K followed by a concatenation with a memory-free scalar nonlinearity S. In this case the mapping Γ in (6.75) and (6.76) is given by a so-called modified Prandtl-Ishlinskii creep extension M_{K}. The corresponding reconstruction model is then given by (6.77) and (6.78) with the compensator $\Gamma_{\mathrm{S}}^{-1} = M_{\mathrm{K}}^{-1}$ defined by

$$\begin{aligned} y(t) &= M_{\mathrm{K}}[x](t) := S(H[x](t) + K[x](t)) \\ &\Uparrow \Downarrow \\ x(t) &= M_{\mathrm{K}}^{-1}[y](t) \Leftrightarrow x(t) = H^{-1}[S^{-1}(y) - K[x]](t) \,. \end{aligned} \tag{6.79}$$

The inverse modified Prandtl-Ishlinskii creep extension M_{K}^{-1} results from solving the implicit operator equation in (6.79). A suitable approach to solve the operator equation, which requires no more steps than the calculation of the operator M_{K}, can be derived analogous to the approach used in [357], and requires the inverse Prandtl-Ishlinskii hysteresis operator H^{-1} and the inverse memory-free nonlinearity S^{-1} in explicit form. Since H and S are of the Prandtl-Ishlinskii type H^{-1} and S^{-1} can be derived analytically with the knowledge of H and S.

6.9.6 Application Example: 1-DOF Piezoelectric Positioning System

The self-sensing solid-state actuator concept illustrated in Fig. 6.139b was implemented into a commercially available positioning system driven by a low-voltage piezoelectric stack transducer [332, 360].

The additional feedback of the reconstructed force F_{r} about the mechanical characteristic Γ_{M} to the compensation filter Γ_{A}^{-1} in the forward path realises the compensation equation

$$X_{\mathrm{i}}(t) = \Gamma_{\mathrm{A}}^{-1}[s_{\mathrm{d}} - \Gamma_{\mathrm{M}}[F_{\mathrm{r}}]](t) \,, \tag{6.80}$$

which follows from the actuator equation (6.76). A scaling of the generalized control signal X_{i} to the control voltage V_{C} leads to a compensation of the hysteretic nonlinearity Γ_{A} in the actuator characteristic and a compensation of the influence of the mechanical load F on the real displacement s at the mechanical output of the self-sensing actuator.

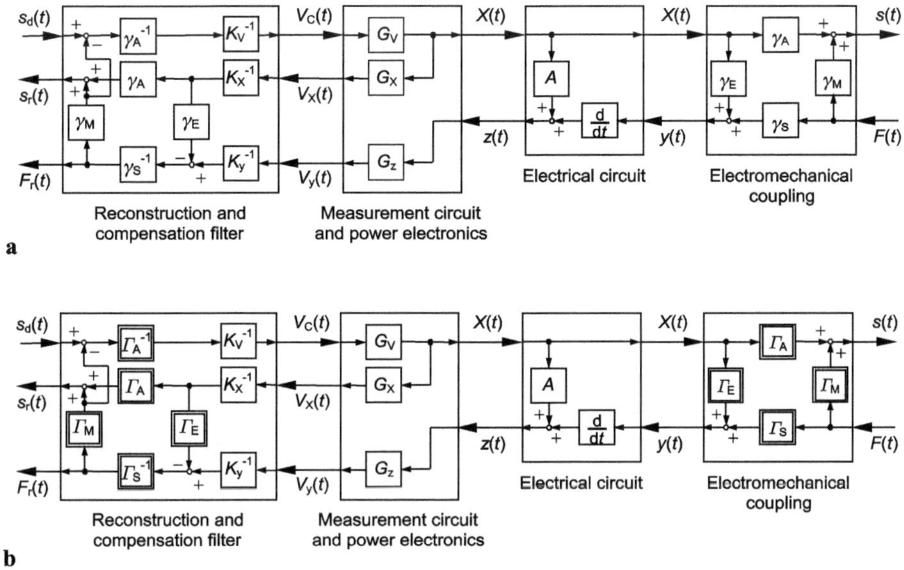

Fig. 6.139. Self-sensing solid-state actuator with reconstruction and compensation filter in the forward path. **a** Linear approach, **b** operator-based approach

The self-sensing solid-state actuator concept illustrated in Fig. 6.139a has the same structure as that in Fig. 6.139b but it is based on the linear system model according to (6.66) and (6.67). The compensation equation follows in this case from the actuator equation (6.67) and results in

$$X_{\mathrm{i}}(t) = \gamma_{\mathrm{A}}^{-1}(s_{\mathrm{d}}(t) - \gamma_{\mathrm{M}} F_{\mathrm{r}}(t)) \,. \tag{6.81}$$

Figure 6.140 shows the measurement results obtained with the two self-sensing actuator principles for electrical large-signal operation. They display the characteristics of the three transfer paths of the bidirectional actuator, illustrated in the form of s–s_{d}, s_{r}–s and F_{r}–F trajectories. In this case, the values X_{i}, X_{m} and y_{m} in the (6.80), (6.77) and (6.78) correspond to the inverse control voltage, the measured control voltage and the measured piezoelectric charge.

In Fig. 6.140a the deviation between the desired displacement s_{d} and the measured displacement s, and between the measured displacement s and the reconstructed displacement s_{r}, are produced, significantly, by the unconsidered hysteresis effect. Moreover a huge deviation occurs between the measured load F and the reconstructed load F_{r} because of the unconsidered hysteresis effect. Using the operator-based filter, the influence of the hysteresis effect is taken into account. As Fig. 6.140b displays, one can hardly recognize the deviations between s_{d} and s and between s and s_{r}. The deviation is seven times smaller as when using a linear reconstruction and compensation model. The deviation between F and F_{r} is now comparatively small too.

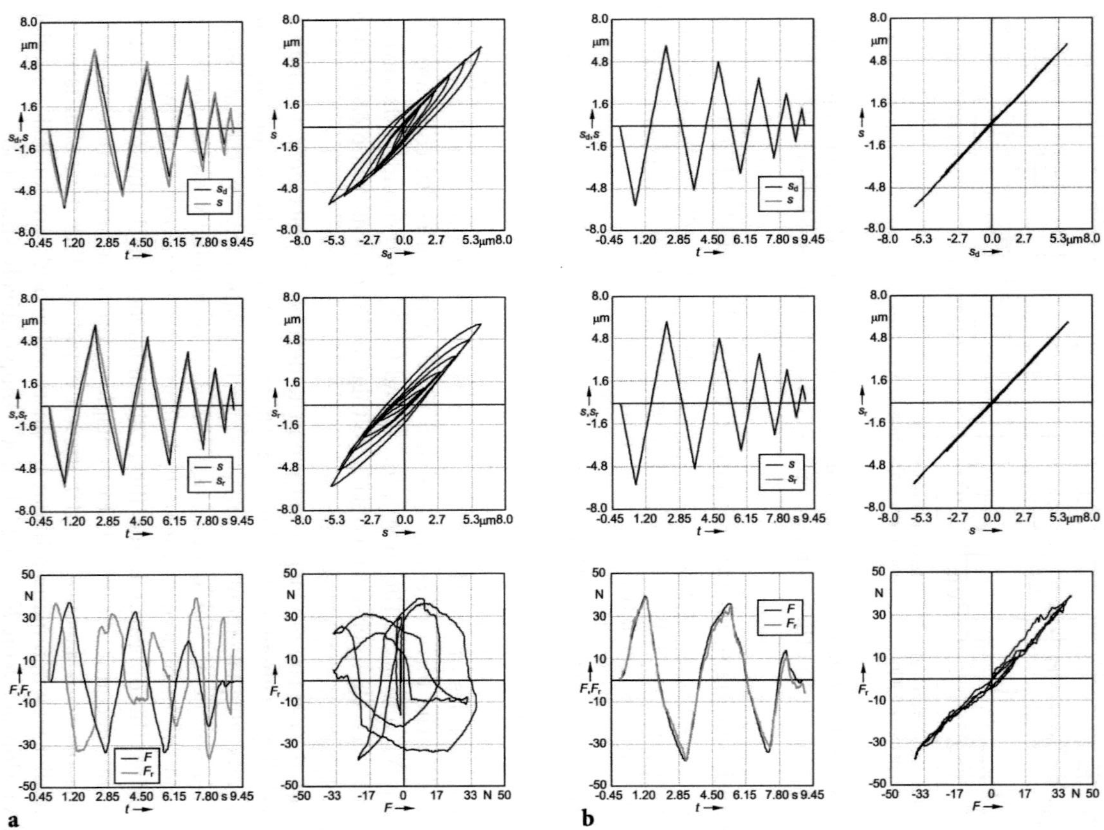

Fig. 6.140. Function of the self-sensing actuator concept according to Fig. 6.139. **a** Linear approach, **b** operator-based approach

6.9.7 Conclusion

The self-sensing actuator concept requires the powerful mathematical machinery of complex hysteresis operators – first for reconstructing the mechanical quantities by means of the measured values of electrical quantities and second for compensating the hysteretic nonlinearities and the load dependency. Whereas robust software tools exist for modeling, identifying and compensating scalar complex hysteretic nonlinearities in practical applications, a considerable amount of research activities is necessary in the field of vectorial hysteresis phenomena to obtain a similar status.

Furthermore, attempts are being made to implement the computationally intensive algorithms of reconstruction and compensation filters within FPGAs, in order that self-sensing solid-state actuators will become available for highly dynamic applications with signal frequencies in the kilohertz range. Recently a built-up realisation of a FPGA-based processor platform which is able to take advantage of the inherent parallel structure of the hysteresis operators used for the modeling and compensation of complex hysteretic nonlinearities in active materials was realised [361]. Hence, it can accelerate the necessary calculations by several orders of magnitude in comparison to conventional DSP-based solutions.

6.10 Power Amplifiers for Unconventional Actuators

H. Janocha, T. Würtz

Power amplifiers are generally designed for driving electrically resistive loads. In contrast, unconventional actuators are mainly electrically reactive loads. This section is dedicated to the interaction between the actuator load and the power amplifier, and we will also treat the common amplifier circuit topologies, providing the reader with valuable information that will help him to design his own power amplifier or choose a suitable commercial product.

Piezoelectric and magnetostrictive actuators in particular, as well as actuators with electrically controllable fluids, are counted among the unconventional actuators. Actuators with shape memory alloys and polymers as well as other, less common actuators sometimes require very simple and sometimes fairly complex amplifiers, which have to be tuned to the actuator and the signals that are to be processed. However, we will not go into the details of such special cases.

Solid-state actuators and actuators with controllable fluids show certain similarities: piezo actuators and electrorheological fluids are driven with electrical fields whereas magnetostrictive actuators and magnetorheological fluids draw their actuation energy from a magnetic field. We will therefore consider the power electronics of these two superordinate groups, but still we will mention the differences between the two different actuator types of each group.

6.10.1 General Information About Power Electronics

There are many possibilities of implementing power stages for unconventional actuators. In the following, we will introduce their fundamental operating principles and some of their combinations.

One, Two and Four-Quadrant Operation

The current and voltage at the output of a power amplifier can be depicted as functions of time or in a voltage-current coordinate system, see Fig. 6.141. Depending on whether the user would like to operate an ohmic, an inductive or a capacitive load, and depending on whether this operation should be unipolar or bipolar, he will require an amplifier with one, two or four-quadrant operation.

If an amplifier used for ohmic loads generates positive current only, one quadrant will suffice. If the load is to be operated with negative voltage as well, two quadrants are required, see Fig. 6.141.

The operation of capacitive loads with positive voltage requires two quadrants: charging requires positive current (quadrant I), whereas discharging requires negative current (quadrant IV), as is illustrated in Fig. 6.142a for an ideal piezo actuator. If the user demands negative voltage as well, e. g. in order to use the full characteristic of a piezo actuator or in order to circumvent electrophoresis within an electrorheological fluid, he will require a four-quadrant amplifier, because all combinations of positive and negative voltages and currents may occur, see Fig. 6.142b.

When operating inductive loads, the current serves as the reference quantity. A positive current will require positive voltages (rise of the magnetic field, quadrant I) as well as negative voltages (decay of the magnetic field, quadrant II). When driving bipolar currents, an amplifier that can pass through

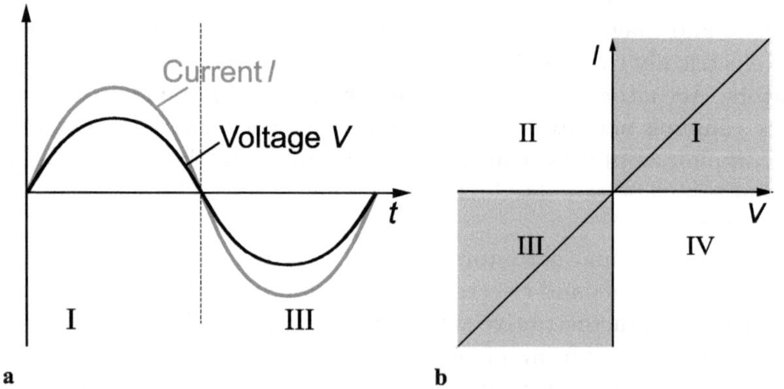

Fig. 6.141. Current and voltage signals of an ohmic load. **a** As function of time, **b** plotted in a voltage-current coordinate system

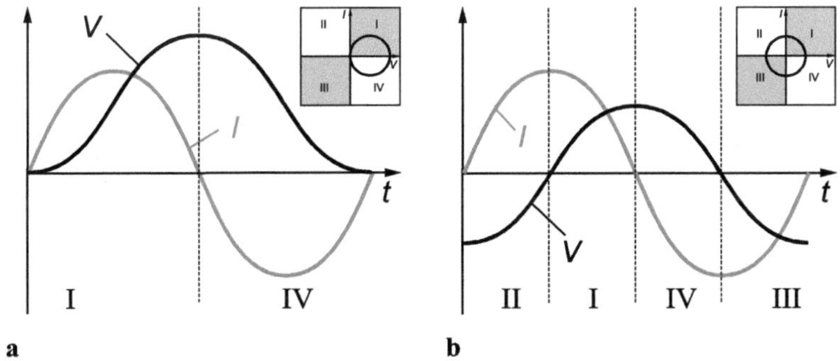

Fig. 6.142. Current and voltage time signals for a capacitive load (e. g. a piezo actuator). **a** Unipolar, **b** bipolar voltage control

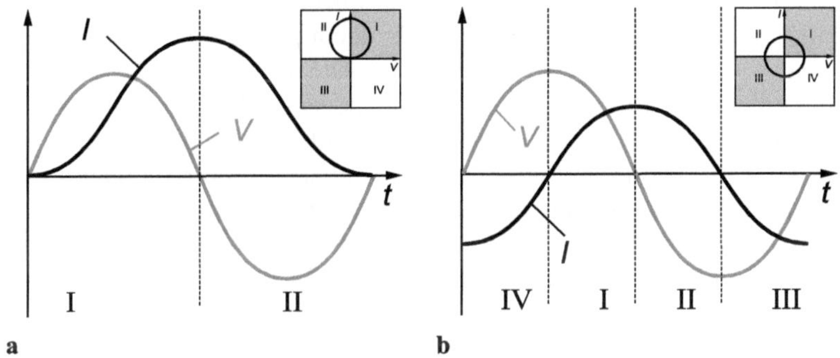

Fig. 6.143. Current and voltage time signals for an inductive load (e. g. a magnetostrictive actuator). **a** Unipolar, **b** bipolar current control

all four quadrants is necessary. These cases are illustrated in Fig. 6.143 for an ideal magnetostrictive actuator.

Switching, Analogue and Hybrid Power Amplifiers

These three circuit concepts differ significantly in terms of their output signal quality, and the efficiency of their energy use.

Switching Power Amplifiers. It is characteristic of switching power electronics that the power semiconductors operate in only two operating modes: they either block or conduct maximally. As long as the corresponding designer guidelines are observed, only minimal losses occur in the semiconductors. Energy is usually stored in a coil and then transferred to a capacitor. Figure 6.144 illustrates several possibilities of connecting actuators and switching amplifiers.

In the lefthand circuit diagram, a piezo actuator is connected as a load. The coil (choke) protects it from rectangular switching voltages, but not from force impulses due to high peaks that can occur in the triangular current-

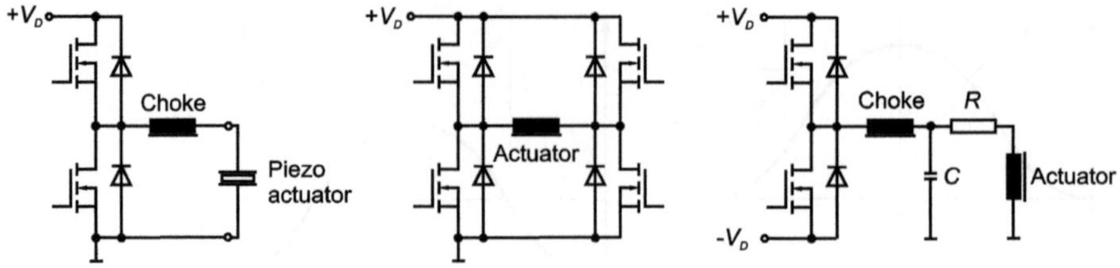

Fig. 6.144. Three principle possibilities of connecting actuators and switching amplifiers

time characteristic. The diagram in the middle shows a magnetostrictive or magnetorheological actuator in connection with a clocked full-bridge; here, eddy currents and parasitic capacitances can cause high current peaks. In the righthand circuit diagram, the actuator is decoupled from the amplifier output by means of an *RLC* filter (which delays the signal).

Depending on the signal quality demanded, the switching frequency must be several orders of magnitude higher than the highest signal frequency. In highly dynamic applications, however, the transmitted power increases linearly with the signal frequency. Furthermore the higher the power that needs to be transmitted, the lower is the working frequency that must be chosen for the switching amplifier. As these two frequencies approach each other, the switching frequency becomes increasingly noticeable in the signal, making it practically impossible to design a filter whose cut-off frequency lies at a clear distance from the signal frequency as well as from the switching frequency.

Subsequently, proper dimensioning of the inductance for energy transmission is crucial for the properties of a switching amplifier. Its design must ensure that the maximum energy required can be achieved without saturation and be transmitted with sufficiently high switching frequency. Another factor that must be taken into account is the greatest energy packet which occurs during the signal period and which must be transmitted in order to provide the load with maximal instantaneous power. Therefore, the coil must be rated to the pulse power requirement of the amplifier.

There are a vast number of different switching amplifier topologies. They can be implemented by means of choke coils or transformers with one, two or several windings, and with very different control concepts. All switching amplifiers can be reduced to two basic types: one type stores all energy in the inductance, and transmits it to the capacitance in a second step. This type is called a flyback converter, and at the capacitance it is able to generate a voltage higher than its own operating voltage, see Fig. 6.145.

In the other type, the coil current runs through the capacitance during charging as well, storing energy in the coil as well as in the capacitance. This

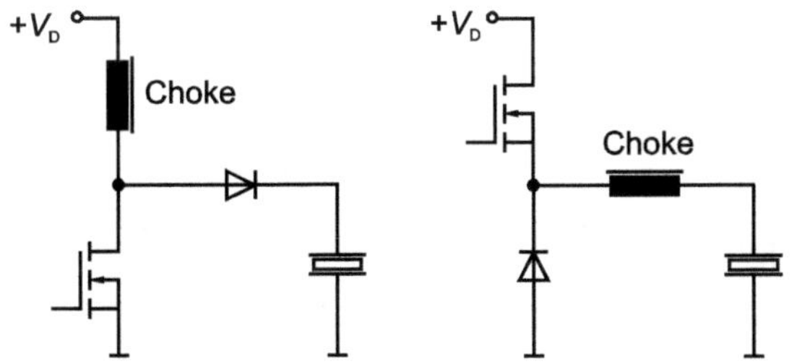

Fig. 6.145. Flyback converter (*left*) and feed-forward converter (*right*) with a piezo actuator as a load

thus called a feed-forward converter. During the next switching operation, the energy stored within the coil is also transmitted to the load. For trouble-free operation, this type always requires an operating voltage higher than the maximum output voltage. Both switching variants are able to feed back the reactive energy from the actuator field to the power supply. Depending on the type and design of the actuator, a great share of the field energy can be recovered as the field decays.

Analogue Power Amplifiers. In contrast to the switching power electronics, with its power transistors that either block or transmit a maximum amount of energy, the analogue switching technology operates its power transistors continuously over their entire operating range as a control element. No energy is stored in reactive elements. Therefore, the analogue switching technology is not able to recover the field energy stored within the actuator, and the field formation does not occur in an energy saving way either, e. g. by storing energy in a reactive element. Instead, (under maximum driving conditions) an amount of energy about as great as the amount to be fed to the actuator is transformed into heat.

When a capacitive load is to be charged to the energy level of $\frac{1}{2}CV^2$, the same amount of energy is transformed into heat in the analogue power stage. This also applies for piezo actuators, which simply speaking can be considered capacitive loads. During discharging, the energy within the actuator is transformed into heat as well. This means that in the analogue circuit technology an entire cycle of charging and discharging takes up $E = CV^2$ or the power $P = fCV^2$, whereby f is the signal frequency or the repetition rate of any periodic signal. Figure 6.146a illustrates the charging process of a piezo actuator (t_1), the powerless holding condition at point t_2, and the discharging process (t_3). Figure 6.146b describes the energy flow at full amplitude during an operation cycle, assuming that hysteresis losses in the actuator amount to 30% of its electrical energy.

In terms of energy efficiency, the analogue circuit technology is less economical than the switching power electronics, but it has some considerable advantages: due to the lack of switching operations, there is no need for extensive signal filtering on the power side, which might cause a serious delay.

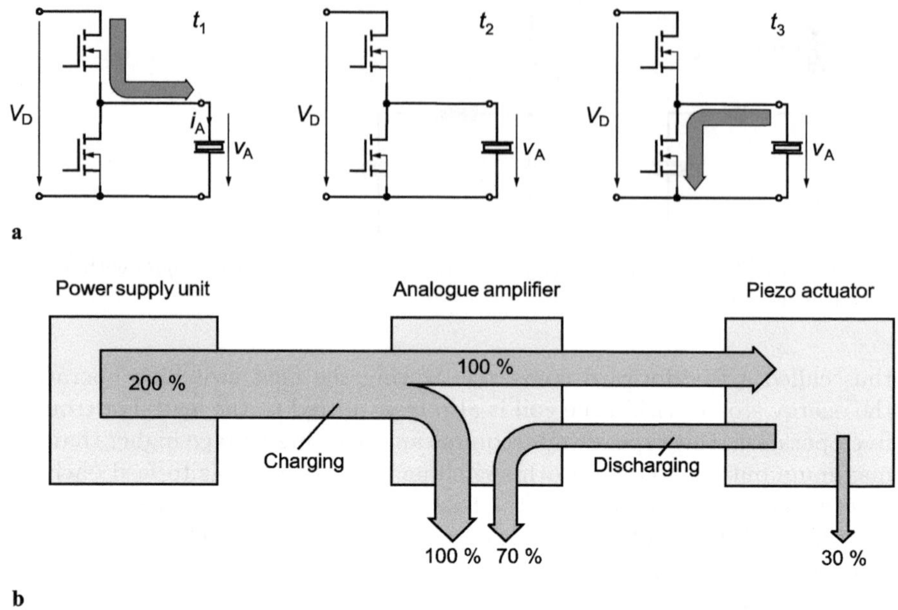

Fig. 6.146. Class C amplifier with piezo actuator as a load. **a** Full charging and discharging cycle, **b** energy flow chart

The analogue amplifier can respond immediately at its output (delay time within the microsecond range), whereas the time required for charging and discharging the coil in a switching amplifier (usually based on a clock frequency) usually causes a signal delay.

The analogue amplifier operates smoothly and de facto without any delays. The most important criteria for its design is its continuous output power as it determines the dimensions of the cooling elements. When the amplifier is operated with pulsating signals of high power a great part of the dissipated energy can be thermally stored. It is possible to achieve a ratio of the pulse power to the continuous power of up to 100.

Hybrid Power Amplifiers. A hybrid power amplifier is a combination of a switching and an analogue amplifier. The switching part transmits the main share of energy from the energy supply to the actuator, and it is able to recover a great share of the stored field energy as the field decays; the analogue part is located between the switching part and the load. The analogue part is fed with approx. 10% of the nominal voltage (see Fig. 6.147), that is, only approx. 10% of the power a purely analogue amplifier would require.

In the circuit, the analogue stage replaces the passive filter, which is often required by switching amplifiers. That is, it performs the tasks of an analogue filter at the power level. The ripple of the output signal of the switching amplifier can be dampened by more than 20 dB. A passive coil filter usually

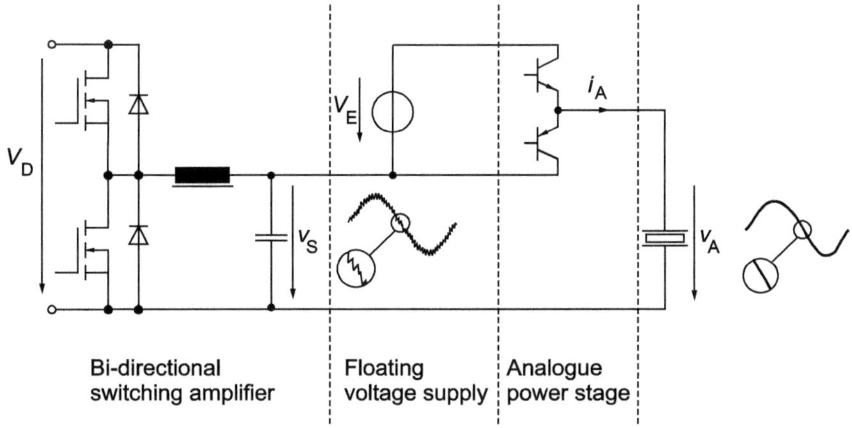

Fig. 6.147. Structure of a hybrid amplifier

causes unacceptably high delay times, however, the analogue power filter can respond to an input signal before the next clock cycle inside the switching amplifier has even started. The analogue stage of the hybrid amplifier is therefore able to compensate most of the disadvantages of a purely switching amplifier.

Comparison of the Circuit Concepts

With analogue circuit technology it is not possible to recover the stored field energy. The continuous output power determines the size of the power supply unit and the dimensions of the heat sink. The continuous output power is therefore the most important criteria in terms of size and weight of the analogue power amplifier. However, this amplifier provides excellent signal quality; very high rates of rise in current and voltage are possible with it, and the amplifier even exceeds the requirements of the HiFi norm (e. g. distortion factor and bandwidth). Analogue amplifiers are usually stable over a wide range of values of the load impedance. From experience, analogue amplifiers are the best choice for universal applications in a laboratory.

The most important advantage of switching amplifiers is the possibility of energy recovery. They operate much more efficiently than analogue amplifiers, and their power components require only about 5% of the energy that an analogue amplifier would transform into heat. This can be very beneficial for mobile systems, because here energy is provided by transportable power sources or must first be generated by other components on board. The power supply unit as well as the heat sink can then be much smaller.

However, one must not underestimate the energy which is absorbed by the much more complex control circuit. It is due to this energy consumption that the energetic advantages of switching amplifiers are practically nullified when the required power output falls below a certain threshold (typically 1... 10 Watt). Since switching amplifiers are usually operated with a fixed clock frequency or with a limited variable clock frequency range, they do not respond spontaneously to changes at the input (command signal) or at the output (load reaction), but react only at certain points in time. The resulting delay can cause undesirable behaviour in the overall system.

Depending on the requirements of output signal quality, one has to connect a filter to the amplifier output, which will even further reduce the already low dynamic response of the switching amplifier. The impedance of the load influences the filter properties and/or the properties (i. e. the cut-off frequency) of the amplifiers switching stage. Subsequently, the admissible range of values of the load impedance is considerably smaller compared to that of an analogue amplifier.

The internal energy buffers of a switching amplifier must be rated to the maximum instantaneous power that can occur. The size and weight of a switching amplifier are determined by the continuous power and by the achieved degree of efficiency (cooling, power supply) as well as by the ratio of the instantaneous power to the pulse power. A factor of 100, which can be achieved in analogue power amplifiers, would require a coil too large and a power switch too complex for the application, so that an analogue amplifier would be the better choice.

Switching amplifiers are well suited to driving a constant load always with the same signal form. An example of this is the fuel injection technology used in the automobile industry. Here, analogue power amplifiers are used in the development laboratories to determine the optimal signal characteristics at the input of the injection valves (see e. g. [362]). Afterwards, switching amplifiers are designed for use in the large series application to generate the optimized signal at the known load, and to achieve efficient performance through recovery of the field energy.

A hybrid power amplifier, as a combination of a switching and an analogue power amplifier (or active signal filter on the power level), combines the advantages of both types. Viewed from the load, the hybrid amplifier behaves almost like a purely analogue power amplifier. The analogue power stage effectively decouples the switching amplifier stage from the load. Moreover, the switching stage does not even work for small signals or control operations that lie within the range of the operating voltage of the analogue power stage. Here especially, the amplifier is a purely analogue amplifier. From the standpoint of the power supply, the hybrid amplifier operates like a switching amplifier. However, the greatest portion of the stored energy is transmitted almost loss-free to and from the load during large-signal operation.

The analogue stage reduces the requirements on the switching stage: the switching stage only has to reach the demanded output value lying within the restricted range of the analogue stages operating voltage because the analogue stage is able to compensate any dynamic or static deviations within

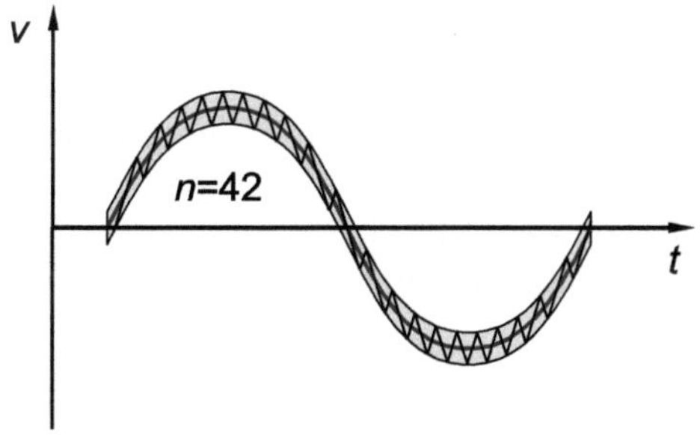

Fig. 6.148. Tolerance band within the switching stage of a hybrid amplifier during dual-state control

this restricted range. In extreme cases it can happen that the ratio of the clock frequency of the switching part to the signal frequency falls from several hundreds for a purely switching amplifier to a few tens for a hybrid amplifier. Figure 6.148 illustrates the output signal of a switching amplifier with dual-state control: since only $n = 42$ switching operations can be executed during one signal period, the residual ripple is too high for direct operation of an actuator, but it is acceptable as a supporting voltage for the analogue filter in a hybrid amplifier (v_S in Fig. 6.147).

Some of the disadvantages of both amplifier types, though diminished, remain nevertheless. Since the output and the load are separated by the analogue filter, the influence of the load impedance on the switching stage has been reduced but not eliminated. The dimensions of the coil continue to determine the dynamic behaviour of the switching stage and thus the large signal dynamics of the overall system. Therefore, the hybrid concept is not necessarily suitable for highly dynamic applications. The switching amplifier must be rated for the entire power that has to be transmitted, and the analogue power stage is comparable to a purely analogue amplifier, even though it requires far less cooling effort and, should the case be, a much smaller number of final stage transistors connected in parallel.

6.10.2 Power Electronics for Piezo Actuators and Actuators with Electrorheological Fluids

Ohmic-capacitive loads mainly require reactive power and only a little active power. The voltage is usually controlled, and the resulting current is the product of the time derivative of the voltage and the load capacitance.

Driving of Piezo Actuators

Piezo actuators are driven with an electrical field strength of up to 2 kV/mm in large signal operation (see Sect. 6.2). The ceramic layers are 30 µm to 0.5 mm thick, leading to a driving voltage in the range of 60 V to 1000 V.

Piezo actuators are mainly operated in the positive voltage range, although, it is in many cases admissible to operate them in the negative range with a maximum of 10% to 20% of the nominal voltage. As a result, the displacement (micrometer per Volt) increases compared to the displacement yielded with positive range operation only, but the area of the hysteresis loop grows disproportionately. Purely positive driving is by far the most frequently used type of operation [363].

Voltage Control. Voltage control is the simplest and most common way of driving an electrical load. Driving a piezo actuator by means of voltage control protects the actuator from uncontrollable undervoltage or overvoltage. The power amplifier has to generate high current amplitudes in order to produce the required signals. Voltage control protects the actuator from voltage drift. In contrast to charge controlled actuators, voltage controlled actuators of equal length and equal layer thickness expand with equal displacement even if they have different capacitances (i. e. different cross-sectional areas). Since the voltage-displacement characteristic shows hysteretic behaviour, precise positioning is only possible with precise displacement control.

A voltage amplifier typically has a low output impedance. When a dynamically excited actuator system oscillates mechanically, the actuator generates positive and negative charges. Due to its low output resistance, the voltage amplifier is able to take energy from the oscillating system. Amplifiers with adjustable maximum current can be tuned so that even when they are driven with a square-wave signal only the current leading to the desired displacement will flow. In this way it is possible to reduce unwanted actuator oscillations to the smallest possible degree.

Current Control. Since the voltage and displacement of a piezo actuator are proportional in their first approximation, the time derivatives of the voltage and the displacement have a similar relationship, that is, current and velocity correspond. This relationship is almost free of hysteresis. Therefore, when an application requires a certain velocity signal at the actuator output, it is possible to drive an actuator by its electrical current. In this case a control loop has to ensure that the operating voltage does not exceed the permissible range.

Charge Control. Integrating the velocity and current over time gives the displacement and charge, respectively. The relationship between these two quantities is also hysteresis-free, when the actuator is not mechanically loaded. However, charge control is accompanied with high demands on the integrating circuit. For controlling the current and charge an amplifier with a high output resistance is used, because this amplifier output does not conduct the charges generated by the piezo actuator. However, without any voltage control, the actuator voltage can drift into undesirable ranges.

Energy Control (Switching Amplifier). Switching amplifiers operate as flyback or as feed-forward converters according to the previously described principles. The flyback converter stores a certain amount of energy in its coil and then transmits this energy to the piezo actuator, which stores the energy $E = \frac{1}{2}CV^2$ after one switching operation. Subsequently, with this type of operation, the piezo actuator is driven with the product of charge $q = CV$ and voltage V.

In the feed-forward converter, the current that charges the coil with energy is conducted through the actuator, generating an additional charge there. Depending on whether the switching is activated by a fixed switching frequency or triggered by exceeding the actuators command value of voltage or current, the system executes either voltage control, current control, charge control or a combination of the previous until the moment of switching. After the switching operation, the energy is transmitted into the piezo actuator, just like in the flyback converter. The energy control description above applies here.

In both cases, the switching transistors are usually blocked after the switching operation, and the amplifier output resistance is high. When not in operation, in order to remove any undesirable charge generated inside the actuator by thermal changes, drift or mechanical load, the amplifier has to have a relatively low output resistance after discharging the actuator. This condition must be fulfilled by means of the circuit.

Control via Inverse Models. The most precise control approach, which also requires no external sensors, make use of an inverse model of the piezo actuator. During a learning phase, the actuator is characterised in a measuring system under the operating conditions awaiting the actuator in future applications. This means measuring the displacement and force on the mechanical side and the current and voltage on the electrical side, and with the acquired data computing a fully inverse actuator model including hysteresis, creep, and external forces. The data is filed in a control unit, which is preconnected to the power amplifier (see Sects. 6.1 and 6.9).

During later operation it suffices to measure the electrical quantities in order to compute the mechanical actuator quantities on the basis of the model, thereby compensating the hysteresis and creep of the piezo actuator. Since the computing process is quite complex and has to be executed in realtime, this control method is presently only implemented for low-frequency operations. However, the applicable frequency range of this inverse control method will grow in conjunction with the technological progress in the field of microelectronics [364].

Retroactive Effects on the Amplifier. Slow thermal changes and mechanical loads can cause a piezo actuator to generate considerably high charges and thus high electrical voltages. These charges should not be allowed to damage the power amplifier. A voltage amplifier should be capable of delivering enough output current to control a piezo actuator that generates charges itself; a high impedance amplifier must be able to cope with the generated voltage peaks without being damaged. When stored, the actuator should be terminated by an appropriate resistor.

Driving of Electrorheological Fluids

Electrorheological fluids (ERF) are operated with field strengths of up to 8 kV/mm (see Sect. 6.5). In order to maintain the required driving voltage as low as possible, the control gap is dimensioned as thin as possible. Depending on the particle size and the hydraulic flow rate, the width of the gap may vary between 0.12 mm and 0.75 mm, while the voltage lies between 1 kV and 6 kV. The polarity of the electrical field has no relevant influence on the formation of the force-transmitting chains within the ERF. However, certain ERFs exist which tend to electrophoresis when driven by a constant voltage, and which therefore must be driven by alternating voltage. This alternating voltage must have a high frequency compared to the signal frequency, and in a straightforward system, it will be amplitude-modulated.

If a (relatively slow) dual-state operation of the actuator suffices for the application, one can use a common power transformer as the power electronic device, which generates the required high voltage directly from the supply voltage, and which can be turned on and off via a solid-state relay.

The response time of electrorheological fluids lies in the range of milliseconds. From an electrical point of view, they are capacitive loads with a parallel conductance. Capacitance and conductance are determined by the geometry of the assembly and by the physical properties of the employed fluids, such as the specific electrical resistance ρ and the permittivity ϵ. When computing the time constant for self-discharge τ, the geometrical parameters cancel each other, so that the time constant is a property of the fluid. Figure 6.149a illustrates a capacitor with the fluid between its parallel plates. Figure 6.149b displays its equivalent circuit diagram.

If the actuator discharges itself too slowly such that it cannot be deactivated at a certain operating frequency, the system requires an additional

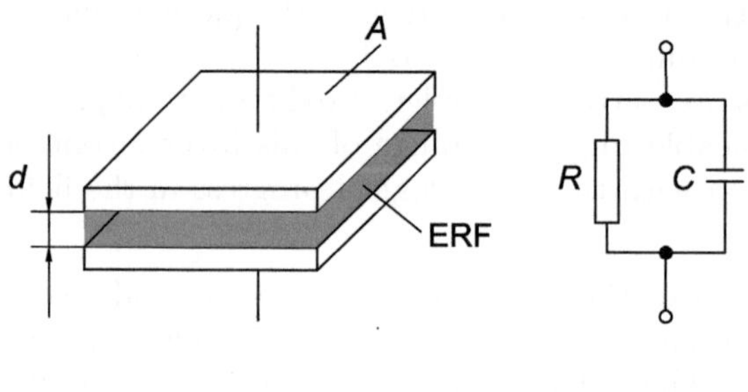

$$R = \rho \frac{d}{A} \qquad C = \epsilon \frac{A}{d} \qquad RC = \tau = \rho\epsilon$$

a **b**

Fig. 6.149. Actuator with electrorheological fluid. **a** Physical arrangement, **b** equivalent circuit diagram

discharging circuit (two-quadrant operation). If the driving frequency is so low that the actuator can discharge itself sufficiently fast, then the charging circuit will suffice (one-quadrant operation).

Retroactive Effects on the Power Electronics. The operating modes of electrorheological actuators are classified into three types: flow mode (or valve mode), shear mode and squeeze mode. The former two modes do not exhibit any retroactive effects, whereas the squeeze mode can generate high voltages when the distance between the capacitor plates are altered quickly. When an actuator is operated in squeeze mode, extra precautions must be taken to protect the amplifier from damage.

Irrespective of the operating mode of the fluid, its maximum field strength (8 kV/mm) is higher than the dielectric strength of air (1 kV/mm or less). If air pockets and contamination within the ER fluid enter the control gap, they can cause a voltage flashover and thus increase the level of contamination. This requires effective measures on the mechanical side to prevent the actuator from damage due to such flashovers, which would cause it to break down fairly quickly.

Since a high operating voltage is required, it is generated by a switching power supply. For the discharging circuit, if necessary, it is possible to use a serial connection of analogue transistors (due to the high voltage).

Important Parameters for the Amplifier Design

Choosing the most appropriate amplifier takes place in several steps. Firstly, one has to identify in which quadrants the amplifier has to operate. Secondly, one has to acquire the values of the voltage, current and power required for the operation.

Nominal Voltage of the Amplifier. The required output voltage of the power amplifier is determined by the parameters of the actuator. When the operating range is not entirely used, an amplifier with a smaller output voltage will suffice.

Average Current, Continuous Output Power. The average of a sinusoidal current signal can be computed through the equation $I = fCV_{\mathrm{pp}}$, and the continuous output power through $P = fCV_{\mathrm{pp}}V_{\mathrm{D}}$, V_{D} being the nominal voltage of the amplifier, see Fig. 6.150a to the right. Complex signals, however, have periods that can contain several charges with varying voltage amplitudes. When calculating the average current which is required to drive a capacitive load, it is not the rate of rise or decay of charging and discharging that matters, but the sum of the individual charging processes per signal period, the actuator capacitance and the repetition frequency.

A piezo actuator that has a large signal capacitance of 10 µF is first to be charged to 120 V, then discharged to 80 V and finally recharged to 180 V. Discharging takes place in the reverse order. The overall cycle is to be repeated at a frequency of 150 Hz. The sum of the charging voltages is then

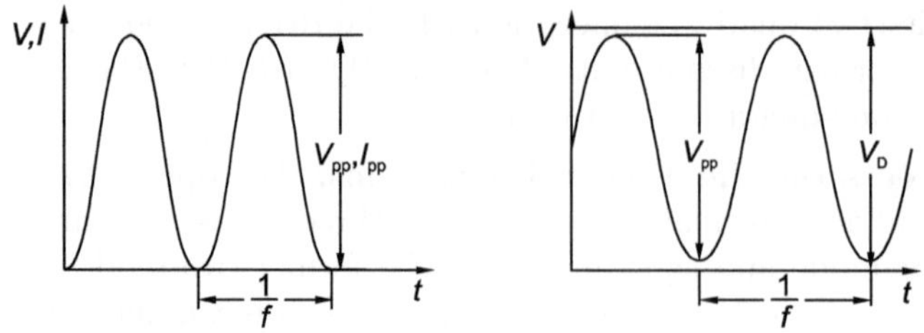

$I_{\max} = \pi f C V_{\mathrm{pp}}$
$V_{\max} = \pi f L I_{\mathrm{pp}}$

$P = f C V_{\mathrm{pp}} V_{\mathrm{D}}$

a

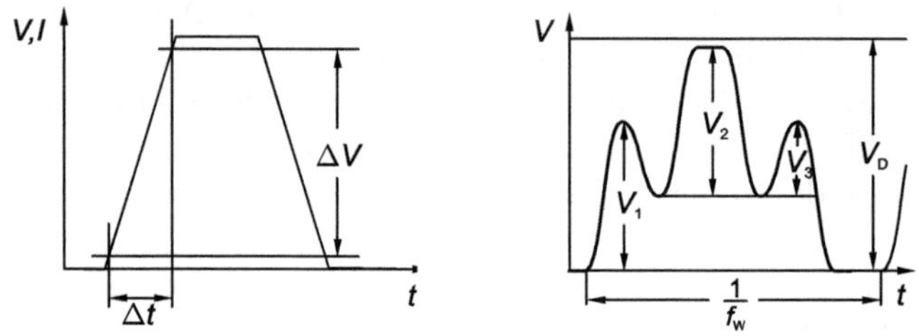

$I_{\max} = C\Delta V/\Delta t$
$V_{\max} = L\Delta I/\Delta t$

$P = f_{\mathrm{w}} C V_{\mathrm{D}}(V_1 + V_2 + V_3 + ...)$

b

Fig. 6.150. Determination of amplifier parameters via the signal form. **a** Monotone signal, **b** complex periodic signal

120 V plus 100 V during the charging process, and 40 V during the discharging process, adding up to a total of 260 V. The average current is computed via the product of 150 Hz · 10 µF · 260 V, resulting in 390 mA. Figure 6.150b to the right displays an example of such a composed signal. The average power is calculated using the 200 V nominal voltage of the amplifier: 0.39 A · 200 V, which equals 78 W. When dealing with ER actuators, one has to consider the additional energy required because of the electrical conductance.

Maximal Current, Pulse Power. Computing the maximal current demands knowledge of the greatest slope in the voltage-time signal. This is gained by means of a curve tangent or by differentiating mathematically. The maximum current results from the charge equation $\mathrm{d}q = C\mathrm{d}V$, which, after rearranging, reads $I_{\max} = C(\mathrm{d}V/\mathrm{d}t)_{\max}$ for the numerical and $I_{\max} = C(\Delta V/\Delta t)_{\max}$ for the graphical solution (C is the so-called large signal capacitance). Figure 6.150 to the left shows two examples. The pulse power is determined via the product of the maximum current and the nominal voltage of the amplifier. When the ratio of the maximum current to the continuous current is high, one can usually neglect the current component related to the conductance of the ER fluid.

6.10.3 Power Electronics for Magnetostrictive Actuators and Actuators with Magnetorheological Fluids

A particularity of ohmic-inductive loads like magnetostrictive or magnetorheological actuators is that they – especially during dynamic operation – mainly require reactive power and only a little active power. Usually, the current is given and the voltage required for operation is determined by the time derivative of the current multiplied with the load inductance. Since the control field is generated by a coil, the copper resistance must also be taken into account.

Both actuator types are controlled by magnetic fields (see Sects. 6.3 and 6.6). The operating point can be established, for instance, using a permanent magnet, whereby the constant field is increased or decreased by means of a control field generated by a coil. This operation requires a four-quadrant amplifier. If pre-magnetisation is not used, the magnetic field is produced entirely by electrical means. This kind of operation requires a two-quadrant amplifier, but one should note that the copper losses are considerably higher.

Driving of Magnetostrictive and Magnetorheological Actuators

Compared to capacitive loads, where the applied voltage can be maintained at a constant level, inductive loads cause the greatest losses in analogue amplifiers when they are driven continuously with maximum direct current. In this case, the coil does not generate any induction voltage, and the operating voltage of the amplifier drops mainly over the final output stage. In this operating state, the losses correspond to the nominal power. Still, in most cases an analogue amplifier will be the better choice for general applications and especially for laboratory applications since operation with a switching amplifier has difficulties of its own.

In contrast to electric motors, where the load is usually driven via a full-bridge as displayed in Fig. 6.144, the vast majority of magnetostrictive and magnetorheological actuators cannot be operated in this way. Electric motors usually require a high share of active power, whereas the actuators treated in this paper, in principle, feature a very high share of reactive power. Their design also differs greatly from that of electric engines, which in some cases can lead to eddy currents in the magnetic circuit. In direct operation via a switching full-bridge, this would lead to great losses due to eddy currents and very high current peaks when switching, so that a well tuned filter is necessary. Such a (signal-delaying) filter would be able to decouple the switching frequency from the load.

The characteristics of magnetostrictive and magnetorheological actuators indicate hysteretic and nonlinear behaviour as well. In contrast to piezo actuators, where properties of the material dominate the operating behaviour, the hysteresis of these actuators is influenced also by design features (partly due to the magnetic circuit). Several control concepts for hysteresis compensation and for inverse modelling seem possible (see Sect. 6.9), but so far they have been subject to research mostly. Here, one mainly applies current control with overlapping feedback control of the displacement and the force, if necessary.

It is worth noting, however, that a high amplifier output voltage is necessary in order to affect fast changing currents, while on the other hand, a quick decay of the dynamic field can induce a high voltage at the inductive load. In order to avoid dangerously high voltage amplitudes, certain measures must be taken depending on the circuit concept at hand. These measures may include recovery diodes between the amplifier output and the operating voltage or device ground as well as circuits in the signal path for limiting the slew rate.

Important Parameters for the Amplifier Design

After identifying the quadrants in which the amplifier has to operate, the required values of current, voltage and power have to be determined.

Determination of Key Data. The parameters of the actuator to be driven form the basis for the choice of the appropriate power amplifier. The amplifier must be capable of providing the required current. The necessary operating voltage is determined by means of the greatest incline in the current-time signal that the amplifier has to produce and the load inductance. To this end, one applies a tangent to the geometric curve or differentiates it mathematically. The procedure is similar to the one described for power amplifiers used to drive capacitive loads: in the examples in Fig. 6.150 to the left simply replace V by I and vice versa and C by L.

If power amplifiers are used that do not make use of variable or switchable operating voltages and which are not tuned to specific loads and signal forms, the continuous power of the analogue amplifier is determined by the greatest incline of the current signal and its corresponding voltage: $P = V_{\mathrm{D}} I_{\mathrm{max}}$.

6.10.4 How to Proceed When Choosing an Amplifier Concept

The need for a power amplifier for driving an unconventional actuator in general laboratory applications will in most cases result in the choice of an analogue amplifier without regard to the fact if there are solid state actuators or actuators with electrically controllable fluids: analogue amplifiers have the highest signal quality, the largest range of frequency and they allow for a wide value range of load impedance. Their comparatively high energy consumption is of minor interest. The analogue amplifier is also useful when a high pulse power is needed to reproduce a signal at the actuator, as is often the case in connection with high dynamic demands.

In systems with a self-contained energy supply and systems in which it is difficult to expel the heat losses, recovery of the field energy is of major interest. The only amplifiers that can perform such a task are hybrid and switching

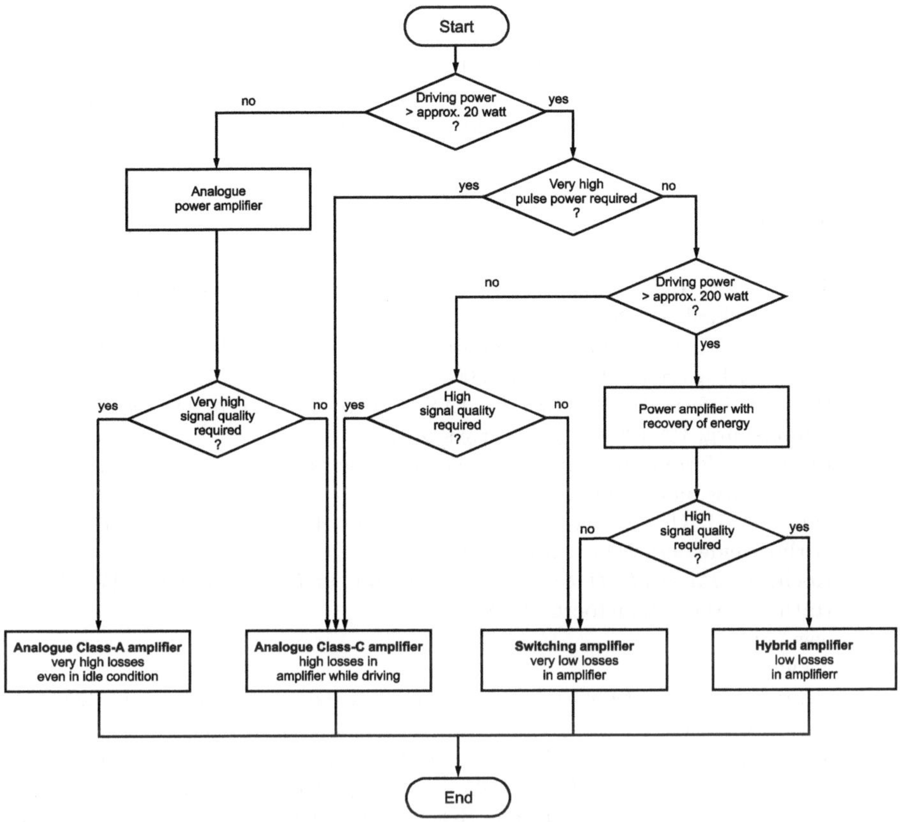

Fig. 6.151. Possible approach for selecting power amplifiers

amplifiers. Compared to analogue amplifiers, these have to be adapted even more closely to the load impedance and to the expected operating signals. Usually, these amplifiers are specially developed for one certain task, and adapting them consists in exchanging certain components of the amplifier or reprogramming the control logic circuit.

The diagram in Fig. 6.151 shows a general approach to finding the ideal amplifier for each application to be specified. For instance, adaptronic concepts might demand the miniaturisation of the power electronics, or even their full integration into a mechanical structure. In this case, the most important goal must be the reduction of the power losses in the amplifier because these determine the complexity of the cooling equipment and thus the amplifiers overall size. Based on these considerations, this type of application would probably require a switching amplifier.

The size of a switching amplifier is not only determined by its efficiency, but also by the applied electrical reactive elements (choke coil, capacitor, filter). So, one will choose a high switching frequency to keep the energy that must be stored and thus the size of the storage elements as small as possible. In the best of cases, the electrical properties of the actuator (storage for electrical or magnetic energy) can be used in place of some of the components the amplifier would normally include (see Fig. 6.144).

References

1. Janocha, H. (ed.): *Actuators – Basics and Applications.* Springer-Verlag, Berlin Heidelberg, New York (2004)
2. Schwinn, A.: *Aktive Vibrationsdämpfung mit verteilten Aktoren.* PhD Thesis, Saarland University, Germany (2004)
3. Product information, ANSYS: *ANSYS 6.1 Full Set of Analysis Guides.* Canonsburg, PA (2002), www.ansys.com
4. Product information, The Math Works: *User's Guide.* Natick, MA (1988–1998), www.mathworks.com
5. Janocha, H. (Ed.): *Actuators – Basics and Applications.* Springer-Verlag, Berlin Heidelberg New York (2004)
6. Koch, J.: *Piezoxide (PXE) – Eigenschaften und Anwendungen.* Dr. Alfred Hüthig Verlag, Heidelberg (1988)
7. www.pi.ws
8. www.noliac.com
9. Uchino, K.: *Ferroelectrics devices.* Marcel Dekker, New York Basel (2000)
10. www.cedrat.com
11. *Wirkungsgradoptimierte Piezoantriebe für hochdynamische Anwendungen in der Flugzeughydraulik (Piezoserv).* Joint project funded by the German Federal Ministry of Education and Research (BMBF), 16SV563, www.tib.uni-hannover.de
12. Sashida, T.; Kenjo, T.: *An introduction to ultrasonic motors.* Oxford: Clarendon Press, vol. 28, Monographs in Electrical and Electronic Engineering (1993), 242p.
13. www.elliptec.com
14. www.exfo.com
15. Six, M.F.; Le Letty, R.; Coste, P.; Claeyssen, F.: *Rotating step by step piezomotor for nanopositioning and space applications.* Proc. 10th Conf. Actuator, Bremen (2006), pp. 353–356
16. Kuhnen, K.; Janocha, H.: *Compensation of the creep and hysteresis of piezoelectric actuators with inverse systems.* Proc. 6th Conf. Actuator, Bremen (1998), pp. 309–312
17. Le Letty, R. (et al.): *Piezoelectric actuators for active optics.* Proc. Conf. ICSO (2004), pp. 717–720
18. Preumont, A.: *Vibration control of active structures, an introduction.* Kluwer (February 2002), ISBN 1-4020-0496-6
19. Sosnicki, O. (et al.): *Active damping of vibrations applied on ski structures.* Proc. 10th Conf. Actuator, Bremen (2006), pp. 932–935
20. Sosnicki, O. (et al.): *Vibration energy harvesting in aircraft using piezoactuators.* Proc. 10th Conf. Actuator, Bremen (2006), pp. 968–971
21. Bozorth, R.M.: *Ferromagnetism.* 2nd, Van Nostrand, NY (1951), p. 980

22. De Lacheisserie, E.: *Magnetostriction: Theory and applications.* CRC, USA (1993), p. 410
23. Clark A.E.: *Magnetostrictive rare earth-Fe_2 compounds, Ferromagnetic materials.* E.P. Wohlfarth, US, Tome 1 (1980), pp. 531–588
24. Verhoeven, J.D.: *The effect of composition and magnetic heat treatment on the magnetostriction of TbDyFe twinned single crystals.* J. Appl. Phys. 66(2) (1989). pp. 772–779
25. Sandlund, L.: *Magnetostrictive powder composite with high frequency performance.* Proc. 3rd Int. Workshop on Power Transducers for Sonics and Ultrasonics, Springer, Fl., May 6–8, (1992), pp. 113–120
26. Quandt, E.: *Magnetostrictive thin film actuators.* Proc. Actuator 94, Axon, Bremen, Germany (1994), pp. 229–232
27. Claeyssen, F.: *Giant dynamic magnetostrain in rare earth-iron magnetostrictive materials.* IEEE Trans. MAG.27, N 6, Nov.1991, pp. 5343–5345
28. Moffett, M.B.: *Characterization of Terfenol-D for magnetostrictive transducers.* JASA, 89 (3) (1991), pp. 1448–1455
29. Kvarnsjö, L.: *On characterisation, modelling and application of highly magnetostrictive materials.* Doct. Thesis TRITA-EEA-9301, ISSN 1100-1593 (1993), p. 173
30. Body, C.: *Non linear finite element modelling of magneto-mechanical phenomenon in giant magnetostrictive thin films.* Proc. CEFC 1996
31. Engdahl, G.: *Loss simulations in magnetostriction actuators.* J. Appl. Phys. 79 (8) (1996)
32. ATILA *– a 3D CAD software for piezoelectric and magnetostrictive structures.* ISEN, Lille (F), Distr. CEDRAT, Meylan (F) & MAGSOFT, Troy New York (US)
33. Debus, J.C.: *Finite element modeling of PMN electrostrictive materials.* Proc. Int. Conf. on Intelligent Materials, ICIM96-ECSSM96, SPIE vol. 2779, Lyon, France (1996), pp. 913–916
34. Claeyssen, F.: *Design and building of low-frequency sonar transducers based on rare earth iron magnetostrictive alloys.* Doct. Thesis, Defence Research Inform. Cent. HSMO, MoD, London, also *Conception et réalisation de transducteurs sonar basse fréquence à base d'alliages magnétostrictifs Terres Rares-Fer.* Thèse INSA Lyon (1989), p. 414
35. Claeyssen, F.: *Modeling and characterization of the magnetostrictive coupling.* Proc. 2nd Int. Workshop on Power Transducers for Sonics and Ultrasonics, Springer (1990), pp. 132–151
36. Claeyssen, F.: *Giant Magnetostrictive Alloys Actuators.* Proc. Magnetoelastic Effects and Applications Conf. L.Lanotte, Pub. Elsevier, Holland, 1993, pp. 153–159, or J. Appl. Electromagnetics in Mater. 5 (1994) pp. 67–73
37. Bouchilloux, P.: *Dynamic Shear Characterization in a Magnetostrictive Rare Earth-Iron Alloy.* M.R.S. Symp. Proc., Vol. 360 (1994), pp. 265–272
38. Bossut, R.: *Finite element modeling of magnetostrictive transducers using Atila.* Proc. ATILA Joint Conf. 2nd Int. Workshop on Power Transducers for Sonics and Ultrasonics, B.F. Hamonic, Isen, Lille, France (1990), pp. 19–26
39. Claeyssen, F.: *Progress in magnetostrictive sonar transducers.* Proc. UDT93, Reed Exhib. UK (1993), pp. 246–250
40. Le Letty, R.: *Combined finite element – normal mode expansion methods for ultrasonic motor modeling.* IEEE Ultrasonic Symp., Proc. (1994), pp. 531–534

41. Claeyssen, F.: *Analysis of magnetostrictive Inchworm motors using f.e.m.* Proc. Magnetoelastic Effects and Applications Conf. L.Lanotte, Pub. Elsevier, Holland (1993), pp. 161–167
42. Claeyssen, F.: *State of the art in the field of magnetostrictive actuators.* Proc. Actuator 94 Conf., Axon, Bremen, Gemany (1994), pp. 203–209
43. ETREMA *Terfenol-D Magnetostrictive Actuators Information.* Etrema Products, USA (1993), p. 6
44. Eda, H.: *Ultra precise machine tool with GMA.* Annals of the CIRP Vol. 41/1 (1992), pp. 421–424
45. Wang, W.: *A high precision micropositioner based on magnetostriction principle.* Rev. Sci. Inst. 63 (1) Jan. 1992, pp. 249–254
46. Cedell, T.: *New magnetostrictive alloy for rapid conversion of electric energy to mechanical motion.* Proc. Actuators 90, Axon, Bremen, Germany (1990), pp. 156–161
47. Suzuki, K.: *Magnetostrictive plunger pump.* Int. Symp. on GMA&A, Japan, Poster 5 (1992), p. 6
48. Hiller, M.W.: *Attenuation and transformation of vibration through active control of magnetostrictive Terfenol.* J. Sound Vib., 133(3), Pap. 364/1 (1989), p. 13
49. Janocha, H.: *Design criteria for the application of solid state actuators.* Proc. Actuator 94 Conf., Axon, Bremen, Germany (1994), pp. 246–250
50. Giurgiutiu, V.: *Solid-state Actuation of Rotor Blade Servo-Flap for Active Vibration Control.* J. Int. Mat. Systems and Structures, vol. 7 (1996), pp. 192–202
51. Lhermet, N.: *Actuators based on biased magnetostrictive rare earth-iron alloys.* Proc. Actuator 92, Axon, Bremen, Germany (1992), pp. 133–137
52. FLUX2D – *a 2D CAD software for electric engineering.* LEG, Grenoble (F), Distr. CEDRAT, Meylan (F) & MAGSOFT, Troy NY (US)
53. Claeyssen, F.: *Design of Lanthanide magnetostrictive sonar projectors.* Proc. UDT91, Microwave Exh.&Pub. (1991), pp. 1059–1065
54. Moffett, M.B.: *Comparison of Terfenol-D and PZT4 power limitations.* J. Acoust. Soc. Am. 90 (2), Letters to editor, August 1991, pp. 1184–1185
55. Jones, D.F.: *Recent transduction developments in Canada and US.* Sonar Transducers Conf., Proc. Inst. of Acoustics, Univ. of Bath, UK, Vol. 17, Pt 3 (1995), pp. 100–106
56. Wise, R.J.: *Ultrasonic welding of plastics by a Terfenol driven magnetostrictive transducer.* Proc. 3rd Conf. on Welding and Adhesive Bonding of Plastic, Düsseldorf, Nov. 1992, pp. 10–12
57. Dubus, B.: *Low Frequency Magnetostrictive Projectors for Oceanography and Sonar.* Proc. 3rd Europ. Conf. on Underwater Acoustics (1996), pp. 1019–1024
58. Kiesewetter, L.: *Terfenol in linear motor.* Proc. 2nd. Int. Conf. on GMA, C.Tyren, Amter, France (1988), Ch. 7, p. 15
59. Vranish, J.M.: *Magnetostrictive direct drive rotary motor development.* IEEE Trans. MAG. 27, N 6, Nov. 1991, pp. 5355–5357
60. Mori, K.: European Patent, N 0155694A2 (1985)
61. Akuta, T.: *Rotational-type actuators with Terfenol-D rods.* Proc. Actuator 92, Axon, Bremen, Germany (1992), pp. 244–248
62. Akuta, T.: *Improved Rotational-type actuators with Terfenol-D rods.* Proc. Actuator 94, Axon, Bremen, Germany (1994), pp. 272–274

63. Claeyssen, F.: *Design and construction of a new resonant Magnetostrictive Motor.* Proc. Intermag 96, Seattle, WA, June 1996, IEEE Trans. MAG also Proc. Actuator 96, Axon, Bremen, Germany (1996), pp. 172–274
64. Claeyssen, F.: *A new multi-mode piezo-electric motor.* Proc. Int. Conf. on Intelligent Materials, ICIM96-ECSSM96, SPIE vol. 2779, Lyon, France (1996), pp. 634-637 also Proc. Actuator 96, Axon, Bremen, Germany (1996), pp. 152–155
65. Fukuda, T.: *GMA applications to micromobile robot as microactuator without power supply cables.* IEEE Microelectromechanical Systems, Proc. (1991), pp. 210–215
66. Honda, T.: *Fabrication of Magnetostrictive Actuators Using Rare-Earth (Tb, Sm)-Fe Thin Films.* J. Appl. Phys. 76 (10), 15 Nov 1994, pp. 6994–6999
67. Betz, J.: *Torsion based, drift-free magnetostrictive microactuator.* Proc. Actuator 96, Axon, Bremen, Germany (1996), pp. 283–286
68. Claeyssen, F.: *Micromotors Using Magnetostrictive Thin Films.* Proc. SPIE Conf. Smart Structures and Materials, San Diego, US (March 98)
69. Engdahl, G.: *A time dependent radially resolved simulation model of giant magnetostrictive materials.* in: Mechanical Modelling of New Electromagnetic Materials, R.K.T Hsieh, Elsevier, pp. 131–138 (1990)
70. Engdahl, G.: *Handbook of Giant Magnetostrictive Materials.* Academic, San Diego, USA (1999), pp. 386
71. Berqvist, A.: *A model for magnetomechanical hysteresis and losses in magnetostrictive materials.* J. Appl. Phys. 79 (8) (1996)
72. Duerig, T.W.; Melton, K.N.; Stöckel, D.; Wayman, C.M.: *Engineering Aspects of Shape Memory Alloys.* Butterworth-Heinemann, London (1990)
73. Funakubo, H. (Ed.): *Shape Memory Alloys.* Gordon and Breach, New York (1984)
74. Hornbogen, E.: *On the Term 'Pseudo-elasticity'.* Metallkunde, Robotica 18 (1995) 1, pp. 341–344
75. Krämer, J.: *Formgedächtnislegierungen in der Automobiltechnik und im Maschinenbau.* Lecture notes of the course 'Konstruieren mit Formgedächtnislegierungen', Technische Akademie Esslingen, course no. 17964/50.051, Ostfildern, Feb. 21–22 (1994), pp. 79–111 (in German)
76. Mertmann, M.; Wuttig, M.: *Application and Hysteresis of Different Shape Memory Alloys for Actuators.* Proc. 9th Int. Conf. on New Actuators, June 14–16, Bremen, Germany (2004), pp. 72–77
77. Olier, P.; Tournie, Y.; Roblin, C.: *Shape Memory Effect and Recovery Stress Measurements in High Temperature Ti-Ni-Hf Shape Memory Alloys.* Proc. 6th Int. Conf. on New Actuators, June 17–19, Bremen, Germany (1998)
78. Gümpel, P.: *Formgedächtnislegierungen.* Expert-Verlag, Renningen (2004)
79. Furuya, Y.; Matsumoto, M.; Matsumoto, T.: *Mechanical Properties and Microstructure of Rapidly Solidified TiNiCu-Alloy.* Proc. Int. Conf. on Shape Memory and Superelasticity, March 7–10, Pacific Grove, CA, USA (1994), pp. 905–910
80. Holleck, H.; Kirchner, S.; Quandt, E.; Schlossmacher, P.: *Preparation and Characterization of TiNi SMA Thin Films.* Proc. 4th Int. Conf. on New Actuators, June 15–17, Bremen, Germany (1994), pp. 361–364
81. Ishida, A.; Takei, A.; Miyazaki, S.: *Formation of Ti–Ni Shape Memory Films by Sputtering Method.* Proc. Int. Conf. Martensitic Transformations ICOMAT-92, July 20–24, Monterey, CA, USA (1992), pp. 987–992

82. Kristen, M.: *Untersuchungen zur elektrischen Ansteuerung von Formgedächtnis-Antrieben in der Handhabungstechnik.* Braunschweiger Schriften zur Mechanik, No. 15–1994, Mechanik-Zentrum der TU Braunschweig (1994), (in German)
83. Hesselbach, J.; Stork, H.: *Simulation and Control of Shape Memory Actuators.* Proc. 5th Int. Conf. on New Actuators, June 26–28, Bremen, Germany (1996), pp. 396–399
84. Hesselbach, J.; Hornbogen, E.; Mertmann, M.; Pittschellis, R.; Stork, H.: *Optimization and Control of Electrically Heated Shape Memory Actuators.* Proc. 4th Int. Conf. on New Actuators, June 15–17, Bremen, Germany (1994), pp. 337–340
85. Russel, S.; Sczerzenie, F; Clapp, P.: *Engineering Considerations in the Application of NiTiHf and NiAl as Practical High-Temperature Shape Memory Alloys.* Proc. Int. Conf. on Shape Memory and Superelastic Technologies, March 7–10, Pacific Grove, CA, USA (1994), pp. 43–48
86. Touminen, S.M.: *High Transformation Temperature Ni–Ti–Hf Alloys.* Proc. Int. Conf. on Shape Memory and Superelastic Technologies, March 7–10, Pacific Grove, CA, USA (1994), pp. 49–54
87. Hesselbach, J.; Kristen, M.: *Shape Memory Actuators as Electrically Controlled Positioning Elements.* Proc. 3rd Int. Conf. on New Actuators, June 24–26, Bremen, Germany (1992), pp. 85–91
88. Pitschellis, R.: *Mechanische Miniaturgreifer mit Formgedächtnisantrieb.* PhD thesis, TU Braunschweig, Fortschritt-Berichte VDI (1998)
89. Bütefisch, S.; Pokar, G.; Büttenbach, S.; Hesselbach, J.: *A New SMA Actuated Miniature Silicon Gripper for Micro Assembley.* Proc. 7th Int. Conf. on New Actuators, June 19–21, Bremen, Germany (2000), pp. 334–337
90. Bütefisch, S.: *Entwicklung von Greifern für die automatisierte Montage hybrider Mikrosysteme.* PhD thesis, TU Braunschweig, Shaker-Verlag (2003)
91. Raatz, A.; Wrege, J.; Plitea, N.; Hesselbach, J.: *High Precision Complaint Parallel Robot. Production Engineering Research and Development.* WGP, Issue XII/1, Berlin (2005), pp. 197–202
92. Hesselbach, J.; Raatz, A.; Kunzmann, H.: *Performance of Pseudo-Elastic Flexure Hinges in Parallel Robots for Micro-Assembly Tasks.* Annals of the CIRP, Vol. 53/1, (2004), pp. 329–332
93. Pickard, W.F.: *Electrical Force Effects in Dielectric Liquids.* Progress in Dielectrics. Vol. 6, Temple, London (1965), p. 3
94. Bullough, W.A.; Stringer, J.D.: *The Incorporation of the Electroviscous Effect in a Fluid Power System.* Proc. 3rd Int. Fluid Power Symp., Turin, B.H.R.A. (May 1973), pp. F3–37
95. Winslow, W.M.: *Induced vibration of Suspensions.* J. Appl. Phys. Vol. 20 (December 1949), p. 1137
96. Whittle, M.; Peel, D.J.; Firoozian, R.; Bullough, W.A.: *Dependence of E R. Response Time on Conductivity and Polarisation Time.* Phys. Rev. E. Pt6A, Vol. 49 (1994), pp. 5249–5259
97. Philips, R.W.; Auslander, D.M.: *The Electro Plastic Flow Modulator.* Source unknown (May 1971)
98. Tsukiji, T.; Utashiro, T.: *Flow Characterisation of E. R. Fluids Between Two Parallel Plate Electrodes.* Proc. ASME Int. Congress and Expo., San Francisco, Developments in Electrorheological flows FED Vol. 235, MD Vol. 71 (Nov. 1996), pp. 37–42

99. Bullough, W.A.; Foxon, M.B.: *The Application of an Electroviscous Damper to a Vehicle Suspension System.* Proc. 3rd. Int. Conf. on Vehicle System Dynamics, Blacksburg, Virginia, Swets and Zeitlinger, Amsterdam (August 1984), p. 144
100. Janocha, H.; Rech, B.; Bölter, R.: *Practice-Relevant Aspects of Constructing ER.* Actuators. Proc. 5th Int. Conf. on ERF/MRS held SMMART Sheffield. World Scientific (July 1995), pp. 435–447
101. Atkin, R.; Xiao, S.; Bullough, W.A.: *Solutions of the Constitutive Equations for the Flow of an Electro-Rheological Fluid in Radial Configurations.* J. of Rheology, Vol. 35 (1991), pp. 1441–1461
102. Winslow, W.M.: *Field Responsive Force Transmitting Compositions.* United States Patent Specification No. 3, 047, 507 (July 31st 1962)
103. Ellam, D.J.; Bullough, W.A.; Atkin, R.J.: *Analysis of smart clutch with cooling flow using 2D Bingham plastic analysis and CFD.* Proc IMechE part A. J. of Power and Energy, Vol. 219, Nr. 8 (2005), pp. 639–652
104. Bullough, W.A. et al.: *The Electro-rheological Clutch: Design, Performance, Characterisation and Operation.* Proc. I Mech E. Vol. 207 (1993), pp. 82–95
105. Ellam, D.J.; Atkin, R.J.; Bullough, W.A.: *An electro structured fluid in transient operation.* Proc IMechE part A. J. of Power and Energy, Vol. 210 (2005), pp. 61–76
106. Tian, Y. et al: *Mechanical Properties of ER Fluids.* Proc 9th Int. Conf. ER Fluids and MR suspensions, World Scientific (2005), pp. 1328–1334
107. Sheng, P. et al: *ER fluids* US Pat 2004/0051076
108. Wu, C.W.; Chen, Y.; Tang, X.; Conrad, H.: *Conductivity and Force between particles in a model ERF.* I – Conductivity II – Force. Proc. 5th Int. Conf. ERF/MRS held Sheffield (1995), pp. 525–536
109. Boissy, C.; Atten, P.; Foulc, J.N.: *The Conduction Model of Electro Rheological Effect Revisited.* ibid pp. 156–165 – see also pp. 710–726 and pp. 756–763 by same authors
110. Peel, D.J.; Stanway, R.; Bullough, W.A.: *The Generalised Presentation of Valve and Clutch Data for an ER Fluid and Practical Performance Prediction Methodology.* ibid pp. 279–290
111. Tozer, R.; Orrell, C.T.; Bullough W.A.: *On-Off Excitation Switch for E.R. Devices.* Int. J. Mod. Phys. B. Vol 8, No. 20 and 21 (1994), pp. 3005–3014
112. Bullough, W.A.: *Smart Fluid Machines.* Chapter 9 of Smart Technologies, World Scientific (2003), pp. 193–218
113. Sianaki, A.H.; Bullough, W.A.; Tozer, R.; Whittle, M.: *Experimental Investigation into Electrical Modelling of Electro-rheological Fluid Shear Mode.* Proc. I.E.E, Sci. Meast. & Tech. Vol. 141, No. 6 (1994), pp. 531–537
114. Bullough, W.A.; Makin, J.; Johnson, A.R.; Firoozian, R.; Sianaki, A.H.: *ERF Shear Mode Characteristics: Volume Fraction, Shear Rate and Time Effects.* Trans. ASME, J. Dyn. Syst. Meast. and Control (June 1996), pp. 221–225
115. Johnson, A.R.; Makin, J.; Bullough, W.A.: *E. R. Catch/Clutch Simulations.* Int. J. Mod. Phys. B. Vol. 8, No. 20 and 21 (1994), pp. 2935–2954
116. Wolfe, C.; Wendt, E.: *Application of ERF in Hydraulic Systems.* Proc. AXON-VDI/E Actuator 94 Conf., Bremen (1994), pp. 284–287
117. Bloodworth, R.; Wendt, E.: *Materials for ER Fluids.* Proc. 5th Int. Conf. ERF/MRS, held SMMART Sheffield 1995. World Scientific (July 1995), pp. 118–131

118. Bullough, W.A.; Makin, J.; Johnson, A.R.: *Requirements and Targets for ER Fluids in Electrically Flexible High Speed Power Transmission.* Am. Chem. Soc. Fall Meeting Washington DC., Plenum (1995), pp. 295–302
119. Naem, A.S.; Stanway, R.; Sproston, J.L.; Bullough, W.A.: *A Strategy for Adaptive Damping in Vehicle Primary Suspension Systems.* 1994 Proc. ASME. Winter Annual Meeting Chicago (1994), pp. 395–399
120. Peel, D.J.; Stanway, R.; Bullough, W.A.: *A Design Methodology based upon Generalised Fluid Data.* Int. Conf. Intel Materials, Lyon. Engineering with ERF (1996), pp. 310–316
121. Hosseini-Sianaki, A.; Makin, J.; Xiao, S.; Johnson, A.R.; Firoozian, R.; Bullough, W.A.: *Operational Considerations in the Use of an Electro-Rheological Catch Device.* Proc. Soc. Fluid Power Transmission and Control, lst Fluid Power Trans. & Control Symp., Beijing, Beijing Inst. Tech. (1991), pp. 591–595
122. Block, H.; Kelly, J.: *Electro-rheology.* J Phys D 21 (1988), p. 1661
123. Peel, D.J.; Bullough, W.A.: *The Effect of Flow-rate, Excitation Level and Solids Content on the Time Response of an Electro-rheological Valve.* J. Intel. Matl. Systems and Structures, Vol. 4, No.l (1993), pp. 54–64
124. Dwyer-Joyce, R.; Bullough, W.A.; Lingard, S.: *Elastohydrodynamic Performance of Unexcited Electro-Rheological Fluids.* Proc 5th Int. Conf. ERF/MRS, held SMMART Sheffield. World Scientific (July 1995), pp. 376–384
125. Yates, J.R.; Lau, D.S.; Bullough, W.A.: *Inertial Materials: Perspective, Review and Future Requirements.* Proc. AXON-VDI/E Actuator'94 Conf., Bremen (1994), pp. 275–278
126. Akio Inoue, Ushio Ryu, Syohji Nishmura: *Caster Walker with Intelligent Breaks employing ER fluid composed of liquid crystal polysiloxane.* Proc 8th Int. Conf. ER Fluids and ER/MR suspensions (2001), pp. 23–29
127. Kiyohito Koyama: *Performance of Electrical Sensitive Fluid.* Proc 4th ESSM and 2nd MIMR Conf (July 1998), pp. 241–253
128. Bullough, W.A.: *Smart Machine Systems.* Chapter 3 in Smart Structures Applications and Related Technologies, Springer (2001)
129. Winslow, W.M.: *Induced Fibration of Suspensions.* J. Appl. Phys., 20 (1949), pp. 1137-1140
130. Phillips, R.W.: *Engineering Applications of Fluids with a Variable Yield Stress.* Ph.D. Thesis, University of California, Berkeley (1969)
131. Carlson, J.D.; Catanzarite, D.M. and St. Clair, K.A.: *Commercial Magneto-Rheological Fluid Devices.* Proc. 5th Int. Conf. on ER Fluids, MR Fluids and Assoc. Tech., Sheffield (July 1995), W.A. Bullough, ed., World Scientific, Singapore (1996), pp. 20–28
132. Carlson, J.D.: *The Promise of Controllable Fluids.* Actuator 94, 4th Int. Conf. on New Actuators, H. Borgmann and K. Lenz, eds, AXON Technologie, Bremen (1994), pp. 266–270
133. Shtarkman, E.M.: U.S. Patent 4,942,947 (1990)
134. Shtarkman, E.M.: U.S. Patent 4,992,190 (1991)
135. Kordonsky, W.: *Magnetorheological Effect as a Base of New Devices and Technologies.* J. Magnetics and Magnetic Materials 122 (1993), pp. 395–398
136. Kordonsky, W.J. and Jacobs, S.D.: *Magnetorheological Finishing.* Proc. 5th Int. Conf. on ER Fluids, MR Suspensions and Assoc. Tech. (July 1995), Sheffield, W.A. Bullough, ed., World Scientific, Singapore (1996), pp. 1–12

137. Carlson, J.D. and Weiss, K.D.: *A Growing Attraction to Magnetic Fluids.* Machine Design, Aug. 8 (1994), pp. 61–66
138. Weiss, K.D.; Duclos, T.G.; Carlson, J.D.; Chrzan, M.J. and Margida, A.J.: *High Strength Magneto- and Electrorheological Fluids.* Soc. Automotive Engineers, SAE, Paper #932451 (1993)
139. Carlson, J.D.; and Sproston, J.L.: *Controllable Fluids in 2000-Status of ER and MR Fluid Technology.* Actuator 2000, Proc. 7th Int. Conf. on New Actuators, Bremen, H. Borgmann ed., Bremen: Messe Bremen (2000), pp. 126–130
140. Rabinow, J.: *The Magnetic Fluid Clutch.* AIEE Trans. 67 (1948), pp. 1308–1315
141. National Bureau of Standards : *Magnetic Fluid Clutch.* National Bureau of Standards Tech. News Bull., 32 No.4 (1948), pp.54–60
142. Rabinow, J.: U.S. Patent 2, 575,360 (1951)
143. Carlson, J.D.: *Innovative Devices That Enable Semi-Active Control.* Proc. 3rd World Conf. on Structural Control, Como, Italy, April 2002, F Casciati, ed., John Wiley, Chichester (2003), pp 227–236
144. Carlson, J.D. and Weiss, K.D.: U.S. Patent No. 5, 382; 373 (1995)
145. Ginder, J.M.; Davis, L.C. and Elie, L.D.: *Rheology of Magnetorheological Fluids: Models and Measurements.* Proc. 5th Int. Conf. on ER Fluids, MR Suspensions and Assoc. Tech., Sheffield (July 1995), W.A. Bullough, ed., World Scientific, Singapore (1996), pp. 504–514
146. Ginder, J.M.: *Rheology Controlled By Magnetic Fields.* Encyclopedia of Appl. Phys., 16 (1996), pp. 487–503
147. Margida, A.J.; Weiss, K.D. and Carlson, J.D.: *Magnetorheological Materials Based on Iron Alloy Particles.* Int. J. Mod. Physics B, 10 (1996), pp. 3335–3341
148. Crosby, M.J.; Harwood, R.A. and Karnopp, D.: *Vibration Control Using Semi-Active Force Generators.* Trans. of ASME, 73-DET-122 (1973)
149. Karnopp, D. and Crosby, M.J.: U.S. Patent No. 3,807,678 (1974)
150. Ivers, D.E. and Miller, L.R.: *Semi-Active Suspension Technology: An Evolutionary View.* Lord Corp., Pub. No. LL-6005 (1994)
151. National Bureau of Standards: *Further Development of the NBS Magnetic Fluid Clutch.* Nat. Bureau Standards Tech. News Bull., 34 No.12 (1950), pp. 169–174
152. Lord Corporation: *MotionMaster Ride Management System.* Lord Corp., Pub. No. PB8008a, (1998)
153. Yanyo, L.C.: *Magnetorheological (MR) Fluid for Automotive Damping Systems.* Proc. IIR Suspension and Damping Conf., Germany (2004)
154. Carlson, J.D.; et al.: U.S. Patent No. 5,878,851 (1999)
155. Guilford, D.: Automotive News. 22 Dec (2003) , p. 10
156. Carlson, J.D.: *What Makes a Good MR Fluid.* J. Intelligent Mater. Systems and Structures, 13 (2003), pp. 431–435
157. Duclos, T.G.: *An Externally Tunable Hydraulic Mount Which Uses ER Fluid.* Soc. Auto. Engineers, SAE Paper #870963 (1987)
158. Duclos, T.G.: *Design of Devices Using Electro-rheological Fluids.* Soc. Auto. Engineers, SAE Paper #881134 (1988)
159. Carlson, J.D.: MR Technology Workshop, Lord Corp., Cary (2004)
160. Lord Corp.: *Rheonetic MRF-122-2ED.* Product Bull. No. 2002-41-0 (2002)
161. Lord Corp.: *Rheonetic MRF-132AD.* Product Bull. No. 2003-15-1 (2003)

162. Lord Corp.: *Rheonetic MRF-336AG*. Product Bull. No. 2003-16-0 (2003)
163. Goncalves, F.D.: *Characterizing the Behavior of Magnetorheological Fluids at High Velocities and High Shear Rates*. Ph.D. thesis, Virginia Polytechnic Institute and State University, Blacksburg (2005)
164. Spencer Jr., B.F.; Dyke, S.J.; Sain, M.K. and Carlson, J.D.: *Phenomenological Model of a MR*. J. Eng. Mechanics, ASCE, 123 No. 3 (1997), pp. 230–238
165. Design News Editorial: *Brake Cuts Exercise-Equipment Cost*. Design News, Dec. 4 (1995), p. 28
166. Anderson, R.: *Carrera MagneShock*. Arre Industries, Atlanta, GA (1998)
167. Delphi Energy & Chassis Systems: *Pub. DE-00-E-019 02/02*, http://www.delphi.com/pdf/vpr/Magnaride.pdf (2002)
168. General Motors Corp., www.cadillac.com/cadillacjsp/cpo/highlights.jsp?section =features&model=seville&year=2002 (2002)
169. Halverson, H.: *The Idaho Corvette Page*. CV World Internet Publishing, http://www.idavette.net/_magride/ (2003)
170. General Motors Corp.: www.cadillac.com/cadillacjsp/cpo/highlights.jsp?section =Specs&model=xlr&year=2004, (2004)
171. New York Times: CLIII, No. 52,810, 5 April (2004), pp. A14–A15
172. Keegan Jr., W.J.: *AutoBlog*. www.autoblog.com/entry/1234000637030854/ (2005)
173. Stewart, J.: *New Car Test Drive.Com*. www.nctd.com/sneakpreview.cfm?Vehicle =2006_Buick_Lucerne&ReviewID=80¿ (2005)
174. Sodeyama, H.; Sunakoda, K.; Suzuki, K.; Carlson, J.D. and Spencer, B.F.: *Development of Large Capacity Semi-Active Vibration control Device Using Magneto-Rheological Fluid*. Proc. ASME Pressure Vessel and Piping Conf., PVP-428-2, Atlanta (2001), pp. 109–114
175. Duan, Y.F.; Ni, Y.Q. and Ko, J.M.: *Cable Vibration Control Using Magneto-Rheological (MR) Dampers*. Proc. 9th Int. Conf. on ER Fluids and MR Suspensions, September 2004, Beijing, K. Lu, R. Shen and J. Liu, eds., World Scientific, Singapore (2005), pp. 829–835
176. Chen, Z.Q. et al.: *MR Damping System on Dongting Lake Cable-Stayed Bridge*. Smart Structures and Materials 2003: *Smart Systems and Nondestructive Evaluation for Civil Infrastructures*. Shih-Chi Liu, ed, proc. SPIE vol. 5057 (2003), pp. 229–235
177. Lansing Linde Ltd. UK, http://www.linde-mh.co.uk/02_products/02_xrange/06_steering/index.html (2005)
178. Carlson, J.D.; Matthis, W.; Toscano, J.R.: *Smart Prosthetic Based On Magnetorheological Fluids, Industrial and Commercial Applications of Smart Structures Technologies*. Proc. SPIE 8th Annual Symp. on Smart Structures and Materials, Newport Beach, 2001 March 5–8, 4332 (2001), pp. 308–316
179. Biedermann, L.: Int. Patent Appl. WO9929272A1 (1999)
180. Biedermann, L.; Matthis, W. and Schulz, C.: Int. Patent Appl. WO0038599A1 (2000)
181. Biedermann, L.; Leg: Eur. Patent Appl. EP 0957838B1 (2000)
182. Bar-Cohen, Y. (Ed.): *Electroactive polymer (EAP) Actuators as artificial muscles. Reality, potential, and Challenges*. Second edition, SPIE, Bellingham, Washington (2004)
183. Madden, J.D.W. et al.: *Artificial muscle technology: physical principles and naval prospects*. IEEE J. Oceanic Eng., 29, 3 (2004), pp. 706–728

184. Kuhn, W.: *Reversible Dehnung und Kontraktion bei Änderung der Ionisation eines Netzwerks polyvalenter Fadenmolekülionen.* Experimentia, 5 (1949), pp. 318–319
185. Katchalsky, A .: *Rapid swelling and deswelling of reversible gels of polymeric acids by ionization.* Experimentia, 5 (1949), pp. 319–320
186. Kuhn, W.; Hargitay, B.; Katchalsky, A. and Eisenburg, H. : *Reversible dilatation and contraction by changing the state of ionization of high-polymer acid networks.* Nature, Vol. 165 (1950), pp. 514–516
187. Osada, Y.: *Advances in polymer sciences.* Springer, Berlin 82, 1 (1987)
188. De Rossi, D.; Kajivira, K.; Osada, Y.; Yamauchi, A. (eds): *Polymer gels.* Plenum-New York (1991)
189. Osada, Y. and Matsuda, A.: *Shape-Memory Gel with Order-Disorder Transition.* Nature, 376, 219 (1995)
190. Osada, Y.; Murphy, R. S.: *Intelligent Gels.* Scientific Amer. 268 (1993), pp. 82–87
191. Wasserman, A.: *Size and shape changes of contractile polymers.* Pergamon, New York (1960)
192. Okuzaki, H. and Osada, Y.: *Electro-driven chemomechanical polymer gel as an intelligent soft material.* J Biomater Sci. Polym Ed.; 5(5) (1994), pp. 485–496
193. Gong, J.P.; Osada, Y.: *Gel actuators.* In: Polymer Sensors and Actuators, Osada Y and D De Rossi (Eds), Springer, Berlin (2000), pp. 272–294
194. Tanaka, T.; Nishio, I.; Sun, S.-T. and Ueno-Nishio, S.: *Collapse of gels in an electric field.* Science 218 (1982), pp. 467–469
195. Tanaka, Y.; Kagamo, Y.; Matsuda, A.; Osada, Y.: *Thermoreversible transition of tensile modulus of hydrogel with ordered aggregates.* Macromolecules, v. 28, 2574–257 (1995)
196. Gong, J.P.; Matsumoto, S.; Uchida, M..; Isogai, N.; Osada, Y.: *Motion of polymer gels by spreading organic fluid on water.* J. Phys. Chem., 100 (26), 11092 (1996)
197. De Rossi, D.; Parrini, P.; Chiarelli, P. and Buzzigoli, G.: *Electrically induced contracatile phenomena in charged polymer networks: preliminary study on the feasibility of muscle-like structures.* Trans. Am. Soc. Artif. Internal Organs, 31 (1985), pp. 60–65
198. De Rossi, D.; Chiarelli, P.; Buzzigoli, G.; Domenici, C.; Lazzeri, L.: *Contractile behaviour of electrically activated mechanochemical polymer actuators.* Trans. Am. Soc. Artif. Internal Organs, 32 (1986), pp. 157–162
199. Osada, Y.: *Chemical valves and gel actuators.* Advanced Mater. 3 (1991), pp. 107–108
200. Narita, T.; Gong, J.P.; Osada, Y.: *Enhanced Velocity of Surfactant Binding after the Volume Collapse of an Oppositely Charged Gel.* Macromol. Rapid Comm. 18 (1998), pp. 853–854
201. Osada, Y.; Onkuzaki, H.; Hori, H.: *A Polymer Gel with Electrically Driven Motility.* Nature 253, 242 (1992)
202. Chiarelli, P.; De Rossi, D.: *Modelling and mechanical characterization of thin fibers of contractile polymer hydrogels.* J. Intell. Mater. Sys. Struct., v3 (1992), pp. 396–417
203. Itoh, Y.; Matsumura, T.; Umemoto, S.; Okui, N.; Sakai, T.: *Contraction/elongation mechanism of acrylonitrile gel fibers.* Polymer Preprints – Japan (English Editions), 36(5–10): E184–E186 (1987)

204. Oguro, K.; Asaka, K.; Takenaka, H.: *Polymer Film Actuator Driven by Low Voltage.* Proc. of 4th Int. Symp. on Micro Machine and Human Science at Nagoya (1993), pp. 39–40
205. Asaka, K.; Oguro, K.; Nishimura, Y.; Mizuhata, M.; Takenaka, H.: *Bending of polyelectrolyte membrane-platinum composites by electric stimuli I. Response characteristics to various waveforms.* Polymer J., 27 (1995), pp. 436–440
206. Asaka, K.; Oguro, K.: *Bending of polyelectrolyte membrane platinum composites by electric stimuli. Part II. Response kinetics.* J. Electroanal. Chem., 480 (1–2) (2000), pp. 186–198
207. Asaka, K.; Oguro, K.: *Bending of polyelectrolyte membrane-platinum composite by electric stimuli. III: Self-oscillation.* Electrochimica Acta, v 45 (27) (2000), pp. 4517–4523
208. Abe, Y.; Mochizuki, A.; Kawashima, T.; Yamashita, S.; Asaka, K.; Oguro, K.: *Effect on bending behavior of counter cation species in perfluorinated sulfonate membrane-platinum composite.* Polymers for Advanced Technologies, v 9, n 8 (1998), pp. 520–526
209. Onishi, K.; Shingo, S.; Asaka, K.; Fujiwara, N. and Oguro, K.: *Morphology of electrodes and bending response of the polymer electrolyte actuator.* Electrochimica Acta 46(5) (2001), pp. 737–743
210. Onishi, K.; Sewa, S.; Asaka, K.; Fujiwara, N.; Oguro, K.: *Effects of counter ions on characterization and performance of a solid polymer electrolyte actuator.* Electrochimica Acta, v 46(8) (2001), pp. 1233–1241
211. Shahinpoor, M.: *Ion-exchange polymer-metal composites as biomimetirc sensors and actuators.* In Polymer Sensors and Actuators, Osada, Y. and De Rossi, D. (Eds), Springer, Berlin (2000), pp. 325–360
212. Oguro, K.; Fujiwara, N.; Asaka, K.; Onishi, K.; Sewa, S.: *Polymer electrolyte actuator with gold electrodes.* Proc. SPIE – Int. Soc. Opt. Eng., v 3669 (1999), pp. 64–71
213. Onishi, K.; Sewa, S.; Asaka, K.; Fujiwara, N.; Oguro, K.: *Bending response of polymer electrolyte actuator.* Proc. of SPIE – Int. Soc. Opt. Eng., v 3987 (2000), pp. 121–128
214. Tadokoro, S.; Yamagami, S.; Takamori, T. and Oguro, K.: *Modeling of Nafion-Pt composite actuators (ICPF) by ionic motion.* Proc. SPIE – Int. Soc. Opt. Eng., v 3987 (2000), pp. 92–102
215. Bar-Cohen, Y.; Xue, T.; Joffe, B.; Lih, S.-S.; Shahinpoor, M.; Simpson, J.; Smith, J.; Willis, P.: *Electroactive polymers (IPMC), low mass muscle actuators.* Proc. 1997 SPIE Smart Mater. Struct. Conf., San Diego California, SPIE 3041–76 (1997)
216. Salehpoor, K.; Shahinpoor, M.; Razani, A.: *Role of ion transport in dynamic sensing and actuation of ionic polymeric platinum composite artificial muscles.* Proc. 1998 SPIE Smart Mater. Struct. Conf., San Diego California, SPIE 3330–09 (1998)
217. Shahinpoor, M.; Bar-Cohen, Y.; Simpson, J.O.; Smith, J.: *Ionic polymer-metal composites (IPMCs) as biomimetic sensors, actuators and artificial muscles – a review.* Smart Mater. Struct., 7, 6 (1998), pp. R15–R30
218. Nemat-Nasser, S.: *Micromechanics of actuation of ionic polymer-metal composites.* J. Appl. Phys., 92, 5 (2002), pp. 2899–2915.
219. Smela, E.; Gadegaard, N.: *Surprising volume change in PPY(DBS): an atomic force microscopy study.* Adv. Mater. 11 (11) (1999), pp. 953–957

220. Mazzoldi, A.; Carpi, F.; De Rossi, D.: *Polymers responding to electrical or electrochemical stimuli for linear actuators.* Ann. Chim.-Sci. Mat., 29 (6) (2004), pp. 55–64
221. Ding, J.; Liu, L.; Spinks, G. M.; Zhou, D.; Wallace, G.G.; Gillespie, J.: *High performance conducting polymer actuators utilising a tubular geometry and helical wire interconnects.* Synth. Met., 138 (2003), pp. 391–398
222. Ding, J.; Zhou, D.; Spinks, G.; Wallace, G.G.; Forsyth, S.; Forsyth, M.; MacFarlane, D.: *Use of ionic liquids as electrolytes in electromechanical actuator systems based on inherently conducting polymers.* Chem. Mater., 15, (2003), pp. 2392–2398
223. Bay, L.; West, K.; Sommer-Larsen, P.: *A conducting polymer artificial muscle with 12% linear strain.* Adv. Mat., 15 (4) (2003), pp. 310–313
224. Otero, T.F.; Cortes, M.T.: *Artificial muscles with tactile sensitivity.* Adv. Mater., 15 (4) (2003), pp. 279–282
225. Lu, W.; Fadeev, A.G.; Qi, B.; Smela, E.; Mattes, B.R.; Ding, J.; Spinks, G.M.; Mazurkiewicz, J.; Zhou, D.; Wallace, G.G.; MacFarlane, D.R.; Forsyth, S.A.; Forsyth, M.: *Use of ionic liquids for conjugated polymer electrochemical devices.* Science, 297 (2002), pp. 983–987
226. Spinks, G.M.; Liu, L.; Wallace, G.G.; Zhou, D.: *Strain response from polypyrrole actuators under load.* Adv. Funct. Mater., 12 (6–7) (2002), pp. 437–440
227. Bay, L.; Jacobsen, T.; Skaarup, S.; West, K.: *Mechanism of actuation in conducting polymers: osmotic expansion.* J. Phys. Chem. B., 105 (36) (2001), pp. 8492–8497
228. Lewis, T.W.; Spinks, G.M.; Wallace, G.G.; Mazzoldi, A.; De Rossi, D.: *Investigation of the applied potential limits for polypyrrole when employed as the active components of a two-electrode device.* Synth. Met., 122 (2001), pp. 379–385
229. Mazzoldi, A.; Della Santa, A.; De Rossi, D.: *Conducting polymer actuators: properties and modeling.* In Polymer sensors and Actuators, Osada, Y. and De Rossi, D. Ed., Berlin: Springer (2000), pp. 207–244
230. Mazzoldi, A.; Degl'Innocenti, C.; Michelucci; De Rossi, D.: *Actuative properties of polyaniline fiber under electrochemical stimulation.* Mat. Sci. Tech. C, 6 (1998), pp. 65–72
231. De Rossi, D.; Della Santa, A.; Mazzold, A.: *Performance and work capacity of a polypyrrole conducting polymer linear actuator.* Synth. Met., 90 (1997), pp. 93–100
232. Della Santa, A.; De Rossi, D.; Mazzoldi, A.: *Characterization and modelling of a conducting polymer muscle-like linear actuator.* Smart. Mater. Struct., 6 (1) (1997), pp. 23–34
233. Baughman, R.H.: *Conducting polymer artificial muscles.* Synth. Met., 78 (1996), pp. 339–353
234. Gandhi, M.R.; Murray, P.; Spinks, G.M.; Wallace, G.G.: *Mechanism of electromechanical actuation in polypyrrole.* Synth. Met., 73 (1995), pp. 247–256
235. Chiarelli, P.; Della Santa, A.; De Rossi, D.; Mazzoldi, A.: *Actuation properties of electrochemical driven polypyrrole free-standing films.* J. Int. Mat. Syst. Struc, 6 (1) (1995), pp. 32–37
236. Kaneto, K.; Kaneko, M.; Min, Y.; MacDiarmid, Alan G.: *Artificial muscle: Electromechanical actuators using polyaniline films.* Synth. Met., 71 (1995), pp. 2211–2212

237. Baughman, R.H.; Shacklette, L.V.: *Science and Application of Conducting Polymers.* Salaneck, W. R.; Clark, D. T.; Samuelsen, E. J. Ed.; New York: Adam Hilger (1990)
238. Otero, T.; Sansinena, J. M.: *Soft and wet conducting polymers for artificial muscles.* Adv. Mater., 10 (6) (1998), pp. 491–494
239. Otero, T.F. et al.: *Artificial Muscles From Bilayer Structures.* Synth. Met., 55–57,pp. 3713–3717 (1993)
240. Pei, Q.; Inganäs, O.: *Electrochemical muscles: Bending strips built from conjugated polymers.* Synth. Met., 57 (1), 3718–372 (1993)
241. Pei, Q.; Inganäs, O.; Lundström, I.: *Bending bilayer strips built from polyaniline for artificial electrochemical muscles.* Smart Mater. Struct., 2 (1993), pp. 1–6
242. Smela, E.; Inganäs, O.; Lundström, I.: *Conducting polymers as artificial muscles: challenges and possibilities.* J. Micromech. Microeng. 3 (1993), pp. 203–205
243. Della Santa, A.; Mazzoldi, A.; De Rossi, D.: *Steerable microcatheter actuated by embedded conducting polymer structures.* J. Intell. Mater. Sys. Structures, vol. 7, n3 (1996), pp. 292–300
244. Web site: http://ndeaa.jpl.nasa.gov/nasa-nde/lommas/eap/EAP-material-n-products.htm.
245. Chou, C.P.; Hannaford, B.: *Static and dynamic characteristics of McKibben pneumatic artificial muscles.* Proc. 1994 IEEE Int. Conf. on Robotics Automation, San.Diego, CA, USA, vol. 1 (1994), pp. 281–286
246. De Rossi, D.; Lorussi, F.; Mazzoldi, A.; Rocchia, W.; Scilingo, E.P.: *A strain amplified electroactive polymer actuator for haptic interfaces.* Proc. EAPAD-SPIE, Newport Beach, CA, March, (2001)
247. Smela, E.: *Conjugated Polymer Actuators for Biomedical Applications.* Adv. Mater., 15 (6) (2003), pp. 481–494
248. Takashima, W.; Kaneko, M.; Kaneto, K.; MacDiarmid, A. G.: *The electrochemical actuator using electrochemically-deposited poly-aniline film.* Synth. Met., 71 (1995), pp. 2265–2266
249. Heinze, J.: *Electrochemistry of conducting polymers.* Synth. Met., 41–43 (1991), pp. 2805–2823
250. Wang, H.L.; Romero, R.J.; Mattes, B.R.; Zhu, Y.; Winokur, M. J.: *Effect of processing conditions on the properties of high molecular weight conductive polyaniline fiber.* J. Pol. Sci. Part B, 38 (1) (2000), pp. 194–204
251. Mattes, B.R.; Wang, H.L.; Yang, D.; Zhu, Y.T.; Blumenthal, W. R.; Hundley, M.F.: *Formation of conductive polyaniline fibers derived from highly concentrated emeraldine base solutions.* Synt. Met., 84 (1994), pp. 45–49
252. Chinn, D.; Janata, J.: *Spin-cast thin films of polyaniline.* Thin Solid Films, 252, 145 (1994)
253. Lee, J. Y.; Kim, D. Y.; Kim, C. Y.: *Synthesis of soluble polypyrrole of the doped state in organic solvents.* Synth. Met., 74, 103 (1995)
254. Kim, I.W.; Lee, J. Y.; Lee, H.: *Solution-cast polypyrrole film: the electrical and thermal properties.* Synth. Met., 78, 177 (1996)
255. Miller, E.K.; Lee, K.; Heeger, A. J.; Lee, J. Y.; Kim, D. Y.; Kim, C. Y.: *Reflectance studies of soluble polypyrrole doped with dodecylbenzene sulfonic acid.* Synth. Met., 84, 821 (1997)

256. Pyo, M.; Bohn, C.C.; Smela, E.; Reynolds, J.R.; Brennan, A.B.: *Direct strain measurement of polypyrrole actuators controlled by the polymer/gold interface.* Chem. Mater., 15 (4) (2003), pp. 916–922
257. Jager, E.W.H.; Smela, E.; Inganäs, O.: *Microfabricating Conjugated Polymer Actuators.* Science, 290 (2000), pp. 1540–1545
258. Smela, E.: *Microfabrication of PPy microactuators and other conjugated polymer devices.* J. Micromech. Microeng., 9 (1998), pp. 1–18
259. Smela, E.; Inganäs, O.; Pei, Q.; Lundström, I.: *Electrochemical muscles: micromachining fingers and corkscrews.* Adv. Mater., 5 (1993), pp. 630–632
260. Smela, E.; Kallenbach, M.; Holdenried, J.: *Electrochemically driven polypyrrole bilayers for moving and positioning bulk micromachined silicon plates.* J. Microelectrom. Syst, 8 (1999), pp. 373–383
261. Jager, E.W.H.; Smela, E.; Inganäs, O.; Lundström, I.: *Application of polypyrrole microactuators.* Proc. SPIE Int. Soc. Opt. Eng. 3669 (1999), pp. 377–384
262. Smela, E.; Inganäs, O.; Lundström, I.: *Controlled folding of micrometer-size structures.* Science, 268 (1995), pp. 1735–1738
263. Jager, E.W.H.; Inganäs, O.; Lundström, I.: *Perpendicular Actuation with Individually Controlled Polymer Microactuators.* Adv. Mat., 13 (1) (2001), pp. 76–79
264. Jager, E.W.H.; Inganäs, O.: Lundström, I.: *Microrobots for Micrometer-Size Objects in Aqueous Media Potential Tools for Single-Cell Manipulation.* Science, 288 (2000), pp. 2335–2338
265. Pede, D.; Serra, G.; De Rossi, D.: *Microfabrication of conducting polymer devices by ink-jet stereolithography.* Mater. Sci. Eng. part C, 5 (1998), pp. 289–291
266. Chang, S.C.; Bharathan, J.; Yang, Y.; Helgeson, R.; Wudl F.; Ramey M. B.; Reynolds, J.R.: *Dual-color polymer light-emitting pixels processed by hybrid inkjet printing.* Appl. Phys. Lett., 73 (1998), pp. 2561–2563
267. Hebner, T.; Wu, C.C.; Marcy, D.; Lu, M.H.; Sturm, J.C.: *Ink-jet printing of doped polymers for organic light emitting devices.* Appl. Phys. Lett., 72 (1998), pp. 519–521
268. Whitesides, G.M. et al: *Soft lithography in biology and biochemistry.* Annual Rev. Biomed. Eng, 3 (2001), pp. 335–73
269. Bao, Z.N.; Rogers, J.A.; Katz, H.E.: *Printable organic and polymeric semiconducting materials and devices.* J. Mater. Chem., 9 (1999), pp. 1895–1904
270. Granlund, T.; Nyberg, T.; Roman, L.S.; Svensson, M.; Inganas, O.: *Patterning of Polymer Light-Emitting Diodes with Soft Lithography.* Adv. Mater., 12 (2000), pp. 269–273
271. Vozzi, G. et al: *Microfabricated PLGA scaffolds: a comparative study for application to Tissue Engineering.* Mat. Sci. Eng. C, 20 (1–2) (2002), pp. 43–47
272. Vozzi, G. et al.: *Microsyringe-Based Deposition of Two-Dimensional and Three-Dimensional Polymer Scaffolds with a Well-Defined Geometry for Application to Tissue Engineering.* Tissue Eng., 8 (6) (2002), pp. 1089–1098
273. Lewis, T.W.; Spinks, G.M.; Wallace, G.G.; De Rossi, D.; Pachetti, M.: *Development of an all polymer electromechanical actuator.* Polymer Preprints, 38 (2) (1997), pp. 520–521
274. Iijima, S.: *Helical microtubules of graphitic carbon.* Nature 354 (1991), pp. 56–58

275. Dalton, B.; Collins, S.; Muoz, E.; Razal, J. M.; Ebron, V. H.; Ferraris, J. P.; Coleman, J. N.; Kim, B. G.; Baughman, R. H.: *Super-tough carbon-nanotube fibers.* Nature, 423, 703 (2003).
276. Baughman, R.H.; Changxing Cui; Zakhidov, A.A.; Iqbal, Z.; Barisci, J.N.; Spinks, G.M.; Wallace, G.G.; Mazzoldi, A.; De Rossi, D.; Rinzler, A.G.; Jaschinski, O.; Roth, S.; Kertesz, M.: *Carbon nanotube actuators.* Science, vol. 284, 1340 (1999)
277. Spinks, G.M.; Wallace, G.G.; Fifield, L.S.; Dalton, L.R.; Mazzoldi, A.; De Rossi, D.; Khayrullin, I.I.; Baguhman, R.H.: *Pneumatic carbon nanotube actuators.* Advanced Mater., 14 (23) (2002), pp. 1728–1732
278. Ahluwalia, A.; Baughman, R.; De Rossi, D.; Mazzoldi, A.; Tesconi, M.; Tognetti, A.; Vozzi, G.: *Microfabricated electroactive carbon nanotube actuators.* Proc. EAPAD-SPIE, Newport Beach, CA (2001).
279. Zhang, M.; Atkinson, K.R.; Baughman, R.H.: *Multifunctional Carbon Nanotube Yarns by Downsizing an Ancient Technology.* Science 306 (2004), pp. 1358–1361
280. Pelrine, R.E.; Kornbluh, R.D. and Joseph, R.D.: *Electrostriction of polymer dielectrics with compliant electrodes as a means of actuation.* Sensors and Actuators A-Phys., 64 (1998), pp. 77–85
281. Pelrine, R.; Kornbluh, R.; Pei, Q. and Joseph, J.: *High-speed electrically actuated elastomers with strain greater than 100%.* Science, 287 (2000), pp. 836–839
282. Pelrine, R.; Kornbluh, R. and Kofod, G.: *High-strain actuator materials based on dielectric elastomers.* Advanced Mater., 12 (16) (2000), pp. 1223–1225
283. Pelrine, R.; Kornbluh, R.; Joseph, J.; Heydt, R.; Pei, Q.; Chiba, S.: *High-field deformation of elastomeric dielectrics for actuators.* Mater. Sci. Eng. C. 11 (2000), pp. 89–100
284. Carpi, F. and De Rossi, D. : *Improvement of electromechanical actuating performances of a silicone dielectric elastomer by dispersion of titanium dioxide powder.* IEEE Trans. on Dielectrics and Electrical Insulation, 12 (4) (2005), pp. 835–843
285. Pei, Q.; Pelrine, R.; Stanford, S.; Kornbluh, R.; Rosenthal, M.: *Electroelastomer rolls and their application for biomimetic walking robots.* Synth. Met. 135–136, 129–131, (2003)
286. Pei, Q.; Rosenthal, M.; Stanford, S.; Prahlad, H.; Pelrine, R.: *Multiple-degrees-of-freedom electroelastomer roll actuators.* Smart Mater. Struct. 13 (2004), pp. N86–N92
287. Kofod, G.; Sommer-Larsen, P.; Kornbluh, R. and Pelrine, R.: *Actuation response of polyacrylate dielectric elastomers.* J. Intell. Mater. Syst. Struct. 14 (12) (2003), pp. 787–793
288. Carpi, F.; Chiarelli, P.; Mazzoldi, A.; De Rossi, D.: *Electromechanical characterisation of dielectric elastomer planar actuators: comparative evaluation of different electrode materials and different counterloads.* Sensors and Actuators A-Phys., 107 (2003), pp. 85–95
289. Carpi, F. ; De Rossi, D. : *Dielectric elastomer cylindrical actuators: electromechanical modelling and experimental evaluation.* Mater. Sci. Eng. C, 24 (2004), pp. 555–562
290. Jungmann, M.; Matysek, M.; Schlaak, H.F.: *Electrostatic solid-state actuators with elastic dielectric and multilayer fabrication-technology.* Proc. Actuator 2004, 14–16 June 2004, Bremen (2004), pp. 686–689

291. Carpi, F.; Migliore, A.; Serra, G. and De Rossi, D.: *Helical dielectric elastomer actuators.* Smart Materials and Structures, 14 (2005), pp. 1210–1216
292. Carpi, F.; Salaris, C. and De Rossi, D.: *Folded dielectric elastomer actuators.* Smart Mater. and Structures, 16 (2007), pp. S300–S305
293. Scilingo, E.P.; Lorussi, F.; Mazzoldi, A.; De Rossi, D.: *Strain-sensing fabrics for wearable kinaesthetic-like systems.* IEEE Sensors J., vol. 3, no. 4 (2003), pp. 460–467
294. De Rossi, D.; Carpi, F.; Lorussi, F.; Mazzoldi, A.; Paradiso, R.; Scilingo, E.P.; Tognetti, A.: *Electroactive fabrics and wearable biomonitoring devices.* AUTEX Research J., vol. 3, no. 4 (2003), pp. 180–185
295. De Rossi, D.; Carpi, F.; Lorussi, F.; Paradiso, R.; Scilingo, E.P.; Tognetti, A.: *Electroactive fabrics and wearable man-machine interfaces.* In: Wearable electronics and photonics, Tao, X.-M. Ed.; Cambridge: Woodhead (2005)
296. De Rossi, D.; Carpi, F.; Lorussi, F.; Mazzoldi, A.; Scilingo, E.P.; Tognetti, A.: *Electroactive fabrics for distributed, conformable and interactive systems.* Proc. IEEE Sensors 2002, Hyatt Orlando, Florida, (October 2002)
297. Lorussi, F.; Rocchia, W.; Scilingo, E.P.; Tognetti, A.; De Rossi, D.: *Wearable, redundant fabric-based sensor arrays for reconstruction of body segment posture.* IEEE Sensors J., vol. 4, no. 6 (2004), pp. 807–818
298. Singh Nalwa, H. Ed.: *Ferroelectric polymers – Chemistry, physics and applications.* New York: Marcel Dekker, 1995.
299. De Rossi, D.; Lazzeri, L.; Domenici, C.; Nannini, A.; Basser, P.: *Tactile sensing by an electromechanochemical skin analog.* Sensors Actuators, 17 (1989), pp. 107–114
300. De Rossi, D.; Nannini, A.; Domenici, C.: *Artificial sensing skin mimicking mechanoelectrical conversion properties of human derims.* IEEE Trans. Biomed. Eng., 35 (2) (1988), pp. 83–92
301. Guckel, H.: *Progress in magnetic microactuators.* Microsystem Technol. 5 (1998), pp. 59–61
302. Tang, W.C.; Nguyen, T.C.H. and Howe, R.T.: *Laterally driven polysilicon resonant microstructures.* Sensors and Actuators A 20 (1992), pp. 25–32
303. Lutz, M.; Golderer, W.; Gerstenmeier, J.; Marek, J.; Maihofer, B.; Mahler, S.; Munzel, H. and Bischof, U.: *A precision yaw rate sensor in silicon micromachining.* Procs. Solid State Sensors and Actuators, 1997, vol. 2, Transducers '97 Chicago (16–19 Jun. 1997), pp. 847–850
304. DeVoe, D. L.: *Piezoelectric thin film micromechanical beam resonators.* Sensors and Actuators A 88 (2001), pp. 263–272
305. www.terfenoltruth.com
306. Quandt, E. and Seemann, K.: *Fabrication of giant magnetostrictive thin film actuators.* Proc. IEEE MEMS 1995 (1995), pp. 273–277.
307. Freygang, M.; Haffner, H.; Messner, S. and Schmidt, B.: *A New Concept Of A Bimetallically Actuated, Normally-Closed Microvalve.* Proc. Transducers '95, Stockholm (1995), pp. 73–74
308. Aikele, M.; Bauer, K.; Ficker, W.; Neubauer, F.; Prechtel, U.; Schalk, J. and Seidel H.: *Resonant accelerometer with self-test.* Sensors and Actuators A: Physical, 92 (2001), pp. 161–164
309. Kohl, M.: *Shape memory microactuators.* 247 p., Springer, Berlin (2004)
310. Hamberg, M.W.; Neagu, C.; Gardeniers, J.G.E.; Ijntema, D.J. and Elwenspoek, M.: *An Electrochemical Micro Actuator.* Proc. MEMS '95, Amsterdam, The Netherlands (1995)

311. Bosch, D.; Heimhofer, B.; Mück, G.; Seidel, H.; Thumser, U. and Welser, W.: *A silicon microvalve with combined electromagnetic /electrostatic actuation.* Sensors and Actuators A 37–38 (1993), pp. 684–692
312. Fu, Y.; Ghantasala, M.K.; Harvey, E.C. and Qin, L.: *Design and fabrication of a hybrid actuator.* Smart Materials and Structures 14 (2005), pp. 488–495
313. Hue, P.-Le: *Progress and trends in ink-jet printing technology.* J. Imaging Sci. and Technol., 42 (1998), pp. 49–62
314. Wirtl, J.: *Die Piezokeramische Pille öffnet den Weg zur leistungslosen Ventilansteuerung.* Firmenschrift der Hoerbiger Fluidtechnik GmbH, D-86956 Schongau
315. Jerman, H.: *Electrically-Activated, Normally-Closed Diaphragm Valves.* Proc. MEMS '91, Nara, Japan (1991)
316. Zdeblick, M.J.; Anderson, R.; Jankowski, J.; Kline-Schoder, B. and Christel, L.; Miles, R.; Weber, W.: *Thermopneumatically Actuated Microvalves and Integrated Electro-Fluidic Circuits.* Proc. Actuator 94, Bremen (1994), pp. 56–60
317. Mettner, M.; Huff, M.; Lober, T. and Schmidt, M.: *How to design a microvalve for High pressure Application.* Robert Bosch GmbH, 70469 Stuttgart (1990)
318. Shikida, M.; Sato, K.: *Characteristics of an Electrostatically driven Gas Valve under High Pressure Conditions.* Proc. MEMS '94, Osio, Japan (1994)
319. Zengerle, R.: *Mikro-Membranpumpen als Komponenten für Mikro-Fluidsysteme.* Verlag Shaker, Aachen (1994), ISBN 3–8265-0216–7
320. Zengerle, R.; Ulrich, J.; Kluge, S.; Richter, M. and Richter, A.: *A Bidirectional Silicon Micropump.* Proc. MEMS '95, Amsterdam (1995), pp. 19–24
321. Gerlach, T.; Wurmus, H.: *Working principle and performance of the dynamic micropump.* Proc. MEMS '95, Amsterdam, The Netherlands (1995), pp. 221–226
322. Olson, A.; Enoksson, P.; Stemme, G. and Stemme, E.: *A valve-less planar pump in silicon.* Proc. Transducers '95, Stockholm (25–29 Jun. 1995), pp. 291–294
323. Keefe, D.O.; Herlihy, C.O.; Gross, Y. and Kelly, J.G.: *Patient-controlled analgesia using a miniature electrochemically driven infusion pump.* British J. Anaesthesia (1994), pp. 843–846
324. Stehr, M.; Messner, S.; Sandmaier, H. and Zengerle, R.: *The VAMP – A new device for handling liquids or gases.* Sensors and Actuators A–Physical 57 (1996), pp. 153–157
325. Howitz, S.; Wegener, T.; Fiehn, H.: *Mikrotropfeninjektor.* FZ Rossendorf e.V., GeSiM mbH Dresden
326. Hornbeck, L.J.: *Digital Light Processing and MEMS: Timely Convergence for a Bright Future.* Plenary Session, SPIE Micromachining and Microfabrication '95, Austin, Texas (October 24, 1995)
327. Gerlach, T.; Enke, D.; Frank, Th.; Hutschenreuther, L.; Schacht, H.-J. and Schüler, R.: *Towards an integrated microsystem for the automatic adjustment of mono-mode optical waveguides.* Workshop MicroMechanics Europe MME '96, Barcelona (21–22 Oct. 1996)
328. Fan, L.; Tai, Y. and Muller, R.S.: *IC-processed electrostatic micromotors.* Sensors and Actuators 20 (1989), pp. 41–47
329. Kämper, K.-P.; Ehrfeld, W.; Hagemann, B.; Lehr, H.; Michel, F.; Schirling, A.; Thürigen, Ch. and Wittig, Th.: *Electromagnetic permanent magnet micro-*

motor with integrated micro gear box. Proc. Actuator 96, Bremen (1996), pp. 429–436

330. Janocha, H. (Ed.): *Actuators – Basics and Applications.* Springer, Berlin Heidelberg New York (2004)
331. Newnham, R.E.; Ruschau, G.R.: *Smart Electroceramics.* J. Am. Ceram. Soc. 74, 3 (1991), pp. 463–480
332. Kuhnen, K.: *Inverse Steuerung piezoelektrischer Aktoren mit Hysterese-, Kriech- und Superpositionsoperatoren.* Doctoral Thesis, Saarland University, Shaker, Aachen (2001)
333. Gabbert, U.; Tzou, H.S. (Eds.): *Smart Structures and Structronic Systems.* Proc. IUTAM Symp., Magdeburg (2000); Kluwer Academic, Dordrecht/Boston/London (2001)
334. Bergqvist, A.: *On magnetic hysteresis modeling.* Royal Inst. Technol., Electric Power Eng., Stockholm (1994)
335. Föllinger, O.: *Regelungstechnik.* Oldenbourg-Verlag, München, Wien (1990)
336. Kuhnen, K.; Janocha, H.; Schommer, M.: *Exploitation of inherent sensor effects in magnetostrictive actuators.* Proc. 9th Int. Conf. on New Actuators, Bremen (2004), pp. 367–370
337. Kuhnen, K.; Janocha, H.: *Self-sensing Solid-state Actuators.* In: 6.43 Control Systems, Robotics and Automation, edited by H. Unbehauen, In: Encyclopedia of Life Support Systems (EOLSS). Developed under the Auspices of the UNESCO, Eolss Publishers, Oxford, UK (2004), [http://www.eolss.net]
338. Clephas, B.: *Untersuchung von hybriden Festkörperaktoren.* Doctoral Thesis, Saarland University, Herbert Utz, München (1999)
339. Clephas, B.; Janocha, H.: *Simultaneous sensing and actuation of a magnetostrictive transducer.* Proc. SPIE Smart Structures and Mater. 3329, San Diego (1998), pp. 174–184
340. Schommer, M.; Janocha, H.: *Rekonstruktion der Belastung eines magnetostriktiven Aktors durch Signalanalyse.* ETG-Fachbericht of the ETG-/GMM-Fachtagung 2004, Darmstadt (2004), pp. 189–194 (CD-ROM)
341. Brokate, M.; Sprekels, J.: *Hysteresis and phase transitions.* Springer, Berlin Heidelberg New York (1996)
342. Jones, L.; Garcia, E.; Waites, H.: *Self-sensing control as applied to a stacked PZT actuator used as a micropositioner.* Smart Structures and Mater., 3 (1994), pp. 147–156
343. Dosch, J.J.; Inman, D.J.; Garcia, E.: *A self-sensing piezoelectric actuator for collocated control.* J. Intell. Mater. Syst. and Struct., 3 (1992), pp. 166–185
344. Vallone, P.: *High-performance piezo-based self-sensor for structural vibration control.* SPIE Smart Structures and Mater. Conf., 2443, SPIE (1995), pp. 643–655
345. Cole, D.J.; Clark, R.L.: *Adaptive compensation of piezoelectric sensoriactuator.* J. Intell. Mater. Syst. and Struct., 5 (1994), pp. 665–672
346. Vipperman, J.; Clark, R.: *Implementation of an adaptive piezoelectric sensoriactuator.* AIAA J., 34 (1996), pp. 2102–2109
347. Ko, B.; Tondue, B.H.: *Acoustic control using a self-sensing actuator.* J. Sound and Vibration, 187 (1995), pp. 145–165
348. Pratt, J.; Flatau, A.B.: *Developement and analysis of a self-sensing magnetostrictive actuator design.* Proc. SPIE Smart Structures and Mater. 1917, Albuquerque (1993) pp. 952–961

349. Banks, H.T.; Smith, R.C.: *Hysteresis Modeling in Smart Material Systems.* J. Appl. Mechanics and Eng. 5, 1 (2000) pp. 31–45
350. Visintin, A.: *Differential Models of Hysteresis.* Springer, Berlin Heidelberg (1994)
351. Mayergoyz, I.D.: *Mathematical Models of Hysteresis.* Springer, Berlin Heidelberg New York (1991)
352. Krasnosel'skii, M.A.; Pokrovskii, A.V.: *Systems with Hysteresis.* Springer, Berlin (1989)
353. Kuhnen, K.; Janocha, H.: *Inverse feedforward controller for complex hysteretic nonlinearities in smart material systems.* Control and Intell. Sys. 29, 3 (2001), pp. 74–83
354. Kuhnen, K.: *Modeling, Identification and Compensation of complex hysteretic Nonlinearities – A modified Prandtl-Ishlinskii approach.* Eur. J. of Control 9, 4 (2003), pp. 407–418
355. Webb, G.V.; Lagoudas, D.C.; Kurdila, A.J.: *Hysteresis Modeling of SMA Actuators for Control Applications.* J. of Intell. Mater. Sys. and Structures, 9 (1998), pp. 432–448
356. Krejci, P.: *Hysteresis, Convexity and Dissipation in Hyperbolic Equations.* Gakotosho, Tokyo, Gakuto Int. Series Math. Sci. & Appl. 8 (1996)
357. Krejci, P.; Kuhnen, K.: *Inverse Control of Systems with Hysteresis and Creep.* IEE Proc. Control Theory Appl. 148, 3 (2001), pp. 185–192
358. Jones L.; Garcia, E.: *Novel approach to self-sensing actuation.* SPIE Smart Structures and Mater. Conf. 3041, SPIE (1997), pp. 305–314
359. Kortendieck, H.: *Entwicklung und Erprobung von Modellen zur Kriech- und Hysteresiskorrektur.* VDI, Düsseldorf (1993)
360. Kuhnen, K.; Janocha, H.: *Inverse Steuerung für den Großsignalbetrieb von Piezoaktoren.* at-Automatisierungstechnik 50, 9 (2002), pp. 439–450
361. Janocha, H.; Pesotski, D.; Kuhnen, K.: *FPGA-Based Compensator of Hysteretic Actuator Nonlinearities for Highly Dynamic Applications.* Proc. 10th Int. Conf. on New Actuators, Bremen (2006), pp. 1013–1016
362. www.dass.de
363. Janocha, H.; Stiebel, Ch. and Würtz, Th.: *Power Amplifiers for Piezoelectric Actuators.* In: Preumont, A. (ed.) Responsive Systems for Active Vibration Control, Kluwer (2002), pp. 379–391
364. Janocha, H.; Pesotski, D. and Kuhnen, K.: *FPGA-Based Compensator of Hysterestic Actuator Nonlinearities for Highly Dynamic Applications.* Proc. 10th Int. Conf. on New Actuators, Bremen, Germany (14-16 June 2006), pp. 1013–1016

7 Sensors in Adaptronics

7.1 Advances in Intelligent Sensors

N.M. White, P. Boltryk

7.1.1 Introduction

The emergence of intelligent sensors arises from the fortunate conjunction of technological demands and technological feasibility. There was a time when engineers made do with a few basic measurements of physical quantities they knew they could measure, rather than seek sensors that could accurately convey the information they really needed. As society and industry become more complex this option becomes increasingly less realistic. There is a growing need to determine precise values of physical and chemical measurands independently of any other variables present. Large scale integration has appeared just in time to provide a solution to the major problems posed by such needs.

In the days of linear continuous electronics the available sensors were limited by stringent requirements on linearity, cross-sensitivity, freedom from drift etc. This meant that most of the vast panoply of possible sensor mechanisms had to be rejected out of hand. The magnitude of change wrought by the appearance of digital electronics would be difficult to overstate. The existence of a drift-free storage mechanism alone provides a solution to many problems, but coupled with an increasing availability of processing power it diminishes once insurmountable barriers almost to the point of negligibility.

Of equal importance with the steadily increasing power of devices is the remarkable decrease in cost. Not only has the density of transistors been doubling every two years, but the cost of a logic gate has been halving every two years [1] and there is no sign of this trend abating. We have related elsewhere [2] how the first suggestions for intelligent sensors were ridiculed on the grounds of the high cost of microprocessors. Now a microprocessor is simply a library element that can be incorporated in an application-specific integrated circuit (ASIC) design and manufactured on a large scale for a few cents.

The term intelligent sensor has been used over the past twenty years or so to refer to sensors having additional functionality provided by the integration of microprocessors, microcontrollers or ASICs with the sensing element (or even adaptronic material) itself. For consistency in this text, we will adopt the term intelligent sensor to refer to a microsensor with integrated microelectronic circuitry and digital processing capability. Intelligent sensors offer a number of advantages for sensor system designers. The integration of sensor (which may itself be adaptronic) and electronics, allows the intelligent sensor to be treated as a module, or black-box, where the internal complexities of the sensor are kept remote from the host system. Hence the intelligent sensor allows the designer to address the issue of an adaptronic system, whereby the electronic hardware and software can be combined with a multifunctional material to create a system that can adapt its behaviour in accordance with its surroundings.

The concept of having a wireless, distributed network of intelligent sensors comprising low-power communications and localised processing has now become a reality [3]. Applications in the areas of environmental monitoring, structural monitoring, surveillance, condition-based equipment maintenance and ubiquitous computing are currently being examined.

7.1.2 Primary Sensor Defects

Before undertaking a brief review of some fundamental principles of sensing, we need to define terms. Searching through the literature in the area of measurement, the reader is faced with many different and sometimes conflicting definitions of transducer, sensor and actuator. Some authorities contend that a transducer should only be applied to energy conversion devices and that sensors are something different. We have chosen to define the terms with reference to the measurement (or control) system. Thus transducers divide into two sub-sets, sensors which input information to the system from the external world and actuators which output actions into the external world.

The intelligent sensor approach means that sensors that were initially thought to be unusable because of fundamental flaws such as non-linearity, cross-sensitivity etc., are now realisable. Before proceeding, it is worth noting the five major defects found in primary sensors [4]. They are:

- non-linearity;
- cross-sensitivity;
- time (or frequency) response;
- noise; and
- parameter drift.

In the days of linear, continuous electronics non-linearity was a major problem. Such compensation techniques as were available were based on diode networks having reciprocal characteristics, but by their nature these were relatively crude. As a result all non-linear primary sensor mechanisms tended to be ignored. Now, linearisation processes such as look-up tables or polynomials are easily realisable with digital electronics.

No primary sensor is sensitive to a single physical variable, and this fact give rise to the important defect known as cross-sensitivity, the dominant form of which is with temperature. Virtually all physical and chemical processes are temperature dependent and hence so are all our uncompensated primary sensors.

The materials and structures associated with primary sensors contain dissipative, storage and inertial elements. These translate into the time derivatives appearing in the differential equation that models the sensor system. Hence another major defect is represented by the time (or frequency) response. The means to neutralise this imperfection involves filtering, which may be thought of in terms of pole-zero cancelation. If the device has a frequency response $H(s)$ then a cascaded filter of response $G(s) = 1/H(s)$ will compensate for the non-ideal time response. The realisation of such a filter in analogue form presents a major obstacle that is greatly diminished in the digital case.

Noise is generally any unwanted signal, but the term is often used to imply random signals. Random noise will always be present, if only because the universe is in a state of continuous agitation. The almost ubiquitous low frequency ($1/f$) noise can cause great difficulties with primary sensors. The nature of $1/f$ noise is not well understood, but, by definition, its amplitude per unit bandwidth is inversely proportional to frequency. Hence measurements of signals down to zero frequency are particularly difficult.

Having looked at the various defects in sensors, we will now address four fundamental techniques of compensation [5]:

- structural compensation;
- tailored compensation;
- monitored compensation; and
- deductive compensation.

Structural compensation refers to the most traditional form. It concerns the way in which the material forming the sensor is physically organised to maximise the sensitivity of the device to the target variable and to minimise the response to all other physical variables. A good example here is the load cell (see later). Not only is the mechanical structure of the device symmetrical, but so is the electrical structure (i. e. Wheatstone bridge), and this illustrates the fundamental manifestation of structural compensation which is design symmetry. The target variable is thus arranged to produce a difference signal, while all other physical variables produce a common mode signal.

Inevitably, there will be a residual effect after applying structural compensation techniques for which it cannot cater, and this residual effect will vary between nominally identical sensors. Further techniques of minor adjustment are thus needed to minimise the residue. The term *tailored compensation* refers to trimming techniques that require action determined by the individual sensor and not the overall design, a major cost item in the traditional industry.

The third class, *monitored compensation*, relies upon taking a measurement of the cross-sensitive variable and compensating computationally, either by reference to a model of the sensor, or by making use of data obtained from a calibration cycle. The tool for monitored compensation is the sensor-within-a-sensor. In the extreme case, such as chemical sensors, where the cross-sensitivity is so severe that it becomes one of lack of specificity, the sensor array approach is preferred.

The final class of compensation is *deductive compensation*. This is resorted to in special circumstances when, for one reason or another, the test object is not physically accessible. Examples of such objects would include a nuclear reactor, the human brain or the cylinder chamber in an internal combustion engine. Deductive compensation requires reference to a model, and because all models are imperfect, it is only used as a last resort.

7.1.3 Hardware Structures

Figure 7.1 shows an example of a generalised hardware structure of an intelligent sensor. Specific examples may include all, or some, of these elements. The sensing element is the primary source of information into the system. The intelligent sensor may also have the ability to stimulate the sensing element to provide a self-test facility, whereby a reference voltage, for example, can be applied to the sensor in order to monitor its response. Some primary sensors, such as those based on piezoelectrics, convert energy directly from one domain into another and therefore do not require a power supply. Others, such as resistive-based sensors, may need stable DC sources, which may benefit from additional functionality such as pulsed excitation for power-saving reasons. So excitation control is another distinguishing feature found in intelligent sensors.

Amplification is usually a fundamental requirement, as most sensors tend to produce signal levels that are significantly lower than those used in the digital processor. Resistive sensors, such as strain gauges in a bridge configuration, often require an instrumentation amplifier; piezoelectric sensors, on the other hand, will require a charge amplifier. If possible, it is advantageous to have the gain as close as possible to the sensing element.

Examples of analogue processing include anti-aliasing filters for the conversion stage. In situations where real-time processing power is limited, there may also be benefits in implementing analogue filters.

Data conversion is the module between the continuous (real world) signals and the discrete signals associated with the digital processor. Typically, this stage comprises an analogue-to-digital converter (ADC). Inputs from other sensors (monitoring) can be fed into the data conversion sub-system in order to implement various forms of compensation. Note that such signals may also require amplification before data conversion. Resonant sensors, whose signals are in the frequency domain, do not need a data conversion stage because their outputs can be fed directly into the digital system.

The digital processing element mainly concerns the software processes within the intelligent sensor. These may be simple routines such as those re-

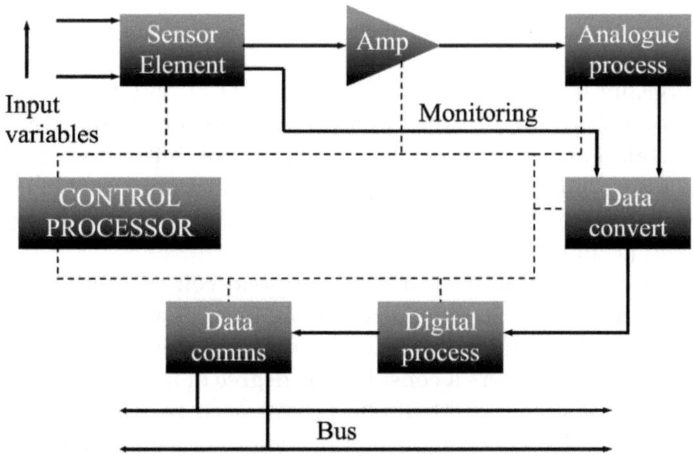

Fig. 7.1. Hardware structure of an intelligent sensor

quired for implementing sensor compensation (linearisation, cross-sensitivity, offset etc.) or may be more sophisticated techniques such as pattern recognition methods (such as neural networks) for sensor array devices.

The data communications element deals with the routines necessary for passing and receiving data and control signals to the sensor bus. It is often the case that the intelligent sensor is a single device within a multi-sensor system. Individual sensors can communicate with each other and also to the host system. There are many examples of commercial protocols that are used in intelligent sensor systems, but we will not cover these here. It is sufficient to be aware that the intelligent sensor will often have to deal with situations such as requests for data, calibration signals, error checking, message identification etc.

The control processor often takes the form of a microprocessor or microcontroller. It is generally the central component within the intelligent sensor and is connected to the other elements. The software routines are implemented within the processor and these are stored within the memory unit.

Illustrative Examples

In this subsection, our main objective is to provide examples of sensors that can benefit directly from the intelligent sensor approach. It is not feasible to cover even a significant fraction of the range, so we have chosen two illustrative examples in the forms of a load cell and gas sensor.

Perhaps the most common element found in mechanical sensors, such as load cells, is the strain gauge. This may take a variety of forms; semiconductor, thick-film or thin-film, but the most readily available is the metal foil gauge. This is attached to the structure by means of an adhesive. The positioning of the gauges is often critical, and great care must therefore be taken to ensure correct positioning. This labour-intensive task is one factor that accounts for the relatively high cost of sensors based on foil gauges. The metal foil strain gauge typically has a resistance of either 120 Ω or 350 Ω, thereby limiting the excitation voltage to about 10 V in order to prevent self-heating effects. Thick-film strain gauges, on the other hand, do not suffer from this problem as they exhibit a high resistance, typically greater than 10 kΩ. The use of microelectronic fabrication techniques also permits such sensors to be deposited quickly and accurately [6–8].

Figure 7.2 shows a representation of a precision load cell together with the electrical configuration of the strain gauges in the form of a Wheatstone bridge arrangement. Metal foil strain gauges are normally used with these devices. The mechanical structure offers a considerable degree of immunity from errors due to eccentric loads. The residual effects still need to be removed, however, and traditionally this is accomplished by tailored compensation in the form of trimming. An eccentric load is applied by attaching a beam to the load cell with a fixed mass at the free end. This is then rotated and small areas are manually filed off the structure to optimise the immunity to eccentricity. First order temperature compensation of the device is traditionally achieved by adding a length of resistance wire, of known temperature coefficient, to one arm of the bridge.

Chemically sensitive resistors are devices comprising a planar electrode pattern deposited onto an insulating substrate. The electrodes are then coated with a suitable chemically sensitive layer. The basic idea is that the conductivity or permittivity of the layer changes in the presence of a chemical measurand and this is measured by monitoring the impedance change between the electrodes. Unfortunately a single sensing element will not only respond to the desired quantity, but will also exhibit a marked cross-sensitivity to other variables including temperature, humidity and different chemical species within the local environment.

Figure 7.3 shows a gas sensor array fabricated using thick-film technology [6]. The chemically sensitive layer is an organic semiconductor. The device also has a platinum heating element situated underneath the electrode pattern. By supplying current to the heating element, the localised sensor site can

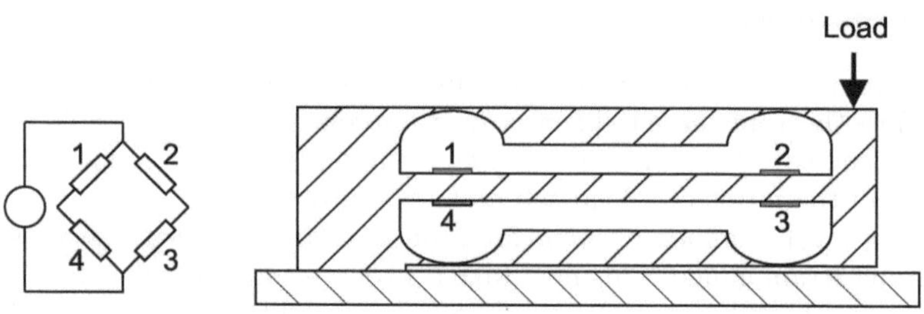

Fig. 7.2. Diagrammatic representation of the electrical and mechanical structure of a precision load cell, illustrating structural compensation by design symmetry

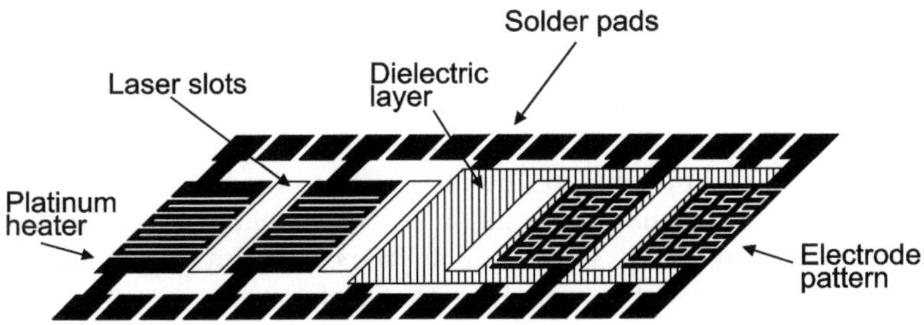

Fig. 7.3. An example of an array of chemically sensitive resistors fabricated using thick-film techniques. The slots are cut by a laser and help to isolate each sensor site from its neighbours (after Brignell et al. [6])

be heated in order to promote the chemical reaction between the organic layer and the sample gas. The resistance of the platinum heater can be monitored to infer the temperature of each sensing site. The cross-sensitivity to other gases is significant and a sensor array is needed to increase the specificity of the device. Within the array, each element is coated with a different reactive organic layer. Elaborate pattern recognition techniques, implemented in software, are required to establish quantitative analysis of a mixture of gases flowing over the sensor array. Research at the universities of Southampton and Warwick has addressed the production of arrays of gas sensing elements on silicon substrates. The operation of the devices is similar to that described above. Polymeric materials are used as the gas sensitive layers and the front-end electronics were fabricated as an ASIC.

The devices described above are examples of intelligent sensor systems that would not have been realisable in the early days of analogue electronic systems. The response of each individual sensor element exhibits a non-linear characteristic and is also cross-sensitive to a variety of other variables.

7.1.4 Software Processes

For compatibility with existing infrastructure, the output from an intelligent sensor should conform to IEEE standard 1451.4 for smart transducers [9]. Adherence to the structure of this standard allows the sensor to interface with the legacy of different communication network protocols and, in particular, enables compatibility with both digital and analogue communications, catering to the needs of networks still employing 4... 20 mA analogue communications and those operating using a digital communication bus.

For integration with data-fusion processes the sensor should provide its management system with an estimate of measurement uncertainty in addition to its measurand estimate. Statistically this information is completely described by the probability density function (PDF) of the process, with the mean and variance of the PDF equating to the measurement value and the uncertainty respectively. Probabilistic techniques such as data-based modeling can be used to estimate the underlying statistical properties directly, avoiding reliance on sensor models derived from first principles.

Conversion of the sensory data into forms compliant with the above specifications requires a range of software modules that range from simple linearisation to sophisticated signal processing methods for onboard sensor and electronics condition monitoring. We can divide the overall intelligent sensor software scheme into a series of sub-modules that include:

- pre-processing;
- signal conditioning;
- feature extraction;
- fault detection; and
- recalibration/reconfiguration.

Figure 7.4 shows a block diagram that illustrates the process. The initial phase, which is required before software can perform calibration of the raw sensory data, is a pre-processing module that converts the signal (which may be in the sensor modality, for example acoustic intensity), into a more universal engineering unit such as volts (or amps). The pre-processing may include basic filtering algorithms for anti-aliasing, rejection of $1/f$ noise, and for signal to noise ratio improvement, together with algorithms for calibration, normalisation and temperature compensation. The calibration process may include signal linearisation using a simple look-up table approach using coefficients stored in the transducer electronic data sheet (TEDS) [9]. An alternative linearisation technique involves summation of the sensors reciprocal characteristic with the signal. Additional functions provided by the calibration procedure include removal of sensor bias effects such as the DC component of the signal and scaling of the output using a technique such as normalisation.

The calibrated signals next pass through a signal conditioning software module to extract a series of features that characterise the data. Feature extraction is a process that is used to derive obscured information from the time history of the sensor signal that is useful both as useful sensor output information and as part of the onboard fault detection strategy. For example, the signal produced by a wireless pressure sensor embedded in a tyre might be corrupted by additive random noise processes, include small cyclic fluctuations caused by regular road surface defects, and fluctuations resulting from temperature. It is unlikely that the vehicles drive-train management system would benefit from the sensor transmitting the entire, high sample-rate time history, and a reduced set of features extracted from the signal such as the mean tyre pressure over a suitable rolling time window would provide the system with sufficient information to monitor the tyres health and pressure. Reduced data-transfer across the wireless network is also preferable for reducing the power consumed by the sensor. More formal feature extraction techniques such as principle component analysis [10] automatically extract

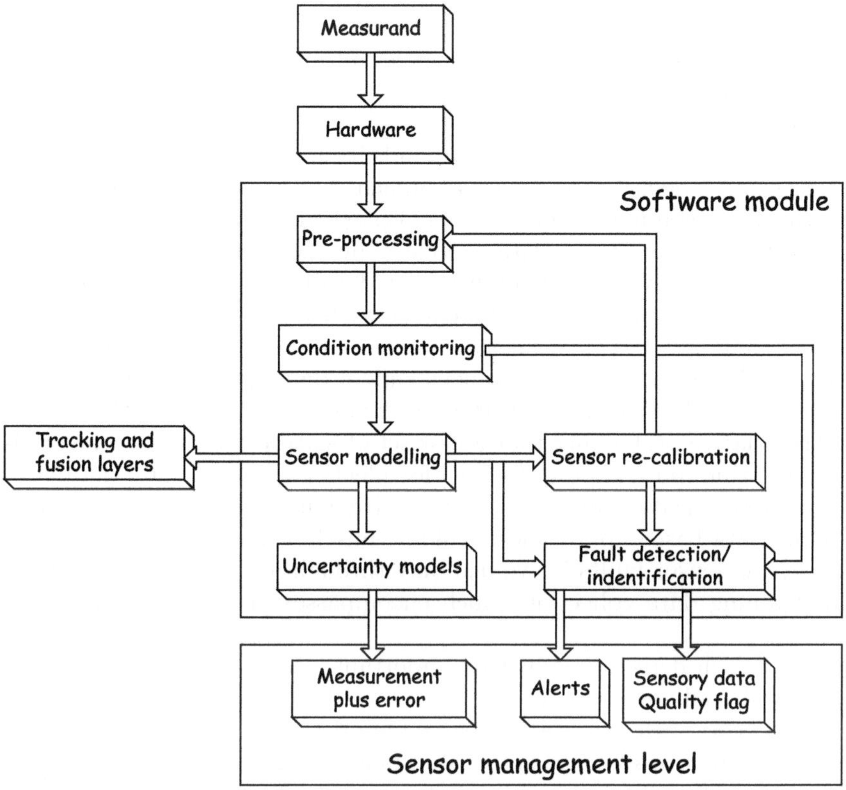

Fig. 7.4. Block diagram describing the software processes in intelligent sensors

a reduced set of uncorrelated statistical features that characterise the underlying process.

The derived features are an essential component for onboard self-diagnostics and fault detection. Onboard fault detection of the sensor condition, based on the sensor data itself, is subtly different to more traditional condition monitoring scenarios such as the monitoring of rotational bearings using accelerometers. Such devices may produce a signal after linearisation that is proportional to the amplitude of the vibration source. Derived features that may be more useful for diagnosing bearing faults may include the maximum acceleration (i. e. the maximum amplitude), and the spectral characteristics since wear conditions may be diagnosed early by identifying specific frequency components in the spectrum. Detecting faults in the sensor itself, however, requires differentiation between the actual sensor defect conditions and changes in sensor signals due to genuine changes in the environment. This point is well explained by considering the same accelerometer when it is suddenly removed from the bearing housing and placed in an environment exhibiting flat, wideband vibration spectra. Whilst it might appear from the accelerometer output signal that the electrical signal has been saturated by extraneous white noise and a fault should be suspected, an intelligent sensor should be able to autonomously detect that the sensor data is reasonable to avoid false alarm conditions.

An approach to fault detection that lends itself to this scenario is novelty detection using a density estimation approach [11]. Novelty detection determines whether incoming feature vectors derived from sensor data belong to the same distribution as the data produced by the sensor when it was operating in a verifiably healthy condition. A data-based modelling approach is used to estimate the PDF of the extracted features for the sensor data when operating as a healthy sensor. If the sensor is moved to a novel environment, or the primary sensor element suffers damage it is probable that there will be resultant changes in the underlying distribution of the output data. The relative probability that a new set of features belongs to the original data distribution is a powerful tool for identifying novel data: data with low estimated density may be indicative of a fault condition. In common with many data-based approaches, avoiding misclassification of environmental changes as sensor faults is avoided only through use of a thorough strategy for training data collection which encompasses all expected operating regimes.

Onboard fault detection is such an important facet of an intelligent sensor that density-based novelty detection may be used in parallel with more traditional approaches such as a residual-based fault detection approach [12]. Here, time series predictions from a data-based model using recent measurements retained in a buffer are compared with the actual current measurement provided by the sensor, to calculate a residual error between the two estimates. Significant discrepancy highlighted by a large residual error is indicative of an error condition.

Once a fault has been detected, the intelligent sensor should attempt to isolate the nature and cause of the fault, and communicate this information to the sensor management system using a set of error codes based on a priori knowledge about likely primary sensor or electronics failure mechanisms. Furthermore, where possible, the intelligent sensor should attempt to remain operational using recalibration and reconfiguration software approaches, and sensor calibration data contained in the TEDS should be updated. Data-based model approaches again provide advantages over physical models derived from first principles for this application, because the sensor models used in the novelty detection and the residual calculation can autonomously retrain using algorithms contained in the sensor software, based on new incoming data. Optimisation techniques for estimating the model parameters are a subject of ongoing research [13], to ensure fast reconfiguration of the sensor, whilst maintaining excellent generalisation capability.

7.1.5 Case in Point: Load Cell

We have already referred to the precision load cell, and this has been a useful test-bed for many of the ideas for the implementation of intelligent sensor processes. It is a good example of a class of devices that pose a particular difficulty in that the measurand is also an important parameter of the physical sensor system. One of the problems posed by mechanical sensors is that they tend to exhibit the oscillatory frequency response characteristic of second order systems. In the load cell, the load being measured contributes significantly to the inertial parameter of the system. The old fashioned way of dealing with the response was to provide massive damping, either mechanically or electrically, but this did nothing to improve the response time; indeed, it only made it worse. By means of digital filtering we could remove the response precisely, but there is an interesting paradox. As the load increases the resonant frequency and the damping of the system both decrease: so, in order to measure a given load rapidly, we have to know the load before we can produce the correcting filter that corresponds to it. The way this chicken-and-egg conundrum can be solved provides a powerful illustration of the capabilities of the intelligent sensor approach [2].

The locus taken by the roots of the characteristic differential equation of the load cell as the applied mass changes can be determined by automatic system identification techniques. Such a locus is illustrated in Fig. 7.5, and the roots of the compensating filter need to follow it. For each value of mass there is a corresponding final output of the compensating filter once oscillation has ceased. The trick is to make the parameters of the filter vary with its own output as dictated by the locus.

When a load is first applied, the compensating filter is set to the parameters for zero load and as the signal begins to rise the parameters follow it. As the output signal crosses the correct value the compensating filter is exactly

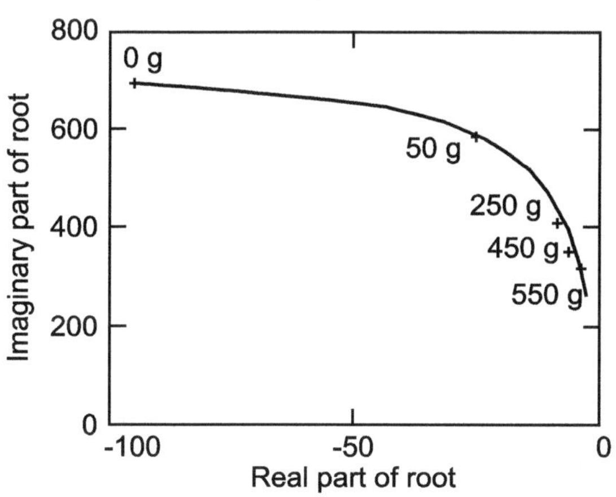

Fig. 7.5. The locus of the roots varies as a function of the applied mass

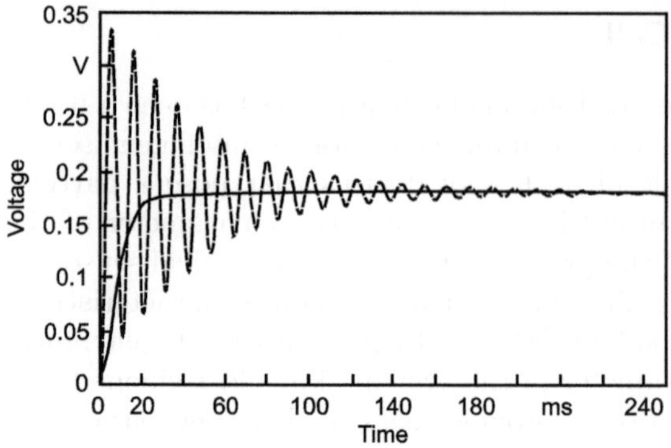

Fig. 7.6. Experimental results showing the effect of adaptive filtering of the response of a load cell. The *broken line* is the uncompensated output and the *solid line* shows the output from the adaptive filter

right and the system locks in at the steady value. A typical output is shown in Fig. 7.6. The precision of such an approach, compared with that of using massive damping, means that overall response times can be improved by at least an order of magnitude.

7.1.6 The Impact of ASICs

Techniques such as those discussed above were first developed on large computers and ultimately implemented on microprocessors. These were still comparatively cumbersome, requiring a circuit board to be associated with each sensor. At this stage it is worth emphasising why the compensation needs to be done at the sensor site. In a large industrial instrumentation system the central computer could be overburdened with sensor compensation processing, while the communication system could be overloaded by raw uncompensated sensor data. Ideally the compensation and communication electronics should be contained in the sensor housing and should be functionally invisible to the user.

Now a substantial analogue sub-system can be accommodated on the same chip as an embedded microprocessor, so it is conceivable that the entire compensation and communications system can be realised in a single chip form. It is important, however, not to understate the scale of the problem of developing and debugging such a system, and unless resources are very substantial it is preferable at the present stage of technology to keep the processor as a separate programmable device. Not least of the problems is the fact that analogue simulators are not nearly as realisable as digital ones.

Also at this point we come up against one of the major problems and a source of cogent criticism of the very concept of intelligent sensors. It has always been a truism that the more complex a system is the less reliable it is. Fortunately this principle can be reversed by the introduction of two

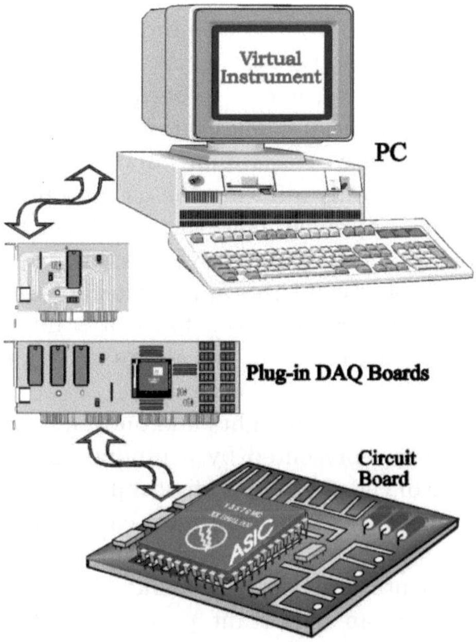

Fig. 7.7. Hardware used with an intelligent sensor ASIC (after Taner and Brignell [15])

concepts – self-test and auto-calibration – and there is one simple component of ASICs that make these realisable; the digitally controlled analogue switch. By means of such switches the analogue sub-system can be made to reconfigure itself to perform various checks (gain, offset, linearity etc.) as well as monitoring possible disturbing variables, such as temperature. In a typical design [14] there are 16 such switches.

The development process on such systems could be fraught with complexity, so it is important to establish methods that give the designer maximum support. A very useful technique is to embed the ASIC in a PC as shown in Fig. 7.7. Data acquisition boards are used to provide intimate access to the functions of the chip. Software is developed in a portable language, such as C, which allows it to be ported onto a suitable microprocessor once it has been developed and tested [15]. This leaves the question of support software which is discussed in the following section.

7.1.7 Reconfigurable Systems

From the above the importance of advances in electronic hardware, in particular ASICs, on the development of intelligent sensors is clearly evident. The role of software drivers is equally essential, as these control and perform the necessary tasks in test, calibration and operating modes [16]. Additionally, the software is responsible for ensuring correct communication between the sensor and the host system, and can also be used to ensure that hazard conditions are eliminated during hardware development.

Returning to the ASIC chip mentioned above, it can be seen that with the 16 analogue switches there are 65,536 potential configurations. Many of these are forbidden conditions which would cause catastrophic failures if they were to arise. The problem could be solved by only allowing the use of a pre-defined set of standard combinations. This approach, however, is extremely inefficient in terms of storage and operating speed, and also restricts the user to the pre-set list which may not be desirable for futures applications of the ASIC. The solution is to provide a software driver in the form of a filter that prevents any destructive configuration being set up, but allows all other combinations.

The switch configuration is stored as a vector of noughts and ones in two, 8-bit bytes. Each configuration is therefore represented by a unique 16-bit word which is stored in the memory space of the controlling digital processor. The sub-system can be switched into a specific self-check or auto-calibration mode with only a few control instructions [17].

Figure 7.8 shows the virtual instrumentation panel for controlling the ASIC. Note that the layout of the ASIC is an important part of the display. The switches can be activated on screen using a pointing device such as a mouse. The values are then passed to the software filter, which initially searches for forbidden settings. A process of binary masking is used to detect the forbidden conditions. The control word from the switch settings is logi-

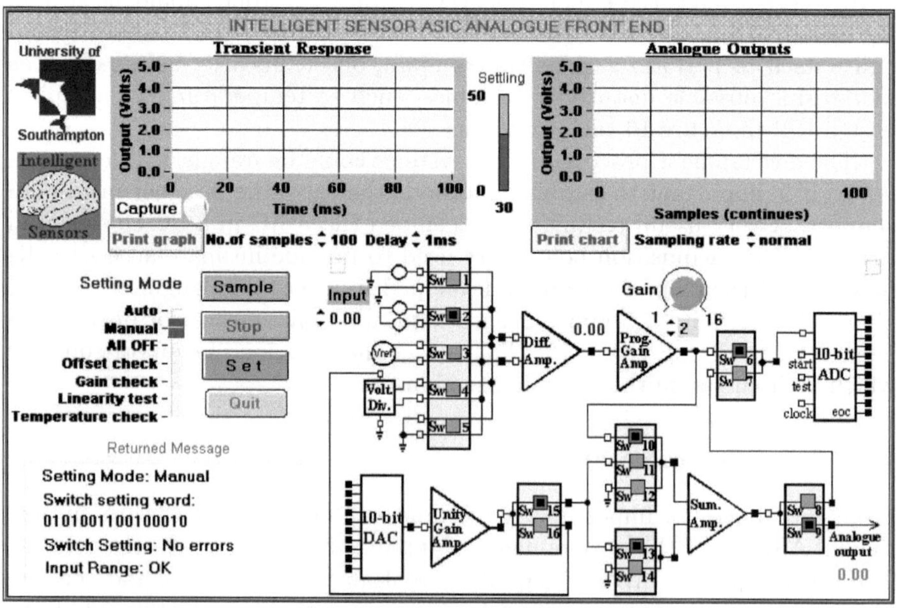

Fig. 7.8. Virtual instrumentation panel after Taner and Brignell [17]

cally ANDed with a mask. The number of bits in the result is counted and this is used to detect an invalid condition.

As the complexity of the intelligent sensor hardware increases, there is a requirement for more sophisticated software for testing, calibration and modes of operation. For example, quantitative analysis of gas mixtures can be performed using the so-called electronic nose [18] utilising an array of chemically sensitive resistors as in Fig. 7.3. The pattern recognition techniques needed for such systems are becoming increasingly complex. Approaches such as neural networks and fuzzy logic mean that there will be additional emphasis placed on the importance of the associated software for sensor applications.

7.1.8 Communications

When we begin to consider multisensor systems, one of the most important factors is the nature of the topology of the network connecting the sensors to the central processor. Figure 7.9 shows four possibilities of methods of networking sensor systems together. For the star topology, each sensor is connected to the center by at least a pair of wires. There are a number of disadvantages associated with this approach. Firstly, a great deal of cabling is required and this could easily become the dominating cost for a large industrial system. Secondly, as more sensors are added a bottleneck occurs at the center where all the cables arrive.

A more attractive idea is based on the bus topology. The transducers share a common pair of wires. We now have the requirement that each device must have a unique address to distinguish it from its neighbours. Another potential problem is that if the shared data highway is severed at any point, all devices beyond that point are disconnected from the system. The third problem is that, as the number of sensors increases, their share of the bus, under time division multiplexing, decreases, though this is not a new consideration as input/output (I/O) resources have always had to be shared. The vulnerability

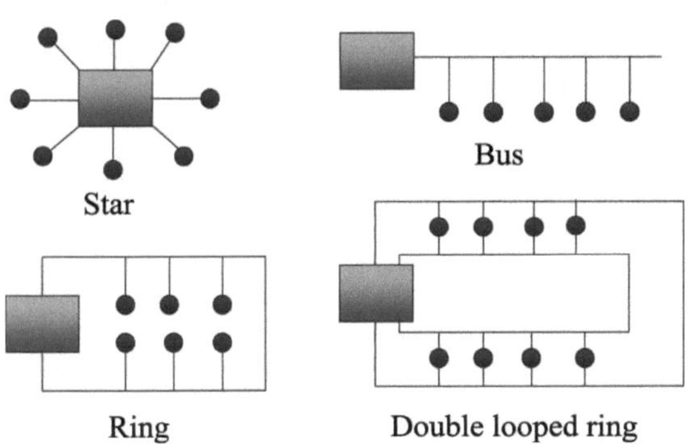

Fig. 7.9. Examples of network topologies

of the simple bus to cable severance can be overcome by the ring topology. Here the bus is arranged in a complete loop and provision is made for it to be driven from either end. This means that if the cable is severed at one point, the system can carry on by addressing both ends separately. This also allows the position of the fault to be determined by observing which devices fail to respond from either end. In cases where there is a particular danger of disruption, for example where there is an explosion hazard, the double looped ring is used. The bus is addressed by four separate drivers. The bus separation, and hence the length of the stubs connecting the transducers, is made large enough to minimise the probability of both buses being disrupted by a catastrophic event.

General Requirements for a Low-Level Protocol

There is an obvious requirement for a procedure which maintains and initiates communication throughout the overall system. Consider a continuous stream of bits being received by a station on a bus. In the absence of a protocol, a number of questions need to be asked concerning the nature of the digits:

- Where does a message begin or end?
- Is the message for me or another station?
- What is the actual information contained in the message?
- How is the message formatted?
- Has the message been transmitted correctly?

It is clear then that a minimal number of fields are required to establish a working protocol. Figure 7.10 illustrates a well known protocol, high level data link control (HDLC). The first field is the opening flag which is a unique signal that cannot occur by accident anywhere else in the message. The bit pattern is 01111110, and in order to preserve the uniqueness, bit-stuffing is used in the non-flag section of the message. A logic zero is added whenever a sequence of five logic ones occurs. The next field is the address field, 8-bits in length thereby allowing up to 256 devices to be uniquely addressed. This is an essential requirement for a shared bus system. The control field makes a statement about the purpose and nature of the message. For example, it could convey a series of instructions such as:

- carry out a self-test;
- set the amplifier gain to 10;
- transmit values of temperature;
- the following field comprises 32 bits divided into four octal sub-fields; etc.

The all important information field is of variable length in the HDLC protocol, although other protocols use a fixed length. The condition of this field is highly conditioned by the fields that have already gone before. The length of the field is contained in the control field and may vary from packet to packet. The penultimate field is the frame check sequence which is a number

Flag	Address field	Control field	Information field	Frame check sequence	Flag
01111110	8 BITS	8 BITS	VARIABLE	16 BITS	01111110

Fig. 7.10. The HDLC protocol

derived from a process in the preceding field. The process is repeated in the receiver and checked for a match. If no match occurs a request is issued for re-transmission of the message. Finally, the end of packet flag 01111110 is transmitted.

Wireless Sensor Networks

There has recently been a great deal of interest in the development of wireless networks of sensor nodes having the ability to collect and disseminate environmental data. Each node is an individual intelligent sensor having the ability to communicate via radio transmission. There are many scenarios in which these networks might find uses. Examples include environmental control in office buildings, robot control and guidance in automatic manufacturing environments, interactive toys, pollution mapping and intelligent buildings. The individual devices in a wireless sensor network (WSN) are inherently resource constrained. They are subject to limited processing speed, storage capacity, and communication bandwidth. The nodes have substantial processing capability overall, but not individually. In most applications, the network must operate for long periods of time and so the available energy resources (batteries, energy harvesting systems, or both) limit the overall functionality. To minimise energy consumption, most of the components, including the radio, will need to be turned off most of the time. The nodes are closely coupled to a changing physical world, and will therefore experience wide variations in connectivity and will, potentially, be subject to harsh environmental conditions. The dense deployment generally means that there will be a high degree of interaction between nodes. Many researchers have been developing low-power radio modules for WSN applications. PicoNodes [19] are small, lightweight and low-cost network elements specifically developed for wireless sensor networks. Owing to the low-power nature of each node, the communication distance between adjacent devices is generally quite small and hence a multi-hop approach is used to achieve communication over larger distances. The network is generally ad hoc, because the number and location of available nodes can vary.

7.1.9 Trends

The reduction in size and cost of the transistors that make up our electronic sub-systems has continued apace over the last four decades, and there is every reason to suppose that it will continue for many years more. Talk is beginning about possible size limitations of a quantum nature, but it has to be remembered that we have only exploited planar structures and the possibilities of three dimensional structures are beginning to emerge. We are entering an era in which the silicon is of negligible cost. In these circumstances, development costs become even more important. Aids, such as those described in the preceding sections, which enable development to be carried out on the device itself via an on-line computer will make a major contribution in this area. We are now used to computer design aids, and digital simulators are now so good that a device that works in simulation can almost be guaranteed to work in practice, but unfortunately the same cannot yet be said of analogue simulation.

Design and development costs will be moderated by the availability of tried and tested sub-systems, so it is important that a systematic approach is adopted rather than a piecemeal case-by-case one. One of the most exciting of recent developments has been microengineering, which turns the photolithographic techniques of circuit production to the manufacture of mechanical systems of micron dimensions. Sub-systems as complicated as working millimeter sized electrostatic motors have been demonstrated, which leads to the possibility of micro-robots working in environments such as the human body. The combination of microengineering and microelectronics on a single structure conjures up all sorts of possibilities, such as self-flushing gas microsensors. Recent developments within the area of wireless, distributed sensor networks have led to the realisation of vast numbers of sensor nodes having localised intelligence. The advantage of such an approach is that retrofitable devices can be installed, without the need for additional wiring. A drawback is that a localised energy source, such as a battery, is needed and these have a limited lifetime and require periodic replacement. A possible solution to this problem is so-called energy harvesting, where ambient energy in the form of solar, thermal, radio frequency, mechanical vibrations etc. is locally converted into electrical energy [20]. Such techniques are in the early stages of development, but will undoubtedly become significant for intelligent sensors over the coming years.

In little over two decades, intelligent sensors have progressed from being a ridiculed academic pipe-dream to an essential component of modern technology, and there is much more to come.

7.2 Fiber Optic Sensors

W.R. Habel

7.2.1 Introduction

At the beginning, optical fibers (lightwave guides) were created to transmit optical pulses over long distances with high transmission rates. Simple test systems consisting of a light-emitting diode, a fiber and a photodetector have been investigated. Since those days, fiber optic sensing techniques have grown significantly in number and type, and fiber optic sensors (FOSs) increasingly have become significant as smart sensing technology. The reasons are:

- their capability of being very sensitive, small, lightweight and chemically inert, and with no perturbing structural properties when embedded;
- their capability of being highly distributed;
- they withstand a few hundred degrees C during the curing process of composites;
- they are electrically passive and not disturbed by electromagnetic fields or by parasitic currents;
- they are network-compatible and amenable to multiplexing;
- they have small interface requirements (the opto-electronic elements and demodulation electronics are confined in the reading unit);
- there is a low risk of sparking because of the very low radiant energy emerging from the fiber optic system; and
- they are almost exclusively driven by standard photonics components.

A complete FOS system consists of two parts:

1. the *sensing unit* contains the fiber optic sensing element equipped with a protective coating and/or an additional protective element (such as a pipe or a small tube) together with attachment material/clinge components
2. the *opto-electronic unit*, which contains the radiation source and a photodetector. Depending on the sensor type and on the size compatibility with the fiber, a semi-conductor laser diode (LD) or a luminescence diode (LED) is used. As photodetectors, PIN diodes or avalanche photodiodes (APD) can be used.

Basically, two fundamental classes of FOS can be distinguished – the intrinsic fiber optic sensor and the extrinsic fiber optic sensor, see Fig. 7.11. An intrinsic FOS takes advantage of measurable changes in the transmission characteristic of the optical fiber itself; that means the sensing element is, at one and the same time, the carrier of information from and to the reading unit. Sensor types of this class are predestined for use in smart components because they avoid additional elements. Some extrinsic sensor types (such as micro strain sensors, see Sect. 7.2.2), where the fiber is not used as a sensor

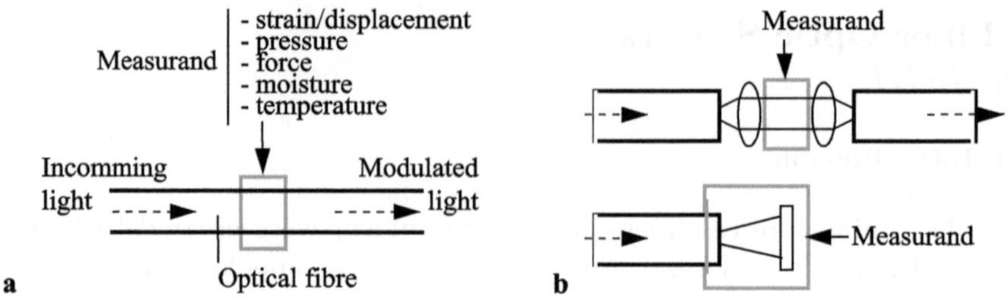

Fig. 7.11. Two main classes of fiber optic sensors: **a** intrinsic type, **b** extrinsic type. The picture above shows the transmission mode; the picture below shows the reflection mode

element but merely as a light guide to and from the sensing area, can also easily be used in smart structures.

The quantity to be measured causes a variation of one or more physical parameters in the sensor. This variation must be detected, recorded, processed and should be re-transformed into a scaling unit of the measured quantity. The great challenge for the engineer is to separate the variations induced by the measured object from any variations induced by some other internal or external effects. Often faulty measurements are produced by an inappropriate application of the sensing element. Some aspects involved with these problems will be discussed in Sect. 7.2.4.

Parameters to be varied by the measurand are the intensity, the wavelength or phase, and the polarization state of light. Additionally used is, by means of optical time-domain reflectometry (OTDR) technique, the measurement of the travel-time of a light pulse launched at one fiber end and backreflected at markers. From measurement of the time of transit, the shortening or extension of the optical path length (contraction or extension of the sensor fiber) can be assessed. However, any effects influencing the fiber can be located. It should be noted that a considerable number of fiber optic sensor types has been created in the past decades for measurement of almost all physical, and a lot of chemical, quantities.

In this section the examples are particularly focused on FOS types for measurement of external disturbances such as strain, displacement, pressure, vibration, acoustics, and for determining the location of damage along a fiber. Sensors for chemical and other physical quantities are briefly mentioned.

7.2.2 Basic Principle of Operation

The basic element of a fiber optic sensor is a thin wire of glass or of plastic (polymeric) material. When light is transmitted into one end of the fiber surrounded by a fiber cladding with a lower refractive index than the core ($n_{\text{core}} > n_{\text{cladding}}$), it propagates through the fiber to the other end corresponding to the physical effect of total internal reflection [21]. Figure 7.12

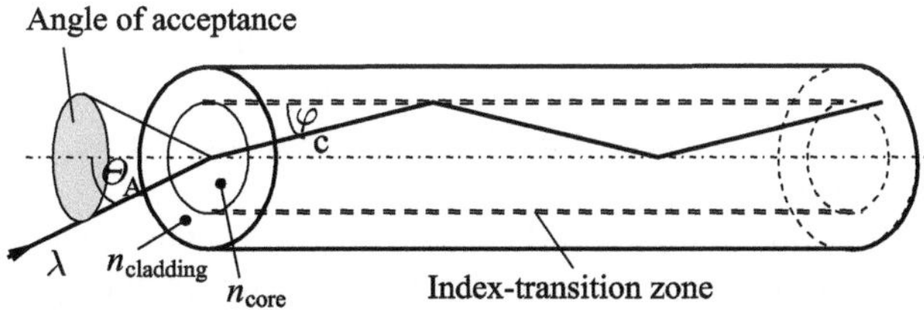

Fig. 7.12. Light beam propagation in a multimode-fiber (λ = wavelength of the source)

shows this effect. The acceptance angle Θ_A of the fiber (Θ_A is the maximum value of the range of the accepted angles Θ) defines the portion of light input at the fiber end. Only light that is input for $0 < \Theta < \Theta_A$ can be guided down the fiber. It is continuously reflected at the interface between the core and the cladding; the critical angle φ_c must not be exceeded.

Depending on the diameter of the core, modes (the interference pattern within the core) are developed. Very small core diameters (< 10 µm) allow only one mode to travel through the fiber (called single-mode fibers). The

Table 7.1. Overview on the most common types of silica optical fibers used for sensors

Fiber type	Multimode step-index fiber	Multimode graded index fiber	Single-mode step-index fiber
Light propagation (schematic)	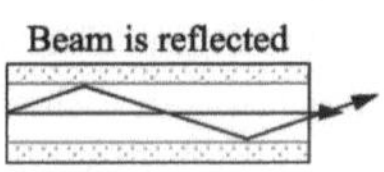	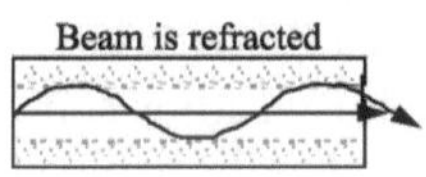	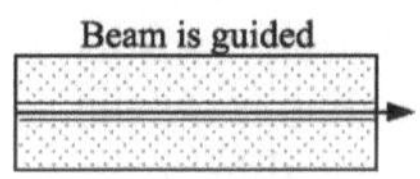
Geometry	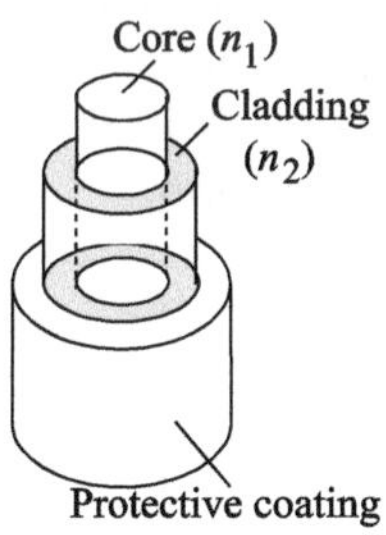	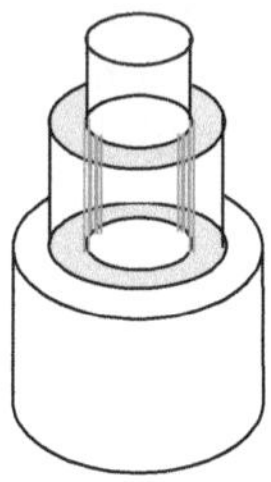	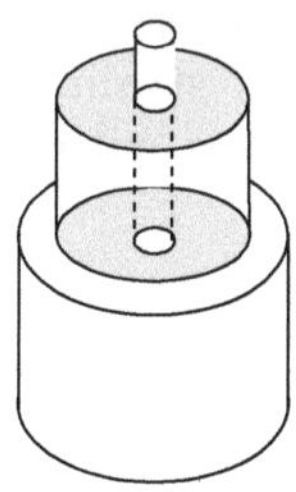
Typical diameters ⇒	core: 50 µm cladding: 125 µm coating: 140 µm ... 250 µm	core: 50 µm cladding: 125 µm coating: 140 µm ... 250 µm	core: 6 µm (870 nm) 9 µm (1300 nm) cladding: 125 µm coating: 140... 250 µm
Refractive index profile	steplike from cladding to core	continuously from cladding to core	steplike from cladding to core

core-to-cladding index transition can change abruptly (step-index fiber type) or gradually with a parabolic profile (graded index fiber type). Thus, there are three main types of optical fibers used for fiber optic sensors; Table 7.1 shows typical distinctions in their geometry.

Due to of the mass-production of single-mode fibers, they are cheaper than other types and therefore often preferred for sensor purposes. For special purposes, such as pressure or current measurements, and sometimes for impact detection, high-birefringent (Hi-Bi) polarization-maintaining (PM) fibers are used because the polarization state of the output signal is definitely affected by external perturbations.

Other specially designed fibers form evanescent sensors. The core of such sensors can (locally) be coated with a cladding that modifies the refractive index in the core-cladding interface region. In the case of variable environment (e. g. a change in the index of refraction between the uncured and the cured state of composites), the absorption coefficient of the fiber can alter. Such sensors are widely used for the detection of changes of chemical or biological environmental parameters.

7.2.3 Commonly Used Sensor Types for Deformation Measurement

From the users point of view, an essential point in smart sensing is the length of the region to be evaluated. In order to record deformations of extended structure components as well as to detect cracks or other damage, long sensor fibers and/or long-gauge-length sensors (area averaging sensors, fully-distributed sensors or quasi-distributed sensors such as segmented sensor fibers) are required. Distributed fiber sensors have the very desirable feature of being able to measure not only a physical quantity influencing the fiber, but also the position where the measurand is acting. The scan frequency of such sensors is limited to a few Hz or less. However, it can be necessary to measure strain, strain distribution or acoustic signals in very limited areas of a few cm^2 with high static or high dynamic resolution.

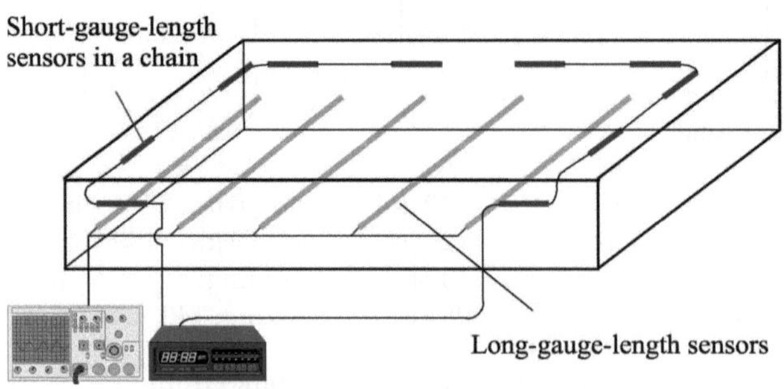

Fig. 7.13. Different fiber optic sensors (embedded or attached to the surface) as an integrated part of a smart structure

A very common measurement task is to detect the long-term and/or load-depending behaviour of composites or laminated materials in highly stressed zones. In such cases, local fiber optic sensors (sometimes denoted as short-gauge-length sensors or point sensors) are used. Commonly used types are Fabry-Perot interferometer (FPI) sensors and fiber Bragg grating (FBG) sensors. In the next subsections, the particular sensor characteristics are considered. Figure 7.13 summarizes possible arrangements of fiber sensors for evaluation of the integrity and the shape or stiffness parameters of a smart structures component. These sensors can be embedded into the material or attached onto the surface of components.

Fiber Sensors Based on Extended Optical Fibers (Long-Gauge-Length Sensors)

The simplest fiber optic sensor uses the light intensity in the fiber. Changes in the intensity signal represent changes in the materials properties culminating in cracks or deterioration of components. This simple principle does not provide an intrinsically absolute measurement value and can thus be prone to errors due to unexpected affects on the leading cable or due to loss of the zero-point information. These intensity-based sensors should be preferred for rather short-term measurements (construction-accompanying and proof loading monitoring, etc).

For long-term monitoring tasks, line-neutral methods are beneficial. Suitable techniques are based on low-coherence interferometry and backscattering. A commercially available long-gauge-length interferometric strain sensor present for long-term measurements on large structures acts as a double Michelson interferometer [22]: a sensing interferometer uses two fiber arms – a measurement fiber that is in mechanical contact with the structure, and the sensors reference, which acts as reference and compensates for the temperature dependence of the measuring fiber. The reference fiber must not be strained and needs to be installed loose near the first fiber. When the measurement fiber is contracted/elongated, deformation of the structure results in a change of the length difference between the two fibers. By the second interferometer contained in the portable reading unit, the path length difference of the measurement interferometer can be evaluated. This procedure can be repeated at arbitrary times and, because the displacement information is encoded in the coherence properties of the light and does not affect its intensity, not only the precision but also the repeatability of measurements is high for this sensor type. The measurement system can be switched off between two measurement events or components such as connectors or cable can be exchanged without zero-point data loss.

Typical parameters of commercially available line-neutral long-gauge-length sensor are:

- Measuring length: 50 cm to several 10 m
- Measuring range: 0.5% in shortening, 1.0% in elongation (for $<170\,^{\circ}$C)

- Precision in measurement: 2 µm (error of measurement: $\Delta\epsilon = \pm 1.25 \cdot 10^{-5}$)
- Proportionality factor between the measured delay and the applied deformation: (128 ± 1) µm/ps.

Such sensors can be provided as tube sensors or as flat tape sensors. Tape sensors allow its integration into composites or into the interface zone of multi-layer materials. Several sensors can be interrogated by multiplexing.

The necessity of an additional unstrained reference fiber could be problematic in smart structures; one-arm sensors should therefore be preferred. Hence, an alternative to two-arm long-gauge-length interferometer sensors is one long optical fiber containing fiber sections separated by reflectors (see Fig. 7.14). By measurement of the time of flight of a short pulse transmitted into the fiber and backscattered on markers (splices, photoinduced reflectors, or squeezing points) at the end of these sections, the measurand can be determined at definite locations along the fiber. An elongation (compression or contraction) of a measuring section, determined by two reflector sites on the fiber, changes the travel-time of the pulse: $\Delta\epsilon \approx \Delta t_{\mathrm{p}}(c/2L_{\mathrm{o}} \cdot n)$; c is the speed of light, n the index of refraction. Based on this relationship, the changes in the average strain of a chain of marked sections along the fiber can be interrogated by an OTDR device.

This method allows the evaluation of strain profiles in large components without using sensor fibers containing discrete sensors along the fiber. The OTDR device used determines the strain resolution achievable. A high-resolution picosecond-OTDR device enables the resolution of elongation to 0.2 mm, assuming the minimum distance between two reflectors in the measuring section is not less than 100 mm [23]. Using this TDM method, a reflector shift of 0.35 mm can be resolved, however only long-term reproducibility of reflector shifts of 0.85 mm can be achieved. This value is sufficient to recognized dangerous changes in the material or loss of bonding integrity. An automatic scanning run takes between one and some ten seconds depending on the desired precision. The sensor sections are interrogated one after another. The position of each reflector can be referred to one stable reference reflector, thus, there is no propagation of error. Off-line measurements are preferred because of the rather expensive OTDR device.

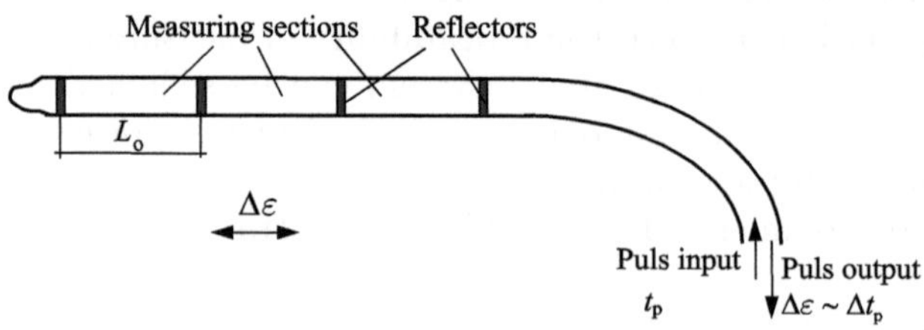

Fig. 7.14. Quasi-distributed fiber sensor based on backscattering signal evaluation

The examples considered above were focused on strain or deformation measurement in the direction of the fiber axis. However in composites, transversely applied pressure, arising forces or beginning delamination might be of interest. For such purposes, single-mode birefringent fibers can be embedded. Internal birefringence can be induced by using non-circular core geometry of the fiber or by introduction of stress anisotropy around the core such as done in panda or on bow-tie fibers. The refractive index difference of the two orthogonal polarization modes produces a differential propagation velocity. Any damage or parameter change in the composite or material structure will perturb the birefringence parameters in the sensor fiber. Using the time delay measurement technique, intensity and position of the perturbation can be located with an uncertainty of about 10 cm [24]. However, a reproducible correlation between affecting external events and optical effects in the fiber is quite difficult because the interface zone of the sensing fiber strongly influences the response of the sensor. Nowadays, fiber Bragg gratings will be written into birefringence fibers to make multiple parameter sensing with the capability of discriminate between them [25].

Fiber Sensors Based on Discrete Sensing Elements (Short-Gauge-Length Sensors)

Numerous types of short-gauge-length optical fiber sensors for strain measurements in materials research and structure evaluation have been proposed, but only a few sensor techniques are commercially available. In contrast to fiber sensors with long gauge lengths, short-gauge-length fiber optic sensors, based on interferometric and spectrometric principles allow the measurement of local deformations with a very high resolution. The most well-known microstrain sensor configurations are the Mach-Zehnder, the Michelson, the Fabry-Perot and the fiber Bragg sensors. In this section, the most widely used sensor types from that list are described.

Fiber Fabry-Perot Interferometer Sensors. This sensor type comprises a cavity defined by two mirrors that are parallel to each other and perpendicular to the axis of the optical fiber. There are two arrangements of Fabry-Perot interferometer (FPI) sensors: first, the (intrinsic) in-fiber FPI sensor, where the cavity is formed by two mirrors at locations in the length of the fiber. The maximum distance of the mirrors (cavity length) can reach some mm and defines the gauge length. The second type is the extrinsic FPI sensor (EFPI). The cavity is produced by positioning a fiber end-face opposite to another, with a small gap of usually some microns. Figure 7.15 shows such an EFPI sensor. The most widely used design is to fix into position the two fiber ends in a hollow tube. The fiber end-faces act as mirrors and produce interference fringes. In general, both fibers with the reflecting end-faces are brought into position by fixing them at the hollow tube, e. g. by fusion splicing if it is a glass tube. The gap between the fibers inside the tube contains

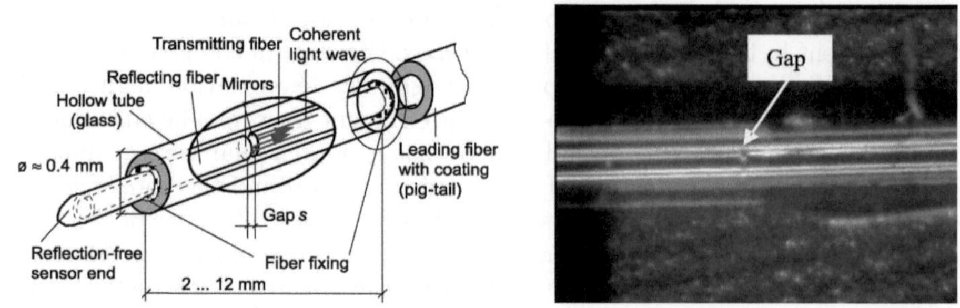

Fig. 7.15. Extrinsic type of fiber Fabry-Perot interferometer (gap $s = 4 \ldots 100\,\mu\text{m}$)

usually air. The very small gap between the mirrors (about 10 to 100 µm) causes, in contrast to an intrinsic FPI sensor, a very low transverse influence.

The functional principle of an FPI sensor is as follows. The incoming light reflects twice: at the interface glass/air at the front of the air gap (the reference [Fresnel] reflection) and at the air/glass interface at the far end of the air gap (sensing reflection). Both reflections interfere in the input/output fiber. The sensor effect is induced by force-induced or temperature-induced axial deformation of the hollow tube. This leads to a shift of the fiber end-faces inside the tube (because they are only fixed at the ends of the tube), which results in changes on the air gap length (gap width s). From this follows a phase change between the reference reflection and the sensing reflection that is detected as an intensity change in the output interference signal.

Interferometer sensors are commercially available, e. g. from FISO [26] for strain, temperature and pressure measurements. They allow local measurements of strain in a range between −5000 µm/m (shortening) and +5000 µm/m (elongation) with a resolution of up to 0.1 µm/m. Available gauge lengths are in the range from 1 mm up to 20 mm. Due to of their excellent response time behaviour of up to 2 MHz, they can also be used for detection of mechanical vibrations and acoustic waves. However, the interrogation unit used defines the dynamic behaviour. With regard to manufacturing and applicability, fiber Fabry-Perot interferometer (FPI) sensor is the most often-applied short-gauge-length interferometric sensor type for structure assessment. It does not need a reference arm and sophisticated stabilization techniques as the Mach-Zehnder or Michelson types do.

For this two-wave interferometer configuration, the observed output intensity I_{out} is a sinusoidal function of the gap width s. Small values of strain variations (less than 300 nm end-face displacement related to the measuring base of about 10 mm) can be measured directly, because the output signal can be defined as linear between the peaks and troughs of the sinusoidal function. The measurement of larger end-face displacements, when a lot of periods are cycled, requires the counting of the interference fringes because the sensitivity decreases near, or becomes zero, at the maxima and minima values of the sinusoidal output signal. Commercially available devices use at least two different wavelengths in the interrogation unit to overcome these insensitive ranges in the sinusoidal function [26]. It is also possible to use a two-gap sensing element or to combine a FPI sensor with fiber Bragg grating elements (see next subsection).

The two-gap sensor contains two input fibers positioned side by side in the hollow tube with a different end-face separation relating to the reflecting fiber. This double-sensor configuration also ensures that at least one of the sensing units is sensitive and the direction of displacement change can be recognized. Although lateral as well as axial strain can be measured with this configuration, the manufacturing process is quite expensive. Such special FPI designs are only used for specific measurement tasks. Examples of applications are described in [27].

A specific flexible Fabry-Perot interferometer sensor type allows almost stress-free deformation measurement with high static and dynamic resolution because one fiber end is able to slide inside the sensor tube. This sensor type can be used to measure deformations of soft materials, of materials bonding zones or of curing materials without reaction to the measuring object, e. g. [28].

Fiber Bragg Grating Sensors. When ultraviolet (UV) light is incident upon such a fiber, the refractive index n of the fiber increases. Meltz et al. [29] demonstrated that gratings can also be formed in the core of an optical fiber by illuminating it from the side by overlapping a pair of coherent UV beams (typical wavelength is less than 250 nm). In the meantime, grating manufacturing as an integrated component of the fiber at special wavelengths with a given periodically changing refractive index and spacing between the individual grating planes (grating period or pitch Λ) well established. Fiber Bragg gratings are usually between 1 mm and 25 mm long.

The distance Λ between the grating planes can vary; the common FBG satisfies the condition $\Lambda < \lambda$ where Λ is less than 1 µm (in contrast to so called long period gratings with $\Lambda \gg \lambda$ where Λ is in the 100 µm up to 500 µm range). Bragg gratings for sensor purposes are primarily referred to as uniform grating: the grating along the fiber has a constant pitch and the planes are

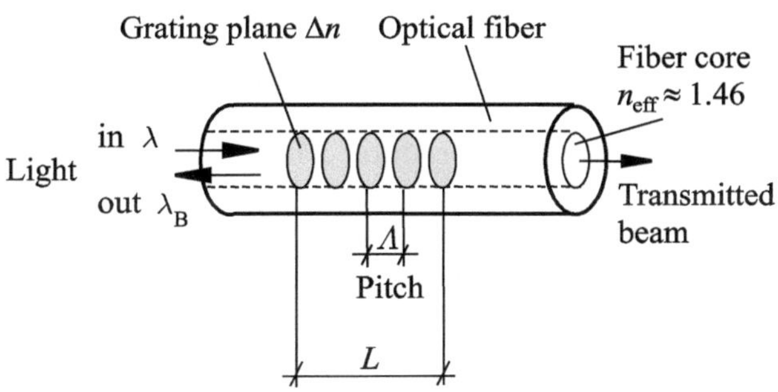

Fig. 7.16. Bragg grating sensor

positioned normally to the fiber axis (as shown in Fig. 7.16). There are other types of Bragg gratings where the grating planes are tilted at an angle to the fiber axis (blazed gratings) or grating planes have a monotonically varying period (chirped gratings). The latter gratings are primarily used in long-haul telecommunication transmission lines.

The principle of function is as follows: when a broadband light signal passes through the fiber Bragg grating, only a narrow wavelength range λ_B, which satisfies the Bragg condition

$$\lambda_B = 2\, n_{\text{eff}} \Lambda \tag{7.1}$$

is reflected back due to interference between the grating planes (n_{eff} is the effective refractive index of the fiber core and Λ is the grating period). The value of the Bragg resonance wavelength λ_B is determined by the grating pitch Λ manufactured and corresponds to twice the period of the refractive index modulation of the grating. The grating periodicity is relatively small, typically less than 1 µm.

From (7.1) it follows that the Bragg resonance wavelength λ_B will change when n_{eff} changes (for example by temperature variation) or Λ changes (due to pitch changes by fiber-grating deformation). That means changes in strain or temperature (or both together) will shift the reflected center wavelength. In general, λ_B increases when the fiber is strained ($\Delta\epsilon > 0$) and decreases when the fiber is compressed ($\Delta\epsilon < 0$). A spectrum analyzer can monitor this wavelength shift; in this way, one can determine strain variations (for constant temperature) or temperature variations (without any deformation of the grating).

When a Bragg grating sensor is to be used as a strain sensor and when the temperature varies under normal operation, only the measurement of the λ_B makes it impossible to differentiate between strain or temperature changes. This undesirable temperature-sensitivity of fiber grating sensors requires taking special measures in order to achieve a separation of the strain and temperature results. Assuming uniform axial strain changes in the grating area and the absence of lateral deformation of the grating, the strain seen by a grating can be computed by a simple linear equation:

$$\epsilon = K \cdot \frac{\Delta\lambda_B(\epsilon_z)}{\lambda_B} + \xi \cdot \Delta T\,. \tag{7.2}$$

K has to be estimated by a calibration procedure. The strain sensitivity depends on the wavelength used; under the condition of constant temperature, the wavelength-strain sensitivity values are written in Table 7.2. The same dependency on wavelength can be observed for the thermal response. In silica fibers, the thermal response in wavelength change is dominated by the temperature-induced change of the refractive index. Only a very small thermal-induced wavelength change comes from the thermal expansion of the glass material (coefficient of expansion of optical fiber glass is $0.55\,\text{K}^{-1}$.

Table 7.2. Strain and temperature sensitivities of FBG for typical wavelengths (approx.)

Wavelength	Wavelength-strain sensitivity [pm/(μm/m)]	Wavelength-temperature sensitivity [pm/K]
800 nm	0.63 to 0.64	5.3 to 5.5
1300 nm	1.0	8.67 to 10.0
1550 nm	1.15 to 1.22	10.0 to 13.7

Table 7.2 also shows the wavelength-temperature sensitivities. However, the pressure-sensitivity for a grating is very low (approx. −3 pm/MPa) so that this sensitivity cannot be exploited for pressure sensing without any transducer elements.

In order to conclude reliable strain and temperature results from the corresponding sensitivity factors, the manufacturer of FBG has to provide these values in the specification table. Investigations with FGB sensors permanently strained by approximately 0.25% (2 500 μm/m) have shown that the strain sensitivity factor can increase by 5% over a period of 6 months [30]. Table 7.2 also makes clear which wavelength resolution has to be reached to resolve a strain change of 1 μm/m (about 1 pm) or a temperature change of 1 K (about 10 pm). This obtainable resolution determines the monitoring method for the wavelength shift.

Bragg grating sensors possess a number of advantages that makes them attractive compared with other microsensor arrangements:

- *Linear response.* The Bragg wavelength shift is a simple linear response to the sensor deformation as shown in (7.1).
- *Absolute measurement.* The strain or temperature information obtained from a measurement system is inherently encoded in the wavelength (strain and/or temperature, index changes due to cladding affection).
- *In-fiber manufacturing.* In-fiber manufacturing enables low-cost fabrication of a large number of gratings.
- *Line neutrality.* The measured data can be isolated from noisy sources, e. g. bending loss in the leading fiber or intensity fluctuations of the light source.
- *Disconnecting the interrogation unit from sensor.* Removal of the reading unit or exchange of leading cable using special connectors with polished angled end-faces do not influence the signal response.
- *Potential for quasi-distributed measurement with multiplexed sensing elements.* As a number of gratings (sensor array) can be written along the fiber and be multiplexed, a quasi-distributed sensing of strain and temperature is possible by serial interrogation of a limited number of gratings.

A few disadvantages should not be missed but they can be overcome by using special sensor arrangements and special demodulation techniques:

- *Relatively short gauge length.* Bragg grating sensors with gauge lengths in the range from 1 mm up to 25 mm are sensitive in the direction of the fiber axis. A special grating sensor, long-period fiber grating (LPG), can be used for the measurement of strain, temperature as well as of bend, transverse load, and torsion [39, 40].
- *Small measurand-induced optical signal changes.* As the strain-induced shift of the Bragg wavelength λ_B can be quite small, transducing elements for amplification of the signal response are sometimes necessary.
- *Temperature-sensitivity in case of field applications.* Strain measurements on-site are perturbed by temperature variations. In order to compensate for this, two superimposed grating elements with different periods Λ_1 and Λ_2 can be used [33]. The static strain sensitivity of this method is reported to be $0.8\,(\mu m/m)/\sqrt{Hz}$. Another method is to combine a FBG with a FPI sensor. The FBG is used as strain-free temperature sensor whereas the FPI sensor acts as strain sensor [34].
- *Weakening of the sensor area by manufacturing.* As the fiber coating must be removed at the location where a grating is to be created, and due to irradiation with UV-beams and the following annealing of the grating, the properties of the glass material that determine strength and fatigue can be expected to change.
- *Vulnerability of the sensors during application.* In applying the gratings, the sensing zone recoated after completion of the gratings creation, must be decoated again.
- *Stiffness.* The stiffness of the fiber – and the corresponding grating area – makes it impossible to measure curing processes at very early ages. For such purposes, stress-free extrinsic Fabry-Perot sensors prevail against Bragg gratings [35].

There are different techniques to read the grating response under the influence of a measurand. The basic operation principles of fiber grating-based Bragg grating sensors are monitoring either the shift in the wavelength or change in intensity of the return signal due to measurand-induced changes. In order to get high-precision monitoring of wavelength shift, laboratory-grade instrumentation based on highly resolving monochromators or optical spectrum analyzers (OSA) have to be used. Laboratory-grade instrumentation often uses instrumentation which is quite expensive, not very robust and unwieldy. Some real-world applications do not have high requirements on strain resolution in the sub-micron or micron range so that small, cost effective and portable reading units are then recommended. There are a number of commercially available interrogation units that fit the laboratory as well as on-site requirements. Table 7.3 shows a rough selection of devices available on the market with the most important specifications.

Table 7.4 gives an overview on the relation between strain resolution and frequency range for a strain measurement task. From the users point of view, for most applications the tunable filter technique is a popular choice. If,

Table 7.3. Interrogation devices for FBG sensors on the market (selection)

Optical specs	si 720 Micron Optics [36]	StrainaTemp JENOPTIK [37]	Spectraleye^{TM} SE600 FOS&S [38]
Wavelength range	1510 nm...1590 nm	850 nm	1527 nm...1565 nm
Resolution	0.25 pm wavelength	about 1 µm/m strain, 0.2 K temperature	1 pm
Uncertainty in wavelength scanning	1 pm	1 µm/m strain repeatability	±10 pm
Max. scan frequency	5 Hz	50 Hz RS 232 1200 Hz Ethernet	1 Hz
Number of channels	2 (8 optional)	max. 16	
Weight	22 kg		1.3 kg
Specialty	Fabry-Perot sensors, long-period gratings	9 V Power supply	90 min battery operation
Preferred use (operating temp.)	Laboratory, no harsh environments	Industrial use −20°C + 40°C	Handheld system 0°C + 40°C

Table 7.4. Sensing characteristics of interrogation methods used

Interrogation technique	Frequency range	Strain resolution
Direct spectroscopy (CCD spectrometer)	DC to 1 kHz	0.1 (µm/m)
Tunable filter	DC to several 100 Hz	1 pm
Interferometric receiver	0.5 Hz to several MHz	0.005 pm

however, high-frequency signals are to be detected, interferometric detection is the most appropriate demodulation technique for FBG sensors to reach the MHz scan range. More details can be found in [39] and [Chapter 18].

In order to exploit the multiplexing capability of FBG sensor, two different methods can principally be used. Due to the wavelength-encoded nature of a grating, each sensor in the fiber can be designed to have its own wavelength within the available source spectrum. Then, using wavelength multiplexing, a quasi-distributed sensing of strain, temperature or other measurands associated with spatial location of the measurand is possible. The number of sensors depends on the bandwidth of the source (typically about 70 nm), on the Bragg reflection bandwidth (typically 0.4 nm) and on the wavelength range needed for pulse shifting due to measurand changes (sometimes up to ±3.5 nm). Using this method, fewer than 20 sensors can be interrogated in series. If a very large number of sensors is to be interrogated in series, time division multiplexing (TDM) method can be used. This enables the use of identical FBG sensors with the same nominal central wavelengths, which are interrogated by a resonant cavity TDM interrogator containing e. g. a diffractive element spectrometer. This method allowed the interrogation of 35 sensors per fiber with an interrogation rate of 2100 Hz per sensor [40].

7.2.4 Fiber Sensors for Physical and Chemical Parameters

Among deformation sensors, fiber optic sensors for measurement of physical and chemical parameters or for detection of substances are gaining an increasingly important role. Especially in biotechnology and in industrial process monitoring as well as for clinical applications, fiber sensors are highly sought after. Important tasks include sensing of temperature, moisture, oxygen, hydrogen, toxicological substances and pH values. Such fiber sensor types are based on evanescent wave-type fibers coupled with the surrounding material or on the evaluation of backscattered signals (see Sect. 7.2.1). There are a number of different both discrete and distributed sensor concepts for the above-mentioned measurands. It is possible to design distributed sensors for measurement of physical parameters or chemical substances. They often use the microbending effect in the optical fiber, which is initiated by the expansion of a chemically reacting or water-swellable polymeric layer (such as a special hydrogel, see Sect. 6.7) on the fiber surface (see below).

Fiber Optic Sensors for Temperature

Temperature sensors form a large class of commercially available fiber optic sensors. Refraining from the advantage of electromagnetic immunity – the main reason for its use – some of these sensors can easily be embedded in or attached to tiny samples without perturbing or heat sinking them. A number of examples can be found in [26, 41–48].

For measurement of *temperature distribution* in composite structures, a backscattering-based sensor is used. This method of distributed temperature measurement can also be used for continuous detection of hot-spots along extended lines like high voltage lines and pipelines [49, 50].

Fiber Optic Sensors for Moisture and Chemical Parameters

Fiber optic sensors can be made sensitive for monitoring of moisture ingress or chemical species. There are both local sensors and distributed sensors [51–

54]. There are special sensor designs for local or distributed measurement of corrosion [55, 56].

7.2.5 Particular Aspects of Sensor Application

Sensor Selection According to the Measurement and Monitoring Tasks

In order to solve a measuring task, an appropriate sensor concept including demodulation techniques has to be selected. Attention has also to be paid on reaching the appropriate sensor characteristics. For example, microstrain sensors have different gauge sensitivity. Fabry-Perot interferometer sensors show the highest sensitivity (in terms of phase changes). However, FBG strain sensors can be perturbed in their signal response by transverse influences [30]. All types of stiff fiber sensors have a limited range of deformability. EFPI sensors normally survive strain values of about 10 000 μm/m. The strength values decrease when dynamic loading with high amplitudes appears. Extrinsic FPI sensors enable more flexibility because one fixing point can be placed outside the tube (by adhering the fiber to the material to be measured) in order to allow free movement of the fiber inside the tube. In this way, sufficiently large displacement of the fixed areas is possible. The resolution of the sensing arrangement can be matched by variation of the gauge length.

Mounting of the Sensor

Depending on the fiber optic sensor type, different kinds of application are used.

- *The sensor is fixed at structure components.* Sensor fibers, which e. g. measure strain, can be surface-mounted on a structure component or fixed inside materials at definite points. The installation process is rather simple; however, special attention must be given to long-term stable fastening to well-defined gauge length. Furthermore, it must not appear to creep or suffer mechanical changes to the fixing components during service.
- *The sensor is attached (glued) on the surface.* Single sensor elements (FBG, fiber Fabry-Perot sensors, short strain-sensitive fibers) can be glued to any kind of surfaces. Two cases have to be distinguished. The measuring area of the sensor (gauge length) is fixed on its ends and the measurand deforms the sensor only by shifting the fixing points. Contrary to the former case, the fiber sensor is fixed to the measuring object along the whole gauge length. In the case of strain sensing, strain transfer, from which arises the strain value at the measurement device, essentially depends on the quality of bonding (suitability of the adhesive) between the fiber sensor and the measuring object.

- *The sensor is integrated into a material.* Fibers with intrinsic sensing areas can be intimately embedded into polymeric complex materials (e. g. composites) as well as into mineral or low-melting materials. The same two cases as above can be distinguished. However, the quality of measurand transfer into the sensor is more difficult to evaluate because there is no possibility for visual inspection. It should be emphasised that the characteristic curve of an applied fiber sensor can strongly differ from the sensor characteristic when not applied.

Since reliable measurand transfer into embedded or surface-attached fiber sensors is the core problem, some more details are to be discussed as follows. The thin polymeric or metallic coatings of applied deformation sensors have to be optimized for reliable sensor/matrix interaction. This concerns sufficient coating strength as well as long-term bond strength to the matrix material. In the case of appropriately bonded sensors, elastic stress transfer will be the dominant mechanism at the interface up to a definite strain level. Assuming that there are no irregularities in the sensor coating (e. g. for recoated FBG) and no irregularities in the matrix microstructure, elastic shear stress distribution along the grating (sensing element!) is then constant (with the exception of its ends). When the sensor/matrix bond exceeds a certain load level, the increasing shear stress at the interface leads to debonding, and reliable measurements are no longer possible.

In order to evaluate the actual bonding behaviour between fiber coating and matrix material as well as the load transfer limit of embedded sensing elements, the micro indentation test method can be used [57, 58]. A thin slice of material containing the embedded sensing element (dimensions, coating, and position) is deformed; this method delivers the shear stress behaviour at the interface when the material is deformed. The recorded force-deformation curve allows the estimation of the limit of reliable sensor operation.

Special Operation-Related Problems

Although the optimum method for installation has been used, in a number of cases, especially when the sensor system has already to work during manufacturing of the structure, or when the object to be evaluated is modified within the period of measurement, the sensor system is subjected to changes. Such changes could include fiber-cabling change, cutting of leading fibers and reconnecting, switching off or disconnecting the power supply. In all of these cases, a line-neutral sensor principle has to be used. Moreover, in order to avoid loss of the bias value (initial value as zero-point reference), the fiber sensor should deliver absolute measurement values. Whenever strain measurements with very high resolution have to be carried out under the influence of frequently changing temperature profiles (e. g. one part of the fiber is installed indoors, another part is installed outdoors under all the climatic conditions along the sensing or leading fibers), the temperature influence has to be clearly identified.

Proof of the Reliability of Fiber Optic Sensor System – Validation

An important detail concerns the re-calibration of installed sensors. If the sensor can be removed from the structure and re-calibrated in the laboratory, there is an opportunity to compensate for unstable characteristic curves, drifts, and aging effects or signal perturbations. If there is no possibility to calibrate the sensing part of the system from time to time, the measurement uncertainty increases over time and the whole system becomes unreliable. In only a very few cases, irretrievably installed sensors for long-term measurement tasks can be forced to traverse the characteristic curve and be compared with a stable reference function. When creating a long-term stable and reliable sensor system, the optimal way would be to design a sensing element which enables an access to the characteristic data and provides a definite zero-point position (zero-point reference) to which all following measurements can be related. Drifts or environmental influences on the sensing part can then be evaluated.

However, all necessary system components such as cables, couplers, sources, demodulation units, require reflection with regard to reliability and stability, especially, when standard fiber sensor components are modified according to specific requirements or for critical application [59–62]. Users of measurement systems want to be definitely sure that a chosen sensor system is suitable and reliable for the specific intended use. A very useful method is the validation procedure of a measurement system or of components of it because they then get assured information about performance and limitations of a sensor system. Then they are able to draw feasible conclusions. Validation is explained in the international standard ISO/IEC 17025 of the International Standardization Organization (ISO) [63]. According to this standard, validation is the confirmation by examination and the provision of objective evidence that the particular requirements for a specific intended use are fulfilled.

7.2.6 Application Examples

Measurement of Strain and Strain Profiles

Strain or strain profiles in structure components can be measured by attaching sensor arrays onto surfaces of components or by embedding it between layers in a composite. One simple example of surface-mounted fiber strain sensors concerns the strain monitoring in a wing spar of an air glider during flight loading. The sensor arrays each consisting of four draw-tower FBG gratings (IPHT Jena) in series with higher tensile strength were attached in the tensile zone to evaluate the axial strain distribution under loading [64].

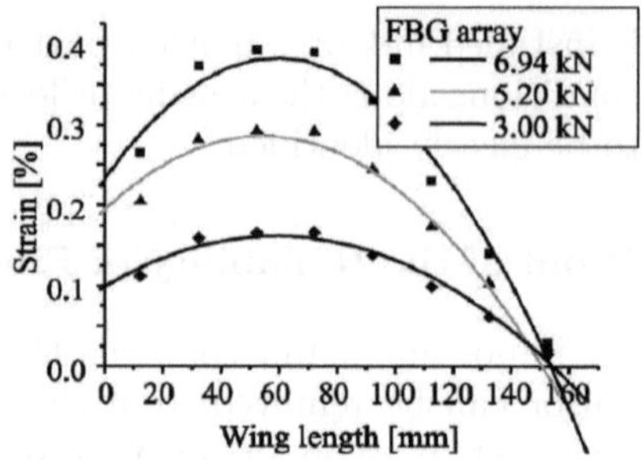

Fig. 7.17. Attaching the FGB array to the upper side of the wing spar (*left*) and results from the loading test (*right*) [64]

Figure 7.17 shows the glued fiber with the sensor gratings and shows the signal response after loading.

Two-dimensional surface-mounted strain rosettes were used to measure planar strain [24]. The sensor patches consist of three independent FBG strain sensors with sensitivities aligned along axes separated by 120°. This 3-axis strain rosettes (Fig. 7.18) temperature-compensated by a separate strain-isolated FBG sensor, enables resolving of the principle directions of planar strain, and hence characterizing of the strain at a location on a structures surface. The optical fiber containing the sensor is constrained into a triangular loop geometry by a process of lamination between a thin polymer film. Use of these materials ensured that the same surface bonding procedures and materials developed with many years of experience for electrical strain gauges could be used unchanged.

In particular, when structure components are simultaneously stressed by climatic and mechanical influences, incorporation of fiber optic sensors be-

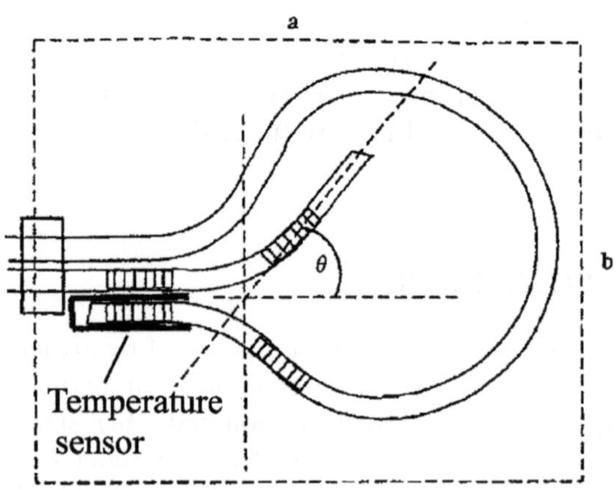

Fig. 7.18. Sensor patch with thermally compensated strain sensor rosette [24]

tween composite layers is more reliable than bonding to the surface of the structure. Apart from additional work and expense during the manufacturing the sensor equipped structure components, and from lack of reparability, the long-term stability of embedded sensor arrays will be higher. Rotor blades of wind turbines equipped with incorporated fiber sensors are one example of smart structures. Stability tests showed that for the dynamic loading of 10^7 cycles with a strain limit of 0.6% embedded FBG sensors show reliable strain response. However, drift effects could be observed likely induced by coating degradation under permanent dynamic stress [65].

Measurement of Vibrations and Acoustic Emissions (AE)

Distributed dynamic measurements, which deliver input signals for active vibration suppression, is one of the important areas of interest in smart structure engineering. Other no less important efforts are the detection of local or partial loss of integrity and the evaluation of the state of curing, e. g. of composite materials. By using embeddable sensors for continuous or periodic AE detection, fatigue cracks or overloading-induced cracks can be detected early and reliably so that a repair can be made at minimum cost, and routine inspections can be reduced. Increasingly, the future health monitoring of structures by smart systems will use acousto-ultrasonic techniques. By introducing ultrasonic stress waves into the structure and detecting stress waves at definite points of the structures, changes in material damping characteristic due to damage can be recognized by using structure-integrated fiber sensors. A similar excitation method can be exploited by using movable fiber optic microphones to detect structural inhomogeneities or to produce proof of structure integrity. Another important potential application is the measurement of the velocity of acoustic waves transmitted through curing materials by a set of sensors. Depending on the state of curing, embedded sensors are able to measure different wave spectra, and after completion of the curing process, the same sensors can be used for determining the in-service strain and vibration state of the structure.

A promising technique for vibration and acoustic emission measurement is that of interferometry. Apart from several sensor arrangements for non-contact interrogation of vibrating surfaces, which are based on reflective types of fiber sensors by means of a focused laser beam (utilized for surface velocity and length measurements), usually two types of short-gauge-length sensors have been used for acoustic emission detection: Fabry-Perot sensors are preferred for highly precise dynamic strain measurements, but FBG has the potential for measuring the distribution of dynamic strain reactions.

In order to measure vibrations or acoustic wave propagation, an interferometer sensor can be embedded or surface-attached, and then interrogated by using a vibration or AE detection system. When fiber optic sensors are used for measurement of AEs, disadvantages of traditionally used PZT transducers for AE sensing such as their large size and their susceptibility to electromagnetic interference are avoided. Fiber optic sensors additionally have the potential for multiplexing a number of sensors.

When the level of elasticity and plasticity is exceeded, fracture of the material emits energy in the form of bursts of (transient response) or continuous acoustic signals. The frequency range of AE is usually between 10 kHz and 1 MHz. Classic interferometer sensors as well as FBG sensors have been applied to measure AE signals. These sensing methods rely on measuring the strain in the fiber sensing area and, thus, the sensor performance is given in terms of strain resolution. For example, FBG sensors can be embedded in a composite structure to detect and localize damage by sensing ultrasound, which is created from Lamb waves [24]. Such Lamb waves are reflected at defects and the maximum strain is parallel to the acoustic wave propagation direction. Using two FBG strain rosettes, damage can be localized and its position can be evaluated.

Among the variety of fiber optic sensors, Fabry-Perot interferometer (FPI) sensors have shown the best performance in the frequency range up to 100 MHz because of their high sensitivity, broad bandwidth and excellent tolerance to low-frequency ambient vibration. Several FPI designs are used whereas intrinsic FPI sensors with flat mirrors show the best performance and are quite compatible with the mechanical structure of composites. Using this type of sensor for measurement of AEs, the detection bandwidth is 15 MHz to a few GHz. The minimum detectable phase of the current system was mainly limited by electronic noise: $4 \cdot 10^{-8}$ rad/Hz$^{1/2}$ [66].

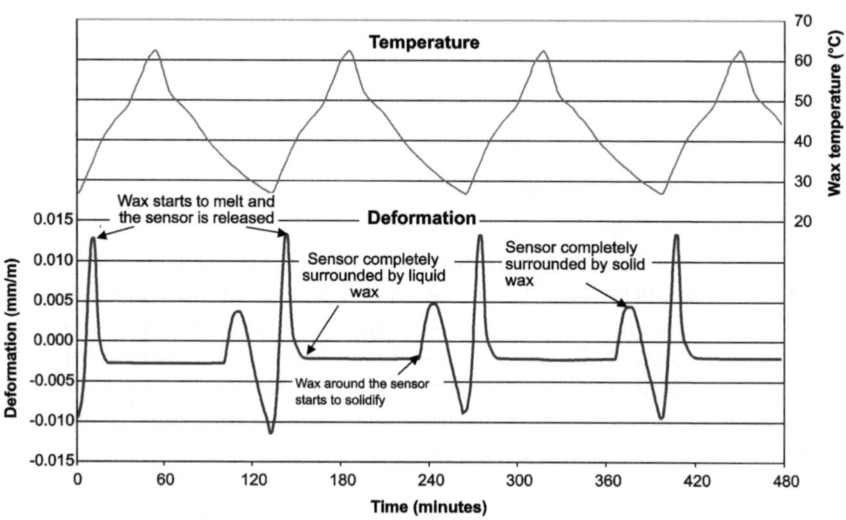

Fig. 7.19. Reproducibility investigations of an EFPI microstrain sensor embedded in wax during cyclic heating and cooling [68]

Measurement of Deformation in Soft Materials

If strain and deformations of curing materials or in rubber-like materials have to be evaluated, sensors are needed that do not react to the measurement object. They must not be stiff like FBG sensors or other sensor fibers. For such measuring purposes, a sliding fiber optic sensor such as flexible Fabry-Perot (EFPI) sensors can be used. In this case, the leading fiber end with one of the reflectors is able to slide inside the capillary. In this way, the necessary force to deform the sensing element is minimized and the sensor does not develop strain over the gauge length when the material to be evaluated deforms. This type of sensor was used to evaluate the deformation behaviour of specific mortar and cement pastes with low water/cement ratios at an early stage of deformations as well as of a repair mortar specimen in the interphase between rheology and solid state [61, 67]. The optically active space inside the tube was protected against water ingress. The measuring range of the sensor is $-2000\ldots+2500\,\mu\text{m/m}$, the strain resolution in combination with the recording device in the order of 10^{-7} to 10^{-8}.

In order to be sure that such a flexible sensor measures reliably, a number of EFPI sensors have been embedded into wax and their deformations during cyclic heating have been measured. An excellent reproducibility could be demonstrated.

Measurement of Force/Stress/Pressure

Fiber optic pressure probes are well-established on the market, mainly driven by oil industry, engine monitoring and medical applications such as pressure gradient measurements in the heart, in the circulatory system and in visceral cavities [69]. These measurement tasks require pressure probes for local pressure sensing. Figure 7.20 shows one possible design. Depending on the pressure value, the movement of a sensing element (reflector) changes the fringe contrast of a white-light fringe pattern at the end of the optical fiber. Another commercially available pressure sensor is based on a Fabry-Perot cavity attached to an optical fiber [70]. Pressure signal changes deflects a membrane,

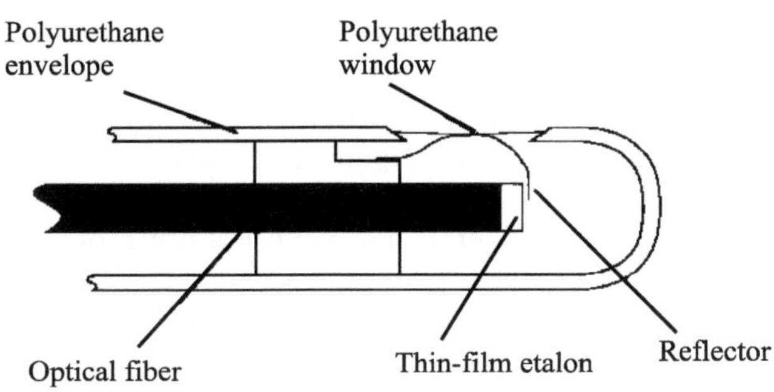

Fig. 7.20. Sketch of a pressure sensor [39]

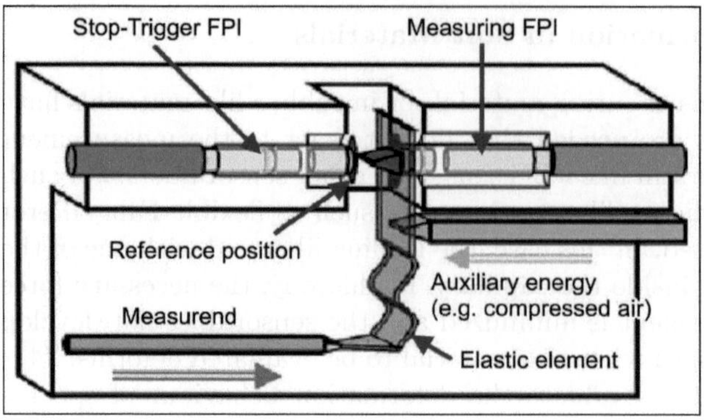

Fig. 7.21. Sketch of a pressure sensor with zero-point reference (BAM/Glötzl) [71]

which results in a change of the depth of the cavity and thus in a change of the reflected light intensity.

A similar but more complex design of a sensor probe allows the interrogation of hydraulic pressure, e. g. used in stress cells or as a force transducer [71]. This probe, currently being prepared for commercial use, is also based on the scanning of a membrane by using a Fabry-Perot interferometer sensor. Additionally, a specially designed second absolute interferometer sensor is used to correct drifts and possible changes of the zero-point reference from time to time. This pressure sensor probe allows measuring of long-term reliable pressure changes with high precision, especially when the power supply is switched off or if components of the measurement system have to be exchanged. The diaphragm deflection can be resolved with 60 nm, the long-term scan drift is smaller than 16 mbar (mean deviation: $< 1.6\%$) and the validated zero-point reproducibility (reference uncertainty) is ± 42.5 mbar. All described pressure probes can easily be designed for a wide range of pressure.

Distributed pressure sensing is more difficult than local sensing. Standard optical fibers usually used for strain or temperature sensing show small pressure sensitivity. A reliable correlation between pressure and inducing events is difficult, even if appropriate coatings that enhance the disturbing effect are used. In contrast to this, polarimetric fiber optic sensors, based on high birefringent (Hi-Bi) fibers, respond to pressure with a change in their polarization state of their output light. Although the use of high birefringence for distributed measurements in Hi-Bi fibers is accompanied by some difficulties (high precision-alignment requirements when splices have to be made, and the high cost of polarization-preserving fibers), new concepts are proposed for the distributed measurement of pressure acting on an optical fiber. Using a side-hole fiber, the distribution of isotropic pressure, e. g. in a fluid can be measured by application of backscatter polarimetry [72]. Such measurement can be made with a resolution of about 1 m in a time of about 1 min. Other sensing arrangements allow the detection of the position of a force and an estimation of its intensity [73].

7.2.7 Research Tasks and Future Prospects

Experience in the past revealed that not all application-related problems are solved. Basically, the application of cylindrical highly sensitive elements must be capable of being manageable under construction and production conditions. Some instructions for use of fiber optic sensors are developed [74]. More effort is still necessary to develop guidelines for reliable application of different fiber optic sensors and for validation of complete sensor systems. These aspects are essential, when fiber sensors are embedded in a laminate material such as glass fiber reinforced plastic (GFRP) or carbon fiber reinforced plastic (CFRP). A close interaction between sensing, control and actuation units creates really adaptive structures. However, because the fiber diameter is considerably larger (by up to ten times) than that of the reinforcement material in composites, they could reduce the tensile (or compression) and fatigue strength of the composite. In order to minimize the possibly reduction of strength parameters of the laminate due to integrated optical fibers, a further miniaturization is desired.

Another open question concerns the actual long-term behaviour of surface-applied or embedded strain/deformation sensors. Future research work should be more intensively focused on optimal design of the interface zone sensor – coating – host material. In adaptronic systems, reliable data must be delivered from sensors over a long period of time. The user has to pay attention to three main points:

- a durable coating or covering material has to be chosen;
- a reliable load transfer from measurement object into the sensing element has to be arranged, that is free of perturbing effects (e. g. temperature, transverse pressure); and
- the installation method must not obstruct the construction process and the long-term functionality of the object being interrogated.

It has been experienced that coating materials usually used can fail under raw environmental conditions. The load transfer characteristics can be perturbed or a long-term bonding to the measuring object cannot be reached. Alternative paths must be trodden.

Worldwide research activities focusing on these problems have been carried out. The next steps should concentrate the worldwide research experience on the still open application-related problems, e. g. establishing of user-friendly evaluation techniques and validation methods to know the long-term sensor characteristics, development of guidelines as well as standards for practical use of available sensors. Fruitful output is expected from current European COST actions, e. g. COST 270 (reliability of optical components and devices in communication networks) [75] and COST 299 (optical fibres for new challenges facing the information society) [76]. Leading the way could also be an international consortium like the International Society for Structural Health Monitoring of Intelligent Infrastructures (ISHMII) [77].

Concerning fiber optic sensors as the heart of a sensing system, two main future trends can be observed: the use of plastic optical fiber (POF) as sensors to an increased extend, and the adaptation of microstructured fibers for the use of sensors. POF sensors have found increasing use in different fields of application, e. g. as chemical, medical and bio-sensors. Due to their significant mechanical properties over glass fiber sensors, new developments such as fiber Bragg gratings in POF and microstructured POF are optimistically considered, unless other limitations such as a maximal operation temperature of about 8 °C or the link length of a few tens of meters preclude their use [78]. Despite the fact that FBG sensors need single mode POFs with sufficient photosensitivity, POF sensor systems would have cheaper interface costs (e. g. low tolerance moulded connectors).

However, the most exciting innovation will be expected from new types of fiber optic sensors based on microstructured materials and/or photonic crystal fibers (PCF). PCF has a lattice of air holes or microstructured areas along a certain length of the fiber. Since the appearance of photonic crystal materials in 1987, a number of remarkable application examples have been published [79]. Two features of PCF are of special importance: a) very small volumes of gases or liquids positioned in the air holes of the fiber can intensively interact with the light propagating in the fiber; b) the distribution and size of air holes, and thus the optical properties of PCFs, can be designed over a wide range. These specific features make PCF particularly interesting for sensor application because the propagation and coupling conditions in optical waveguides can be easily influenced. Several sensor concepts, e. g. the design of PCF as gas sensors [80, 81] or for the use as two-dimensional bend sensor [82] have been investigated and reported. Very promising sensing features show long-period gratings (LPG) made on a silica-based PCF. Results are reported, on first investigations into its use as a sensor, that the sensing effect can be enhanced compared to LPG inscribed in conventional single mode fibers [83].

7.3 Piezoelectric Sensors

R. Petricevic, M. Gurka

7.3.1 Introduction

The direct piezoelectric effect, via mechanical deformation of the piezo crystal lattice, causes an electric polarization by charge displacement. Vice versa, the effect of an electric field will cause a deflection of the crystal lattice and therefore of the whole crystal (inverse piezoelectric effect). Both effects are linear for small field strengths or deflections.

Piezoelectricity appears in natural crystals such as quartz, tourmaline, rochelle salt as well as in artificially produced ceramics and polymers such as e. g. nylon or copolymers of vinylidenefluoride (VDF) with trifluoroethylene (TrFE) or with tetrafluorethylene (TeFE). Most of the piezoelectric materials used for commercial sensor applications are synthetically produced polycrystalline ferroelectric ceramics such as e. g. lead-zirconate-titanate (PZT).

Ferroelectric materials show a spontaneous polarization that can be aligned by an external electric field (>1 kV/mm). Originally, polycrystalline ferroelectric ceramics such as PZT contain statistically polarized regions whose smallest grain areas with unique polarization are called domains or Weiss areas. Above the Curie temperature PZT has a cubic (m3m) lattice whose charge centers coincide and thus the corresponding crystal has no electric dipoles (paraelectric behaviour). On cooling down below the Curie temperature the crystalline structure of the PZT passes through a lattice distorting phase transformation which causes the formation of an electric dipole in each unit cell.

Within single crystals and ceramic crystallites, respectively, the dipole moments of neighbouring domains are either perpendicular or anti-parallel to each other. For polycrystalline materials the orientation of the crystallites and thus of the domains is randomly distributed. In the original state these materials do not exhibit a macroscopic polarization and thus no piezoelectric effect. However, the latter can be induced by applying a static electric field below the Curie temperature where the domains of uniform dipole moments arrange towards the polarization field (paraelectric polarization). The field strength applied should be between the saturation and the breakdown range. Due to this polarization the ferroelectric material becomes piezoelectric.

A part of the domains will turn back into the original state after switching off the electric field while the major part will remain remanently oriented (polarized). By the application of an electric field with reverse polarity the dipoles from a specific threshold of the so-called coercive field strength E_c start to turn over into the opposite direction, and the polarization is reversed.

If the values of the dielectric displacement D or the electric polarization P are plotted as a function of the field strength E the described processes are shown in a hysteresis curve (Fig. 7.22) that is characteristic for the piezoelectric material. When applying an increasing field to a not yet polarized material below the Curie temperature the polarization follows the so-called virginal curve. The saturation polarization P_s is reached for high field strengths. It is kind of identical with the spontaneous polarization in the domains. If the electric field is then reduced to zero, the so-called remanent polarization P_r remains and will be about $0.3\,C/m^2$ for PZT. Finally the entire hysteresis curve can be traversed by applying an electric field ramp with reverse polarity, returning to zero, and reapplication of the original field ramp.

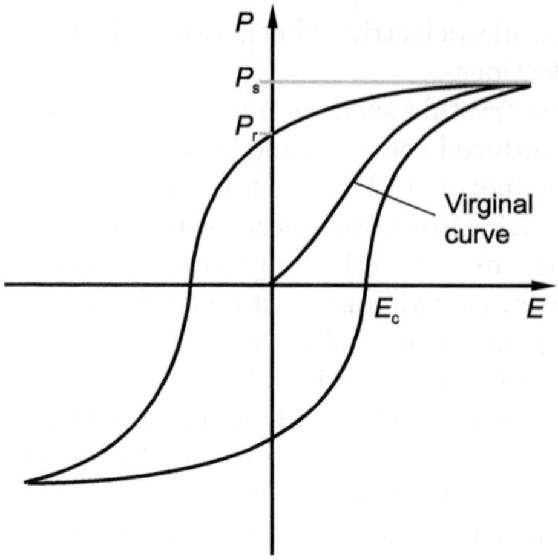

Fig. 7.22. Ferroelectric hysteresis loop

Due to the polarization being orientated into a preferred direction, and its elastic coupling via the crystal lattice, piezoelectric composite materials have a strongly anisotropic character. Simultaneously, a linearization of the electrostrictive features is achieved. Graphically, electrostriction means the directional orientation of the present dipole moments from their statistical disorder which normally leads to an extension or strain of the material in the field direction that is proportional to the square of the field strength ($\epsilon \sim E^2$). Due to the polarisation remaining remanently in the field direction, an inner electric field E_0 is induced in the material itself which means that an additional external field ΔE ($\Delta E < E_0$) can only have an effect at the absolute value of the dipole moments i. e. the increase of the distance between charge centers of polar molecules. In this way, the excited deflection goes linear with the electric field strength ($\epsilon \sim E$).

The piezo effect produced after the poling is quantified by the tensor coefficients of the piezoelectric charge coefficients d_{33}, d_{13} and d_{15}. For a clear indexing the Cartesian x_3-coordinate (i. e. the z-axis) is applied as a reference axis in parallel direction to the polarization vector in general [90–92].

7.3.2 Sensor Relevant Physical Quantities

Piezoelectric Charge and Voltage Coefficient/-Constant. For a piezoelectric material interactions between the electrical field and mechanical quantities have to be considered. In a good approximation this can be described via the linear context

$$D_{\mathrm{i}} = d_{\mathrm{ij}}^{\mathrm{sens}} T_{\mathrm{j}} + \epsilon_{\mathrm{in}}^{\mathrm{T}} E_{\mathrm{n}} \tag{7.3}$$

$$S_{\mathrm{k}} = s_{\mathrm{km}}^{\mathrm{E}} T_{\mathrm{m}} + d_{\mathrm{jk}}^{\mathrm{act}} E_{\mathrm{j}} \; . \tag{7.4}$$

Here D is the vector of the dielectric displacement (size: 3×1, unit: $\mathrm{C/m^2}$), S is the strain (size: 6×1, dimension 1), E is a vector of the electric field strength (size: 3×1, unit: V/m) and T is a vector of the mechanical tension (size: 6×1, unit: $\mathrm{N/m^2}$). As the piezoelectric constants depend on the direction in space they are described as tensors: $\epsilon_{\mathrm{in}}^{\mathrm{T}}$ is the permittivity constant also called dielectric permittivity at constant mechanical tension T (size: 3×3, unit: F/m) and $s_{\mathrm{km}}^{\mathrm{E}}$ is the elastic compliance matrix (size: 6×6, unit: $\mathrm{m^2/N}$). The piezoelectric charge coefficient $d_{\mathrm{ij}}^{\mathrm{sens}}$ (size: 6×3, unit: C/N) defines the dielectric displacement per mechanical tension at constant electrical field and $d_{\mathrm{jk}}^{\mathrm{act}}$ (size: 3×6, unit: m/V) defines the strain per electric field at constant mechanical tension [84]. The first equation describes the direct piezo effect (sensor equation) and the second the inverse piezo effect (actuator equation).

An equivalent formulation would be

$$E_{\mathrm{i}} = -g_{\mathrm{ij}}^{\mathrm{sens}} T_{\mathrm{j}} + \frac{D_{\mathrm{k}}}{\epsilon_{\mathrm{ik}}^{\mathrm{T}}} \tag{7.5}$$

$$S_{\mathrm{k}} = s_{\mathrm{kj}}^{\mathrm{D}} T_{\mathrm{j}} + g_{\mathrm{km}}^{\mathrm{act}} D_{\mathrm{m}} \, , \tag{7.6}$$

where $g_{\mathrm{ij}}^{\mathrm{sens}}$ and $g_{\mathrm{km}}^{\mathrm{act}}$ are the piezoelectric voltage coefficients. It is important to consider that the quantities s_{kj} and ϵ_{ik} strictly speaking depend on the electrical field E or on the mechanical stress T. In the equation system above the upper index indicates that in the present case a certain value for s_{kj} or ϵ_{ik} is meant for the constant upper index.

For short-circuited electrodes E is held constant at zero (upper index E), for open electrodes the dielectric displacement D remains constant.

The (7.3), (7.4), (7.5) and (7.6) show that the piezoelectric coefficients g and d can be defined in two ways. In the hydrostatic mode the piezoelectric coefficients are represented by the effective quantities $d_{\mathrm{h}} = d_{33} + 2d_{31}$ and $g_{\mathrm{h}} = g_{33} + 2g_{31}$. For hydrophone materials the product $d_{\mathrm{h}} g_{\mathrm{h}}$ is often reported as a measure of quality [85].

Sensitivity. The sensitivity of a piezoelectric material is taken to be equal to the generated open-circuit voltage that drops across to the contact with the distance t (= thickness) divided by the applied stress or the product $g \cdot t$, where g is the relevant piezoelectric voltage coefficient. The voltage coefficient g is connected with the charge coefficient d via the dielectric permittivity $\epsilon = \epsilon_{\mathrm{r}} \epsilon_0$ according to

$$d = \epsilon_{\mathrm{r}} \epsilon_0 g \, . \tag{7.7}$$

For a sufficient sensitivity possibly a high permittivity or capacitance of the sensor is required to compensate electrical losses via the cables. However, it is important to consider that a higher permittivity according to the relation above, implies a decrease of the voltage coefficient.

Coupling Factor and Energy Efficiency. The electromechanical coupling coefficient is an important quantity for piezoelectric sensor materials in the resonant operation mode. The square of the coupling coefficient k is a measure for the conversion of electrical energy into mechanical energy and vice versa:

k^2 = stored mechanical energy/applied electrical energy

k^2 = stored electrical energy/applied mechanical energy.

If the coupling factor relates to a piezoelectric element with optional dimensions it is also referred to as the effective coupling factor k_{eff} considering the energies appearing in all directions [90]. If the electrical and mechanical quantities of a piezoelectric element appear in certain directions the coupling factor k_{ij} is provided with the corresponding indices analogically to the piezoelectric coefficients. Special cases are the planar coupling factor k_{p} and the thickness coupling factor k_{t}. Formally, k_{p} would correspond to k_{31}, and k_{t} would correspond to k_{33}. For k_{p} and k_{t} however, the influence of the other direction components in contrast to k_{33} and k_{31} are not contained.

In contrast to the coupling factor the total efficiency is defined as

η = converted effective energy/energy consumed by the transducer.

Temperature Drift. For application over a wide temperature range, knowledge of the temperature coefficient is required for the signal that acquires the measuring quantity. In general this is the relevant charge or voltage coefficient. By registration of the sensor temperature the signal can be corrected online or later on correspondingly. It is more comfortable to minimize the temperature drift by a capacitance without a temperature coefficient which is additionally connected to the measuring circuit. So therefore, besides the total capacitance even the temperature coefficient will be reduced. For voltage measurements a parallel capacitance is connected in, and for charge measurements the capacitance is connected in series. Thus one can achieve that the temperature coefficient for the output quantity can be minimized [91].

Pyroelectricity. A sudden modification of the environmental temperature of the crystal (or ceramic) causes a modification in the length (thermal expansion) of the crystal axis whose direction matches with the polarization direction. Due to the piezo electric effect charging occurs. However, the permanent polarization changes with the temperature as the dipole moments in piezoelectric materials depends on the temperature.

The polarization is:

$$P_{\mathrm{py}} = p \cdot \Delta T \,, \tag{7.8}$$

where p is the so-called pyroelectric constant.

Both effects are in the same direction and lead to an external charging of the crystal. Therefore changes in temperature are accompanied by changes of the relevant measuring signal (charge and voltage) without having an external mechanical reason for it.

This pyroelectric effect can be utilized for sensors such as e. g. infrared cameras. However, in sensors that use the electromechanical effect pyroelectricity can be disturbing. The disturbing effects arise especially in low-frequency or quasi-static applications as the temperature drift is often a slow process. A one ohm resistance is applied in parallel to suppress this effect. That way, the pyroelectric induced charges are deflected and the cut-off frequency of the sensor is raised.

Nonlinear Behaviour. The linear relation between deformation and electrical field strength or charge is only valid for a limited range that can be determined via the hysteresis curve. The nonlinear behaviour is caused by domain reorientations in poled materials. The extension of the linear range depends on the magnitude, direction and frequency of the generated or applied field strength in relation to the coercive field strength.

A reorientation or depolarization of the domain is also effected by mechanical stress (e. g. 20 ... 50 N/mm^2 for PZT). Influencing factors besides the stress magnitude are its direction and frequency as well as the kind of electrical circuit (e. g. open circuit, load or short circuit). If the electrical field induced by a force is in the polarization direction, the nonlinearities are essentially smaller than those of a generated field in the opposite direction or in the case of short circuit.

If the material is heated up to the Curie point T_c a complete depolarization follows where the domains become randomized upon thermal motion. For a long-term operation without significant depolarization $T_c/2$ should not be exceeded.

Due to high power requirements the nonlinearities of actuators and ultrasonic transducers are accepted despite the accompanying dissipative losses.

7.3.3 Materials and Designs

Sensor Materials

Crystals. Naturally appearing crystals such as quartz and Rochelle salt can be mostly substituted by synthetically produced alternatives. For an optimized piezoelectric performance the crystals must be adjusted and tailored along specific crystallographic directions.

Currently, quartz is often utilized in accelerometers. Due to their high piezoelectric voltage coefficient g_h lithium sulfate and tourmaline are often applied in commercial hydrophones especially to measure shock and pressure waves. Rochelle salt can be found in acoustic pickups and special devices to measure acoustic pressure. Due to their long-term stable piezoelectric properties natural crystals are in particular perfect for sensor applications where the monitoring of a quantity has to be made over long periods [85].

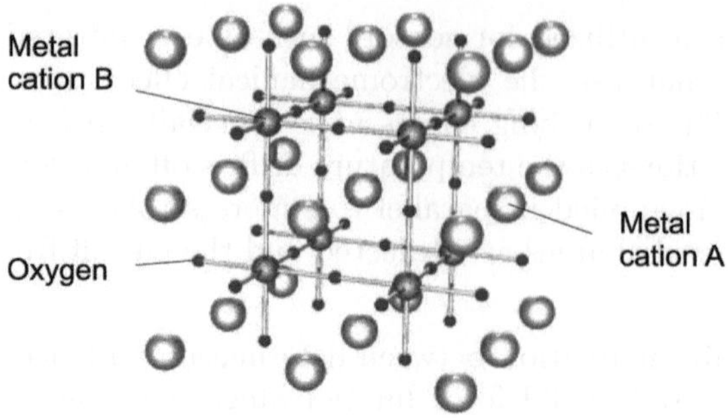

Fig. 7.23. Perovskite structure (Source: Wikipedia)

Ceramics. Barium titanate ($BaTiO_3$) was discovered in 1943 independently from American, Japanese and Russian scientists and was thus the first polycrystalline ferroelectric ceramic with Perovskite structure (Fig. 7.23).

The advantages over natural crystals are the subsequent polarizability, very high permittivity, chemical resistance, free possibility of forming and low-cost manufacturing by the ceramic manufacturing process.

Before Jaffe et al. discovered lead-zirconate-titanate (PZT) in 1954, barium titanate with its excellent features was the piezoceramic of choice. Compared to barium titanate the Curie temperature ($\approx 360°C$) as well as the coupling factor ($k_{33} \approx 0.7$) for PZT is considerably higher.

Due to their versatile producibility and processability and the good piezoelectric performance PZT ceramics in a diversity of designs are quite appropriate for the implementation of sensors and actuators in adaptronic systems.

Adjusting the mixing ratio of the components and doping the ceramic in a special manner is a way to influence the lattice structure of PZT. A detailed description of the effects of doping to the different features of PZT is given in diverse publications and in the information material of the manufacturers. An overview can be found in [85].

Soft PZT ceramics are characterized by high piezoelectric coefficients, high relative permittivity, high dielectric losses, high electromechanical coupling factors, very high insulating resistance, low mechanical quality factor and low coercive field strength. Corresponding application fields are electroacoustic devices (sound generator and receiver), metrology (sensors), ultrasonic diagnostics and static or quasi-static deformation elements as actuators.

Hard PZT ceramics are characterized by low piezoelectric coefficients, a smaller relative permittivity, minor dielectric losses, lower insulating resistance, high mechanical quality factor and high coercive field strength. Corresponding applications fields are ultrasonic generators with the highest required output powers such as ultrasonic cleaners or transducers for sonar applications.

Polymers. The best known piezoelectric polymer is polyvinylidene difluoride (PVDF) discovered in 1969. PVDF is a thermoplastic consisting of long chains of repeating monomers ($-CH_2-CF_2-$). The PVDF film sensors are fabricated via film drawing from the melt with unidirectional stretching and subsequent polarization. During the viscose melting phase the dipoles are randomly oriented and so the melt does not show any polarization. After the coagulation and unidirectional stretching the polymer chains are preferably justified along the stretching direction. During the contemporaneous polarization process a permanent dipole moment is impressed and the PVDF film subsequently shows piezoelectric properties [84].

Due to the unidirectional chain orientation the material becomes piezoelectrically orthotropic, i. e. $d_{31} \neq d_{32}$. The draft direction is defined as the 1-direction. For very small strains, however, the material is widely isotropic. Due to its Youngs modulus which is essentially smaller compared to that of PZT, the influence of the stiffness of PVDF on the dynamics of the host structure in most cases is negligible. That is why PVDF films are especially appropriate for sensory applications.

Its good elasticity and mechanical flexibility as well as the simple processing together with low costs and an excellent adaptability give PVDF films a certain attractiveness for a wide range of applications, especially those where the low acoustic impedance of PVDF (comparable with water or organic materials) is useful. Among those are transducers for acoustic sound (hydrophones), ultrasonic signals (up to 24 GHz) as well as electromechanical and pyroelectric applications. PVDF with some kV/mm exhibits an extremely high coercive field strength compared to crystals and ceramics.

Disadvantages of PVDF are the low piezoelectric charge coefficient (about 1/10 of PZT, but comparable with quartz) as well as the strong temperature depending performance (temperature drift) due to the pyroelectric properties and the low thermal stability. As a result of their viscoelastic behaviour (like all polymers) temperature and frequency have a strong influence on the mechanical and electrical properties of PVDF. The maximum tolerable working temperature is 100°C. The piezoelectric features, however, in a permanent application already diminish significantly above the room temperature because of relaxation processes. The small relative permittivity constant of $\epsilon_r \approx 12$ can be a disadvantage for the application as a sensor (see Sect. 7.3.2, Subsect. Sensitivity).

As an actuator PVDF foils are unsuitable for most adaptronic applications due to the small forces and damping losses.

Sensor Designs

Plates, Disks, Cylinders, Globes. Plates, disks and cylinders are the simplest geometries and are often applied in electroacoustic sensors. These geometries are either formed by monolithic piezoelectric materials or are correspondingly arranged in segments. The fundamental resonances of the components are defined by the corresponding geometric dimension that is responsible for the effect. For omni-directional characteristics even spheri-

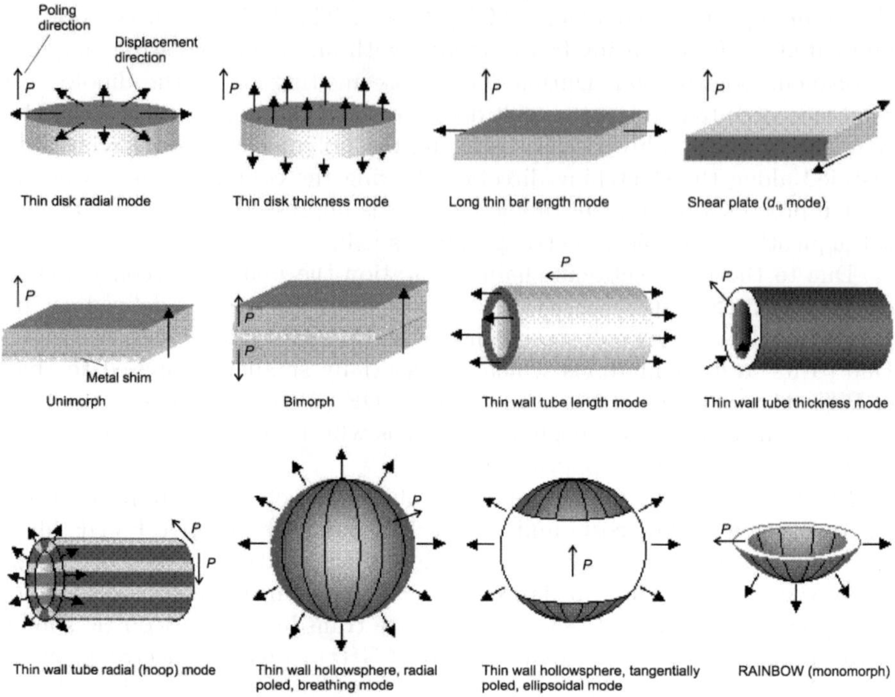

Fig. 7.24. Sensor configurations [85]

cal geometries can be produced e. g. by adhesion of bent triangular ceramic segments (half melon pieces) [85].

Bending Transducers. Bending transducers are produced by sticking together two reverse polarized piezoceramic plates via a common electrode surface (bimorph arrangement). This results in an addition of the signals from both plates, due to a deflection of one plate as well as a compression of the other one. This is a widespread geometry for ultrasonic sensors and accelerometers and it is quite appropriate for applications working in the low ultrasonic frequency range. The combination of a ceramic element with a thin metal plate (used as an electrode) is designated as an unimorph arrangement (see Fig. 7.24).

Conventional bending transducers are to be found as bimorph and unimorph arrangements. The so-called monomorph transducer is a more exotic device with single RAINBOW (reduced and internally biased oxide wafer[1])

[1] A lead containing a piezoceramic disk (e. g. PZT) is reduced on one side by high temperature treatment in direct contact with a carbon block. This reduced layer is no longer piezoelectric but therefore a good electric conductor. Due to the thermal expansion mismatch between the reduced and oxide layers, a curvature develops in the structure, giving it a dome (or rainbow) shape.

ceramic plates that are especially applied as low pressure sensors (<100 kPa) or as acoustic transducers. For efficient force coupling RAINBOW sensors have to be attached to a ground plate. By virtue of the geometry, RAINBOW transducers are extremely robust, but their signals do not depend on the pressure in a linear manner (the more pressure the more flattened the transducer) [85].

Piezoelectric Composite Sensors. The concept of piezoelectric composites comes from the idea to connect an active piezoelectric phase with a passive matrix phase in a way that the best features from both components can be enhanced and the shortcomings can be minimized correspondingly. Common examples are composites from a stiff ceramic with a soft polymer.

Newnham et al. [86] established a notation that indicates the number of dimensions in which each phase is continuously connected to itself. There are ten possible combinations of two different components in one composite, that is to say 0–0, 1–0, 2–0, 3–0, 1–1, 2–1, 3–1, 2–2, 2–3 and 3–3. In the case of piezo composites the first number refers to the connectivity of the piezoelectric (active) phase and the second one refers to the connectivity of the passive phase.

Figure 7.25 shows an array of composites that are realized with various piezoceramic modifications. The diverse designs serve the purpose of decoupling the d_{33} and d_{31} coefficients in order to influence the directional char-

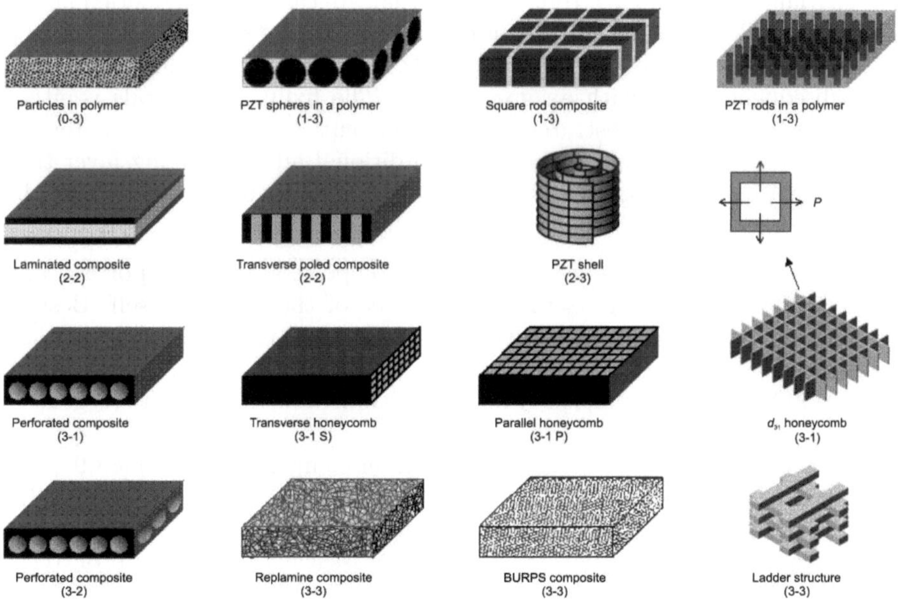

Fig. 7.25. Connectivity of constituent phases in piezoelectric ceramic-polymer composites [85]

acteristic. Furthermore certain composite geometries are used to optimize resonant and non-resonant bandwidths.

Due to improvements of mechanical properties and higher damage tolerances, piezoelectric ceramic-polymer composite materials show interesting aspects with regard to adaptronic applications. Some brief approaches to fabricate these composites based on the connectivity of their constituent phases are shown by the following examples [85].

The 3–3 composites are porous piezoceramics with interconnected pores. They can be made for example by the sintering of PZT powder mixed together with small spheres made of a volatile polymer and subsequent filling the remaining pores with silicone rubber. Another possibility is the inner surface coating of open porous organic foams by infiltration with piezoceramic slurry and a subsequent pyrolysis and sintering.

The 1–3 composites are the most examined and applied ones. They consist of individual PZT rods or fibers embedded in a polymer matrix and oriented parallel to the poling direction. Fiber diameter and spacing, composite thickness, volumetric PZT content, aspect ratio (radius/length) of the fibers and the stiffness of the polymer matrix have an influence on the composite performance. The force transfer between the rods or fibers and the polymeric matrix is due to shear-coupling at the polymer-fiber interface or due to compressive coupling at the front end of the fiber bundle.

Basically there are two different manufacturing technologies:

- spatially adjusted PZT rods or fibers that are cast in a polymeric matrix.
- the so-called dice-and-fill technique: with a diamond wafer saw rod-shaped pillars are cut free from a massive PZT block, or from thin plates and subsequently cast with an epoxy resin. The fixing of the rods is either made by the PZT substrate itself (by only cutting a part of the material), or in the case of thin plates by an additional adhesive fixing layer that is not or only incompletely cut through and can be removed after the casting.

The polymers and adhesives used in the composites are as important and performance-determining as the performance of the ceramic itself. Besides processing features (e. g. viscosity during fabrication) the used polymers must exhibit a good dielectric strength (20... 30 kV/mm), high shear strength (20... 30 MPa) and a sufficiently high glass transition temperature (e. g. $T_g = 150\,°C$).

Active fiber composites (AFC), macro fiber composites (MFC) [88,97] and piezo fiber composites (PFC) [87,98] are advanced variants of 1–3 composites that are particularly designed for adaptronic applications. Those variants are based on a basic design developed at the Massachusetts Institute of Technology (MIT) [89, 93–96]. For that purpose uniaxial arranged piezoceramic fibers from PZT are embedded in a polymer matrix and contacted via interdigital electrodes at the surface. Such composites have a high flexibility and robustness and can therefore be easily applied to components or structures

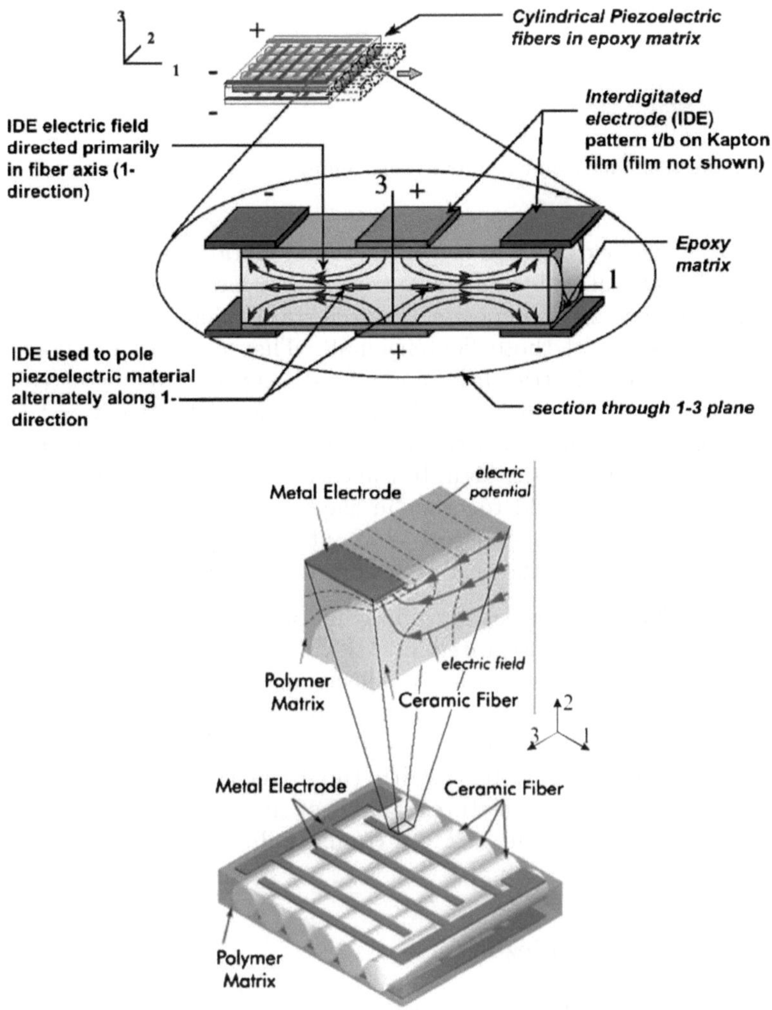

Fig. 7.26. Schematic of the cross section of a 1–3 piezo fiber composite [88, 89]

with a given (e.g. bent) shape. Figure 7.26 shows the details of this fiber composite concept [89].

Ceramic Metal Composites. Ceramic metal composites are characterized by a simple design and extreme robustness. This is achieved by the combination of an active ceramic with metal clamping plates (shells or caps). The metal plate is used to achieve the coupling of the active ceramic to the surrounding medium. The metal plate is a mediator or coupler between the operating force and the ceramic. The best ceramic metal composite sensors are the flextensional type transducers. For this construction the flexural modes

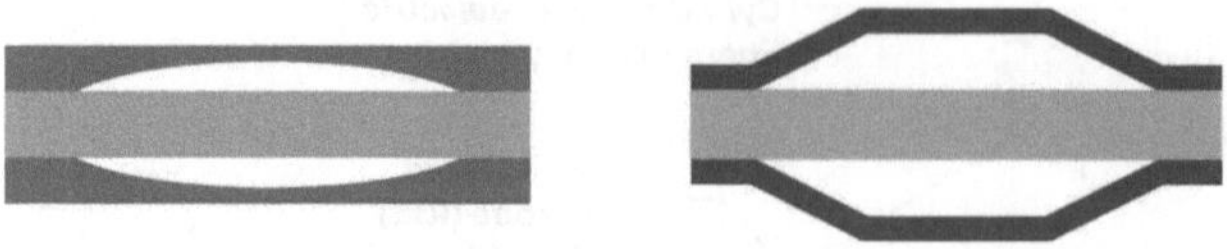

Fig. 7.27. Moonie (*left hand*) and Cymbal (*right hand*) 2–2 composites

of the metal shell causes extensional or contractional vibrations of the piezoelectric element. In general such an arrangement is very large and heavy [85].

Moonie and Cymbal 2–2 composite transducers, depicted in Fig. 7.27 are miniaturized versions of those flextensionals. They consist of a poled and face electroded piezoelectric disk sandwiched between two metal end caps with air-filled cavities above the electrodes. Due to the cavities the metal caps act as transformers of axial compressive stress into tangential and radial components of opposite sign. In the case of a hydrostatic load the contributions of d_{31} and d_{33} add together in the effective d_{h} of the device. The d_{33} coefficient of a cymbal structure is about 70% higher than that of a Moonie configuration but at the cost of nonlinear load sensitivity [85].

Piezoelectric MEMS. Piezoelectric thin layers or films (most frequently made of PZT) that can be integrated in MEMS offer a broad range of advantages. In contrast to electrostatic or electromagnetic MEMS-devices piezoelectric MEMS are especially characterized by large deflections at small hysteresis. They exhibit a high energy density as well as high sensitivities with a very broad dynamic bandwidth (up to GHz) low power consumption. Further outstanding features are: easy integration, high temperature stability, good scalability, CMOS compatibility and simple signal processing.

7.3.4 Passive and Active Piezo Sensors

Sensors transform strains or movements or their derivatives in electric signals (e. g. an electrical field). Piezoelectric strain sensors should be easy to handle and simple to apply. Sensor properties such as sensitivity (strain, movement or acceleration), bandwidth and geometric size are of particular importance for dimensioning. Features such as temperature sensitivity, linearity, hysteresis, repeatability, electromagnetic compatibility, and integration capability as well as peripheral electronics (size and power requirement) determine the sensor performance. Typically the sensitivity for a resistor gauge is about $30\,\mu\mathrm{V/ppm}$, for semiconductor gauge is $10^{-3}\,\mathrm{V/ppm}$ and for piezoelectric or piezoceramic DMS $10^{-2}\,\mathrm{V/ppm}$. The sensitivity of fiberoptic sensors is defined differently and corresponds to $1^{\circ}/\mathrm{ppm}$ [100].

The piezoelectric measuring methods can be roughly divided in two groups, corresponding to the fact of whether the sensor works with a passive or an active principle.

Passive Sensors

Passive sensors directly transform a mechanical quantity into an electrical signal. The direct piezoelectric effect is used e. g. to detect or quantify structural deformations or deflections caused by pressure, tensile loads or motions. These deformations are transferred to the piezoelectric material via a force or friction locked bonding (as strain, bending, shearing or compression) where they will be transformed into an electrical signal. Typical applications are based on detecting amplitude variation of the charge signal at constant frequency or on analysis of the frequency spectrum (and its change) generated by the sensor.

There are two types of sensors to be distinguished: axial and bending sensors. In the case of an axial sensor the force works in the polarization direction, while for the bending sensor the force works perpendicularly to it. Concerning the latter, emerging tensile or pressure forces depend on the distance of the active material from the neutral fiber [91].

The main advantage of piezoelectric sensors in contrast to conventional strain gauges is their higher signal-to-noise ratio and their high-frequency-noise suppression. In applications with low strain levels, piezoelectric sensors require significantly less signal conditioning. Other advantages are their compact design, high sensitivity over a broad strain and frequency range as well as their simple integration capability. Most frequently applied sensors are based on piezoelectric polymers (e. g. PVDF) and ceramics (e. g. PZT). Piezoceramic sensors are often chosen for applications that simultaneously require sensory and actuating capabilities.

PZT sensors exhibit a high Youngs modulus, are very brittle and have only a low tensile strength. Leakage currents appear at dc voltage, depolarization under high mechanical loads as well as deviations from linearity at high strains. Despite those problems there are numerous examples of how piezoelectric ceramics can be applied successfully in adaptive structures as well as in health-monitoring systems. A comparison of different PZT sensors can be found in [101]. However, piezoelectric polymers such as PVDF films show themselves as orthotropic but mechanically isotropic for small strains.

Independently from the material the following equations describe the sensor effect [84]:

$$D_{\mathrm{i}} = d_{\mathrm{ij}} T_{\mathrm{j}} + \epsilon_{\mathrm{ik}} E_{\mathrm{k}} + \alpha_{\mathrm{i}} \Delta T \tag{7.9}$$

$$d = \begin{bmatrix} 0 & 0 & 0 & 0 & d_{15} & 0 \\ 0 & 0 & 0 & d_{25} & 0 & 0 \\ d_{31} & d_{32} & d_{33} & 0 & 0 & 0 \end{bmatrix} \tag{7.10}$$

$$\epsilon = \begin{bmatrix} \epsilon_{11}^{\sigma} & 0 & 0 \\ 0 & \epsilon_{22}^{\sigma} & 0 \\ 0 & 0 & \epsilon_{33}^{\sigma} \end{bmatrix} \tag{7.11}$$

where α is the thermal constant vector. These sensor equations are based on the direct piezoelectric effect, i. e. the sensor is exposed to a mechanical stress that generates an electrical field. Monolithic PZT sensors are transversally isotropic, i. e. $d_{31} = d_{32}$ and $d_{15} = d_{25}$.

If the electric field is zero and no thermal expansion is predominant above, the equation reduces to

$$D_{\mathrm{i}} = d_{\mathrm{ij}} \cdot T_{\mathrm{j}} \,. \tag{7.12}$$

A mechanical tension vector T causes a dielectric displacement D that generates a charge q which is

$$q = \int\int D \cdot \mathrm{d}A \,, \tag{7.13}$$

where dA contains the three components of the electrodes differential surface area in the 2–3, 1–3 and 1–2 plane directions.

Charge q and voltage V_{c} are related to each other via the sensor capacitance C_{p} according to

$$V_{\mathrm{c}} = q/C_{\mathrm{p}} \,. \tag{7.14}$$

For a certain voltage the force and thus the strain can be determined. In the case of a thin rectangular sensor plate whose main surfaces are contacted (polarization direction perpendicular to the surface = 3-direction) and uniaxially loaded in the 1-direction the capacitance is

$$C_{\mathrm{p}} = \epsilon_{33}^{\sigma} l_{\mathrm{c}} b_{\mathrm{c}} / t_{\mathrm{c}} \,, \tag{7.15}$$

where l_{c}, b_{c} and t_{c} are length, width and thickness of the sensor. For example, for a strain in the 1-direction the voltage results in

$$V_{\mathrm{c}} = \frac{d_{31} Y_{\mathrm{c}} b_{\mathrm{c}}}{C_{\mathrm{p}}} \int_{l_{\mathrm{c}}} \epsilon_1 \mathrm{d}x \,. \tag{7.16}$$

In this equation Y_{c} is the Youngs modulus of the sensor and the strain ϵ_1 is averaged over the measuring length. Transforming the latter results in

$$\epsilon_1 = \frac{V_{\mathrm{c}} C_{\mathrm{p}}}{d_{31} Y_{\mathrm{c}} l_{\mathrm{c}} b_{\mathrm{c}}} \,, \tag{7.17}$$

assuming that strain is applied only in the 1-direction and that there are no strain losses in the interface layer. Considering the Poissons ratio (Poissons number ν) the following relation results [84]

$$\epsilon_1 = \frac{V_{\mathrm{c}} C_{\mathrm{p}}}{d_{31} \left[1 - \nu (d_{32}/d_{31})\right] Y_{\mathrm{c}} l_{\mathrm{c}} b_{\mathrm{c}}} \,. \tag{7.18}$$

For PVDF foil sensors and piezo fiber composites with 1–3 or 2–2 connectivity, the transversal sensitivity is very small and can be neglected under certain circumstances.

With only one piezoceramic plate it is not possible to separate different strain directions from each other. If the transversal strain is not known already, it is not possible to determine longitudinal strains with one piezoelectric sensor plate only. This can be more easily achieved by piezoelectric composites with 1–3 or 2–2 connectivity.

The shear lag effect caused by a finite thickness of the bond layer can be taken into account starting from strain beam theory. The shear lag in the bond layer causes a reduction of the effective length and width of the sensor considered in the following equation [84]:

$$\epsilon_1 = \frac{V_c C_p}{d_{31}\left[1-\nu\right] Y_c l_{c_{\mathrm{eff}}} b_{c_{\mathrm{eff}}}} \,. \tag{7.19}$$

The effective length and width of a piezoceramic sensor depends on both the sensor and layer properties. Sirohi and Chopra [84] describe how to determine those effective quantities. In general this is to say that the smaller the thickness and stiffness of a sensor, the smaller the shear lag losses.

Electromechanical Behaviour. The properties of piezoelectric sensors depend on the mechanical boundary conditions as well as on the peripheral electronics. In principle, the gain response of the sensor signal (i. e. the sensitivity) as described schematically in Fig. 7.28 consists of a quasi-static range that is frequency independent (at low frequency) and a resonant range with a frequency depending sensitivity (at higher frequencies). In the quasi-static range the sensor behaves like a capacitance with a voltage source connected in series or with a current source connected in parallel, respectively. Correspondingly, the sensor signal at the extremities can be measured as an open-circuit voltage or a short circuit charge. Other loading cases can easily be determined accordingly.

For very low frequencies the capacitive inner resistance, however, is problematic since the signal is reduced by parasitic bleeder resistors in the sensor material (i. e. the PZT ceramic or the polymer matrix within a composite) and in the circuit. This cut-off frequency, where the sensitivity is strongly diminished (counterdrawn curve), can significantly be moved towards lower

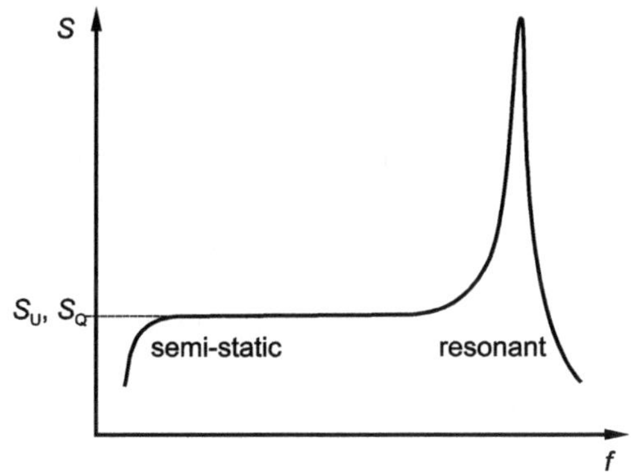

Fig. 7.28. Schematic frequency response of sensitivity S

frequencies by charge amplifiers. But then, the signals are increasingly superimposed or dominated by the pyroelectric properties. The measuring of very slow processes is therefore extremely problematic; static measuring is impossible.

In the resonant domain the sensitivity, frequency and bandwidth depend on the load. With an increasing load resistance the resonant frequency moves from the series resonance to the parallel resonance. The bandwidth or damping of the resonator runs through a maximum if the impedance of load and sensor capacitance exhibit approximately the same values. With additional inductive adjustment a continuous tuning of the resonant behaviour is possible.

For large strains above 150 ... 200 ppm (micro strain) the behaviour of piezo sensors is increasingly nonlinear. Thats why this measuring range should not be exceeded. Pure PZT sensor usually does not require temperature corrections as long as the temperature variation is smaller than ± 40 K.

Although piezoelectric properties such as relative permittivity and piezoelectric coefficients change with the temperature, the total effect for the calibration of PZT sensors far below the Curie temperature is very small. However, PVDF foils show a significant temperature dependence of the pyroelectric properties in addition to their temperature depending piezoelectric features. Thus PVDF sensors are quite temperature sensitive, and in general, an appropriate temperature compensation is necessary. This behaviour, however, can be improved by special forming and spatial distribution of the PVDF sensors [100].

Active Sensors

The principle of active sensors can be described as an indirect reaction of the sensor system to external influences. Here the environment or the object to be measured is influenced actively by a piezo transducer (e. g. by sending defined ultrasonic waves) while the reaction or the signal response is registered at the sensor position. From the signal transfer function conclusions can be drawn about the structure under examination or about external influences.

Active sensors preferably work in the resonant mode and are almost exclusively ultrasonic sensors. Acoustic measurement methods primarily utilize the bi-directional transforming ability of piezoelectric materials. By using acoustic piezo transducers a very high coupling efficiency for the sound signal to liquid or solid media can be achieved. Gaseous media, however, are disadvantageous for the simple piezo transducers but via impedance matching techniques effective sensors can be realized here, too.

As ultrasonic sensors work predominantly in pulse mode, the pulse transmission behaviour of the piezo transducer together with the peripheral amplifier and signal conditioning electronic will determine the signal quality.

A broad bandwidth is important for the shortest possible pulses of the transducer and thus for the resolution of the sensor system. Another important quantity is the directional characteristic of the transducer. Ultrasonic sensors utilize physical correlations that can detect many quantities acoustically [91]. Therefore different procedures are utilized. The most important technologies for adaptronic systems will be described here briefly.

Non-Destructive Techniques. There is a wide range of different variants of non-destructive evaluation (NDE), non-destructive testing (NDT) and non-destructive inspection (NDI) techniques in order to identify local damage and the onset of damage in critical structures. In particular, ultrasonic inspections used over several decades are still most popular [102].

In an infinitely extended solid medium elastic waves can propagate in two basic modes: pressure (P) waves and shear (S) waves. However, if the medium is bounded, wave reflections occur at the boundary and more complicated wave patterns emerge.

Of particular interest especially for adaptronic systems are the guided waves which remain contained in a wave guide and thus can travel over large distances. Depending on the wave guide structure different kinds of waves are distinguished: Lamb waves propagate along thin plates, Rayleigh waves are restricted to the surface of a material, Love waves travel along layered materials, and Stoneley waves travel along their boundaries.

Guided waves can exist in massive and hollow cylinders as well as in shell structures. In flat plates, ultrasonic guided waves travel as Lamb waves and as shear horizontal waves (SH). Lamb waves are vertically polarized while SH waves are horizontally polarized. Both, Lamb waves and SH-waves can be symmetric or antisymmetric with respect to the plate mid-plane [103].

Ultrasonic NDE Methods. Ultrasonic NDE methods rely on the elastic wave propagation and reflection within the material trying to identify wave field disturbances due to local damage and imperfections. In contrast ultrasonic NDT involves the measurement of the following quantities: time of flight (TOF), path length, frequency, phase angle, amplitude, acoustic impedance and angle of wave deflection (reflection and refraction).

Conventional ultrasonic methods include the pulse-echo, the pulse-transmission and the pulse-resonance techniques [104]. Depending on the incidence of the piezo transducer with respect to the structural surface as well as on their design, P-waves, S-waves or a combination of both can be generated within the structure. P-waves are best suited for the inspection of thick components, for through-the-thickness damage detection, and are quite effective for the detection of anomalies along the sound path. By the pulse-echo method, detects are detected in form of additional echoes. In the pulse-transmission method wave dispersion and attenuation due to diffused damage in the material indicate possible defects [103].

A classic thickness-wise inspection with the P-wave method is generally very time-consuming as the transducer must be moved mechanically along the surface to scan and interrogate the entire volume of the material. Guided ultrasonic waves however open up a great potential. With the variable mode structure and mode distribution of the wave fields, a specific sensitiveness for different defect types, propagation over long distances and the guiding character which enables following of curvature and the reaching hidden or buried parts, it is possible to cover a diversity of inspection tasks (e. g. of planes, pressure vessels, oil tanks or pipelines) [103].

Electromechanical Impedance Method. This kind of NDE method enables the direct identification of the local structural dynamics via an electromechanical impedance signature of a piezoelectric sensor permanently attached to the structure. Thereby, the sensor dynamics must be taken into account, too. Structural damages cause changes in the high frequency range (1 ... 500 kHz) of the electromechanical impedance spectrum which will become visible as the frequency shifts, splitting up of resonance peaks or appearance of new resonances. The spectral data can be classified e. g. via statistical methods or stochastic neural networks to determine the grade of the damage.

The high frequency spectrum is neither affected by global structural modes nor by static loads, ambient vibrations or other changes of the boundary conditions. Therefore the impedance method is particularly appropriate for the monitoring of the onset of local damage such as fractures, cracks or delaminations that bear no noticeable changes in the global dynamics of the entire structure. That is why this method is especially appropriate for monitoring local areas of the structure where the beginning of damage is already preassigned and piezo sensors can be installed nearby [105].

7.3.5 Piezo Sensors as Integral Components of Structures

Piezo sensors enable an implementation of smart structure concepts in two ways: the so-called adaptive and the sensory structures [106].

The idea behind the adaptive structures is to implement structural properties (e. g. stiffness, strength) that can be adapted to external influences by means of appropriately integrated sensors and actuators. In principle, a functionally well adapted passive structure is extended by active components (sensors, actuators, microprocessor based controller and computational capabilities).

Sensory structures however, are able to detect and monitor deformations, deflections or even structural conditions and properties (e. g. the growth of damages) via integrated sensors. In the case of a critical condition or property an output of information about the actual condition or just a warning will be generated. This kind of condition monitoring (CM) is also called structural health monitoring (SHM).

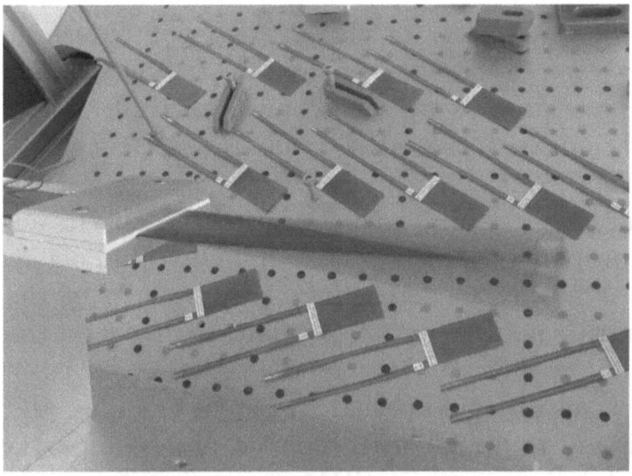

Fig. 7.29. Fully embedded 1–3 piezo fibre composite within a carbon fibre reinforced plastic beam [87]

The following fundamental technology developments have formed the base for the implementation of adaptive and/or sensory functional design:

- fiber reinforced composite construction enables the full integration of active elements such as demonstrated in Fig. 7.29 by means of a carbon fiber reinforced plastic beam. By the plies assembly and the utilization of anisotropic laminate features, properties such as stiffness can be adapted to the specific requirements of active elements.
- more efficient and robust active elements such as piezo composites that can be fully integrated into the structures in order to enable an effective coupling and adaptation of electrical and mechanical features.
- extreme miniaturizations of the electronic and computer technology that can then be integrated closely to the active components. Of course signal processing, artificial intelligence and efficient control strategies are also of great importance [106].

7.3.6 Sensory Structures

Concerning practical applications, especially the damage detection and the active limitation of damage in constructions with very long working life or complex maintenance are to be focused on. By means of SHM the integrity of the structure shall be monitored as a function of time via permanently embedded sensor networks. The SHM methods can be passive or active. Passive SHM methods determine the state of the structure by means of passive sensors that monitor signals versus time and feed them into a structure model. Passive SHM means just listening to the structure without influencing it. Active SHM however, requires scanning of the structure where necessary, in order to detect damage and estimate its dimensions. As permanent integrated piezo sensors for active SHM, work similarly to classic NDE techniques, it is also referred to as embedded non-destructive evaluation (e-NDE). This relatively new method makes it possible to transfer conventional ultrasonic techniques to embedded applications in active SHM systems [103].

In adaptive and sensory structures passive piezoelectric sensors are applied to measure the conditions (e. g. strain, mechanical stress, speed, acceleration, frequency) and the properties (e. g. stiffness, damping and eigenmodes) of the structure. This is a way to determine the deviation of the requested actuating variable and to initiate an appropriate feedback to the actuators.

Passive piezoelectric sensors can also be applied to detect cracking and crack growth via acoustic emission or to identify damage proceedings via impact detection. Active piezoelectric sensors in SHM structures are suited for the detection of remote damage via pulse-echo, pulse-transmission and phased-array methods or for the identification of damages nearby the sensors via high-frequency electromechanical impedance methods.

Due to their high costs, weight and size conventional NDE ultrasonic transducers are not suited for active SHM applications. However, novel 1–3 composite transducers such as piezo fiber composites (PFC) are adapted quite well to SHM applications. They can be used for passive as well as for active methods i. e. as transmitters and receivers for ultrasonic waves. PFCs are small, lightweight and extremely robust and thus can be structurally embedded for e-NDE purposes in a large number.

7.3.7 Adaptive Structures

Two conditions have to be fulfilled for the production of adaptive structures with embedded piezo sensors: a safe contacting method of the piezo elements and force-locked coupling to the structure. In general the question arises if the sensors are to be embedded into the components or to be applied at their surface. The mounting of sensors on the surface has advantages concerning the better accessibility for application and maintenance. The disadvantages are the lack of protection against damaging and the dependence of the performance on the surface condition.

Embedded sensors are relatively inaccessible for inspection but better protected and an interconnection with other sensors can be implemented easier. Integrated piezo sensors must have a Youngs modulus comparable to the base structure to avoid structural discontinuities. The Curie temperature should be higher than the curing temperature of the base component. Furthermore piezoelectric sensors should be electrically isolated from the base structure. The isolation must not reduce the force-lock between the sensors and the structure.

To embed electronic circuits into the structure they have to be isolated electrically, cooled if necessary and isolated mechanically from the load paths.

For the optimal performance of a multilayer structure (e. g. FRP) it is important to minimize the number of ply interruptions. Presently applied piezo ceramics allow a maximum strain of about 1000 ppm and exhibit a stiffness in the range of 50 ... 90 GPa. In order to design an adaptive structure with integrated sensors comprehensive tests are necessary to cover operating conditions and service strengths such as mechanical stresses, strains, temperatures and voltages.

The manufacturing of adaptive structures requires new production methods that need a lot of experience and expertise for the fabrication of complex systems with embedded or applied actuators and sensors. Patch-like sensors such as 1–3 piezo fiber composites are quite appropriate to be attached to the surface or to be integrated into lightweight multilayer constructions. The bond layer between the piezo element and the base structure determines the transfer behaviour of strain, vibrations and acoustic waves from the actuator to the structure and from the structure to the sensor. Local stress distributions can strongly be influenced by the bonding technique.

Up to now piezoelectric composite transducers (e. g. AFC, MFC and PFC) are not widespread in adaptive structures. There exist many publications about the application of piezoelectric ceramic plates or wafers in structures but there is rarely something said about realistic application conditions or the load capacity of such devices. With regard to that piezo composite transducers promise to be robust alternatives to bulk ceramic devices.

In most studies about adaptronic systems piezoceramic wafers have been applied. Piezoceramics, however, are very brittle and the effect of dynamic loads to the piezoceramic is therefore an important issue. The knowledge of the mechanical and electromechanical fatigue behaviour of the ceramic sensors under dynamic loads is thus a basic requirement for the reliable design of such components and should be examined systematically. Most of the applications described in literature use the d_{31} and d_{33} effect. The d_{15} effect (shear mode) is also mentioned. An essential handicap for the latter, however, is the high electrical voltage needed to receive a significant effect. Furthermore the manufacturing of the transducers is quite complex as polarization and operation can not be performed with the same electrode arrangement.

For a direct integration of the sensor elements in the structure (or at least of an outline compliant application to the structure) piezoelectric composites or polymers (e. g. PVDF) are particularly appropriate.

Essential requirements for the embedding capability of sensors are: compatible surfaces (i. e. chemically and physically adhesive), high interlaminar shear strength, electric isolation and an appropriate electromagnetic shielding (especially for the integration in C-FRP due to the fluctuating capacitance between the plies). Especially for the passive sensor operation the influence from the bond layer to the transfer behaviour should be ascertainable. This is essential especially for the utilization of bond layers (e. g. elastic adhesives) that exhibit a frequency depending damping characteristic.

References

1. Brignell, J.E.: *Quo vadis smart sensors?* Sens. Actuators, A37–38 (1993), pp. 6–8
2. Brignell, J.E.; White, N.M.: *Intelligent sensor systems.* Inst. Phys. Publishing, Bristol (1994)
3. Culler, D.; Estrin, D.; Srivastava, M.: *Overview of sensor networks.* IEEE Computer, 37(8) (2004), pp. 41–49
4. Brignell, J.E.: *Digital compensation of sensors.* J. Phys. E: Sci. Instrum., 10 (1987), pp. 1097–1102
5. Brignell, J.E.: *Sensors in distributed instrumentation systems.* Sens. Actuators, 10 (1986), pp. 249–261
6. Brignell, J.E.; White, N.M. and Cranny, A.W.J.: *Sensor applications of thick-film technology.* IEE Proc. I, 135, (4) (1988), pp. 77–84
7. White, N.M.; Brignell, J.E.: *A planar, thick-film load cell.* Sens. Actuators, 26, (1/3) (1991), pp. 313–319
8. White, N.M. and Brignell, J.E.: *Excitation of thick-film resonant structures.* IEE Proc.- Sci. Meas. Technol., 142, (3) (1995), pp. 244–248
9. Mark, J.; Hufnagel, P.: *The IEEE 1451.4 Standard for Smart Transducers* (2004), available online from http://standards.ieee.org/regauth/1451/IEEE_1451d4_Standard_Genl_Tutorial _090104.doc
10. Dunia, R.; Qin, S.J.; Edgar, T.F.; McAvoy, T.J.: *Identification of faulty sensors using principal component analysis.* AIChE J., 42 (1996), pp. 2797–2812
11. Markou, M.; Singh, S.: *A Review, Part I: Statistical Approaches, Signal Processing.* Signal Processing, 83(12) (2003), pp. 2481–2497
12. Gertler, J.: *Model based fault diagnosis.* Control-theory and advanced technology, 9 (1) (1993), pp. 259–285
13. Chen, S.; Wang, X.X.; Brown, D.J.: *Orthogonal least squares regression with tunable kernels.* Electronics Letters, 41(8) (2005), pp. 484–486
14. Taner, A.H.; Brignell, J.E.: *Aspects of intelligent sensor reconfiguration.* Sens. Actuators, A46–47 (1995), pp. 525–529
15. Taner, A.H.; Brignell: *The role of the graphical user interface in the development of intelligent sensor systems.* Man-machine interfaces for instrumentation, IEE Digest No: 1995/175, 3/1–3/6 (1995)
16. Taner, A.H.; Brignell, J.E.: *Aspects of intelligent sensor reconfiguration.* Actuators, A46–47 (1995), pp. 525–529
17. Taner, A.H.; Brignell, J.E.: *A graphical user interface for intelligent sensor ASIC reconfiguration.* Proc. Sensors & their Applications VII, Dublin (10–13 Sept. 1995), pp. 365–369
18. Gardner, J.W.; Bartlett, P.N. (eds): *Sensors and sensory systems for an electronic nose.* Kluwer Academic, Dordrecht (1992)
19. Rabaey, J.M.; Ammer, M.J.; da Silva Jr., J.L.; Patel, D.; Roundy, S.: *PicoRadio supports ad hoc ultra-low power wireless networking.* IEEE Computer, 33(7) (2000), pp. 42–48
20. Glynne-Jones, P.; White, N.M.: *Self-powered systems, a review of energy sources.* Sensor Rev., 21 (2) (2001), pp. 91–97
21. Göpel, W. (ed.), et al.: *Optical-Fiber Sensors.* In: Sensors – A Comprehensive Survey, VCH (1992)
22. http://www.smartec.ch

23. Habel, W.R.: *Long-term monitoring of 4,500 kN rock anchors in the Eder gravity dam using fibre-optic sensors.* Proc. Int. Symp. Geotechnical Measurements and Modelling, Balkema, ISBN 90 5809 603 3 (2003), pp. 347–354
24. Staszewski, W.J.; Boller, C.; Tomlinson, G.R.: *Health Monitoring of Aerospace Structures – Smear Sensor Technologies and Signal Processing.* Wiley (2004)
25. Frazao, O. et al.: *Strain and Temperature Discrimination using a Hi-Bi Grating partially exposed to Chemical Etching.* 17th Intern. Conf. on Optical Fibre Sensors (OFS-17), Bruges, Belgium, SPIE-Vol. 5855 (2005), pp. 755–758
26. http://www.fiso.com
27. Measures, R.M.: *Structural Monitoring with fiber optic technology.* Chapter 9.5, Academic. (2001)
28. Molter, M.; Hegger, J.; Habel, W.R. et al.: *Characterization of Bond Performance of Textiles in Cement-Matrices Using Fiber-Optic Sensors.* Int. Conf. on Smart Struct. and Mater. 2002. SPIE-Vol. 4694 (2002), pp. 253–258
29. Meltz, G.; et al.: *Formation of Bragg gratings in optical fibers by a transverse holographic method.* Optics Lett. 14(1989)15, pp. 823–825
30. Lebid, S.; Habel, W.R. and Daum, W.: *How to reliably measure composite-embedded fibre Bragg grating sensors influenced by transverse and point-wise deformations* Meas. Sci. Technol. 15(2004)8, pp. 1441–1447
31. Yun-Jiang Rao: *Long-Period Fiber Gratings for Low-cost Sensing.* 17th Int. Conf. on Optical Fibre Sensors (OFS-17), Bruges, Belgium, SPIE-Vol. 5855 (2005), pp. 13–16
32. James, S.W. and Tatam, R.P.: *Optical fibre long-period grating sensors: characteristics and application.* Meas. Sci. Technol. 14(2003), pp. R49-R61
33. Farahi, F.: *Simultaneous Measurement of Strain and Temperature Using Fiber Grating Sensors.* Proc. 11th Eng. Mechanics Conf. Fort Lauderdale, Conf. vol. 1 (1996), 351–354
34. Liu, T. et al.: *Simultaneous Strain and Temperature Measurement Using a Combine Fibre Bragg Grating/Extrinsic Fabry-Perot Sensor.* 12th Int. Conf. on Optical Fiber Sensors. Williamsburg, USA (1997), pp. 20–23
35. Habel, W. R.; et al.: *Deformation measurements of mortars at early ages and of large concrete components on site by means of embedded fiber optic microstrain sensors.* Cement & Concrete Composites 19(1997)1, pp. 81–102
36. http://www.micronoptics.com
37. http://www.jenoptik.com
38. http://www.fos-s.be
39. López-Higuera, J.M. (Ed.): *Handbook of Optical fibre Sensing technology.* Wiley (2002)
40. Lloyd, G.D. et al.: *Re-configurable, multi-channel, high-speed FBG strain sensing system for vibration analysis in oil risers.* 17th Intern. Conf. on Optical Fibre Sensors (OFS-17), Bruges, Belgium, SPIE-Vol. 5855 (2005) pp. 218–221
41. http://www.aos-fiber.com
42. http://www.broptics.com
43. http://www.quasys.ch
44. http://www.luxtron.com/product/fluoroptic_thermometry.htm
45. Labs, J.; Rose, K. and Werner, S.: *Kopplung von optischen Komponenten.* Me 7 (1993)1, Fachbeilage Mikrosystemtechnik (1993), pp. IV-V

46. Farahi, F.: *Fiber Optic Sensors for Heat Transfer Studies.* SPIE-Vol. 1584, pp. 53–61
47. Wang, A., et al.: *Sapphire optical fiber-based interferometer for high temperature environmental applications.* Smart Mater. & Struct. 4(1995), pp. 147–151
48. Yibing Zhangö, et al.: *Single-crystal sapphire-based optical high-temperature sensor for harsh environments.* Opt. Eng. 43 (2004)1, pp. 157–164
49. Grosswig, S., et al.: *Pipeline leakage detection using distributed fibre optical temperature sensing.* 17th Int. Conf. on Optical Fibre Sensors (OFS-17), Bruges, Belgium, SPIE-Vol. 5855 (2005), pp. 226–229
50. Inaudi, D.; Glisic, B: *Development of distributed strain and temperature sensing cables.* 17th Int. Conf. on Optical Fiber Sensors (OFS-17). SPIE-Vol. 5855 (2005), pp. 222–225
51. Yeo, T.L., et al.: *Fibre-Optic Sensor for the Monitoring of Moisture Ingress and Porosity of Concrete.* 17th Int. Conf. on Optical Fibre Sensors (OFS-17), Bruges, Belgium, SPIE-Vol. 5855 (2005), pp. 491–494
52. Kunzler, W.; Calvert, S. and Laylor, M.: *Implementing fiber optic sensors to monitor humidity and moisture.* Int. Conf. on Smart Struct. and Mater. 2004. SPIE-Vol. 5384 (2004), pp. 54–63
53. Kronenberg, P.; Rastogi, P.K.: *Relative humidity sensor with optical fiber Bragg gratings.* Optics Lett. 27(2002)16, pp. 1385–1387
54. Khay Ming Tan, et al.: *High relative humidity sensing using gelatin-coated long period grating.* 17th Int. Conf. on Optical Fiber Sensors. SPIE-Vol. 5855 (2005), pp. 375–378
55. Dantan, N.; Habel, W.R.; Wolfbeis, O.S.: *Fiber optic pH sensor for early detection of danger of corrosion in steel-reinforced concrete structures.* Int. Conf. on Smart Struct. and Mater. 2005. SPIE-Vol. 5758 (2005), pp. 274–284
56. Bürck, J.M., et al.: *Distributed fiber optical HC leakage and pH sensing techniques for implementation into smart structures.* Int. Conf. on Smart Struct. and Mater. 2004. SPIE-Vol. 5384 (2004), pp. 1–12
57. Habel, W.R.; Krebber, K. et al.: *Fibre Bragg Grating Sensors to Monitor the Rotor Blades of Wind Turbines – Criteria and Method to put them to the Best Possible Use.* 7th German Wind Energy Conf. DEWEK 2004 (CD-ROM), Wilhelmshaven (2004)
58. Habel, W.R. and Bismarck, A.: *Optimization of the adhesion of fiber-optic strain sensors embedded in cement matrices; a study into long-term fiber strength.* J. Structural Control 7(2000)1, pp. 51–76
59. Berghmans, F.: *Reliability of Components for Fiber Optic Sensors.* Int. Conf. on Smart Struct. and Mater. 2005. SPIE-Vol. 5758 (2005), pp. 417–426
60. Habel, W.R.: *Fiber optic sensors for deformation measurement: criteria and method to put them to the best possible use.* Int. Conf. on Smart Struct. and Mater. 2004. SPIE-Vol. 5384 (2004), pp. 158–168
61. Habel, W.R.: *Stability and Reliability of fiber-optic Measurement Systems – Basic Conditions for Successful Long-Term Structural Health Monitoring.* In: F. Ansari (Ed.): Sensing Issues in Civil Structural Health Monitoring. Springer (2005), pp. 341–351
62. Culshaw, B.; Habel, W.R.: *Fibre sensing: Specifying components and systems.* Symp. on Optical Fiber Measurements SOFM 2004, Session X: Fiber Bragg gratings and fiber sensors. Boulder, Colorado, USA, September 28–30 (2004)
63. DIN EN ISO/IEC 17025:2000 (trilingual version): *General requirements for the competence of testing and calibration laboratories*

64. Internal information from Dr. Krebber and Dr. Trappe, Federal Institute for Materials Research and Testing (BAM) Berlin, (Winter 2005)
65. Krebber, K.; Habel, W.R., et al.: *Fiber Bragg grating sensors for monitoring of wind turbine blades.* 17th Int. Conf. on Optical Fibre Sensors (OFS-17), Bruges, Belgium, SPIE-Vol. 5855 (2005), pp. 1036–1039
66. Park, H.S.; Thursby, G. and Culshaw, B.: *High-frequency acoustic detector based on fiber Fabry-Perot interferometer.* 2nd Europ. Workshop on Optical Fibre Sensors, Santander, Spain 2004. SPIE-Vol. 5502 (2004), pp. 213–216
67. Habel, W.R.; Hofmann, D., et al.: *High-performance Concrete – Lime Optimization with Fiber Optic Sensors.* 14th Eng. Mechanics Conf. (CD-ROM). Austin, Texas/US (2000)
68. Hillemeier, B.; Scheel, H.; Habel, W.R.: *Enhancing Durability of Structures by Monitoring Strain and Cracking Behavior.* In: F. Ansari (Ed.): Sensing Issues in Civil Structural Health Monitoring. Springer (2005). pp. 155–164
69. Pinet, E.; Pham, A. and Rioux, S.: *Miniature Fiber Optic Pressure Sensor for Medical Applications: an Opportunity for Intra-Aortic Balloon Pumping (IABP) therapy.* 17th Int. Conf. on Optical Fibre Sensors (OFS-17), Bruges, Belgium, SPIE-Vol. 5855 (2005), pp. 234–237
70. http://www.samba.se
71. Glötzl, R.; Hofmann, D.; Basedau, F.; Habel, W.R.: *Geotechnical Pressure Cell Using a Long-Term Reliable High-Precision Fibre Optic Sensor Head.* Int. Conf. on Smart Struct. and Mater. 2005. SPIE-Vol. 5758 (2005), pp. 248–253
72. Roger, A.J.; Shatalin, S.V. and Kanellopoulos, S.E.: *Distributed Measurement of Flow Pressure via Optical-fibre Backscatter Polarimetry.* 17th Int. Conf. on Optical Fibre Sensors (OFS-17), Bruges, Belgium, SPIE-Vol. 5855 (2005), pp. 230–233
73. Campbell, M.; et al.: *Optimisation of Hi-birefringence Fibre Based Distributed Force Sensors.* Smart Struct.: Optical Instrumentation and Sensing Systems Conf. 1995, SPIE-vol. 2509 (1995), pp. 57–63
74. *Installation, use and repair of fibre optic sensors.* Design manual. ISIS-M02-00, Canada, 2001
75. http://www.cost270.com
76. http://cost.cordis.lu/src/list_of_mc.cfm
77. http://www.ishmii.org/
78. Kalymnios, D.: *Plastic Optical Fibres (POF) in sensing – current status and prospects.* 17th Int. Conf. on Optical Fibre Sensors (OFS-17), Bruges, Belgium, SPIE-Vol. 5855 (2005), pp. 1–4
79. Pagnoux, D. et al.: *Microstructured fibers for sensing applications.* 17th Int. Conf. on Optical Fibre Sensors (OFS-17), Bruges, Belgium, SPIE-Vol. 5855 (2005), pp. 5–8
80. Wehrspohn, R.B., et al.: *Photonic crystal gas sensors.* 17th Int. Conf. on Optical Fibre Sensors (OFS-17), Bruges, Belgium, SPIE-Vol. 5855 (2005), pp. 24–29
81. Lehmann, H., et al.: *Toward photonic crystal fiber based distributed chemosensors.* 17th Int. Conf. on Optical Fibre Sensors (OFS-17), Bruges, Belgium, SPIE-Vol. 5855 (2005), pp. 419–422
82. Bjarklev, A., et al.: *Photonic crystal structures in sensing technology.* 2nd Europ. Workshop on Optical Fibre Sensors, Santander, Spain, SPIE-Vol. 5502 (2004), pp. 9–16

83. Zhi Wang, et al.: *Properties of PCF-based long period gratings.* 17th Int. Conf. on Optical Fibre Sensors (OFS-17), Bruges, Belgium, SPIE-Vol. 5855 (2005), pp. 298–301
84. Sirohi, J.; Chopra, I.: *Fundamental understanding of piezoelectric strain sensors.* Proc. SPIE – Int. Soc. Optical Eng., Vol. 3668 (1999), pp. 528–542
85. Tressler, J.F.; Alkoy, S. and Newnham, R.E.: *Piezoelectric sensors and sensor materials.* J. Electroceramics, 2 (4) (1998), pp. 257–272
86. Newnham, R.E.; Skinner, D.P. and Cross, L.E.: *Connectivity and Piezoelectric-Pyroelectric Composites.* Mat. Res. Bull. 13 (1978), pp. 525–536
87. Petricevic, R.; Gurka, M.: *High performance piezoelectric composites.* European Space Agency, Special Publication, ESA SP (2005), pp. 763–767
88. Wilkie, W.K.; Bryant, R.G.; High, J.W.; et al.: *Low-cost piezocomposite actuator for structural control applications.* Proc. SPIE – Int. Soc. Opt. Eng., Vol. 3991 (2000), pp. 323–334
89. Janos, B.Z.; Hagood, N.W.: *Overview of Active Fiber Composites Technologies.* Proc. 6th Int. Conf. New Actuators, Bremen, Germany (17–19 June 1998), pp. 193–197
90. Jaffe, B.: *Piezoelectric Ceramics. Non-Metallic Solids.* Ed. J.P. Roberts. Vol. 3. 1971, The University , Leeds, England: Academic, London and New York (1971)
91. Ruschmeyer, K.; Koch, J.; Lubitz, K.; Schönecker, A.; Helke, G.; Petersen, A.; Möckel, T. and Riedel, M.: *Piezokeramik.* Expert Verlag (1995)
92. Xu, Y.: *Ferroelectric Materials and Their Applications.* University of California Los Angeles, CA, USA: Elsevier, North Holland (1991)
93. Hagood, N.W.; Bent, A.A.: *Development of piezoelectric fiber composites for structural actuation.* Collection of Technical Papers – AIAA/ASME Structures, Structural Dynamics and Mater. Conf. (1993), pp. 3625–3638
94. Bent, A.A.; Hagood, N.W. and Rodgers, J.P.: *Anisotropic actuation with piezoelectric fiber composites.* J. Intelligent Mater. Systems and Structures, 6(3) (1995), pp. 338–349
95. Bent, A.A.; Hagood, N.W.: *Improved performance in piezoelectric fiber composites using interdigitated electrodes.* Proc. SPIE – Int. Soc. Optical Eng., Vol. 2441 (1995), pp. 196–212
96. Bent, A.A.; Hagood, N.W.: *Piezoelectric fiber composites with interdigitated electrodes.* J. Intelligent Mater. Systems and Structures, 8(11) (1997), pp. 903–919
97. Williams, R.B.; Grimsley, B.W.; Inman, D.J.; Wilkie, W.K.: *Manufacturing and mechanics-based characterization of macro fiber composite actuators.* Amer. Soc. Mech. Engineers, Aerospace Division (Publication) AD, Vol. 67 (2002), pp. 79–89
98. Petricevic, R.; Gurka, M.: *Extremely Robust Piezoelectric Actuator Patches – Properties and Applications.* Proc. 10th Int. Conf. New Actuators, Bremen, Germany (14–16 June 2006), pp. 62–65
99. Trolier-Mckinstry, S. and Muralt, P.: *Thin film piezoelectrics for MEMS.* J. Electroceramics, 12(1–2) (2004), pp. 7–17
100. Chopra, I.: *Review of state of art of smart structures and integrated systems.* AIAA J., 40(11) (2002), pp. 2145–2187
101. Giurgiutiu, V.; Zagrai, A.N.: *Characterization of piezoelectric wafer active sensors.* J. Intell. Mater. Sys. and Structures, 11(12) (2000), pp. 959–976

102. Krautkramer, J.; Krautkramer, H.: *Ultrasonic Testing of Materials*. Springer-Verlag, Berlin (1990)
103. Giurgiutiu, V.; Cuc, A.: *Embedded non-destructive evaluation for structural health monitoring, damage detection, and failure prevention*. Shock and Vibration Digest, 37(2) (2005), pp. 83–105
104. Blitz, J.; Simpson, G.: *Ultrasonic Methods of Non-destructive Testing*. Springer-Verlag, Berlin (2005)
105. Giurgiutiu, V.; Zagrai, A.: *Damage detection in simulated aging-aircraft panels using the electro-mechanical impedance technique*. Amer. Soc. Mech. Engineers, Aerospace Division (Publication) AD, Vol. 60 (2000), pp. 349–358
106. Elspass, J.; Flemming, M.: *Aktive Funktionsbauweise*. Springer-Verlag, Berlin (1997)

8 Adaptronic Systems in Engineering

8.1 Adaptronic Systems in Aeronautics and Space Travel

C. Boller

8.1.1 Implications and Initiatives

Since the early days of adaptronics aeronautics and space travel has been a most significant player in driving adaptronics ahead. A large amount of work was done along concept studies which ended up in partially even showing hardware demonstration at full scale very recently. Adaptronics in aeronautics can be split into the following areas of activity:

- *Structural health monitoring (SHM)*: the integration of sensing and possibly even actuation devices for either monitoring the operational and/or damaging condition of aerostructures.
- *Shape control and active flow*: the mainly static deformation of space and aerodynamic structures to either improve communication performance of antennas or adapt structures to optimum aerodynamic fluid flow, both achieved by integrated actuation mechanisms.
- *Damping of vibration and noise*: passive and specifically active damping with respect to improving aeroelastic and flutter performance as well as reduction of noise generated through aerodynamics and/or engines, monitored by sensors and alleviated through actuator systems being integrated into structural components.
- *Smart skins*: load-carrying structural elements with integrated avionics (antennae), which can be either a sensory, active, adaptive or even intelligent structure, depending on its term of use.
- *Systems*: Mainly small air vehicles such as micro aerial vehicles (MAV), uninhabited aerial vehicles (UAV) or micro-satellites, which are less constrained with regard to design specifications and regulations and as such more flexible in terms of integrating advanced sensing and actuation technologies.

Compared to the mid 1990s, there has not been too much variation in the sensing and actuation principles to be used. With regard to sensing optical fibre and piezoelectric sensors are dominant, followed by MEMS, which however still rather plays a secondary role. On the actuation side piezoelectric and shape memory alloys (SMA) are dominant, followed by shape memory polymers and foams and possibly electroactive polymers as a further type of actuation material to emerge. Activities related to electrorheological and magnetorheological fluids have been comparatively small over the past and have more moved into ground-based damper applications.

After the initial aerospace related activities such as those in the USA initiated through NASA-JPL on space structures, NASA/Lockheed/Northrop-Grumman on adaptive wings, MURI programmes on adaptive helicopter rotor blades, and possibly others, or those in Europe mainly driven by BAE Systems, DASA (now EADS), DLR, Eurocopter or ONERA and funded by the EU Framework Programmes, defence related EUCLID programmes and national ministries. The last ten years has shown further aerospace related programmes on adaptronics to emerge and those not only in North America and Europe but also in Australia, India and Japan. In the USA the adaptive wing programme funded by DARPA/AFRL/NASA [1–4] has possibly had the most long lasting impact on adaptronics technology in aerospace and has achieved a complete adaptive wing system for UAVs realised in hardware and successfully tested in a wind tunnel. These activities also triggered further programmes such as NASAs Morphing Wing [5] or has been closely linked to DARPA funded programmes such as the Compact Hybrid Actuators Program (CHAP) and the Smart Aircraft and Marine Propulsion System demonstration (SAMPSON) [6] programme respectively. A similar continuation in the USA can be seen in the rotorcraft development where the DARPA-funded Smart Rotor Program or the Aeroelastic Rotor Experimental System (ARES) have been key to this sector. In structural health monitoring (SHM) different US-programmes were funded by the US Air Force and US Army with industry such as Boeing, Honeywell and BF Goodrich taking currently over through own evaluation initiatives.

A large number of topics generated through US initiatives have also been followed up in Europe. Activities on aeroelasticity were very much considered in the DASA/DLR Adaptive Wing Programme and was continued in EU funded programmes such as 3AS and ADAPT while the Eurocopter/DLR Adaptive Rotor Systems Programme AROSYS and the recently started EU funded programme SMARTCOPTER is looking into the rotordynamic, aeroelastic and noise aspects of rotorcraft. The defence related EUCLID programme VIBRANT focussed specifically on the aspect of using commercial-off-the-shelf (COTS) components and ways on how to support them with active damping to allow military specifications to be met with the lower cost COTS components. The SHM for aerospace applications has been funded through different EU (MONITOR, DAMASCOS) and EUCLID (AHMOS) programmes, which has found continuation in the recent EU programmes named TATEM and SMIST.

In Japan the New Energy Development Organisation (NEDO) launched a nationwide adaptronics related programme in 1998 [7], where the aerospace related projects have been looking into SHM of aircraft fuselages and space structures, cabin noise reduction in helicopters through piezoelectric actuators, and adaptive damping in deployable ultra-lightweight satellite antennas.

India has been running large adaptronics related conferences since a decade now [8, 9] with strong implications from the aeronautics side. Much of the effort has been driven from advanced sensing technology triggered by development in MEMS and optical fibre sensors with national programmes now ongoing. Analytical work has also been widely provided through concepts and systems in the actuation field using piezoelectric, magnetostrictive and/or shape memory alloys for the actuation of flaps, as shock mounts, for external store vibration control, as antennae or helicopter rotor blades.

Australias emphasis in aerospace related adaptronics technology is very much driven by the Defence Science and Technology Organisation (DSTO) and different universities around the country. The activities include F/A-18 aircraft fin buffet alleviation in collaboration with Canada, New Zealand and the USA as well as a variety of structural health monitoring projects ranging from monitoring cracks in metals to the integrity of bonded repairs [10]. This is further added by development of various types of sensors as well as an acousto-ultrasonic monitoring system and different technologies for corrosion monitoring [11, 12].

Trying to rate adaptronics in aerospace on the basis of NASAs Technology Readiness Level (TRL) brings it currently no much higher than to TRL 5 or 6 (component validation) on the 9 levels scale to be achieved when considering a system to be flight proven.

Adaptronics research in aeronautics is very much communicated through conferences, symposia and workshops as well as through scientific journals and recently also books. The largest forum is possibly SPIEs annual International Symposium on Smart Structures and Materials [13] followed by the International Conference on Adaptive Structures ICAST and CanSmart [14], also both held annually. The SPIE hosts a variety of further conferences mainly in Australia and India. The SHM is mostly discussed separately such as SPIEs annual Nondestructive Evaluation for Health Monitoring and Diagnostics Symposium [13], the International Workshop on SHM [15] and the European Workshop on SHM [16], the latter both taking place biannually in alternating sequence. There are some further events of a more national character such as AIAAs Structural Dynamics in the USA or the Adaptronic Congress in Germany where adaptronics in aerospace is discussed as well.

A variety of overview papers related to aerospace can be found in 'Smart Materials and Structures' published by the Institute of Physics Publ. since 1992 and the International Journal on Intelligent Material Systems and Structures published by SAGE Publications since 1990. Another important source is the AIAA-Journal. Besides different aerospace related articles spread over the different issues of Smart Materials and Structures, there have been published a number or aerospace related special issues being related to space [17,18] and rotorcraft [19,20] respectively. A number of different overview articles do also report on developments related to rotorcraft or other aerospace related issues [5, 21–23]. There is also a book on health monitoring of aircraft which largely encompasses the adaptronic aspect and which appeared in late 2003 [24].

Further details within the different areas mentioned above can be found in the subsequent paragraphs.

8.1.2 Structural Health Monitoring

Aircraft related structural health monitoring (SHM) has been widely described in [24]. SHM includes loads, condition and damage monitoring respectively. Loads monitoring of aircraft structures dates back to the 1950s. It is either based on monitoring acceleration in time domain multiplied by a structures mass or strain sequences monitored at locations representative for an aircrafts load. While the former uses accelerometers the latter can be satisfied by using conventional electrical strain gauges. In some cases such as the Eurofighter Typhoon [24] different flight parameters are recorded which are then fed into the aircrafts digital loads model that allows the actual loading sequence of the aircraft to be determined and stored in a specific load occurrence matrix used to determine the aircrafts fatigue life (Fig. 8.1).

Compared to the relatively slow development of fixed wing aircraft loads monitoring, helicopters are possibly a step ahead. Helicopters today are only sold with a health and usage monitoring system (HUMS), which is mainly

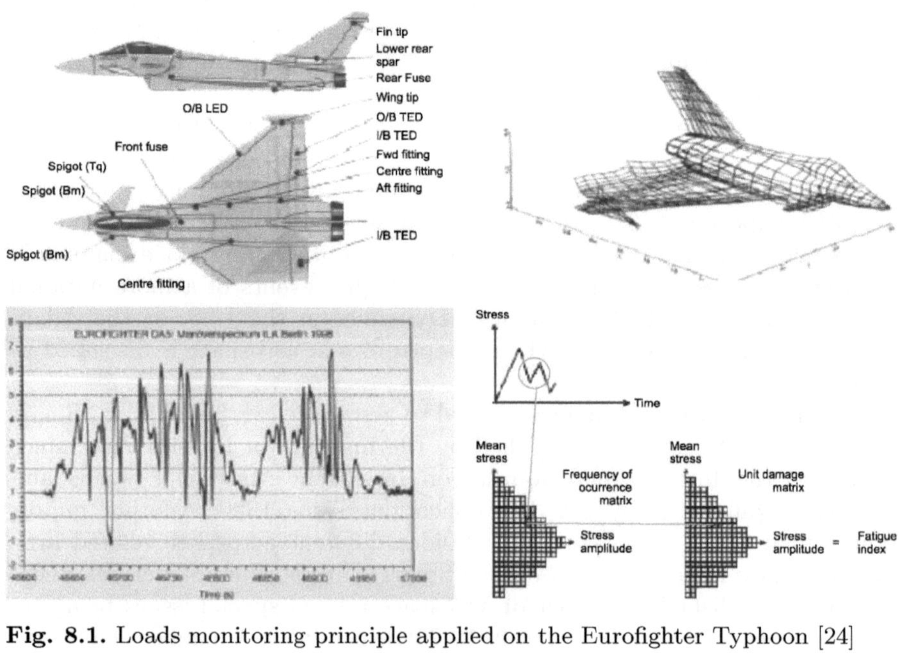

Fig. 8.1. Loads monitoring principle applied on the Eurofighter Typhoon [24]

based on monitoring vibrations generated in components such as gears, shafts and couplings, bearings, rotors, and any further components critical for flight performance [25]. Jet engines are monitored for decades as well. Systems used in that regard include the full authority digital control (FADEC), the remote data concentrator (RDC), the engines spool speed, engine distress monitoring system (EDMS) and the ingested debris monitoring system (IDMS) respectively [26].

Loads monitoring however does not have to be limited to mechanical loads only. It can also include any other environmental loads such as resulting from temperature, humidity or other corrosive gases and liquids that may lead to deterioration of the structural components. Monitoring these phenomena, possibly combined with monitoring classical mechanical loads is certainly a challenge where sensing in the adaptronics sense offers some promising opportunities. Optical fibre Bragg grating sensors is one of the options where sensors do not only allow to measure mechanical strain but can also be applied to monitor temperature and pressure at the same time (see Sect. 7.2). Different studies (i. e. [27,28]) have shown that these sensors can be elegantly integrated into metallic as well as composite full scale aircraft structures and that virtually hundreds of these sensors can be aligned along a single optical fibre. This enormously reduces complexity of the sensing system when compared to the amount of wiring being required with conventional electrical strain gauges. Another potential type of sensor with multi-parameter monitoring capability is MEMS which virtually allows to be configured to monitor any of the loading parameters mentioned above. MEMS has been specifically suitable for monitoring accelerations, humidity and/or different types of gases and has as such been targeted in monitoring corrosive effects [29]. Boeing [30,31] has currently longer lasting tests ongoing where optical fibre and MEMS sensors are positioned in a Boeing 767-300ER and 737-800 aircraft, both with different airlines and in different continents for monitoring different aspects of aircraft and structural performance regarding loads, moisture and corrosion. Following a cost benefit study the locations of the sensors have been well selected with an example of those locations shown in Fig. 8.2.

Besides optical fibre and MEMS sensors there is also a variety of other sensor types emerging for monitoring aircraft structural components. In the context of acoustic monitoring in the ultrasonic range piezoelectric sensors is possibly the type being most discussed (see Sect. 7.3). Smart Layer, Smart Suitcase and ACESS software from Acellent [32] is a technology being currently closest to a product where aerospace related applications have been specifically reported in [33]. Alternative suggestions based on eddy current technology have been made by Jentek [34] where a magnetic field is generated via a number of conductive metallic windings and recorded by another number of windings all being configured as shape-field sensors on a carrier. The system has been proven to work on cracked aluminium panels, around

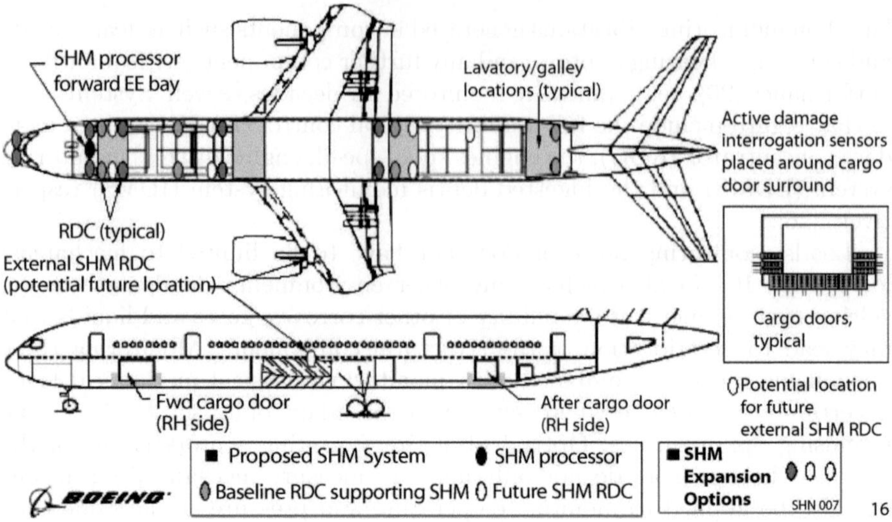

Fig. 8.2. Locations monitored using optical fibre and MEMS sensors on a Boeing 767-300ER during a long term in service performance test [36]

rivet holes and on a flight deck chine plate of a C-130 military transport aircraft. Further to this there have also been suggestions to combine acousto-ultrasonics with eddy current methods in a sensor layer to be applied in composites [35].

Benefits from integrating SHM into current aerostructures have mainly to be seen in the automation of the monitoring process, improvement of aircraft availability and operability and as a consequence in saving cost. Schmidt et al. [37] brought in a new aspect of enhancing damage tolerance through SHM. The opportunity of detecting cracks with SHM much earlier than with conventional means allows maintenance intervals to be extended or allowable stresses to be increased which in the latter case leads to lighter weight. Weight savings of up to 20% per component have been considered likely.

Since Boeing has now committed itself to SHM with the Boeing 787 there is hardly any way back anymore. Adaptronic technologies currently under development will therefore have to demonstrate that they can be operated reliably under real in-service operational conditions. Sensor signal processing, which still shows significant potential for further improvement and where methods and options are described in [24] will help to handle large numbers of sensors and extract the appropriate sensor signal information accordingly (see Sect. 7.1). Another challenge is the proper integration of SHM into the current maintenance process of aircraft such that it leads to the requested savings in direct operating cost. Since aircraft maintenance processes are highly complex it is not very likely that these will be significantly altered due to SHM. An adequate adaptation of SHM to these processes is therefore an essential need with options currently explored within the EU funded integrated project entitled TATEM. In North-America SHM is now finding its way into the traditional activities such as the Aircraft Structural Integrity Program (ASIP) or the different initiatives on ageing aircraft and maintenance, repair and overhaul (MRO). Japan has just completed a major large scale composite aircraft fuselage structure where a variety of optical fibre sensors have been integrated for monitoring impact loads, delaminations and damage propagation in general [38].

8.1.3 Shape Control and Active Flow

Lift and drag as well as the velocity of the airflow and the ambient gas temperature and pressure are very sensitive functions of an aerofoil. Much care is therefore spent on the design and especially the camber of these aerofoils, where drag should always be kept at a minimum. As long as the camber is kept constant, this minimum of drag is only possible for a specific amount of lift. Whenever more lift is required (e. g. for a manoeuvre), drag may increase significantly. Figure 8.3 shows a selection of lift-drag relations for different cambers.

A minimum in any lift-drag relationship of an aerofoil is achieved when camber can be varied according to different operating conditions. This is what birds and insects do permanently and what is done with conventional fixed wing aircraft through a partially complex flap system possibly combined with aeroelastic tailoring. A comparison of shape adaptation in nature and for an artificial flapping mechanism is shown in Fig. 8.4. A major motivation for adaptronics in that regard has therefore been to explore how far sensing and actuation options could be integrated in the wing to achieve a more adaptable, less complex, lighter weight and possibly even more elegant solution.

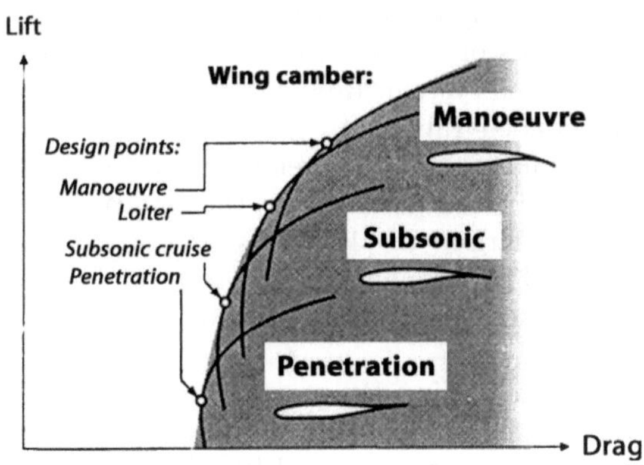

Fig. 8.3. Lift-drag relationship for different cambers

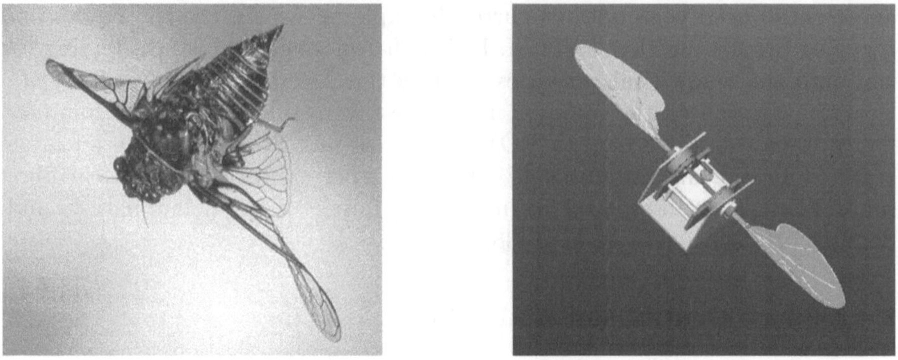

Fig. 8.4. Adaptation of wings in nature and engineering (Source: R. Zbikowski, Cranfield/UK)

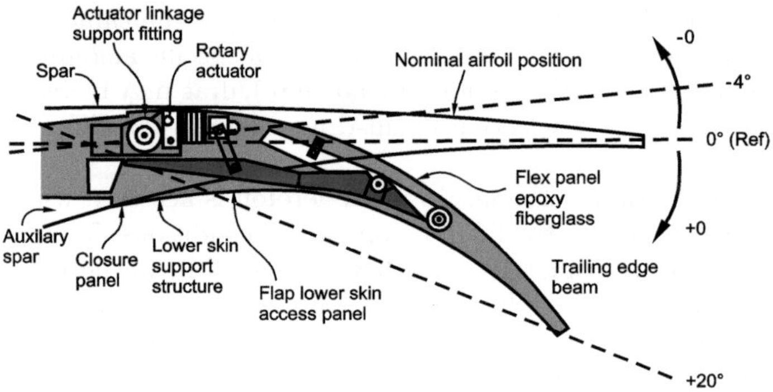

Fig. 8.5. Mission adaptable wing [39]

Early work trying to improve wing performance through adaptation was done on the basis of what was entitled the mission-adaptable wing [39] shown on Fig. 8.5. This wing that contained a fully elastic surface coating could be bent in accordance to the different needs of camber using a rotary actuator and a conventional mechanical leveraging system. Rewards with such a system can be seen in higher aircraft manoeuvrability.

The early steps of adaptive aerodynamic profile development being more based on adaptronics have been run in the USA under the Smart Wing Programme and have been well summarised in [40]. This programme together with further programmes related to aeronautical structures and adaptronics was then merged under the umbrella of the Morphing Wing Program [5].

The ideas generated under the Smart Wing Programme included rods, antagonistic filaments and honeycomb sandwiches made of or including shape memory alloys. The different concepts of what was called Phase I and which are shown in Fig. 8.6 were realised in hardware on a 1/16 downscaled model of a fighter aircraft wing. Validation was done in the wind tunnel of NASA

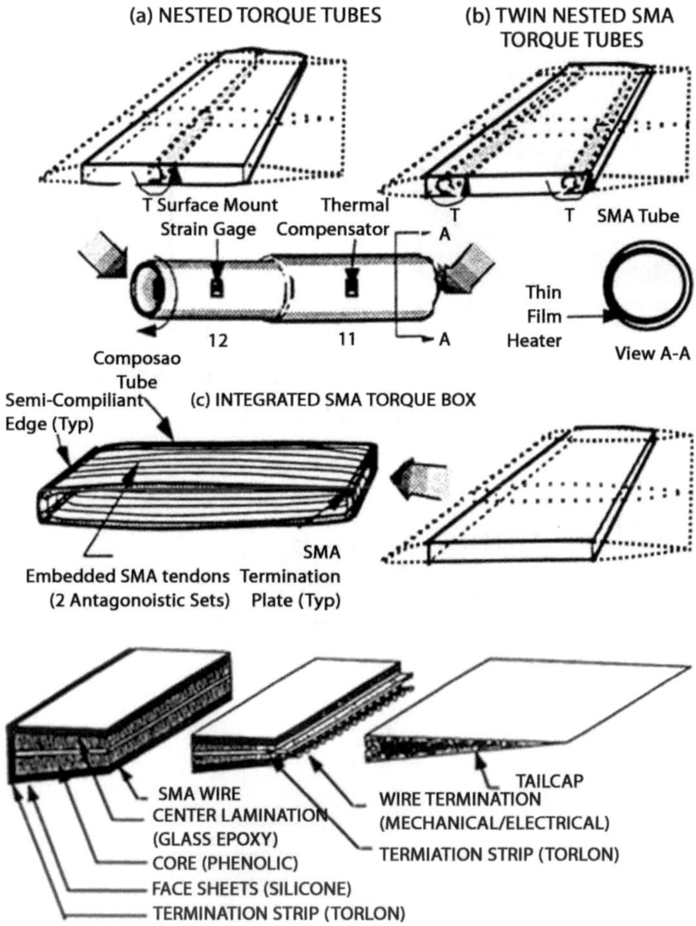

Fig. 8.6. Different concepts of SMA actuated aerodynamic profiles [41]

Langley and allowed to show that the expected improvements in performance could be achieved. A summary of the Phase I achievements has been given in detail in [41–44].

Findings in Phase I were then used to design a 30% full span model of an uninhabited combat air vehicle (UCAV) in Phase II as shown in Fig. 8.7 [45].

The model consists of two different types of wings. The conventional side has control surfaces actuated through electrical motors while the smart side is actuated through the SMA actuation concepts developed under Phase I (Test 1) and a new accentuator concept (Test 2). The smart trailing edge is configured as a hinge-less control surface that consists of segmented actuation elements which are then covered by a silicone skin, a center laminated backbone and a flexible honeycomb core as shown in Fig. 8.8. The two adaptronic actuation concepts, the one being based on SMA wire actuators and

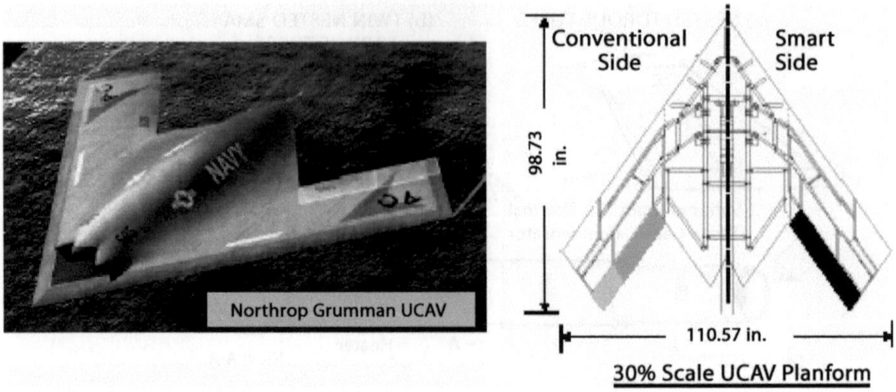

Fig. 8.7. Northrop-Grummans Smart Wing Model Phase II [45]

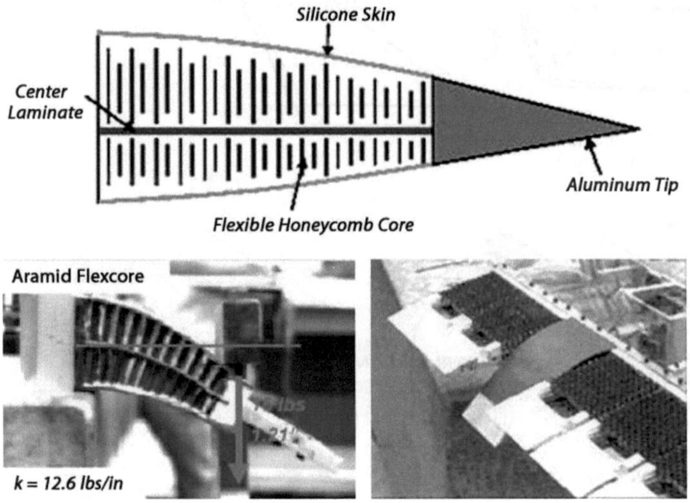

Fig. 8.8. Hingeless control surface segment design principle [45]

the other on eccentuators driven by ultrasonic travelling wave motors are both shown in Fig. 8.9. Each of the concepts was then wind tunnel tested in air and heavy gas at flight Mach numbers 0.8 and dynamic pressures of 300 psf (= 1464 kg/m^2). The hinge-less control surface concept has been able to demonstrate a 17% improvement in rolling moment coefficients at 15 degrees of control surface deflection when compared to the conventional control surface solution [46].

Another remarkable DARPA funded programme driven partially through the CHAP programme on hybrid actuators has been SAMPSON [6] which made the engine air inlet duct of a F-15 fighter adaptive. This improves aerodynamic performance, air intake and thus engine performance in dependence to different flight conditions. The modification consisted of four components allowing the cowl to rotate, the lip to deflect, the air intake wall to deflect

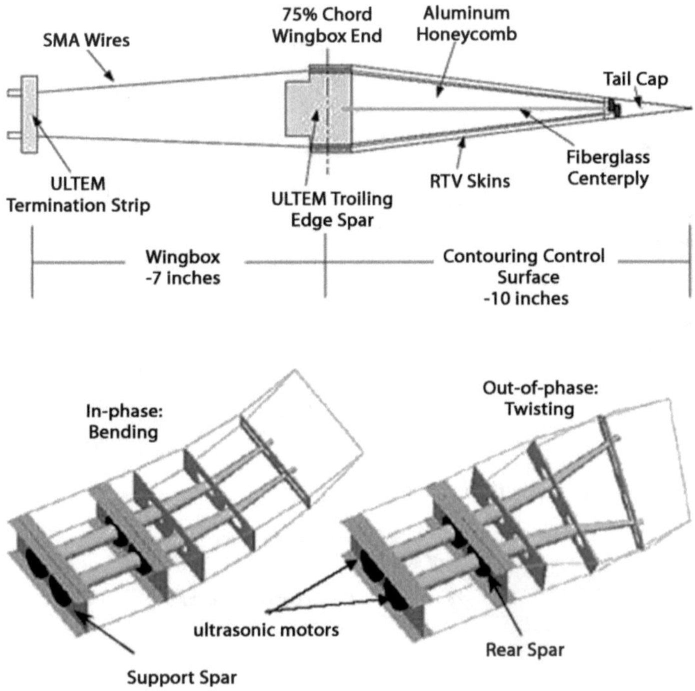

Fig. 8.9. Actuation principles applied for each hinge-less control surface segment: SMA wire based (*left*) accentuator and ultrasonic motor based (*right*) [45]

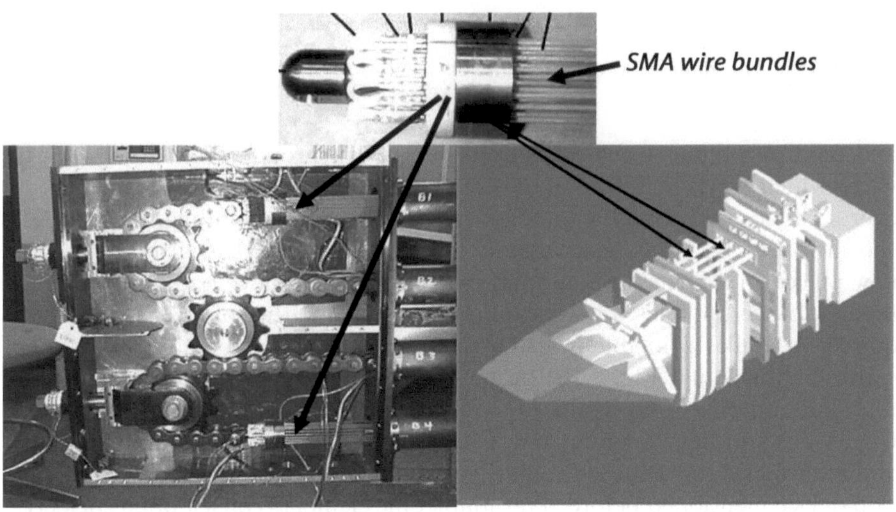

Fig. 8.10. Cowling deflection actuation system for jet engine air inlet [6]

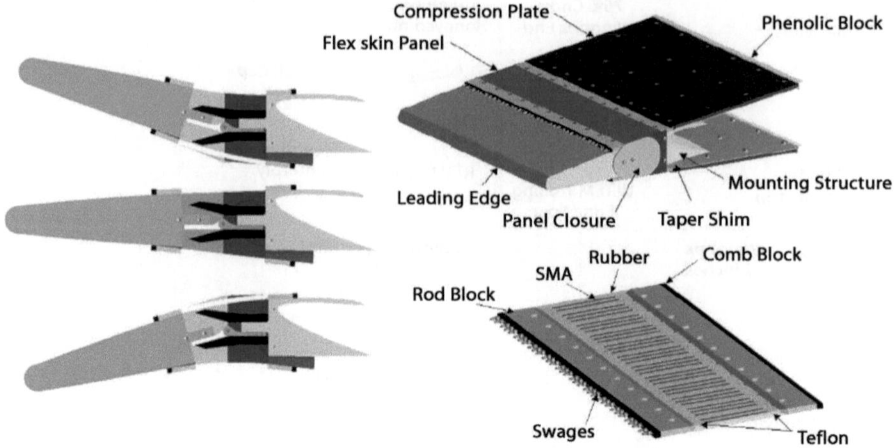

Fig. 8.11. Air intake duct lip deflection mechanism [6]

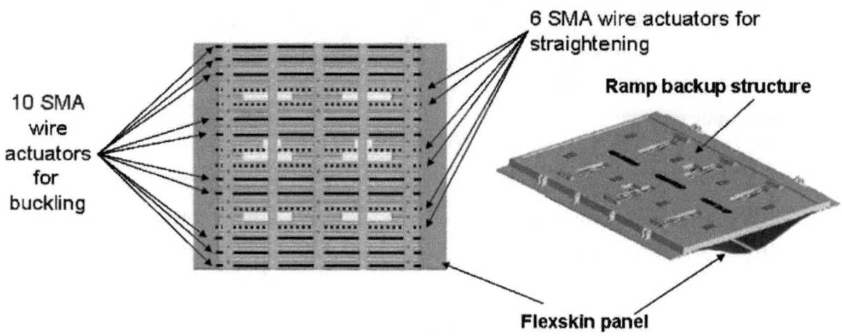

Fig. 8.12. Wall deflection component [6]

and the lip to blunt. Cowl deflection is achieved through bundles of 60 SMA wires which work in an antagonistic fashion and actuated a chain moving over a slider and around a sprocket (Fig. 8.10). The air intake lip deflection shown in Fig. 8.11 has been again actuated through SMA wires which were now placed in a flex skin panel that was positioned above and below a hinge linking to the lip. This allows the lip to deflect in either of the directions required. A similar principle is also used to introduce a wall deflection through controlled buckling (Fig. 8.12). Again SMA wires were integrated into a flexible skin and butted up against the ends which allow the skin to buckle once the SMA wires have been heated. Finally a lip blunting device was designed and realised (Fig. 8.13) that could be used in combination with the lip deflection device mentioned before. The adaptronic elements consist of piezoelectric materials that are used in a motor to rotate a shaft. Superelastic SMA was used to cover the hinged surface at the lip tip. The system was finally tested in full scale in the 16 foot transonic wind tunnel at NASA Langley.

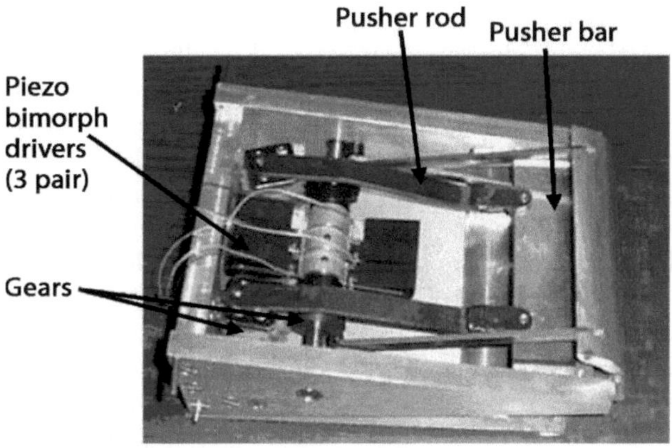

Fig. 8.13. Lip blunting component [6]

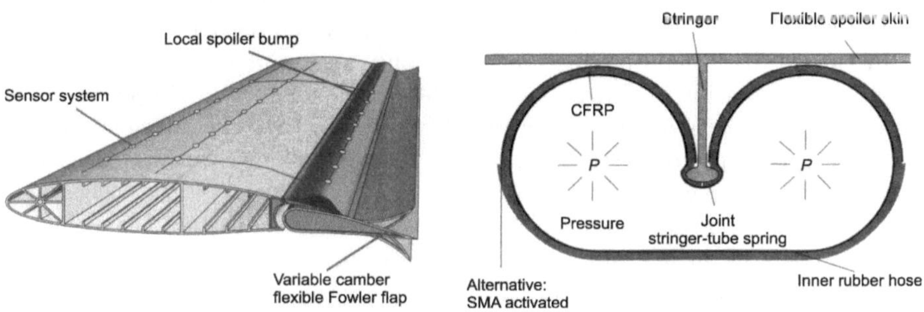

Fig. 8.14. Adaptive bump for the control of transonic shock waves [48]

Another concept considered for noise reduction in jet engines for civil aircraft are aerodynamic devices called chevrons, which are placed along the trailing edges of a jet engine primary and secondary exhaust nozzle. These chevrons need to be fine tuned between noise-benefit and thrust-loss, where a solution has been proposed by using SMA actuators to position the chevron in accordance to the on-ground and cruise level temperatures [47].

Active flow control and separation has been a longstanding issue within aero-structural design of large transport aircraft wings. Two concepts have mainly emerged within the context of adaptronics. The one being a more mechanistic one has been proposed among others in [48]. The idea is to generate a bump (Fig. 8.14) at the location where the supersonic flow on an aerodynamic profile is usually generated and to move the point of flow separation further down the profile, resulting in increased lift. This is specifically useful under those conditions where high lift is required, such as the starting and landing phase of an aircraft. Solutions for generating these bumps have been made such as a twin elliptical rubber hose either pressurised or activated through shape memory alloys.

A more fluid dynamics driven alternative solution is in delaying turbulent boundary layer separation by superimposing forced oscillations on the mean flow being on the verge of separation. This is done by mixing the high momentum fluid outside the boundary layer with the lower momentum fluid near the surface [49]. Oscillatory excitation such as that provided by a synthetic jet was shown to be up to two orders of magnitude more efficient than steady suction or blowing. Further concepts being pursued include travelling waves [50] and near wall vortex generators [51] using mechanical devices and phased plasmas.

As for the fixed wing aerodynamic profiles, similar activities have been ongoing for the rotary wing configurations. The Boeing Active Flow Control Systems (BAFCS) Programme [52, 53] sponsored by DARPA has been looking into the flow around helicopter rotor blades and how to improve them by adaptronic solutions. Suggestions generated have been synthetic jets and active flipperons, both made out of piezoelectric polymer materials. The concepts were then tested on a 0.1 scale 3D V-22 powered model and a 7 to 12% reduction in drag could be observed when only actuating one wing and up to 16% when actuating both wings. Both the active jets and the flipperons have been based on piezoelectric materials. While the jets are operating similar to the principle of piezoelectric ink jets, the flipperons have been configured as sandwiched multilayered PVDF beams or as PZT single crystals. A comparison between the PVDF multilayers and the PZT single crystals have shown increases of 10% in lift and a 20% in angle of attack respectively as well as lower drag for the single crystal option.

Further conceptual studies have dealt with the problem of combined bending and torsion in large backwards swept wings [54] that changes along different operational conditions of an aircraft and can lead to increases in drag in case the wing is not operating in the optimum condition. Solutions made consisted either of discrete actuators operating along the wing in the span direction of the aircraft at locations of reduced stiffness or as distributed actuators in terms of continuous SMA wires being integrated into a composite structure.

Most of the adaptronics work started in the late 80s of the last century with space applications (i. e. [55, 56]). Shape control and vibration damping of antenna structures, mirrors and solar panels have been the major issue. Extensive studies have been dealing with the techniques of adapting and integrating PZT patch actuators into spherically curved antenna structures (Fig. 8.15) [57].

Recent work [58, 59] in that context has been dealing with the use of piezofibre actuators completed as pre-encapsulated patches which have been integrated into a lightweight honeycomb core satellite mirror. The optimum shapes and curvatures of these actuators were determined numerically and then realised and integrated into the real structure.

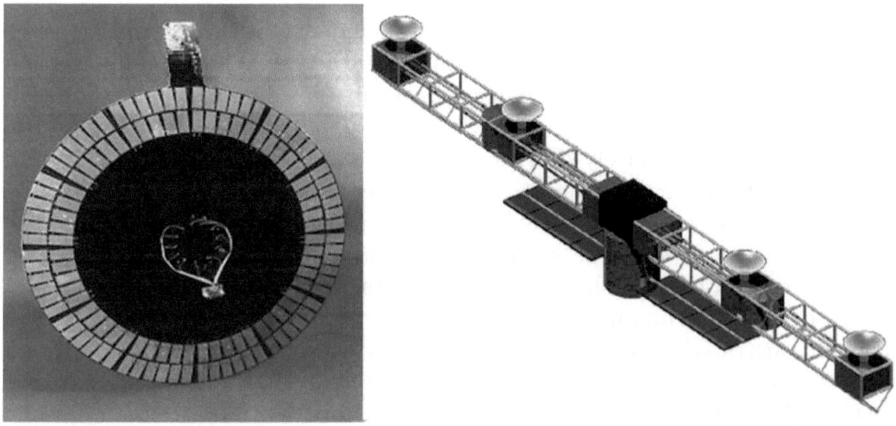

Fig. 8.15. Spherical antenna structure with adapted piezoelectric patch actuators [57]

Although being classified, there is a lot of development ongoing with missiles, where all of the applications mentioned before can be applied as well. Winglets with either preconditioned SMAs [60] or piezoelectric benders [61] have been reported for submarine missiles and ground to air guided missiles respectively. The latter has even shown deflections up to $\pm 16°$ and load increases up to $12.6\,g$ in an experiment which significantly improved manoeuvrability.

8.1.4 Damping of Vibration and Noise

Fixed-Wing Aircraft

Early fixed wing work was related to aeroservoelastic control of wings where mainly piezoelectric actuators were attached to the wing model used [62]. Substantial work towards true application was launched during the early 1990s at NASA Langley where the fin buffet problem of fighter aircraft such as the F/A-18 was tackled on wind tunnel test models and even in real flight tests with the Australian Air Force during later stages [63]. Another sector which has been very much dealt with is noise reduction in aircraft fuselages which was highly driven by turboprop aircraft such as the Saab 2000 or the Dornier 328 and where proof-of-concept has been shown at various occasions [64, 65]. With the decline of turboprop aircraft over the past years and jets taking more and more over, the problem of active cabin noise reduction has currently become less relevant.

Fin buffeting, which is a vortex generated by the leading edge of an aircraft at high angles of attack and which can cause high loads and reduced fatigue life on an aircrafts fin is an issue of broad exploration. Different options have been considered in [63, 66, 67] which included those shown in Fig. 8.16 below. Three of the five options were further explored. Structural integration of piezoelectric elements was realised on a fin box demonstrator shown in Fig. 8.17.

The fin box adequately downscaled from a real fighter finbox to around 2 m in height and around 1 m in width was equipped with 2410 specifically tailored piezoelectric wafers [68] and was run in a vibration test where the first bending mode could be reduced from a maximum of 4.1 g down to 0.6 g with a similar observation made for the torsion mode as well. The second option being based on an additional rudder actuated by a piezoelectric motor and demonstrated in a wind tunnel test on a down scaled model showed that power spectrum density could be reduced by more than 60% for the first bending and torsion moment at angles of attack above 30° [69]. The third option using adaptive control surfaces has been shown analytically to be possibly the most effective solution. However suitable actuators with the required strokes and actuation forces have not shown to be available so far.

Integration of SMA wires into the composite skin of a fin of 0.5 m in height has been realised within the EU-funded project ADAPT [70]. SMA wires of 150 µm thickness were integrated into the glass fibre reinforced composite skin which could then be actuated. An initial simple test showed that tip deflection amplitudes could be reduced by around a half in a vibration test once the SMA wires had been heated up to an austenitic condition (Fig. 8.18).

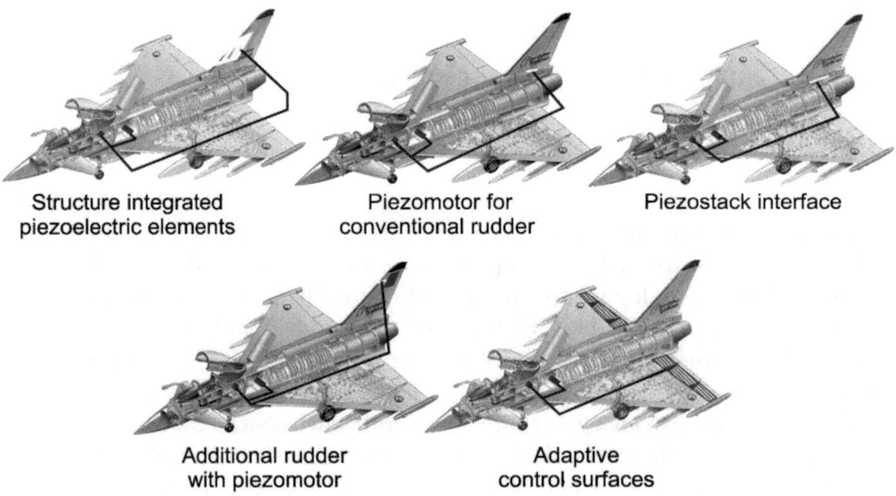

Fig. 8.16. Different options for fin buffet alleviation [66, 67]

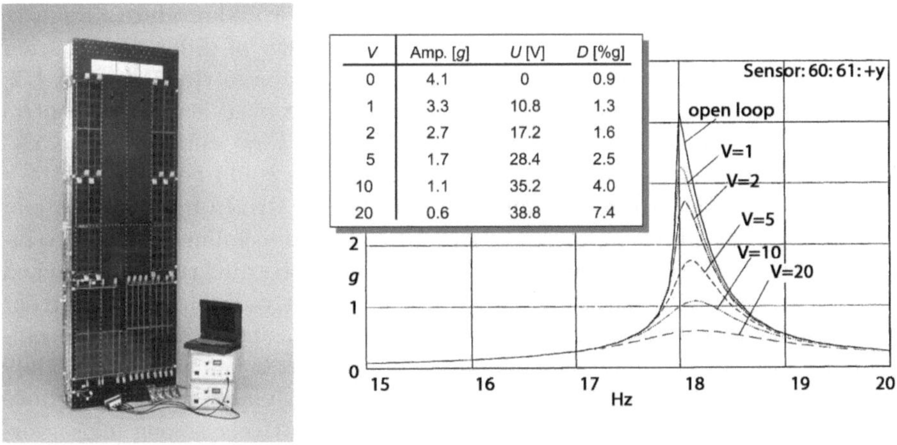

Fig. 8.17. Fin box with adapted piezoelectric actuators [66,67]

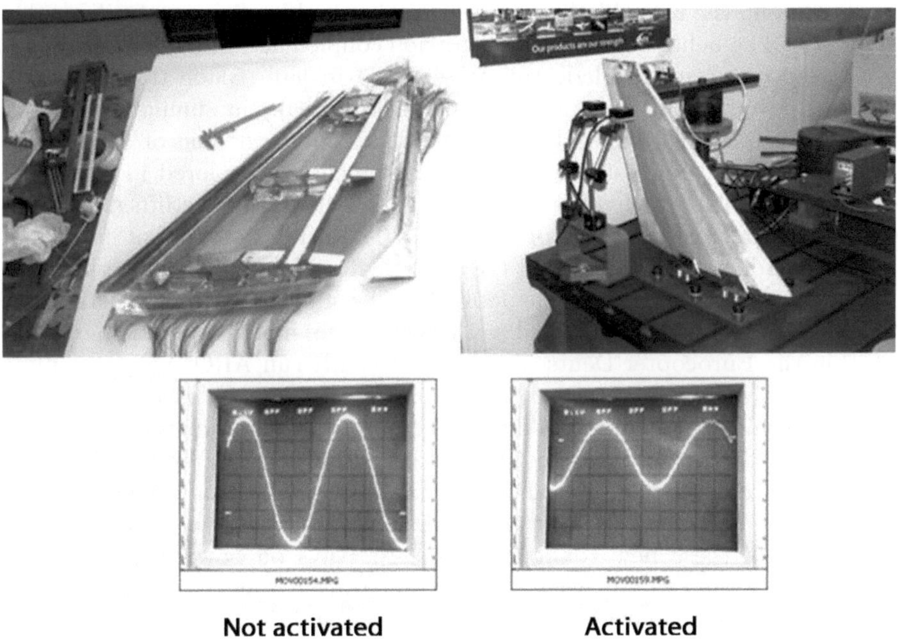

Fig. 8.18. Fin with SMA-reinforced composite skins [70]

Rotorcraft

Adaptronics for rotorcraft was very much driven by concepts in the early 1990s looking at rotor blade twist to improve aerodynamic performance as well as reduction of the negative drawbacks of lead lag damping. Individual blade control has become key in rotorcraft technology and it became quickly apparent that only a hinged flap on each rotor blade would produce a remarkable effect in that regard. As a consequence a variety of different actuation concepts were further explored which included piezoelectric [71], SMA [72] and magnetostrictive Terfenol-D [73] actuators. An overview of the control background in that regard has been given in [74]. First concepts were also given with respect to how cabin noise in helicopters could be reduced [75].

Smart structure rotor dynamics has been highly driven by I. Chopra and his group at the University of Maryland/USA. Initial techniques on how to integrate piezoelectric actuators were reported with the directionally attached piezoelectric (DAP) actuators [76]. Since these early days a lot of further concepts have been developed in simulation as well as in hardware to specifically improve the damping behaviour of rotor blades. It has been analytically shown that DAP-based blade twist technology is a viable means to reduce blade-vortex interaction noise by 2 ... 4 dB for relatively strong, close vortex interactions while 7 ... 10 dB can be expected for the weaker ones [77]. Other principles proposed include segmented constrained layer damping actuators along the rotor blade axis [78] or individual leading and trailing edge flaps [79]. The behaviour of bending-torsion coupled actuators in rotor blades with respect to active blade tips is reported in [80]. Blade tip deflections were achieved in the order of 2° (half peak-to-peak) in simulation as well as in a one-eighth down-scaled rotor model. The integration of active fibre composites has been analytically and experimentally explored in a two-cell shaped rotor blade with respect to 20% increase in torsional stiffness through an increase in twist actuation of 5% [81,82]. The study showed that individual blade control is possible and that the analytical model well allows explanation of the dynamic behaviour of the four blade rotor system.

Hardware has been realised with respect to individual blade control. Within the Eurocopter Deutschland/EADS/DLR run AROSYS programme different solutions with respect to actuation and control have been evaluated [83] which resulted in a leading and trailing edge flaps driven rotor blade where flaps are moved by piezoelectric stack actuators [79]. Figure 8.19 shows the hardware developed by EADS Corp. Research in Germany. Genetic algorithms have been recently used to specifically optimise the aero-servo-elastic behaviour of these new configurations [84]. Ways on how to design these piezostack actuators and their way on integrating them into the different rotor blade designs have been described in [85] and [86].

Validation of such flap actuation solutions have been performed in wind tunnel tests on a one-seventh downscaled Bell-412 Mach-scaled rotor hub [87]. It has been shown that trailing edge deflections of ±4° to ±5° can be achieved at up to 1800 rpm which allowed suppression of vibratory bending moments under an open loop control condition. Even some preliminary closed-loop tests using a neural network controller were performed which however required simultaneous actuation of all four blades. In [88] an induced-shear piezoelectric actuator has been described to actuate trailing edge flaps

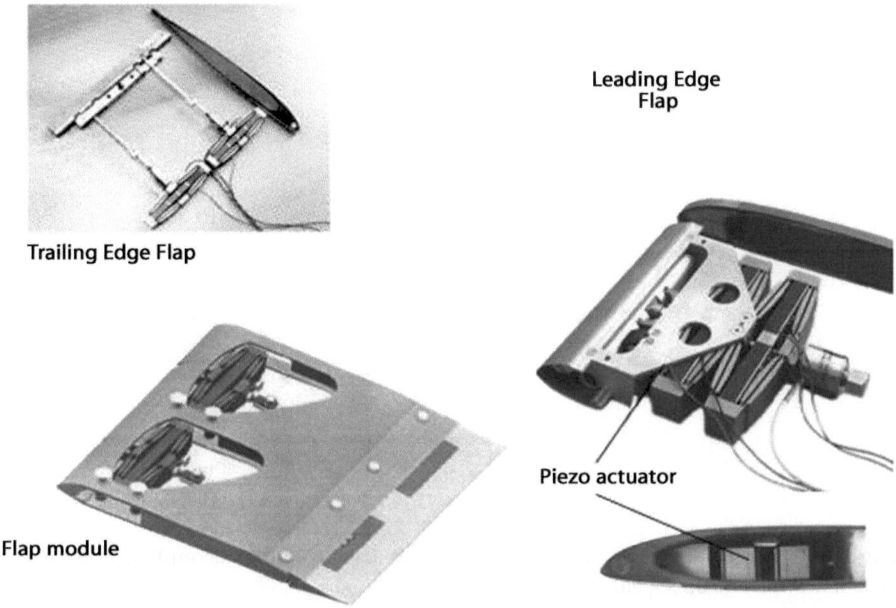

Fig. 8.19. Piezoelectric stack actuator driven mechanisms for rotor blade leading and trailing edges [79]

through induced torsion of the piezoelectric tube actuator. A 12 inch flap was deflected by ±2.8° at 0 rpm and ±1.4° at 400 rpm respectively.

Other adaptronics related damping and actuation mechanisms for the rotor blade include magnetorheological (MR) dampers linked to the rotor blade root. This has been studied analytically for lag damping applications in [89] and shown that uncertainties in the modelling have a significant effect. A concept based on SMA actuation has been proposed in [90] where SMA wires that served to actuate a trailing-edge tab were integrated into a NACA 0012 profile on 12 inch in chord and span. It could be shown that prestraining the SMA-wires to above 3% allows the tab to be deflected by 29°.

Piezoelectric actuators have also been used in a similar collaboration between EADS Corp. Research and Eurocopter both in Germany and France to develop smart struts which serve to decouple the helicopter cabin from neighbouring gear boxes [91] of which the principle is shown in Fig. 8.20. Control mechanisms and ways on how to simulate waves transmitted through these struts have been described in [92].

Space Vehicles

Damping of vibrations is possibly one of the major issues which have triggered adaptronics in aerospace. Early work started in the 1980s with active

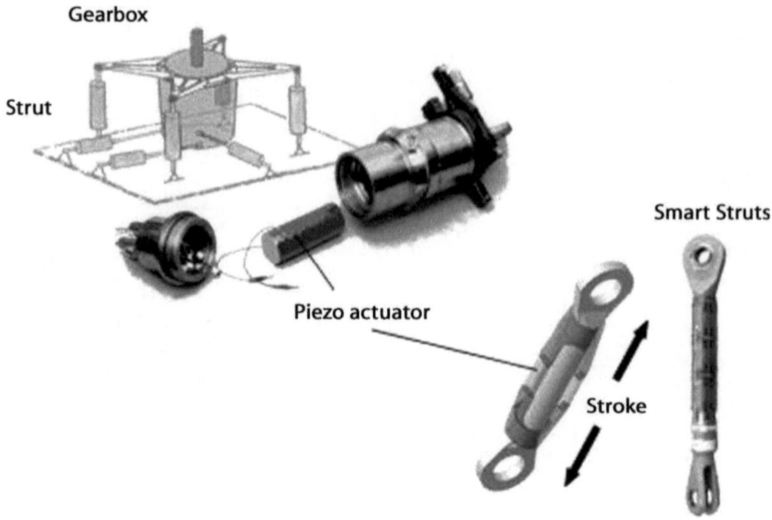

Fig. 8.20. Smart struts driven by piezoelectric actuators for decoupling noise from gearboxes in helicopter cabins [91]

control algorithms that resulted in vibration suppression techniques. Much consideration was also related to large antennas and space platforms. Various demonstrators were built such as the ASTREX and ACTEX smart struts where the background related vibration control analysis and test have been reported in [93–95]. Another application was successfully shown along the articulating fold mirror of the Hubble Space Telescope in the 1990s.

Different of these programmes have been ongoing still far into the mid and late 90s. Further work related to adaptronics for space vehicles has been pursued but at smaller scales. The issues tackled still include vibration suppression [96] and damping [97] mainly based on analytical work. The European Space Agency (ESA) did study some microvibration pointing accuracy platform proof-of-concept studies which were related to six degree-of-freedom passive elastomer isolators, a passive vibration damping system based on distributed piezoelectric wafers, an active damping system based on piezoelectric wafers and positive position feedback control, and a centralised anti-phase control scheme for a distributed sensor actuator system [98]. ONERA in France has been looking on how to isolate vibrations from precision measurement devices on satellites using active struts and has also been able to prove this experimentally [99]. Active magnetic isolation techniques for the sub Hz isolation of equipment racks have been an alternative solution followed up by a variety of space organisations [100]. In [101] a six hybrid isolation strut system configured in terms of a hexapod is described which allows for fluid damper based passive isolation as well as for active damping provided through a linear motor system. Optical fibre sensors have also been explored in the context of ultrahigh sensitivity strain monitoring in complex composites and sandwich structures [102].

Another space related topic is the reduction of noise impacts on payloads through active noise control using technologies such as an acoustic foam with embedded PVDF material, distributed active vibration absorbers (DAVA) made from acoustic foam linked to a metallic plate and in a latest version even added by a very small electrodynamic shaker that allows to cover lower frequencies [103]. This sandwich of acoustic foam, PVDF and electrodynamic shakers is then used as an active coating on the fairings of space vehicles.

8.1.5 Smart Skins

Smart skins in aerospace mainly means the integration of antennae into the structure and/or making the aircraft as much electronically invisible as possible (stealth) by using material being highly electronically absorbent. Structural integration of antennae means that these antennae do also take over a load carrying function. This results in a bivalent relationship such that structural weight can be saved as a result of multi-functionality on the one side but that structural loads and vibrations may affect the functionality of the antennae on the other. Integration of antennae further means an improvement in the aerodynamic performance of an aircraft since a large number of external sword antennae can be avoided. Up to 50% of an aircrafts structural surface can be used for the integration of antennae. Today 66 antennae apertures are located at 37 sites on an F-18 fighter aircraft, covering a frequency band from 200 MHz to 18 GHz [104], and these are intended to be reduced to only nine apertures in nine sights with future designs. This will allow a reduction in weight by a factor of two and in the cost by 30% [105].

Another initiative with the USAF Wright Patterson Laboratory is to design, develop and test a conformal structural load-bearing communication navigation and identification (CNI) antenna in the 0.15 ... 2 GHz range [106]. Further attempts and ideas are thermoadaptive and electrochromic adaptive antennas developed by McDonnell Douglas [107, 108] or a conformal spiral antenna developed at Penn State University [109]. Phased array antennae are another result of miniaturisation efforts, leading to weight savings and structural conformity.

The integration of antennae into an adaptronic system also allows to adaptively steer an antennas beam. Solutions of that kind have been proposed by actuating thick metalised substrates through surface bonded PZT actuators [110] which can be further enhanced when including Rainbow actuators for deflection enhancement [111]. The bandwidth of antennae can however also be increased at smaller scales through electronic measures when using a barium strontium titanate substrate and changing the biased voltage [112].

Smart skins are to be considered in the context of radar absorbing materials where solutions have been given in [113, 114]. Smart microwave windows have been proposed in the context of using poly(aniline)-silver-polymer electrolyte composite materials which have shown a change in microwave reflectivity when a small electrical dc potential is applied among them [115]. An adaptive radar absorbing structure based on the topology of a Salisbury screen combined with an active frequency selective surface controlled by PIN diodes has been described in [116] which allows for superior reflectivity-bandwidth.

MEMS in the sense of adaptronics is playing an increasing role. In their first generation MEMS have been developed to measure parameters such as pressure, temperature, shear, stress, acceleration and rates of those. Their second generation is now looking into using these sensors in terms of arrays and this large scale integration leads them to be considered for smart skins. One of these areas is fluid shear stress measurements on aircraft wings for flow control which have been reported in [117, 118]. For space applications fine-pointing mirrors for inter-satellite optical links have been proposed in [119] which allow for a significant reduction in size, mass, power and cost when compared to state-of-the-art solutions. This has specifically become interesting in the context of microsatellites. Antennae based on new fractal antennae and RF-MEMS further allow for reconfiguration and steering and are considered for communication satellites and electronically scanned arrays for space-based radars [120].

8.1.6 Control

Control plays a significant role within everything being related to flight performance of the air vehicle itself as well as any active vibration control or actuation based structural health monitoring. The various sensor systems being onboard the aircraft are mainly part of the larger flight control system which also includes the ultimate goal of autonomous flight control. Specific situations emerge with fighter airplanes that are required to be controlled after battle-damage and as such under unstable conditions. In all conditions classical and emerging control techniques are applied which are described in Chap. 4 in this book.

8.1.7 Systems

Although adaptronics is a systems approach most of the systems considered for aerospace such as adaptive/morphing wings and aerodynamic profiles in general, adaptive rotors, adaptive cabin noise reduction and many others have so far not bypassed the proof-of-concept stage. Reasons for this can be seen in the lack of financial viability and also in an aircraft systems relatively high complexity. Any change in the design principle – which adaptronics mainly requests – requires a freedom in design which is often not given within our complex and modern aircraft.

A field in aerospace where design is still less constrained and where adaptronics could have a larger and more sustainable impact is Micro Aerial Vehicles (MAV). MAVs are small uninhibited airplanes usually in the range of 600 mm and less in span and 1 kg or less in weight. They are considered to monitor areas which are difficult to access due to contamination (environmental damage), exposed position (towers, chimneys, bridges, rocks, caves, etc.), congestion (traffic), crime (terrorism) or defence. MAVs are available on the basis of a fixed, rotary or flapping wing design principle mainly.

Due to the small size, miniaturisation of the technologies to be implemented is a challenge where the integral approach of adaptronics could be of advantage. Various successful applications have been shown with respect to hardware functionality such as directionally attached piezoelectric (DAP) elements for the control of the servopaddles of the 60 cm span rotor blade of a model helicopter [121]. Piezoelectric actuation has also been used as a piezoelectric flexspar bender element in a fin of a small missile [122]. In a follow-on step this principle was finally used for stabilators of a rotary wing MAV [123]. Adaptronic systems of a similar nature include the adaptive wing concepts explained in more detail along Fig. 8.9. A roll control system using SMA filaments driving wings in balanced, antagonistic pitch has been designed, built and tested [124]. The system mainly consists of two pairs of SMA wire actuators where the first pair of actuators is in charge to produce the pitch while the second pair is in charge of turning the pitch back again to zero. The system was applied to a 2 m wingspan UAV and allowed for wing pitch deflections of ± 3.5 deg with a corner frequency of 1.2 Hz in an airflow of 25 knots. A much smaller wing span for an MAV with a nearly circular aerodynamic shape where the camber has been changed adaptively through the integration of SMA wires has been described in [125]. This concept allows to replace conventional servos which leads to further weight savings. Weight savings are also reported through integration of the battery function into a MAV wing using a rechargeable plastic-lithium-ion battery technology laminated into the wings structure [126]. EAP is another actuation material that has been specifically used for mimicking the muscular system of insects and birds in terms of flapping wing MAVs. In [127] a principle has been proposed where four silicone bowtie actuators drive the mechanism, which is designed such that the optimum flapping frequency of the wings coincides with the resonance of the EAP actuators.

The ability to machine devices at the micro scale on the basis of MEMS technology also drove Ho et al. to explore how far very light weight MAVs such as artificial insects could be manufactured [128, 129]. It has been found that titanium-alloy etched from a bulk material is best for manufacturing the veined structure of the wing. This structure can then be coated with a silicone foil and the process has allowed realisation of capacitor powered MAVs down to a mass of 7.5 grams in total and just 0.3 grams for the wing only [130]. A next step in micromachining has been to look at 'soft' materials such as

Parylene which can be used as a skin for flapping wings [131]. Some recent development has been in exploring options on how to control the airflow on the skin. The solution is an array of valves where each valve consists of a tethered cap, such that it can be actuated individually. The whole system is said to be packaged at 20 µm thickness and to be mass producible.

MEMS is also playing a major role with micro-satellites. Micro-satellites are considered to be satellites in the class of 10 to 100 kg of weight. In [132] a micro-propulsion system is described which uses the catalysed chemical decomposition of high-concentration hydrogen peroxide to produce a 500 µN impulse over 140 to 180 seconds. Another application is a MEMS based actuator array used as a docking system that allows docking of pico-satellites of a few grams of mass [133]. MEMS have also been developed as RF switches implemented on pico-satellites, which allow communication between different micro-satellites as well as between satellites and the ground [134]. A further MEMS application has been suggested for space inflatable structures where a MEMS sensor and actuator system allows the structure to unwrinkle and structural vibrations to dampen once the structure is inflated [135].

8.2 Adaptronic Systems in Automobiles

T. Melz, D. Mayer, M. Thomaier

For automotive applications in particular, there is a rather smooth but increasing transition from mechatronic to adaptronic structures. Once having started with mechatronic systems such as the central car locking system and today realizing adaptive light control (ALC) systems it can be stated that more and more safety critical active components carrying mechanical loads are being introduced into commercial vehicles. Examples are active chassis components which adjust its characteristics depending on external parameters such as mechanical loading, road conditions, speed, lateral and pitching accelerations, and even driver preferences. Some examples are active body control (ABC) systems, semi-active dampers or active stabiliser [136, 137]. This trend will continue whereas a special focus will be on exploiting the well-known multifunctional, intelligent material systems. Even though today there still are no such fully adaptronic structures within commercial vehicles, it is clear that at least most OEMs and Tier 1 suppliers – supported by R&D facilities – are investigating the potential of adaptronic structures for various applications.

8.2.1 Preamble

The focus of todays R&D projects in the automotive industry lies in the improvement of lightweight design, product quality impression, comfort and life cycle cost as well as noise emission, pollution and safety, whereas the latter are especially driven by legal requirements.

Most automotive R&D projects within adaptronics are concerned with active vibration control (AVC) and active structural acoustic control (ASAC) for optimization of NVH (noise, vibration and harshness) characteristics and even sound design. Furthermore, active measures to increase the passive safety of vehicles are being developed. Shape and position control as well as structural health monitoring (SHM) are still of secondary importance.

For automotive applications the following intelligent material systems are of primary interest because of the respective design constraints such as temperature range, humidity or mechanical loads:

- piezoelectric materials;
- electro- and magnetorheological fluids (ERF, MRF);
- shape memory alloys (SMA).

Until now, piezoelectrics have predominated adaptronic system design. In particular, the realization of PZT-based piezoceramic fuel injection systems has had a major effect on manufacturing technology, reliability, availability,

Table 8.1. Overview on potential automotive applications of adaptronics

Technical Approach	Customer Effect
Active vibration control (AVC)	Increased passengers comfort, reduced load/stress level, extended lifetime, weight reduction
Active structural acoustic control (ASAC)	Reduced structure-borne noise and sound emission/imission, increased passengers comfort, weight reduction, interior sound design
Integrated safety, active crash systems	Active pre-crash and crash systems, reversible locking systems, adaptation of car structure, increased deformation zone, load control, active damping
Structural health monitoring (SHM) Structural health and load control (SHC)	Increased reliability reduced maintenance costs, maintenance on demand, adapted safety factors and lightweight design
Active shape and position control	Reduced aerodynamic drag, aerodynamic structural stabilization, extended quality impression
Substitution of conventional actuators by use of smart materials	Reduced complexity, increased reliability, weight reduction
Energy harvesting	Power supply free applications, e. g. energy autarkic SHM systems

and cost on this material [138]. This situation strongly aids the acceptance of this material type for wide use within adaptronic car systems. Due to legal requirements lead free ceramics [140] are being developed. Further R&D activities concern high temperature resistant and transparent ceramics.

Electro- and magnetorheological fluids are mainly used in semi-active damping devices, current car suspension systems in premium class vehicles based on MRF [139]. Furthermore, shape memory alloys are under investigation, especially for active pre-crash systems [141].

Since all these material types cause certain limitations to adaptronic solutions it must be expected that new material developments with respect to stroke, stiffness or robustness will even expand the technical potential of adaptronics for automotives. One example can be magnetostrictive material which could complement or even replace piezoceramics. Today magnetostrictive Terfenol-D with a superior energy density is rather expensive, availability is limited, the reliability unclear, etc..

8.2.2 AVC/ASAC Project Examples

As mentioned in the preamble AVC and ASAC applications prevail in current R&D projects and will most likely be the first commercial adaptronic systems in passenger cars. Some reasons for this are:

- legal requirements for noise emission;
- demand for further weight reduction;
- passengers comfort requirements (NVH);
- design restrictions of purely passive structure means.

The noise emission from road vehicles is limited by legislation in order to protect the environment against high noise pollution. The related noise limits were significantly reduced over the last 30 years (Fig. 8.21) [148]. Predominant noise sources are tires, exhaust systems, intake systems, power-train and combustion engine. Purely passive systems like exhaust mufflers,

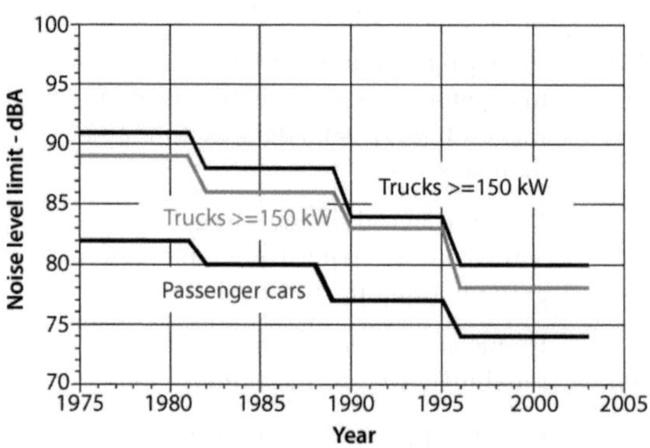

Fig. 8.21. Pass-by noise limits since 1975 [148]

optimized intake systems, special noise damping materials, acoustic partial shielding or sound absorbing linings are reaching technical limitations and often result in an increase of the overall system weight. Adaptronic solutions indicate the potential to optimize both, noise emission and system weight.

Apart from the noise emission into the environment, the interior sound and vibration characteristics are crucial for the passengers comfort, thus representing essential design drivers. Sources for disturbing noise and vibration within the passenger compartment are mostly the same as for the emission into the environment, namely the combustion engine, power-train and intake- and exhaust-systems. Furthermore, HVAC-systems, auxiliary equipment, aerodynamically and road-tire-contact generated noise and vibration are considerable sources for the NVH quality impression [148].

Adaptronic solutions like active mounts, adaptive vibration absorbers, or distributed in-plane actuators can optimize these structural dynamic and vibroacoustic characteristics [149]. Typical target functions are the active control of introduction and transfer of disturbances and/or damping of elastic modes. Basically, three different conceptual approaches can be considered:

- interference at the NVH source (engine);
- interference at the NVH recipient (steering wheel);
- interference within the transfer path between both (car body, stiffening struts).

The type of interference depends on the respective target function, constraints and system design. As an example, if a variety of sources and only a few transfer paths exist, it would seem unreasonable to interfere with each source whereas design constraints might still be required for such an approach.

Typical application scenarios for adaptronic systems in automobiles focusing on AVC/ASAC are (Fig. 8.22):

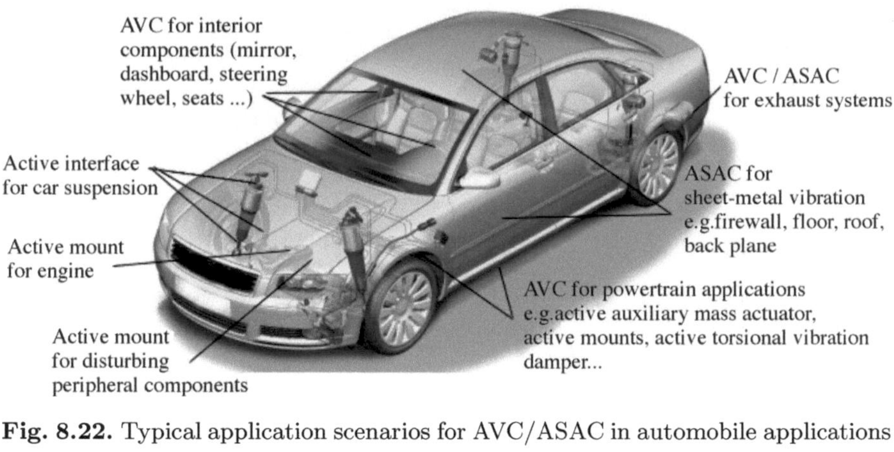

Fig. 8.22. Typical application scenarios for AVC/ASAC in automobile applications

- soft active mounts as engine supports [157];
- stiff active mounts (interface) to reduce road-tire-contact induced vibration [154,161];
- active struts to reduce torsional vibration within convertibles [150,151];
- active add-on systems such as adaptive absorbers or auxiliary mass actuators [152];
- in-plane actuators to reduce sheet-metal vibration of firewall, roof or windshield [163].

These systems can be active or semi-active. Examples for commercially available semi-active systems are 'adaptable' car suspension damping devices which are realized as controlled mechatronic CDC (continuous damping control) systems [161], used within the Lancia Thesis, VW Phaeton, VW Touareg, Audi A8, Porsche Cayenne and even the Opel Vectra or semi-active motor-mounts. Another recent example exploiting intelligent materials is realized as 'magnetic ride' within the current Audi TT [153]. An even more challenging approach was developed by the Bose Corporation utilizing electromagnetic actuators replacing the conventional car suspension systems [158]. This active suspension system is focussing on low frequency driving dynamics and not yet been integrated within series cars. Stiff active interfaces are especially interesting for higher dynamics up to the vibroacoustic frequency range. An exemplary market need for an adaptronic solution results from the increasing application of run-flat tires – first introduced within luxury cars. These are stiffer than conventional tires, thus degraded NVH characteristics within the passengers compartment can be observed. One possible adaptronic solution is proposed below.

Semi-active soft engine mounts have also been investigated which make use of electromagnetic systems or even MRF [155,156]. Active mounts focus on integrating pneumatic, electrodynamic and electromagnetic actua-

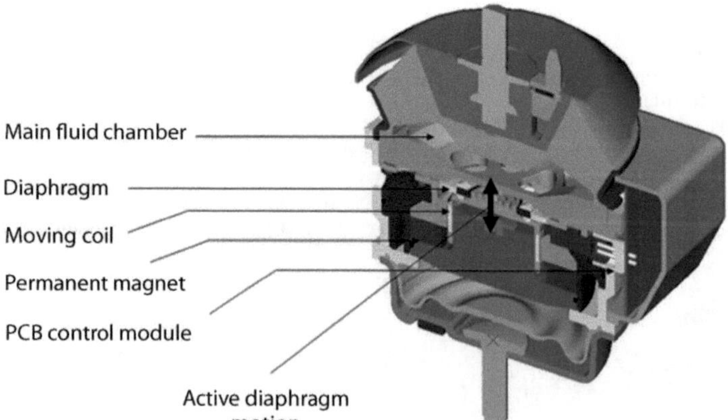

Fig. 8.23. Avon VMS active engine mount [157]

tors [159]. One current commercial system is integrated within the Jaguar XJ6 TDVi [157], (Fig. 8.23).

In addition to the trend to increasingly use soft active engine mounts there is an increasing importance to optimize the NVH-characteristics of the chassis. An example for a semi-active mechatronic system for chassis control based on electrohydraulics was investigated in [160], whereas different active systems recently focussed on AVC within convertibles [151]. The latter will be discussed below.

AVC of Torsional Vibration in Convertibles

The dynamic stiffness of the car body predominates the driving comfort of a passenger car in the low frequency range (<30 Hz). The torsional mode is especially relevant for the drivers comfort impression. For convertibles the corresponding natural frequency is lower than for sedans due to the fact that no stiffening roof structure exists ($f_t \approx 20\,\text{Hz}$ for convertibles compared to $f_t > 50\,\text{Hz}$ for sedans). To prevent high vibration amplifications of the car body caused by the engine it is preferable to realise relatively high natural frequencies of the car body. A low natural frequency results in perceptible vibrations of the rear-view mirror, the seats and even the steering wheel.

To reduce these vibrations usually different kinds of passive means are realized such as stiffening the body by additional sheet-metals, enlarging the cross sections of beams, applying struts in the car underbody or even using the rear panel as an additional load carrying part. Such means increase the total weight of a body-in-white of a convertible compared to a sedan to about 50 kg. Furthermore, heavy passive absorbers are added to reduce disturbing vibrations at the natural frequency of typically 10 ... 20%. Typical system weights for this range from 8.5 to 14 kg in four seaters [150, 151]. Another important design rule is to prevent natural frequencies of subsystems to coincide, such as the suspension, chassis, engine mounting and car body. However, with all these means the dynamic stiffness of a sedan cannot be reached.

One adaptronic approach to reduce torsional vibrations was to integrate actuators within the diagonal stiffening struts of the car underbody (Fig. 8.24) [150,151]. Different actuators have been investigated, piezoceramic stacks, hydraulic cylinders and hydraulic muscles. Control approaches like adaptive feed-forward and feedback were implemented and significant vibration reductions were achieved (Fig. 8.24). To commercialize this affirmative active concept system cost, size, complexity and power consumption must be further reduced.

Active Interfaces for Front and Rear Suspensions

Current car suspensions are designed to realize a compromise between car dynamics, comfort, cost and weight. Typically, their operative frequency range

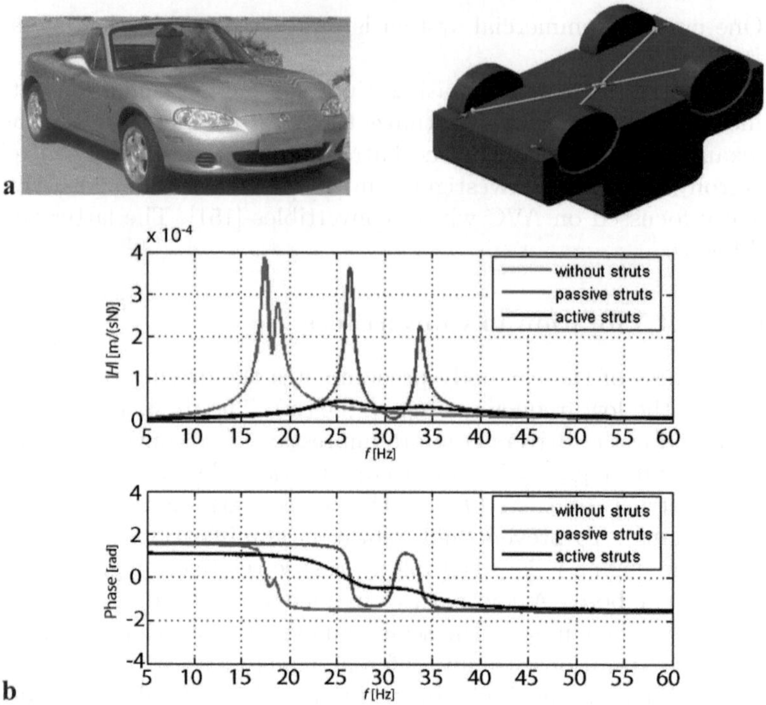

Fig. 8.24. AVC-application within a convertible: **a** convertible and concept of active struts, **b** achievable vibration reduction

is limited to the needs of low frequency car dynamics. Higher frequencies do directly concern the passengers NVH comfort. To optimize NVH characteristics both using high damping rubber material and air springs for upper class cars, is state-of-the-art, although these reach limits with respect to increasing NVH and lightweight design demands of modern vehicles. Further NVH problems come from the new runflat tires which are stiffer than conventional tires in order to prove fail-safe operation with no air pressure [154].

A promising solution is to integrate stiff active mounts into the suspension system and thus to actively prevent structure-borne noise from spreading into the car body (Fig. 8.25). Typical operative frequencies should range from approximately 30 Hz to several hundred Hz.

One design approach for such active interfaces is to integrate stiff piezoceramic stack actuators in an elastic housing and realize a robust, compact and housed system. Using piezoceramic actuators ensures short response times and the ability to withstand high active and passive loads, which is necessary for a direct integration into the load transfer path.

A recent design study based on low cost actuators, which have been developed as a mass product for fuel injection systems [138], is described in [161,

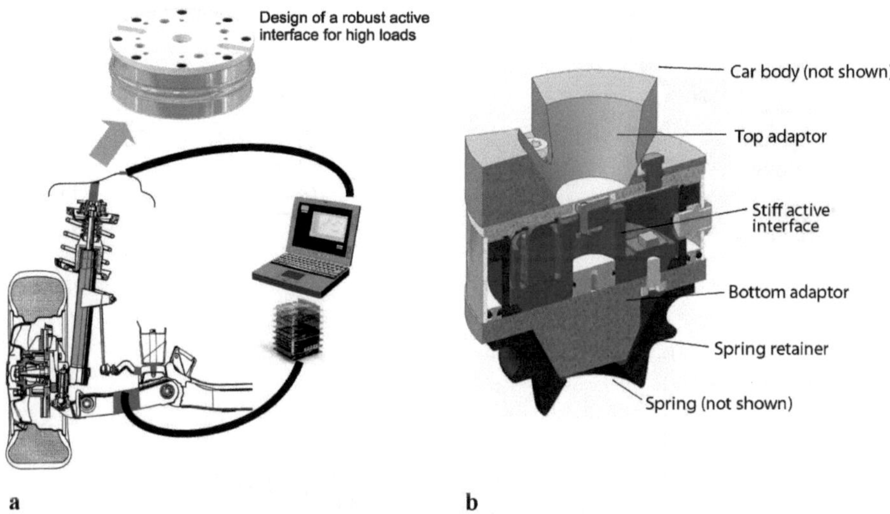

Fig. 8.25. Integration of stiff active mounts into car suspensions: **a** integration into Mc Pherson front suspension, **b** design example for rear suspension spring mounting

164]. It provides a stroke of 70 µm and is designed for typical forces relevant within the front suspension. First tests have successfully been performed with loads of up to 18 kN in z-direction. Furthermore, it has 3 degrees of freedom (translation along z, rotation about x and y) whereas for this application only the translation is relevant. More recent versions are more compact and can be loaded even higher while providing an adapted reduced stroke.

Different control approaches like velocity feedback (VF), integrated force feedback (IFF) and independent modal space control in combination with velocity feedback (IMSC-VF) as well as more advanced IMSC with adaptive filters like internal modal control in modified error configuration (IMSC M.E.IMC) have been implemented. In initial experiments a significant broadband vibration reduction was shown [142, 162]. Furthermore, experiments with different control platforms ranging from rapid control prototyping (RCP) to embedded systems like µC, DSP and FPGA have been done to achieve the required system integration for commercial automotive applications.

One interesting aspect of such stiff active interfaces for automotive applications, is that they can be used for a variety of interconnection points within the car structure, rear suspension, power train, exhaust system, rear axles suspension or engine. This is especially interesting when properly combined with passive, elastomer or hydraulic supports.

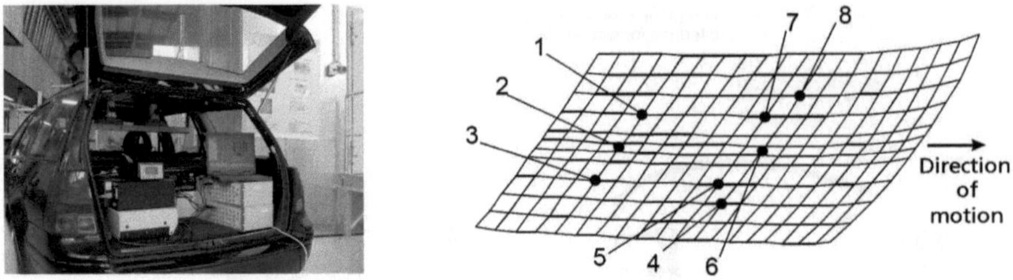

a

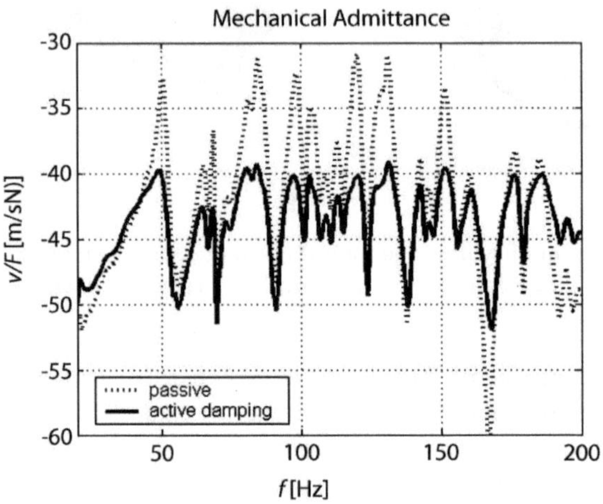

b

Fig. 8.26. Vibration reduction of a VW Bora roof: **a** test vehicle equipped with electronics and actuator locations on the roof, **b** results of control in comparison to passive behaviour

ASAC for the Car Roof

Interior noise is a very important purchase reason for customers. The overall interior noise within a passengers compartment is caused by different sources. One source is the vibration of sheet metal components which are excited by rotating parts (like engine, wheels or powertrain) or even aerodynamics. Typically, the car body is poorly damped. Large sheet metal areas therefore can emit annoying noise. To prevent this, for the high frequency range car manufacturers usually use acoustic insulation like foams, carpets and fabrics whereas for lower frequencies highly damped, heavy weight materials like bitumen are applied to sheet metals. However, these approaches are limited with respect to its effect and cause additional weight.

One approach to optimize weight and acoustic emission is to actively influence sheet metals via ASAC. A corresponding project example is an actively damped car roof structure which was part of the German Leitprojekt Adaptronik [165]. Several distributed piezoceramic actuators have been attached onto the roof of a VW Bora variant [163]. Investigations were done to assess the potential of the ASAC approach.

Within the respective subproject a step-by-step procedure has been chosen. After analysing the passive structure the position of piezoceramic patch actuators were numerically optimized (Fig. 8.26) and attached to the roof. First results showed that the available actuators were not capable of sufficiently exciting the roof structure to counteract the passive vibrations – today wide range optimized patch actuators are commercially available. To compensate the insufficient actuator performance and further validate the ASAC approach electrodynamic actuators were applied to the roof. Different controllers have been tested and an active broadband structural control was successfully proven (Fig. 8.26).

8.2.3 Current Research Topics for Automotive Smart Structures

As mentioned in the preamble AVC and ASAC applications prevail in current R&D projects. To commercialize such adaptronic systems which today can still be characterized as laboratory type solutions, current R&D efforts are made to optimize cost, overall system size and complexity, system integration including power electronics and embedded controllers, component and system reliability as well as compatibility. For automotive applications it has to be considered that typical technical and economic restrictions given by car manufacturers are especially challenging.

Development Methodology for Adaptronic Systems in Automobiles

The development of adaptronic structures must be understood as a rather complex process due to the fact that the functional interdependencies of active and passive components – in which these are integrated – are inherently strong (the actuator performance depends on the passive host structures characteristics and its loading and vice versa). To systematically optimize the adaptronic system realization process it is advisable to establish an efficient development methodology which considers all typical functions of an adaptronic system (mechanical, electronic, signal conditioning, software, control) as well as engineering methods and tools (FEA, SEA, MBS, CAD, CACE, EMA, RCP, etc.) respectively. One such approach is described in detail in [164], (Fig. 8.27).

For the development of active systems, it is vital to carefully analyze the mechanical host structure with respect to the operative boundary conditions to determine requirements and constraints for the active system such as its operative frequency range, deformations, strokes, loads, etc. This analysis is typically a process in which numerical and experimental investigations do complement each other. This means, that experimental data can be used to

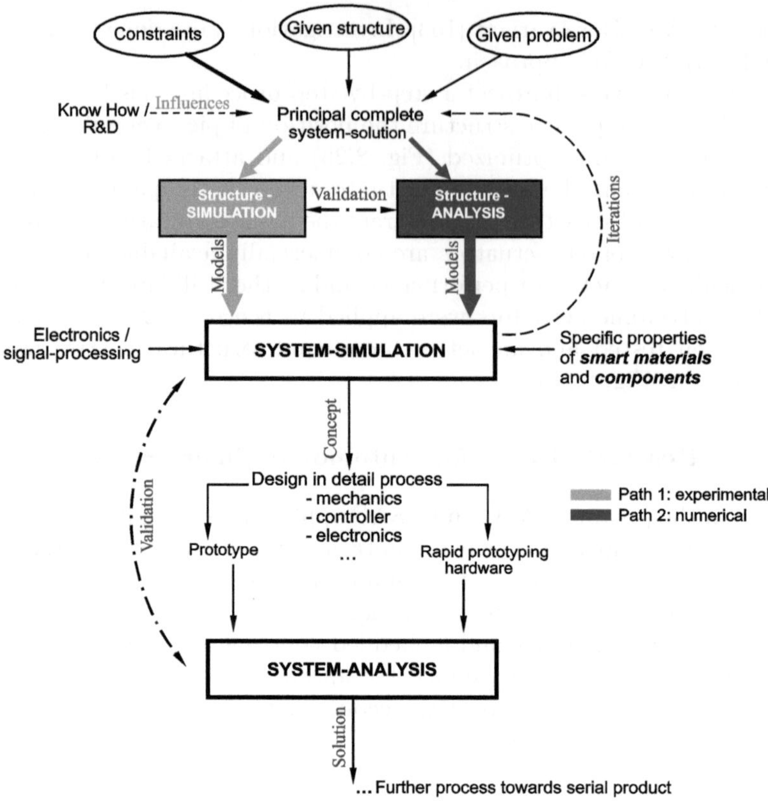

Fig. 8.27. Flow chart of a development method for active systems [142, 162]

generate numerical models and with this reduce modelling efforts wherever reasonable or to verify numerical models by experimental model analysis (EMA). FEA data can be used to feed MBS models or to generate mechanical data for controller design within CACE (computer aided control engineering) which could be excited based on experimental data. It can even be reasonable to use analytical models for simple estimations.

An important step is then to set up a full system simulation which integrates the different functions of the active system, i. e. especially host structure, actuators, sensors, signal conditioning, control and electronics. Depending on the current perspective of the system analysis it is favourable to allow for shifting the simulation focus from CACE to FEA to MBS or even electronic design automation (EDA). To reduce the modelling effort it is desirable that the different models share as much information as possible. This can be achieved by integrating different domains into one simulation environment or by co-simulations controlled by a central tool. Moreover, as mentioned before, it is important to enable combining of the different numerical methods with experimental based methods like the frequency response functions, transfer functions or impedance based formulations.

One typical example for such full system simulation that allows integrating the different engineering discipline models is the controller development using CACE tools such as Matlab/Simulink™ or open source tools like Scilab/Scicos. Within such an environment, the mechanical structure, experimental data, actuator performance, etc. can be modelled as complete systems in state space, allowing to analyse the coupled systems behaviour, to design the controller and even to study aspects of the systems reliability. This includes the system in its passive condition as well as its active performance. These CACE environments additionally offer special advantage to work with toolboxes to archive certain functions such as control algorithms or even actuator behaviour, thus helping to accelerate the system development process for future tasks.

The next step of the development methodology is the system analysis of the overall system equipped with prototypes, investigating the performance of the adaptronic system, ACV/ASAC in the lab or even within operative conditions by exploiting RCP-tools and doing even hardware-in-the-loop (HIL) tests. At this stage further verifications and updating of the prior modelling work is typical. The final step is to then realize and test the prototype system.

As a final remark it should be mentioned that an adequate development methodology for adaptronic systems must comply with these different engineering needs while offering enough flexibility with respect to analysis focus and methodology expendability.

Cost and Size Reduction

In addition to establishing an efficient, accelerative development methodology it is essential to further reduce the cost and size of the adaptronic system significantly. To enhance current laboratory state AVC and ASAC solutions and commercialize them within automotive applications one obvious approach is to utilize low-cost alternatives to customized components for all adaptronic system components – at least for market introduction of these active systems. This would correspond to an approach which can be called design-to-standards in which mass-production compatible components such as MEMS-sensors, fuel injection actuators or market available embedded systems for implementing control algorithms and signal conditioning are being used to set up an active system rather than to derive new customized components with optimized features.

For this, it is first necessary to identify products that can be used as laboratory substitutes, mass-produced components for consumer goods. Subsequently, these products must be analysed concerning the specific demands from the considered applications, for a low cost sensor this would be its bandwidth, linearity, thermal characteristics and signal-to-noise ratio. Examples for this approach are the use of acceleration sensors based on micro-electro-mechanical-systems (MEMS) in AVC systems or use of low cost piezoceramic stack actuators used in fuel injection systems as actuators in active mounts. Some other approaches to reduce cost and size are:

- profound system analysis and optimized design-to-requirements including reduced, i. e. adopted functionality and integration of functions;
- development of new low cost components (power electronics);
- adoption of manufacturing processes for adaptronic components and systems.

The design-to-requirements approach, which could be expected to follow market introduction and which would be feasible for large series solutions, would focus on an extensive knowledge of the given system based on intensive system analysis and which would enable a more precise knowledge on required actuator stroke and hence enable a reduction of safety factors for this. Consequently, this would allow for an adoption of the effected components such as power electronics or controller platforms, thus enabling the design and realization of optimized, custom-made components. Another aspect affected by this approach would be the reduction of functionality meaning that many laboratory type adaptronic solutions tend to offer more functions than required for the given problem. Moreover, this also affects the integration of functions which corresponds to the fact that quite often in laboratory type solutions different functions are realized in separate components. Some simple examples are the mechanical actuator pre-stressing, mechanical missuse and system housing function which could be realized by one component or the actuator drive and the full system diagnosis function which could be combined within the controllers signal conditioning. This approach would have the benefit of leading to the continuous development of new duplicate adaptronic parts which would serve as standard components, and thus relieve cost, and increase the reliability of such active systems.

It is necessary to continuously optimize and adapt manufacturing technologies to utilize them for adaptronic components and system solutions. This comprises new materials processing such as optimized manufacturing of magnetostrictive materials, component manufacturing such as for cost reduction within piezoceramic stack actuators, and system manufacturing. The latter aspect comprises processes compatible with unique copies as well as small to large series in order to increase design flexibility and enable series production with high reproducibility. Examples of current research are projects on metal casting technology compatible with adaptronic devices [169, 170] and rapid manufacturing approaches.

All the aforementioned aspects must not only consider the costs for actuators and sensors and smart materials respectively but at the least, the control unit, the signal processing devices and power amplifiers. These subcomponents do have a significant influence on the overall cost and size. In particular the development of *low cost amplifiers* for piezoceramic actuators, the investigation of *low cost production processes* for mechanical parts but even for electronics and the realisation *low cost controllers based on embedded systems* are subject to continuous research work. Current results indicate the implementation of first adaptronic systems in near- to medium-term.

Systems Integration

In modern vehicles, using mechatronic systems is more or less state-of-the-art. Numerous actuators, sensors and controllers are already integrated with the vehicle. Thus, it is necessary to solve the task of systems integration for adaptronic systems in automotive applications as well.

For example, an adaptive feed forward controller has to be connected to a present engine speed signal to generate a correlated reference signal for the suppression of engine noise and vibration [143].

To reduce wiring effort, a realisation of a multiple channel AVC system with several actuators and sensors with the help of an automotive bus system like CAN, LIN or FlexRay would be favourable. However, current bus systems do not offer enough bandwidth to transmit actuator and sensor data sampled at 1 kHz in real time together with handling other data from existing mechatronic systems. Therefore, decentralized vibration control systems are an interesting alternative, for ASAC systems for large panels. Local vibration control systems with typically one actuator, sensor and controller are networked by a bus system. Together with a central computational unit, a hierarchical control system can be set up: the local controllers run at high sampling rates and transmit processed sensor data at a lower sample rate to the central unit. Based on this information, optimized control parameters for the local units are calculated and transmitted back.

To realise these decentralized control systems, it is necessary to realise small signal processing and actuator driving units. In the case of piezoelectric actuators, the high driving voltage also has to be supplied, preferably generated from the common 12 V DC vehicle electrical system. Concepts based on piezoelectric transformers have been examined as potential solutions for this task [145].

For vibration reduction of panels mostly, semi-active vibration damping systems based on piezoelectric transducers have been studied. The piezos are attached to the structure and electrically shunted with resistor-inductor circuits. By proper tuning of the circuit elements, effects compared to the application of mechanical vibration absorbers can be observed, however only laboratory experiments have been performed yet [147].

Current research work does focus on the implementation of the advanced control algorithms on adequate embedded systems such as microcontrollers, DSPs and FPGAs and finally even ASACs [154]. The corresponding work is called embedded control whereas there is an inherently close link to implementing additional diagnostic and sensor signal conditioning functions on such platforms which might further relieve standardization and cost reduction efforts. For successfully exploiting embedded systems it is necessary to develop tools for target machine independent control encoding which should be interlinked to the aforementioned development methodology and especially the CACE tools. Moreover, tools for code generation for the specific target machines are being developed and even commercially available for some platforms. Some current investigations have been done to test interlinking such code generators with the open source CACE environment Scilab.

With the integration of piezoelectric actuators into the suspension or at other difficult to access locations, the implementation of diagnostic functions becomes an important aspect. Monitoring systems can supply information about the condition of the actuator gained from electromechanical impedance measurements [144]. Since powerful signal processing platforms are becoming smaller and cheaper, even the implementation of structural health monitoring functions may be possible for automotive components. Wireless sensor networks for monitoring tasks are presently under development for civil and aircraft structures [146], but in-situ monitoring of mechanical loads of a high performance vehicle is an attractive application for these networks as well. To implement a fully wireless network, either power supply from a battery or local energy harvesting are alternatives. The latter can also be implemented with the help of smart materials, as a piezoelectric generator system which recovers electrical energy from mechanical vibrations.

8.2.4 Summary and Outlook

Today, mechatronic systems are an integral part of modern vehicles. This trend will continue (Fig. 8.28) and expand to fully active adaptronic systems and the question is which wording – adaptive, active, or some other – will predominate. Executive consultants and the European Commission (EC) very often call such systems smart structures or talk of exploiting smart materials [148, 168]. According to a recent EC study [166] 'smart materials' that adapt to different conditions by changing properties are expected to be in widespread use in the future. Some ideas exist for even self-healing structures and those capable of performing some kind of structural health control. It seems to be less a question of whether these active systems will be commercialized within trucks or passenger cars and non-automotive systems, but more like, when this will happen? Some advanced systems such as Delphis Magnetic Ride, Avons active Vibramount or Boses suspension system indicate the current trend. Moreover, it is clear that Tier 1 suppliers and OEMs are engaged in respective R&D-work. This might indicate that in the coming or next but one vehicle generation there will be some first adaptronic AVC or ASAC system.

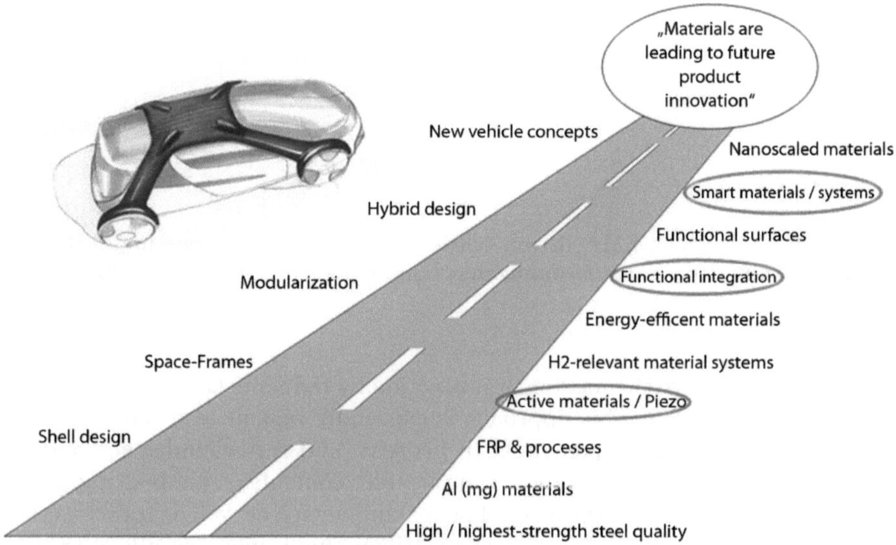

Fig. 8.28. Trends in automotive mechanical engineering [168]

Active Crash Systems

Apart from the aforementioned NVH problems one current rather exotic R&D focus is on using smart materials for adopting the mechanical characteristics of vehicles with respect to crashworthiness. The motivation is that even advanced current approaches of active and passive safety (multiple airbags, pre-safe, active braking, distance control, adaptive belt pretensioning, active bonnets) are not expected to be sufficient to halve the number of road traffic deaths to 25 000 by 2011 as required by the EC whereas the economic damage caused by accidents is roughly 2% of ECs gross domestic product, equalling € 200 Mrd. With respect to side-impact collisions, the motor vehicle is rather weak. The potential to optimize the structure is high although the deformation space available is very low [167]. One new endeavour is to realize structures which make use of pre-crash sensor information and thereby offer an integrated safety approach. The idea is to cope with sensor information – which by the nature of very short term side crash scenarios must be of limited reliability – and to realize active structures that adapt the side vehicle stiffness to different side crash scenarios such as pole, barrier, SUV or motorcycle impact. Due to such sensor information it is not possible to drive nonrecurring actuators such as pyrotechnical but to realize fast, reversible and thus fault tolerant actuators. Whereas a first step could be to have digital actuation at discrete locations for structure locking, the next logical step would then be to continuously adapt the deformation profile.

One current crash study indicates that a selective, reversible interconnection of the doors, Fig. 8.29, with the surrounding stiff car structure can

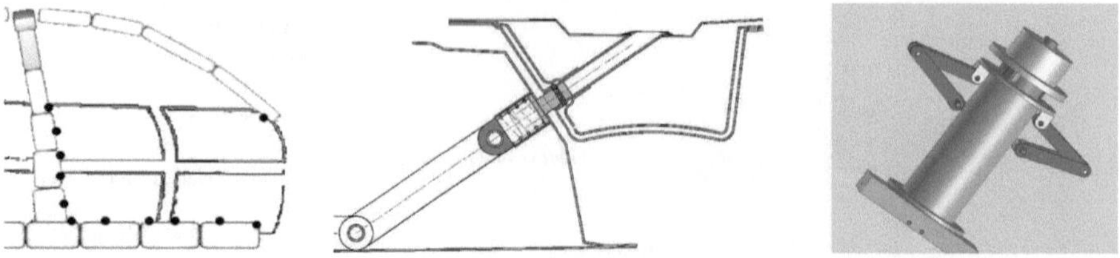

Fig. 8.29. Reversible, SMA driven release, *left*: locking positions under investigation at front door, *middle*: interconnection of door and B-pillar, *right*: release system

significantly improve the passenger protection by reducing the intrusion depth of the door structure. This approach is especially beneficial when combined with means of fixing the passenger to his seat, and if it includes active door paddings to realize a softer and, if applicable, controlled damped impact of the passenger to the car structure. The locking device as well as its retraction mechanism is activated by a capacitor driven shape memory alloys (SMA) and offers activation times of about 5 ms. One derivation of such an unlocking mechanism could be to complement conventional pyrotechnical airbag systems for certain structure parts for the aforementioned crash pad or for seat rests to fix the passenger. This might enable a reduction of peak velocities and lower biomechanical load values for the human body.

Another idea for an active crash system is an adaptive side intrusion beam that can locally change its stiffness and deformation shape. It is designed with several hinges which can vary their stiffness and friction by integrated linear actuators. Many other ideas are currently evolving which are concerned

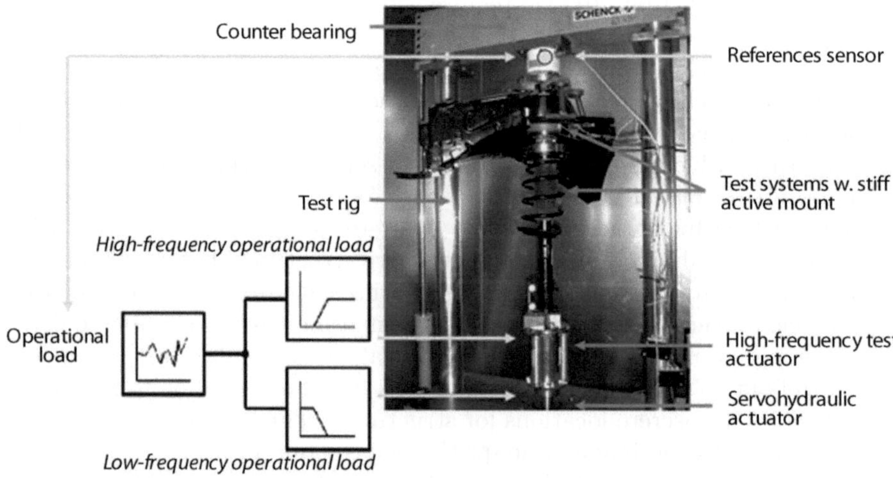

Fig. 8.30. Prototype of hybrid testing facilities

with increasing deformation space, modifying the impact load vector, and stiffening the cars structural parts.

Hybrid High Dynamic Test and Development Racks

AVC and ASAC solutions like active interfaces for car suspensions or active motor mounts require novel equipment for testing as well as for developing the active components within operative conditions. For this, it is necessary to operate these systems in a frequency range of up to at least several hundred Hz while being mechanically loaded, thus extending the conventional test drive solutions. Only then is it possible to assess the reliability and durability of active systems and components under operational lab conditions as well as to develop the controller and assess the system performance.

Conventional test equipment like servo-hydraulics or servo-electrics are limited to driving frequencies of typically 50 Hz with respect to working with arbitrary drive signals. Thus, one approach to extending the frequency range can be to combine low and high frequency actuator devices within one test facility whereas the high frequency actuation could be done by piezoceramic interface structures similar to those described before.

Such hybrid devices could be controlled by filtering the broadband input signal into a low frequency domain for driving the servo-electric motor and a high frequency domain driving the solid state actuators. Hence, it is possible to investigate the active operative system performance while mechanically loading it.

Some R&D projects focus on developing high frequency test stands with highly rigid test frames and actuators designed with electrorheological fluids (ERF) and piezoceramic actuators. Some more recent projects cope with the aforementioned hybrid test stands. All these focus on uniaxial force loading of components. Future applications will demand more complex full vehicle test stands with even more challenging approaches for test environments in which the mechanical characteristics of the active systems host structures – the vehicle body and the suspension for active suspension mounts – are virtually simulated by controlled actuators at the connection points. This new test approach can be called active control of connection impedances.

Outlook

New actuator systems with a larger stroke, reduced price and increased reliability will obviously extend the adaptronic portfolio. New sensors such as piezoresistive DLC layers will enable to realize robust, very small force sensors for integration within safety critical areas and allowing for static to dynamic measurements and thus enabling robust force control of even stiff active structures. As a vision distributed computing for various applications such as sensor signal conditioning or active control might relieve system cost within the future. Moreover, SHM systems to monitor usage and health will help to increase reliability, expand system life, reduce maintenance cost as well as lower safety factors and finally reduce system weight. Exploiting concepts of energy recovery such as using electromechanical transducers can result in energy autarkic senor nodes setting up distributed sensor networks. One further step can be to use wireless communication to once more reduce mass.

8.3 Adaptronic Systems in Machine and Plant Construction

H. Janocha

This section focuses on the application of adaptronic concepts in machine tools and manufacturing plants. Such concepts are applied to improve the performance of the machines. Experts agree that machine tools of high performance stand out because of their excellent product quality, high processing speed, as well as reliability and flexibility when different tools and materials must be processed.

So far, adaptronic systems have mainly been applied in aeronautics and space travel, see Sect. 8.1, but hardly ever in commercial machine tools and manufacturing plants, which means that most of the concepts and experiences described here result from research projects conducted at institutions of higher learning. For many years, these institutions have applied such systems in their experimental set-ups.

In 2002, the German Research Foundation (Deutsche Forschungsgemeinschaft, DFG) started a so-called Priority Programme named 'Adaptronics for Machine Tools' to support and fund such activities conducted by universities and research institutions. A vast number of research groups are collaborating within the scope of approximately 20 projects over a period of 6 years, whereby the groups will mainly focus on the following topics:

Active Damping and Reduction of Structural Oscillations. Under this heading, the research groups develop fundamental concepts of optimized positioning and integrating adaptronic components (actuators, sensors) in machine tool structures. Research focuses on approaches and implementations that will reduce structure-born noise by means of control techniques.

Compensation of (Quasi)Static Deformations. This topic aims to stiffen machine tool structures and thus protect them from thermal and (quasi) static deformations caused by other effects. To this end, researchers must examine sensor-actuator combinations integrated into supporting and coupling components and intelligently modify the feed motion performed by the machine axes.

Autonomous Adaptation of Adaptronic Components During the Production Process. This topic is about the development of adaptive control algorithms that serve to autonomously adapt the properties of adaptronic components. To achieve this, research groups are working out concepts to integrate additional sensors which will acquire relevant process parameters as well as concepts to adapt already existing methods of avoiding errors and breakdowns to the specific requirements and production processes.

Adaptronic Tools and Tool Holders. This is about dimensioning and implementing adaptronic tools and tool holders for the most diverse production technologies, about wear monitoring, and about optimized tool employment to minimize wear. Scientists must also find ways of computing and simulating integrated sensor-actuator combinations close to the tool, taking into account the machine tool structure.

In the following, the author will introduce some implementation examples for each of the above mentioned fields of research and development, illustrating the potential of adaptronic concepts for different machine tools and manufacturing methods. Naturally, this section cannot be exhaustive, but rather must be limited to methods of cutting manufacturing. Since many of the implementation examples are not process specific, this section will provide the reader with a good overview on the state of the art, and he will find many suggestions that will help him to develop his own concepts.

8.3.1 Grinding Machines

External Plunge Grinding

External plunge grinding is a fine finishing technique used for manufacturing cylindrical workpieces of high quality. This technique allows the user to produce workpieces with very little surface roughness and very narrow dimensional tolerances. The metal removing rate of external plunge grinding machines is, among other things, limited by dynamic incidents (e. g. oscillations) that occur between the tool and the workpiece. However, it is possible to compensate for disturbing oscillations by measuring changes in the distance between the workpiece center point and the tool surface (grinding wheel) with sensors, and by positioning the workpiece via the center points with proper phase control. This type of dynamic displacement control is achieved using active damping systems.

Active dampers make use of actuators to produce a force opposing the disturbing vibration at the location of the occurrence. This type of application takes place in a closed-loop control since the disturbance must first be measured at the location of interest. Special process parameters can be identified with the help of a general process model. A control algorithm can be formed that supplies the appropriate control signal to the actuator. The process model and the control algorithm exist in a process computer, which forms part of the active damping system. Active dampers are more complex than passive ones as they can adapt to changing process behaviour and are therefore effective within a much wider band of frequencies.

Based on their ability to generate both high frequency displacements and large forces, solid-state transducers are well suited as actuators for active damping of undesired vibrations in heavy mechanical structures. Figure 8.31 illustrates this feature in the case of external plunge grinding, in which relative displacements between the tool and work piece resulting from dynamic instability (chatter) are compensated for or damped. The additional necessary mechanical energy is generated using a piezoelectric actuator integrated directly into the center points. A system applying this concept demonstrated broad band improvement of dynamic behaviour of the grinding machine [171].

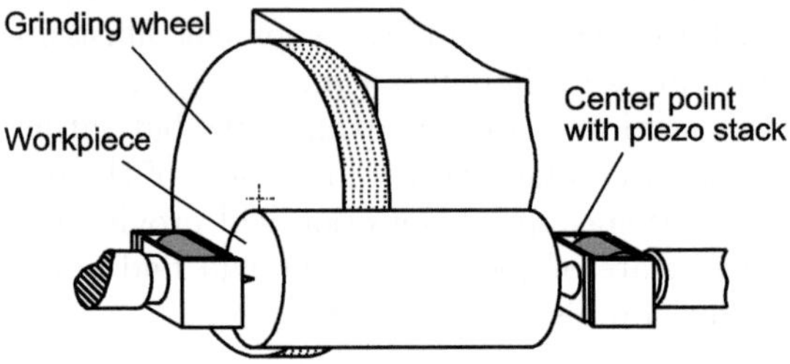

Fig. 8.31. Active vibration damping with piezoelectric actuators in external plunge grinding

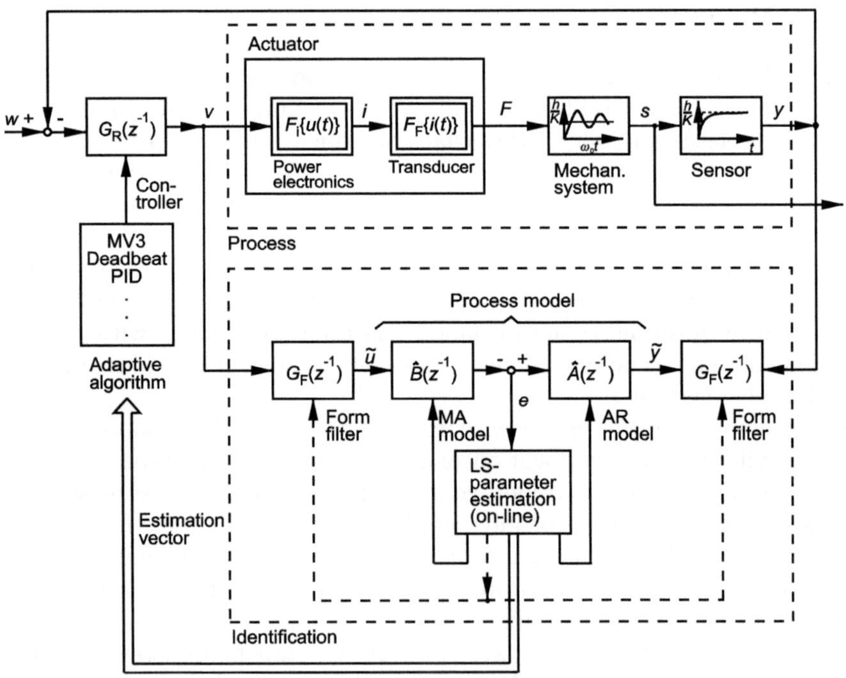

Fig. 8.32. General signal-flow diagram for real-time control of active damper systems

A so-called ARMA process model tracks the real process on-line, with a least-squares (LS) procedure for identifying the unknown parameters of the process – the undesired vibrations. Input and output process parameters are passed through form filters to the individual parts of the model (see Fig. 8.32). The difference e in the model outputs is a generalised error and serves as input to the LS procedure for adapting both model parts to the process. The estimation vector, which contains the identified process parameters, forms the basis of an adaptation algorithm used to adjust the controller according to appropriate strategies (minimal variance, dead-beat, PID, etc.) and follows the changing process behaviour.

Internal Circular Grinding

Internal circular grinding is a production technique that is often used to manufacture roller bearings, see Fig. 8.33. The production costs for such components are mainly determined by the required grinding operations, whereby internal circular grinding is the greatest cost factor. This is one of the reasons why in recent years, users have applied grinding wheels with cubic crystalline boron nitride (CBN) embedded in ceramic material, especially for internal circular grinding.

The wear resistance of extremely hard CBN is several orders of magnitude higher than that of conventional corundum, so that these grinding wheels need to be dressed only after a long period of time. Although CBN grinding wheels are expensive, their use in industry is economical, because the non-productive time associated with dressing decreases considerably. Besides, there are possibilities to protect the expensive CBN grinding wheels against damage due to overstress: adaptive process control is an excellent approach to provide this kind of protection.

With this background, the term 'adaptive internal circular grinding' defines a process control method, for example, in which a process parameter

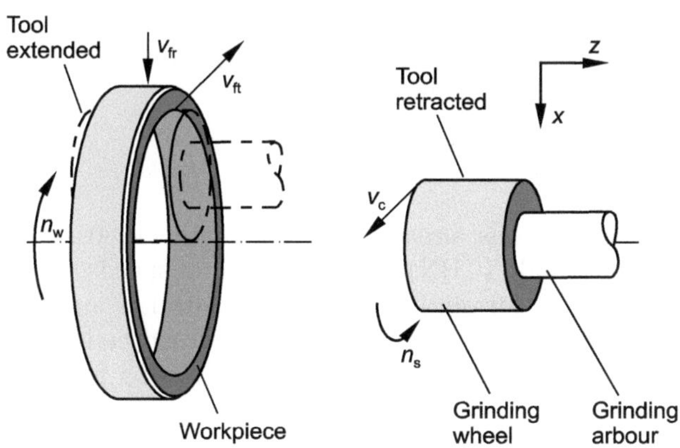

Fig. 8.33. Principle of internal circular grinding (derived from [172])

such as the normal grinding force F_n is constantly measured and maintained at a preset value (controlled variable) within a closed control loop. Any deviation from the command value of the grinding force, e. g. as a result of grinding wheel wear or elastic deformation of the machine structure due to process forces, is compensated by altering the radial positioning velocity v_{fr} (manipulated variable).

There are several possible techniques for measuring the controlled variable. For instance, one can measure the grinding forces in the normal and tangential directions to the surface of the grinding wheel by means of a multi-axis piezoelectric force measurement platform mounted beneath the grinding spindle. Since this is, however, an expensive solution, the active electrical power required by the grinding spindle serves as a control variable, making the spindle a multifunctional element [172].

Internal circular grinding machines are usually equipped with high frequency grinding spindles (three-phase asynchronous motors) and frequency converters for continuously variable speed control. As a measure of the mechanical power emitted by the grinding spindle, one can use the electrical output power delivered by the converter, which can easily be acquired by conventional measurement techniques. The required control variable – one speaks in this case of constant-power grinding – is therefore almost free of charge.

Many commercial applications have proven that adaptive internal circular grinding shortens the phase of rough-cutting considerably compared to the time required when applying conventional grinding. This time savings result from the immediate and quick rise in the normal force to the command value and, consequently, the quickly established machining rate. In addition, if the time characteristic of the control value is known, it is possible to implement a contact sensor to reduce the periods of unproductive air grinding.

The method applied here for measuring the control value requires a fast reacting and soft mechanical structure at the place where the process force acts. Internal circular grinding fulfils this requirement automatically due to the small dimensions of the grinding wheel (i. e. small moment of inertia) and the slim grinding arbour (= elastic beam). In contrast, external plunge grinding will not necessarily fulfil this condition.

Adaptive Dressing

The following grinding application was already patented in the 1980s. The use of cubic crystalline boron nitride (CBN) permits grinding speeds much higher than those achievable with conventional grinding materials such as corundum. However, tighter precision requirements are placed on dressing the grinding wheel. Piezoelectric actuators are suitable for this dressing process. Their smart properties enable them to be used both as sensors and actuators. Figure 8.34 illustrates one configuration for active dressing using piezoelectric actuators. In the first step of dressing, the dressing stone is positioned close to

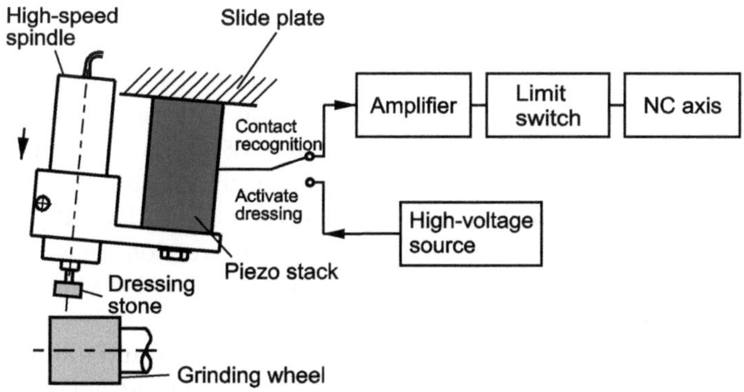

Fig. 8.34. Principle of adaptive dressing with piezoelectric actuators (Source: University of Hannover)

the grinding wheel along the NC axis of the grinding machine. The slide plate stops moving as soon as contact with the grinding wheel has been sensed by the piezoelectric transducer. In the second step, the wheel is dressed with the help of the piezoelectric actuator, making use of its high loading capacity.

8.3.2 Milling and Turning Machines

Milling

Incline Correction Table. Thermal deformations in machine tools – like process forces – lead to undesirable displacements between the tool and the workpiece at the site of action. The degree of structural deformation is determined by the relative position of the heat source to the assembly group, and therefore by the feed axes. Furthermore, the thermal load is characterized by heat sources, which depend on the technological quantities in different ways. This is why deviations of position and orientation above the workspace are variable, i. e. they are dependent on the position.

Control of deformation through intelligent thermal design of the structure (thermo-symmetric layout) and heat flows (displacing of or insulation from heat sources) has already reached a high standard, so not much further improvement is to be expected. Therefore, the effects of remaining thermal deformations must be eliminated by means of focussed correctional movements between the tool and the workpiece. This requires the acquisition of their position deviation at the point of action and superimposing this value to correct the command movement. Since the deviation between the workpiece and the tool is not directly measurable during the removal of material, it is determined indirectly by means of a compensation model via substitute parameters.

Due to the varying time demands, this compensation model is broken up into a thermal state model and a correction model. The former is to provide information about the current thermal machine behaviour, taking into account the process and environment dependent thermal load and the process-dependent structural changes due to feed motion. The state model acquires the deviation of position and incline at the point of action for several discrete points within the workspace. These points are interpolation points, which serve the correction model to determine the required corrections of position and incline of any point within the workspace. The corrections must be computed in time with the position control by the correction model whereas the state model can be updated much slower.

In a concrete case, the thermal deformations in a 3-axis drilling and milling machine were to be compensated [173]. Since the machine tool did not provide a rotation axis, which could have been used for correcting the occurring incline, a separate piece of final controlling equipment was required. Figure 8.35 shows that an incline correction table was mounted on the machine tool bed instead. With four position-controlled piezo actuators it is possible to set inclines in φ_{xz} and φ_{yz}, whereby the correction model provides the required command values of the position. The four linear axes are used under the aspect of the symmetric distribution of stiffness; due to the redundancy within the system, all axes must be driven simultaneously. Figure 8.36 shows the construction of the axis, which can cope with a force of $-10 \ldots +15$ kN in the z direction (preload force 10 kN) and which has an overall stiffness of 360 kN/μm with a regulating range of 60 μm [173].

Milling Spindle. Milling with long, slender tools can result in the milling cutter bending due to the process forces as demonstrated (excessively) in Fig. 8.37 (left). The consequence would be a loss in production quality and productivity. To avoid such disadvantages the process forces might for example be pre-estimated and used to intervene in the machine controller to correct the milling path in-process. Another approach involves an adaptronic milling spindle that, independently of the machine controller generates a correcting movement by inclining the spindle, see Fig. 8.37 (right). Such a spindle system has been implemented within the scope of the previously mentioned DFG priority programme at Hannover University [174].

Figure 8.38 gives a simplified 3D illustration of the adaptronic spindle. The lower clamping ring is assembled at the shell of the milling spindle, the supporting ring is bolted to the machine frame where a torsion-proof membrane leads the torsional moments from the milling process directly into the housing. Between the clamping ring and the supporting ring there are, equally distributed, three piezoelectric stack translators and three bias springs to protect the piezo actuators from damaging tensile stresses. Spherical discs at both ends protect the actuators from shearing stresses. The stiffness of the bias springs is around 0.5 N/μm; this is about 120 times lower than the stiffness of the actuators so their maximum achievable displacement is hardly affected.

With the help of this auxiliary device it is possible to orient the milling tool in three directions in a controlled way. In an experimental set-up the

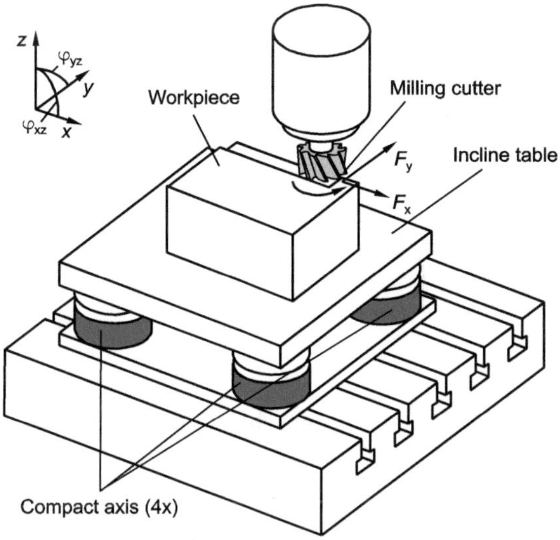

Fig. 8.35. Application of an incline correction table during front milling (derived from [173])

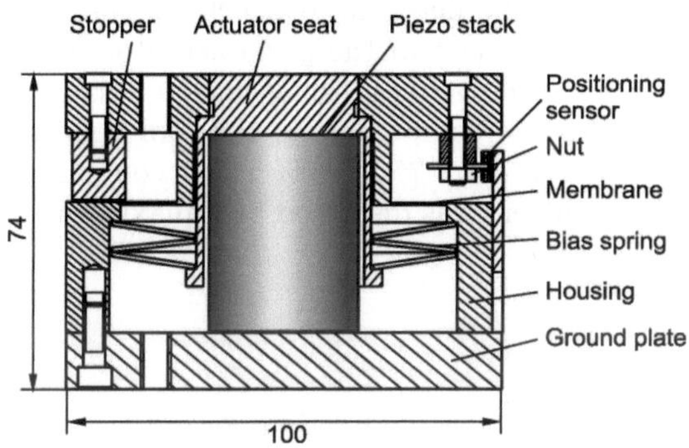

Fig. 8.36. Structure of the axis for the incline correction table (derived from [173])

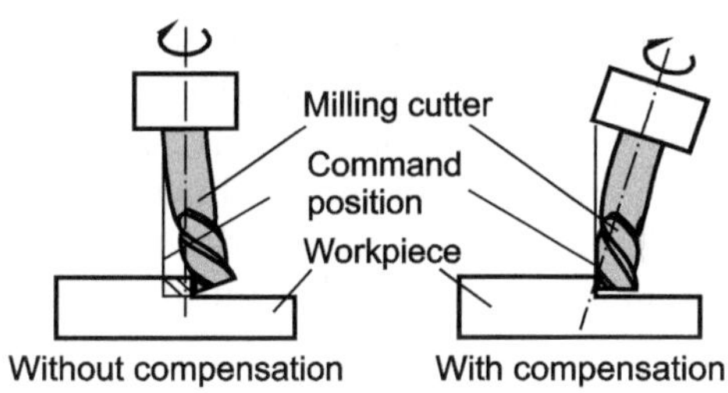

Fig. 8.37. Bending of the milling cutter, exaggerated (derived from [174])

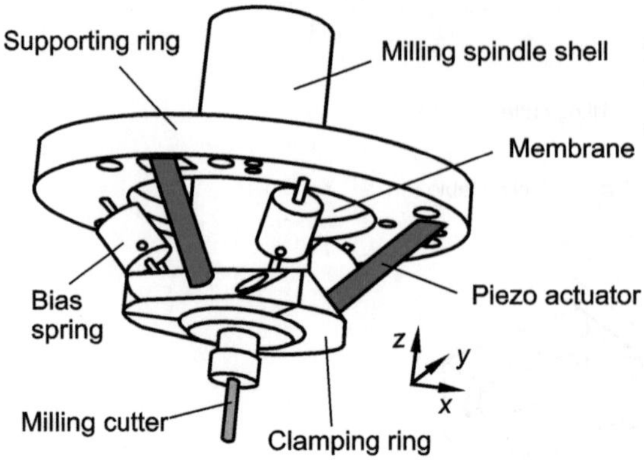

Fig. 8.38. 3D depiction of the adaptronic spindle system, based on a CAD model (derived from [174])

tool was clamped to a load sensing platform to register the force between the tool and the workpiece in-process. With knowledge of the milling cutter stiffness the actual tool bending could be determined. Consequently, the new command position of the tool could be determined which was used to calculate the necessary command displacement of the piezo actuators with help of the inverse spindle kinematics (their actual displacements were registered by strain gauges). Compared to no compensation, the tool bending error when milling aluminum was reduced by 75 ... 90% using compensation, where the tool (100 mm milling cutter) was displaced by a maximum of ±150 μm.

Turning

Complete machining of a workpiece during a single fixation not only reduces the investment costs and the throughput time but also leads to an improvement in the workpiece precision. In the case of large dimension lathes (turned diameter: 1000 ... 3000 mm, z displacement: 4000 ... 8000 mm), however, the machining precision is limited by the fundamental properties of the feed axis. Apart from the guiding properties of the z axis (longitudinal axis), it is the x axis (transverse axis) that influences the precision of the finishing considerably. Its positioning properties limit the reliable correction of deviations of the desired diameter as well as of the required surface quality during single-fixation machining.

The solution to this problem was a numerically controllable precision positioning device, which is parallel to the x axis and which is equipped with an integrated tool holder [175]. Figure 8.39 shows its basic structure: two membranes – each clamped on one side to the frame and on the other end to the

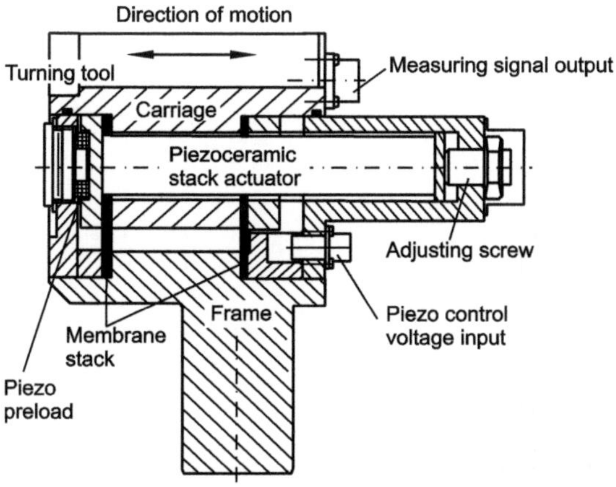

Fig. 8.39. Construction of a precision positioner with a piezo actuator (derived from [175])

side of the tool holder – form a zero-play parallel spring device conveying the carriage with the turning tool (displacement force 10 kN) when driven by the piezoelectric stack actuator. The position of the carriage is measured relative to the frame via an integrated, photoelectric displacement measuring system (not illustrated in Fig. 8.39). The positioning device achieves displacements of 0.1 µm to 0.2 mm.

Figure 8.40 illustrates the control and integration of the positioning device in a CNC machine controller. The digital displacement signal of the micro axis (x'_{act}) is available via a digital measuring system input to the CNC. The deviation from the desired position is then computed in the CNC, using the programmed command value $x'_{\text{com}}(z'_{\text{act}})$ and the actual value x'_{act}; the corresponding digital error value is converted into an analogue signal. This signal is available at the analogue output of the CNCs axis control interface, and is sent to a high voltage amplifier, which contains a PI controller as an integrated component. The amplifier then supplies the piezo control voltage for the position-controlled micro axis.

Experimental and practical tests with such an additional X2-NC micro axis have proven that it is possible to perform absolutely stick-slip free motion, even at extremely slow trajectory speeds, as well as to achieve reproducible positioning in the sub-micrometer range [175]. This enables the user to increase the machining precision, e. g. through correction of the systematic reference trajectory errors as well as deflections of the workpiece due to gravity. Furthermore, this axis enables expanding the process functionality, e. g. with cambering. Thanks to its robust design and the high static and dynamic stiffness of the axis, it is possible to achieve outstanding surface quality.

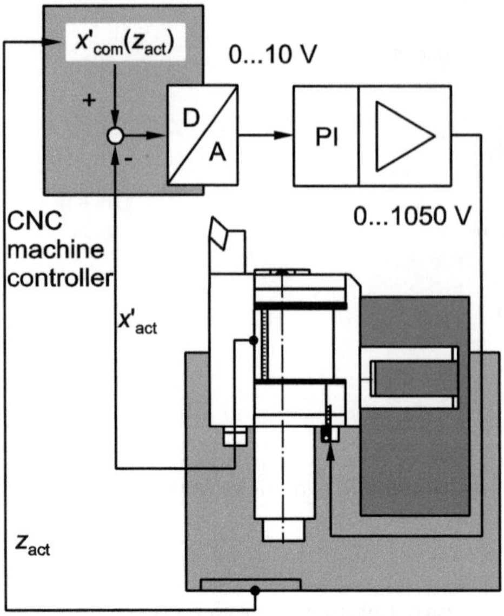

Fig. 8.40. Integration of the micro axis into a CNC machine controller (derived from [175])

8.3.3 Deep Drilling Tools

Inside Turning

Inside turning with a long boring bar fixed on one side is used for processing large housings for turbines and cement ovens, for instance. However as the protruding length of the boring bar grows, it becomes harder to control this manufacturing method: even in the case of small cutting-volume rates chatter vibrations can occur, forcing the user to stop processing long before the installed machine power has been exploited. The strongest, and therefore most relevant dynamic compliance occurs at the first bending eigenfrequency (order of magnitude: some tens of Hertz) of the boring bars free end, and is many times greater than the static compliance.

Apart from design measures during the design phase and special process control strategies for reducing undesirable oscillations, another approach consists of feeding damping-proportional counterforces into the structure via a dynamic add-on system. In principle, such forces can be generated absolutely or relatively. Absolute exciters are supported by a seismic mass, which is accelerated. A reaction force, the dynamic exciter force, is generated, which is fed into the structure to be damped. Relative exciters, in contrast, are able to feed forces into the structure, because they are supported by a fixed reference point, or because they are directly integrated into the force flow of the machine tool.

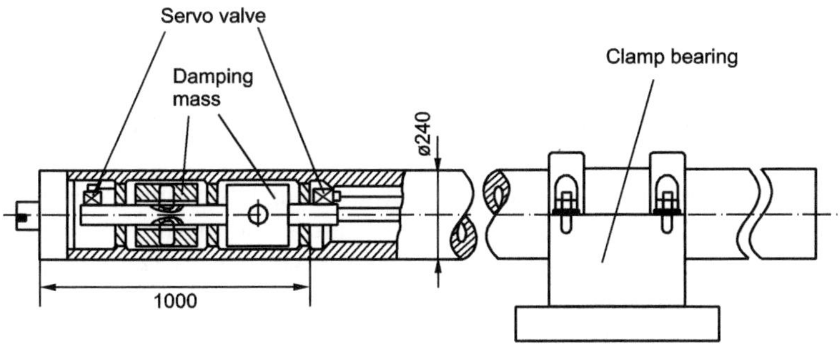

Fig. 8.41. Arrangement of the dampers within the boring bar [176]

Tools with protruding components, such as boring bars, cannot be damped with relatively operating add-on systems since, in this case, it is technically impossible to support the exciter with a fixed machine component. Subsequently, the required damping forces must be generated absolutely. According to Newton, they are proportional to the seismic mass and to the produced acceleration, which for a harmonic motion can be described as $\ddot{x} = -\omega^2 x$. Subsequently, the force depends on the square of the oscillating frequency and on the displacement amplitude of the damping mass. Since the operation of boring bars requires high dynamic forces (kN range) at low frequencies (<100 Hz), the damping mass can weigh several tens of kilograms.

Figure 8.41 illustrates an implemented test tool [176]. The auxiliary system consists of two exciter units, which work perpendicularly to each other, and which are able to generate – when independently controlled – dynamic forces in any radial direction (0... 360°). The damping masses of 50 kg each are shaped like hollow cylinders and are moved by hydraulic drives. Two absolute measuring sensors, situated close to the site of action between the workpiece and the tool, measure the oscillation velocity, because only a velocity-proportional control parameter increases the damping effect. By applying a PD controller, tests were able to decrease the dynamic compliance of a boring bar at the relevant first eigenfrequency of 10 µm/N to less than 1 µm/N (corresponding to −20 dB); the amplitude of the undesirable oscillations in line with the cutting force was reduced to about 20%.

8.3.4 Adaptronic Machine Components

Adaptronic Strut. Process loads as well as errors in manufacturing and assembly of machine components cause the displacement of a machine tools tool-center point, which influences its accuracy of work in a negative way. Since in many cases an improvement of the production accuracy of machine components is not favorable either for economic or for technological reasons, the insertion of active components that autonomously register a state of load or deformation and can perform a corresponding compensation movement, may be a low-cost solution. The active strut described in the following has been developed at the Karlsruhe University and is based on an adaptronic approach and is suitable for compensating for geometric errors as well as static and quasi-static positioning errors of machine tools with parallel kinematics [177, 178].

The core of the strut is a piezoelectric stack translator that produces compensating displacements and forces as an actuator and at the same time measures the load dependent length change of the strut as a sensor (self-sensing principle, compare Sect. 6.9). As a piezoelectric sensor cannot register static loads and displacements (see Sect. 7.3 and Fig. 6.130) a solution has been found: Making use of steel wire (and a lever) the actuator is preloaded mechanically thereby facilitating positive as well as negative displacements and forces (see Fig. 8.42). At the same time the wire functions to produce a load-dependent sensor signal. To this effect it is excited electromagnetically to its natural frequency of oscillation. If the strut is exposed to an extreme load the mechanical wire strain changes and with it the frequency of the force working on the piezo transducer. Due to the direct piezo effect an electrical alternating voltage arises whose frequency is a measure for the external load and that can easily be separated from a quasi-static sensor signal (dotted line in Fig. 8.43) by electronic means [178].

The described strut was implemented in welded steel construction with the dimensions Ø 70 x 700 mm. Here a piezo transducer of Ø 15 x 60 mm was used and the wire in the form of a strip of spring steel oscillating with a fundamental frequency of 1200 Hz. Comprehensive tests with this prototype in an experimental set-up showed that under the maximum load of ±2000 N a stroke of ±25 µm could be reached, while the expectations on the dynamic behaviour of the strut for stepped external loads have been completely fulfilled.

Hydrostatic MR Fluid Bearing. A hydrostatic bearing consists principally of two sliding surfaces that are separated by a thin film of oil. The oil is pressed with a constant flow rate Q by an external pump through the evolving gap, see Fig. 8.44. A disadvantage of hydrostatic bearings is that the

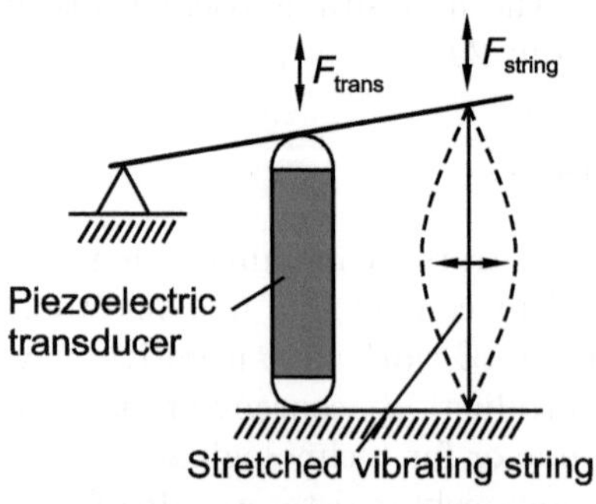

Fig. 8.42. Concept of the adaptronic strut [177]

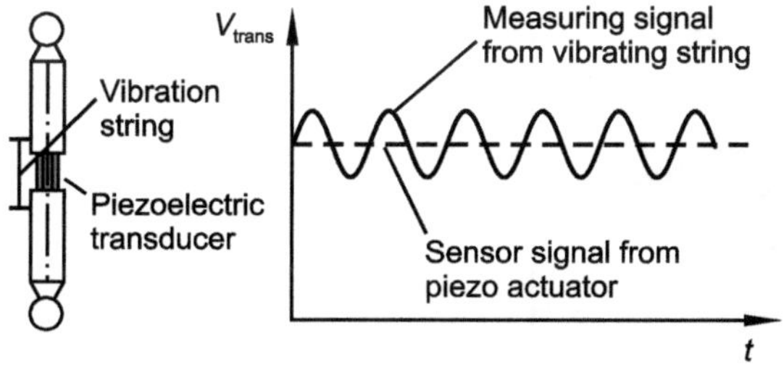

Fig. 8.43. Separation of the sensor signal from the piezo transducer and the vibrating string [177]

bearing gap h varies with the payload F. Conventional systems compensate this dependence with a change in the oil flow rate, e. g. by means of external valves. This leads to a poor response time of the bearing to load changes.

In order to circumvent this disadvantage, from [179] replacing the oil with a multifunctional magnetorheological fluid is suggested, i. e. to employ both its hydraulic and rheological properties, whereby the latter can be controlled through an external magnetic field. Furthermore, the upper sliding surface is put into rotation (constant rotation speed n), which increases the shear rate of the fluid inside the gap (see Sect. 6.6). Due to the fact that the viscosity of the fluid depends on the shear rate, a rotation leads to a change in the gap.

Figure 8.45 indicates the decrease of the bearing gap h with the increase of the payload F. Further on, one can see how a constant bearing gap can be achieved: for a payload, for example $F = 25\,\mathrm{N}$, and a gap of $h = 310\,\mu\mathrm{m}$ a certain value of the magnetic field B is necessary (marker A in Fig. 8.45). When the payload changes to 160 N, the bearing gap remains constant if the magnetic field is changed to another value (marker B). In closed-loop control, a nearly infinite stiffness can be achieved, limited only by the resolution of the system for measuring changes of the gap width.

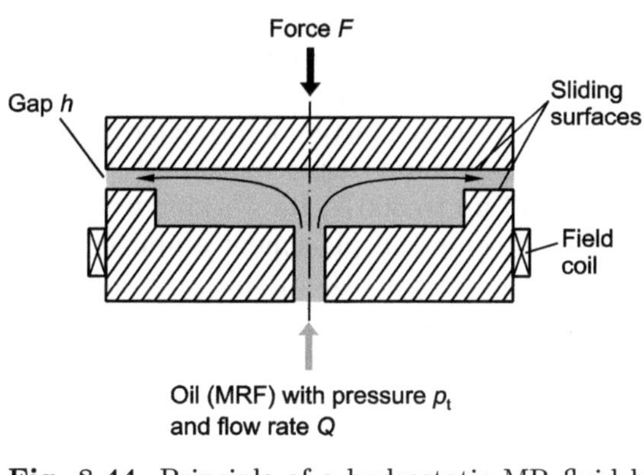

Fig. 8.44. Principle of a hydrostatic MR fluid bearing. The field coil is required for the MRF (derived from [179])

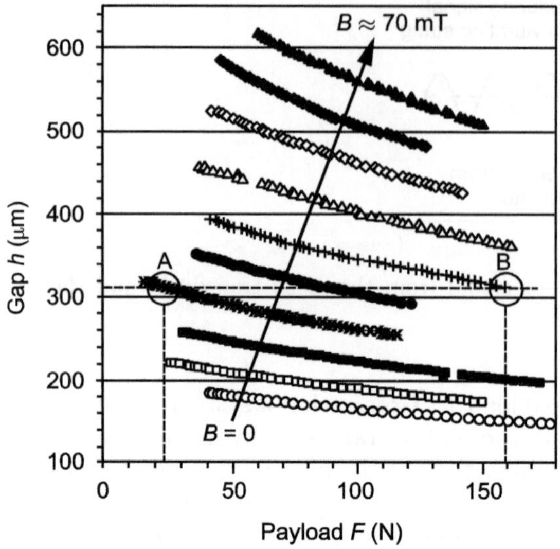

Fig. 8.45. Change of bearing gap h with the payload F for different flux densities B [179]

Positioning Drive. One possible application for an ER fluid valve is the positioning unit shown in Fig. 8.46. The design consists primarily of a hydraulic cylinder, four ER fluid valves arranged in the form of a full bridge, a four-channel high voltage source, and a controller. The position x of the piston is measured by a displacement sensor and the force F working against the piston by a force sensor. A pump operating at a constant speed generates the pressure in the hydraulic circuit. Continuous control of the pressure drop across the ER fluid valves enables continuous control of the pressure on the piston and therefore its direction and speed of travel as well as its holding force.

The above concept was the basis of an ER actuating cylinder which was developed as a highly dynamic power supply for a testing machine. It made use of the ER fluids hydraulic properties, since the electrorheological (as well as the magnetorheological) effect is, as a matter of principle not capable of generating forces directly. For maximum displacements of ±35 mm the actuating cylinder produced forces of up to 300 N. Its working frequency reached up to 100 Hz and for smaller force and displacement amplitudes up to 400 Hz. Similar actuating cylinders were used in other applications in combination with a hydraulic accumulator functioning as a spring-damper element [180]. They produced forces of up to 5 kN (operating pressure 120 bar) and strokes of up to ±85 mm, while the working frequency reached up to several Hertz.

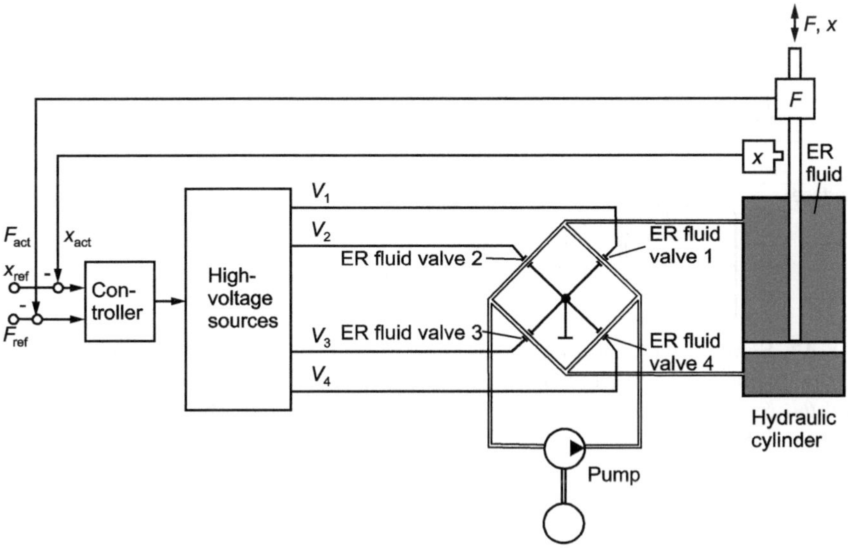

Fig. 8.46. Example of a positioning drive with ER fluid valves

Adaptive Joint Connection. An interesting variation of active dampers is referred to by its creators as a semi-active damper'. Semi-active devices are those in which the passive stiffness and/or damping properties are varied in real time based on a feedback signal. Semi-active elements have, like passive elements, the ability to dissipate system energy. Through implementation of an appropriate adaptive control law, semi-active elements are able to adapt to different vibration environments and/or system configurations. Another advantage over passive damping elements is their ability to utilize sensor information from other parts of the structure to form a so-called non-collocated sensor/actuator architecture. Typically, semi-active elements have low power requirements and are less massive than their active counterparts.

One of the most significant sources of passive damping in large mechanical structures are bolted joint connections. Energy dissipation in joints occurs primarily as a result of impact and dry friction present at the sliding interface. With nonlinear, local control, the energy dissipated by the frictional joint can be maximized, using the normal force as control input. This concept of varying the normal force in a frictional joint to enhance the energy dissipation from a vibrating structure has accomplished using a piezoelectric disc plate (see Fig. 8.47). Voltage applied to the piezo stack tries to extend the piezoelectric material, resulting in an increase of the normal force. If the dynamic behaviour of the piezoelectric material can be neglected, the normal force is proportional to the input voltage. The goal of the controller is to prevent the semi-active friction damper from sticking [181].

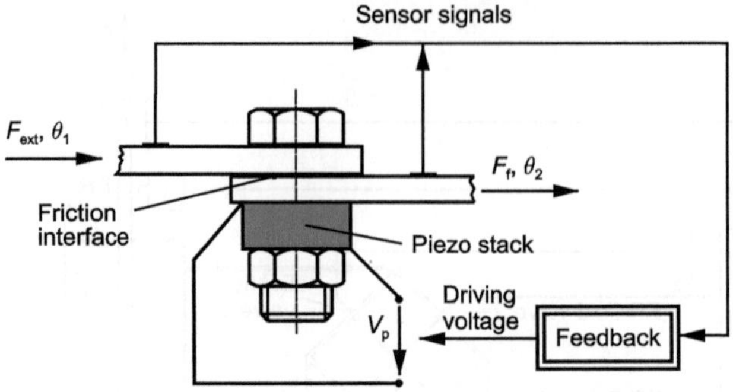

Fig. 8.47. Semi-active joint connection. The system is excited by an external force F_{ext}, the displacement in front of and behind the friction interface are denoted by θ_1 and θ_2, respectively, and F_f is the friction transmitted by the semi-active joint connection [181]

8.3.5 Conclusion

Within the context of machine tools and manufacturing plants, the term adaptronics defines systems which feature a high functional density because they not only introduce the classical load-bearing and shape-defining functions into a static structure, but also sensory and actuator properties. The integration of convenient control algorithms enables such a system, and subsequently the entire structure, to become adaptive, i. e. it can be adapted to changing operating and environmental conditions.

Up to now, adaptronics has mainly been an object of research at universities, and the industry has only just started to apply adaptronic systems; a most recent example is an intelligent mineral casting mould [182]. This machine bed for machine tools is to adapt itself to changing thermal conditions, thereby preventing the structure from any deformations. To accomplish this, temperature sensors are embedded into the mineral casting material (polymer concrete); cooling elements, which are integrated into the bed, serve as actuators.

Finally, it is worth noting that the development of adaptronic components to be used for the most diverse machine tools and manufacturing plants is a multidisciplinary task requiring the collaboration of production, civil, machine and control engineers. A multidisciplinary approach of this task is crucial for the successful implementation of optimised systems with structurally integrated actuators and sensors and adaptive control algorithms in machine tool applications.

8.4 Adaptronics in Civil Engineering Structures

G. Hirsch†

Civil engineering structures are exposed to dynamic loading from several sources, including high winds, earthquake ground motion, rotating and reciprocating machinery. While large deflections of tall buildings do not necessarily pose a threat to the safety of a structure, they can cause considerable discomfort and even illness to building occupants. The requirements for control of civil engineering structures concerning comfort are significantly less demanding than those for safety.

The concept of active control – a first step to adaptronics in civil structures – is an attempt to make structures behaviour more like aircraft, machinery, or human beings in the sense that they can be made adaptive or responsive to external loads.

Kobori and Minai [183] advocated the concept of 'dynamic intelligent buildings' capable of executing active response control when they are subjected to severe earthquakes.

In the U.S., Yao [184] marked the beginning of active control research when he proposed an error-activated structural system whose behaviour varies automatically in accordance with unpredictable variations in the loading, as well as environmental conditions. A remarkable number of different systems, mechanisms and devices has been proposed by researchers in the past 30 years. Although each of them introduces a certain novelty, all the presented systems can be classified in three groups:

- active tendons;
- active mass dampers; and
- pulse control.

It should be noted that adaptronics in civil engineering structures is not making progress at present. Moreover structural control is not the same as control theory, which has been developed in electrical engineering and applied mechanics, or the methods for control of space structures. The essence of structural control is the management of the performance of relative massive civil structures that require the application of large control forces but do not require a high degree of accuracy. Control of space structures have developed knowledge that, to some degree, provides information of value to structural control but does not solve the problems of civil structures control.

The control of earthquake response of structures is only one part of structural response control. The effects of wind, explosive shock, micro-tremors, etc., are also of concern. The control of the response of sensitive items, such as medical equipment, emergency power equipment, etc., must be considered.

The first international conferences on structural control were held at Waterloo University, Canada, in 1979 and 1985 [185, 186]. Although active control has been researched and utilized in many applications throughout the last few decades, it has been only in recent years that applications in civil engineering structures have been contemplated. Progress in development of structural control has been summarized within the scope of the International Workshop on Structural Control, Hawaii 1993 [188]. The First World Conference on Structural Control, Pasadena 1994 was the first official act of the newly formed International Association for Structural Control. State-of-the-art reports, second generation of active control were topics of the conference.

Much progress has been made in research and development of active control technology in the U.S. and in Japan [189]. However, the true potential of active control has not been fully exploited. Concerning adaptronics in civil engineering structures, it should be noted that for the control of large motions, large uncertainties in the structural model exist since tests are not at amplitudes corresponding to building collapse, and time-varying non-linearities may exist in the foundations or soil that are significant to the lower vibration modes of interest [190].

This section is about the state of the art of active control in civil engineering, the second generation of active control, the results of experimental and full-scale tests in Japan and the U.S. and conclusions relating to the realization of adaptronics in civil engineering structures. It doesn't claim to be complete and is a selection of the authors experiences in passive and active control.

8.4.1 State of the Art for Active Control of Civil Engineering Structures

Concerning control of civil engineering structures, we have to distinguish between

- slender, tower-like structures;
- tall buildings; and
- long-span bridges (suspension and cable type).

The structural control depends from the dynamic properties (mass distribution, stiffness and damping) as well as the dynamic loading (wind, traffic, machinery, earthquake). From an engineering point of view, the different realities require adapted measures. Practical guidelines relating to vibration problems in structures are given by Bachmann et al. [191].

Tower-Like Structures

Tower-like structures are understood, in general, to be slender, tall structures (as television towers, lookout towers, chimneys, masts, and bridge pylons). Usually, gust-induced vibrations in the wind direction predominate, especially those at the fundamental bending frequency. The vibrations connected with vortex shedding that is transverse to the wind direction, however, can be more important. Particularly sensitive in this respect are steel chimneys (of welded construction, not insulated or lined with masonry, and with a fixed base). Vibrations of chimneys, masts and other low-damped tower-like structures lead to structural safety (fatigue) and serviceability problems. The occurrence of unacceptable vortex-induced cross-wind vibrations cannot be ruled out. As remedial measures in this case, both passive aerodynamic and mechanical aids should be mentioned. From an economic point of view, tuned mass dampers (TMD) are becoming more and more popular. The mass ratio (the mass of the TMD to the generalized mass of the structure) is often chosen to be 0.05.

An illustrative example (see Fig. 8.48) of the application of remedial measures to an existing group of steel chimneys when another chimney is added is given by Hirsch in [192]. However, in cases of transient loading by wind-gusts or earthquake, the effectiveness of TMDs will be reduced and therefore the active control will be particularly important in these cases of dynamic loading.

In practice, the optimum values of natural frequency and damping are usually not attained precisely. However, the sensitivity of such added systems with respect to deviations from the optimum is relatively small. Moreover, the TMD is not effective from the beginning of critical excitation of the main structure, as Fig. 8.49 shows.

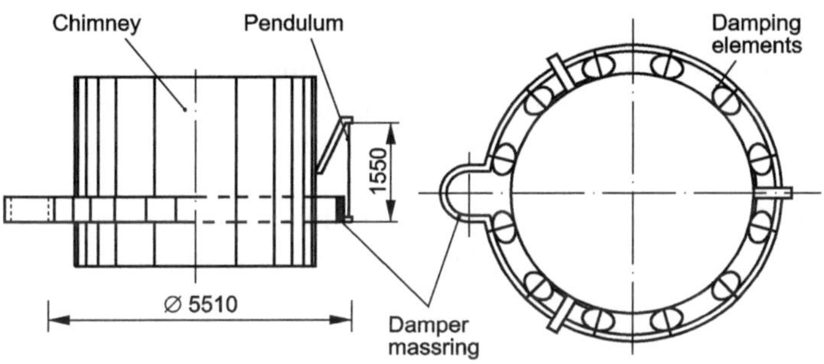

Fig. 8.48. Design of pendulum-suspended TMD for chimneys

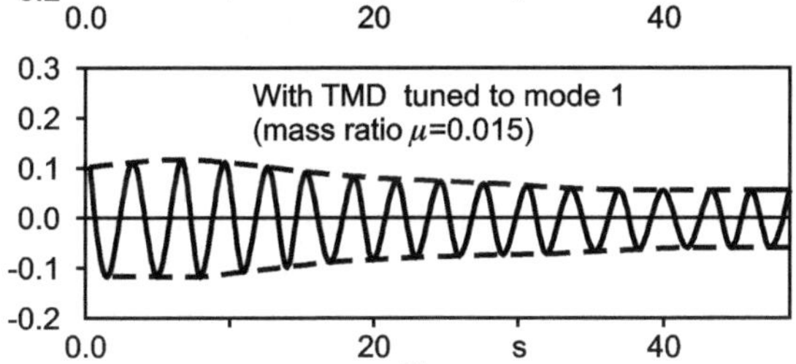

Fig. 8.49. Uncontrolled and controlled top displacement

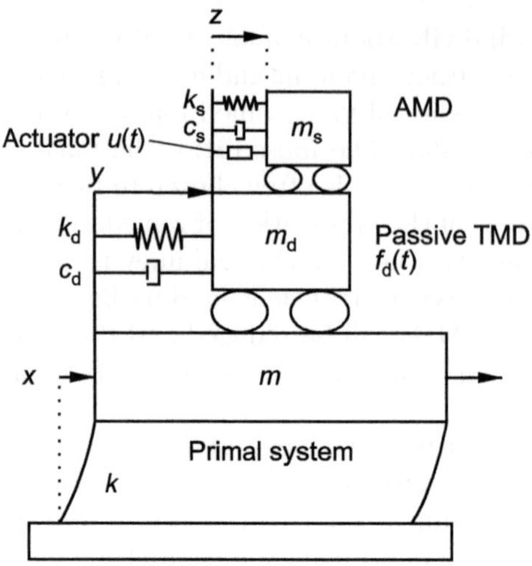

Fig. 8.50. Active-passive composite TMD

The time delay between the beginning of a dynamic response and the counteraction of TMD is one of the most important topics in structural control (passive and active) from an engineering point of view. Therefore, Koshika et al. [193] have proposed an active-passive composite tuned mass damper (APTMD). The feasibility and the practicability of the theory was confirmed by demonstrating its control effect using an experimental model, as Fig. 8.50 shows.

The APTMD is one of the proposed active control device in civil engineering structures, especially of tall buildings against strong-wind effects and earthquake loading, described in the next subsection.

Tall Buildings

Tamura [194] shows several popular mechanisms of active vibration control systems for civil engineering structures. Mass damper systems, which use the inertia force of the auxiliary mass as the reaction force, are most commonly adopted. They need only a small space for installation and they can suppress the response of tall buildings very effectively during strong winds.

The mass damper systems are classified into four types from the energy supply point of view: the passive type (TMD), semiactive type (SAMD), full active type (AMD) and hybrid (HMD). The TMD is effective if the natural period of the device is tuned well to that of the tall building (mostly the fundamental mode). However, if the period of the device or the building gets changed, or the predominant period of the building response differs from the period of the device, the effectiveness of the TMD cannot be maintained. Therefore the other three types are developed to solve this problem. The

SAMD is designed to ensure the optimum adjustment modifying the characteristics of the devices. Both of the AMD and HMD are the systems to use active control forces by supplying the energy from outside for suppressing the vibrations of the buildings. The AMD has no spring and no damping element; whereas the HMD is equipped with adequate values of spring and damping elements. In the case of a power blockout, the effectiveness of TMD will remain, and this is important in relation to reliability of control.

Active bracing or tendon systems can apply the control force directly to structures. This system is one of the most studied mechanisms. The reasons for favouring this mechanism is that the bracings and the tendons are existing structural members. This is important relating to realization of adaptronics in civil engineering structures. This system seems to be effective when the tall building structure is light. However, if the structure becomes larger and the external excitation level is higher, it will be difficult to apply the system because the required control force increases significantly. Moreover, the dynamic behaviour of the bracings or tendons have to be considered.

Hybrid systems combined with base-isolated buildings have been proposed against earthquake loads. They have the possibility of decreasing the vibration of the structures drastically, if the adequate devices for supporting the buildings with low stiffness and the devices for generating the control force accurately are developed. Since the isolation devices have usually nonlinear properties, the control algorithm for such systems has been actively studied recently [183].

Bridges

The dynamic loading of bridges is of different kind:

- traffic load (randomly);
- dynamic wind loads (gusts, buffeting, vortex shedding, galloping, flutter), occurring randomly, harmonically, or in a self-excited manner;
- earthquake (transient).

To avoid or suppress bridge-flutter, know-how transfer from other disciplines of engineering (ship and aircraft engineering) is possible and suitable in order to understand the control application. Domke (in [186]) reported on active deformation control of a 10 m test girder of Aachen University. As Fig. 8.51 shows, pneumatic cushions were supported on steel cables, which were connected to both ends of the girder. Pressures in the cushions were adjusted according to the deflection of the girder until the cable forces counteracted the sum of dead weight and live load.

Moving vehicles induce vibration of bridge girders by two mechanisms: one is due to the moving force acting on the girders in a finite time and the other due to the roughness of the road surface. In the case of cable-stayed bridges, this mechanism will be excited at the higher modes of the bridge deck and the supporting points of the cables. Parametric vibrations

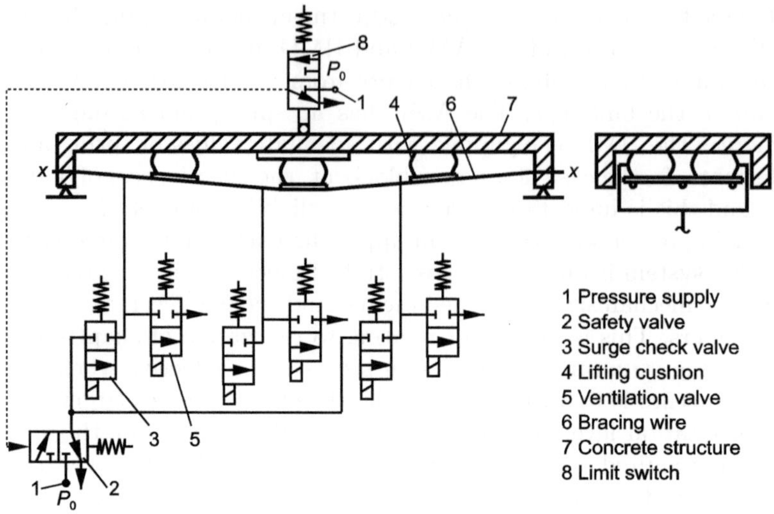

Fig. 8.51. Test girder with control device

of the cables may occur (Bangkok Bridge for example), because vehicles are mass-spring-dashpot systems and the interaction between the girder and vehicles is not negligible in the girder vibration. The traffic-induced vibration causes the following problems: (1) runability of vehicles, especially of high speed trains; (2) fatigue of girders and vibration-induced noise; and (3) excitation of buildings near bridges. The first problem occurred with the Honshu Shikoku Bridge and was reported by Y. Fujino in [188]. The third problem often occurs in elevated urban viaducts. Vibrations of the bridge piers induce waves propagating in the ground and this excited the buildings nearby. As the occupants are sensitive to noise and vibrations, this may entail controversy.

Fujino and co-authors (in [189]) reported about active control of traffic-induced vibrations in highway bridges. Figure 8.52 shows the vehicle, bridge girder with surface roughness, and an active TMD.

This simplified model should be self-explanatory. It is shown that both bridge response and support reaction forces can be significantly reduced. Linear quadratic (LQ) optimal control with full state feedback is used. The passive TMD (around 5% of the beams modal mass) is tuned to the first mode of the beam. Figure 8.53 shows some results. namely the dynamic reaction forces with and without control.

In order to effectively control wind-induced vibrations of long span bridges aerodynamic devices from the viewpoint of required energy are more attractive than mechanical control means. Figure 8.54 shows the simplified model of bridge-section with aerodynamic devices (surfaces) to flutter-control.

Active aerodynamic measures are being used successfully in aeronautical engineering to suppress wing flutter. Using additional control surfaces, con-

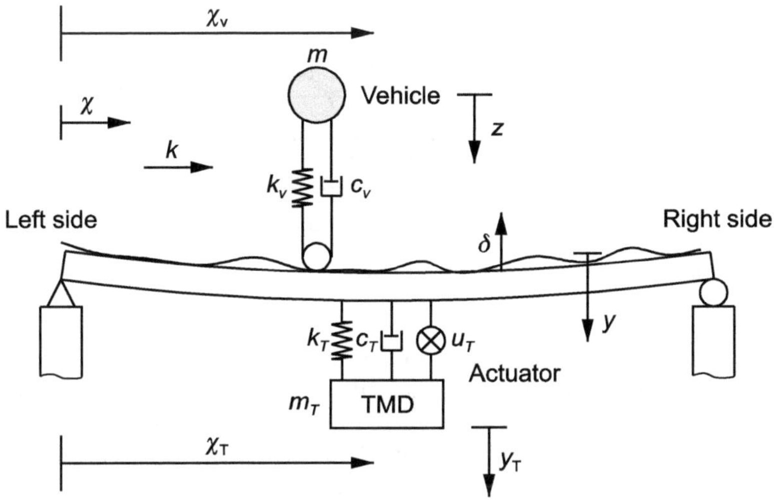

Fig. 8.52. Bridge girder with vehicle and active TMD

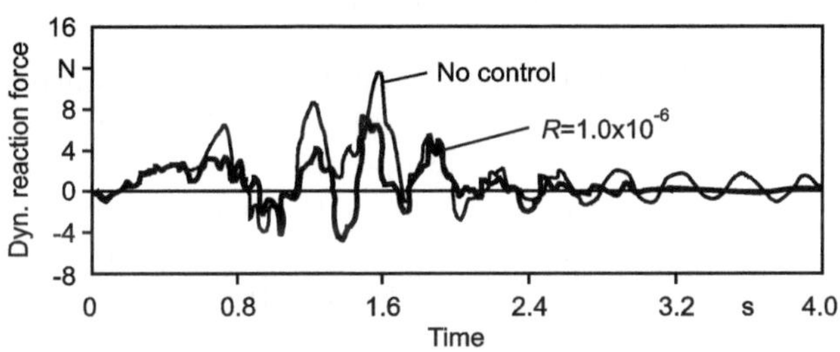

Fig. 8.53. Dynamic reaction forces with and without control

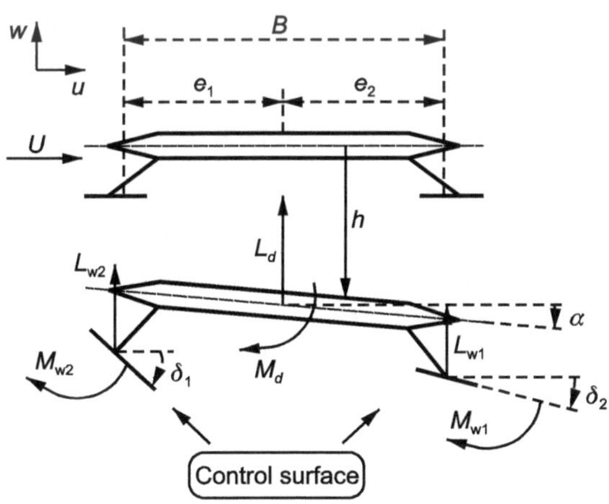

Fig. 8.54. Bridge section with control surfaces

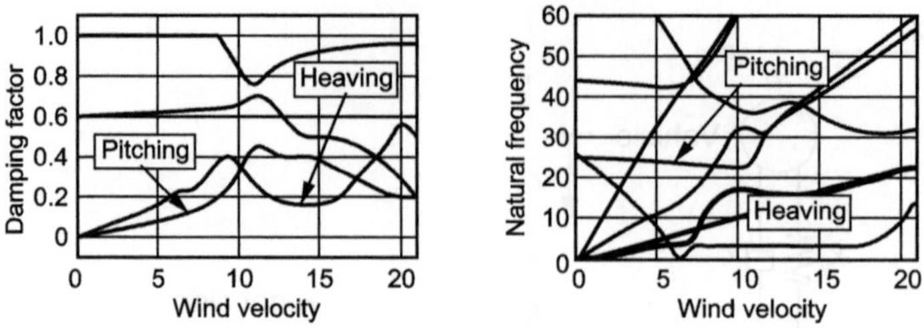

Fig. 8.55. Example of bridge flutter suppression

trol forces that are anti-phase to the inducing forces are implied to prevent the vibration of the structure in the critical flow. Sensburg et al. (in [185]) reported the results of flight tests with active flutter suppression. The suppression of bridge flutter by means of control surfaces will be based on the same idea. Figure 8.55 shows some characteristic results.

8.4.2 The Second Generation of Active Control

Much of the theoretical basis in the development of active structural control over the last thirty years is rooted in modern control theory. For example, most of the control algorithms used in the current operating control systems for large civil structures are based on the principles of the linear quadratic regulator (LQR), as Housner et al. [189] showed. However, it needs to be recognized that control applications to civil engineering structures are unique in many ways and present a set of different challenges. For example, in comparison with conventional control design as practiced during the first generation. Some distinguishing features of civil engineering structural control will be: structure (complex system with more than one eigenmodes, nonlinear behaviour), sensing and actuation (few sensors and actuators, large control forces with higher speed), and control strategies (simple but robust and fault-tolerant control, suboptimal control, implementable control laws).

In incorporating active control into a structure, either as a new design or a retrofit, it is important to consider active control as a member of a family of innovative protection technologies which include, in earthquake loading for example, among others, base isolation and passive energy dissipation. For a specific application, technical merits and cost effectiveness of active control systems can then be evaluated more realistically in this context. Moreover, as stated in the recommendation of a working group on experimental methods [188], the testing of possible control devices that can deliver the required control force (for example) is necessary in order to assess the implementability of theoretical results: practical issues such as time delay and spillover effects can only be addressed after one learns of their magnitude and effects through experiments.

8.4.3 Application of Active Control from Practical Engineering Aspects

Although active structural control has been researched and utilized in many applications throughout the last few decades, it has been only in recent years that applications in civil engineering structures have been contemplated. The acceptance of building applications has similar features in the U.S. and Japan, the leading states in active control of civil engineering structures. However, the acceptance procedures in Japan are expedited by direct involvement of each owner-construction company with the professional community. In the U.S., procedures were identified as to be geared by cost, insurance and performance criteria. While specific performance criteria should be established for individual devices to meet safety and operational requirements, the performance of buildings should meet current safety and service criteria. It was felt that the design engineer should be trusted with this engineering judgement and additional limiting standards should not be established. This will probably encourage and not limit necessary innovation.

The most recent implementations are combinations of add-on devices. While mass dampers provide the majority of implementation (i. e., tuned mass dampers, active mass dampers, hybrid mass dampers), active stiffening or bracing systems, energy dissipation/absorption dampers and hybrid isolations were also implemented in actual applications or full scale experiments. It was noted in [188] that most of the systems were considered as add-ons and the integration of the systems into structural design is not yet completely developed. That is important in relation to the application of active control from practical (engineering) aspects.

The following is a synopsis of the conclusions of several workshops on structural control concerning the development needs in the field of control engineering practice: (1) actuators, including modeling and identification; (2) performance robustness versus stability robustness, improving performance and improving stability for large civil engineering structures; and (3) the development of accurate, reliable and inexpensive displacement sensors would provide a substantial benefit to many approaches in structural control.

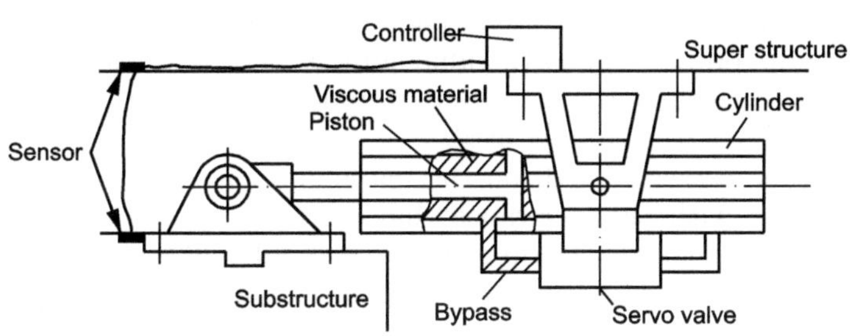

Fig. 8.56. Variable damper

From an engineering practice point of view, the actuators for civil structures are most interesting. Simple and robust actuators are necessary. There are only few practical examples of actuators for civil structures. Figure 8.56 shows an example of an actuator with variable damping showed by Kawashima and Unjoh in [188].

8.4.4 Results of Experimental and Full-Scale Tests (in Japan and the U.S.)

Application of active control of civil structures from practical aspects has been pioneered in Japan. Multiple earthquake loadings and strong-wind effects on tall buildings enforced the realization of new techniques. Synopses of practical examples are given in the Japanese contributions in [188] and [189]. In this section, some practical solutions will be explained.

The Kyobashi Seiwa Building was built in 1989 and is the first building in the world that has an active control system. The active mass driver was installed to suppress dynamic responses caused by earthquakes or strong winds. It was reported from Yutaka Inoue et al. in [188] that the building had experienced several moderate earthquakes and strong winds during which ground acceleration, wind velocities and structural responses had been measured. The measured responses during the earthquakes are compared with the simulated responses by numerical analysis for an uncontrolled structure. Wind-response observations were performed every 30 minutes with and without control. From these comparisons, a remarkable decrease in amplitude due to the active mass driver system was confirmed.

An active tendon system has been examined by using a six-story steel structure [187]. Figure 8.57 shows that a control force is transmitted to the structure through diagonal braces connected to the first floor by servo-controlled hydraulic actuators. The 600 ton symmetric building, as shown, has been erected in Tokyo, Japan. In fact, two control systems has been tested on the structure (a biaxial active tendon system and a biaxial active mass damper system).

In relation to adaptronics in civil engineering structures, the active bracing system represents one of the best possibilities and therefore will be outlined in more detail. Reinhorn et al. [195] reported on the braces, the hydraulic actuators, the hydraulic power supply, the analog-digital converter and the sensors. The observed performance of the system under actual earthquakes and other artificial loadings will be presented.

The design of the braces was based on the maximum control force and the anticipated stiffness with the assurance that buckling will not occur under actuator actions. Circular steel tubes were used with 360 cm length, 165 mm diameter, 4.5 mm thickness and 564 kN strength. The measured stiffness of the braces is 98.4 kN/mm in the x-direction and 73.8 kN/m in the y-direction. Four units of Parker, heavy-duty hydraulic cylinder series 2H-style TC (NFPA style Mx2) were selected as actuators, with the following

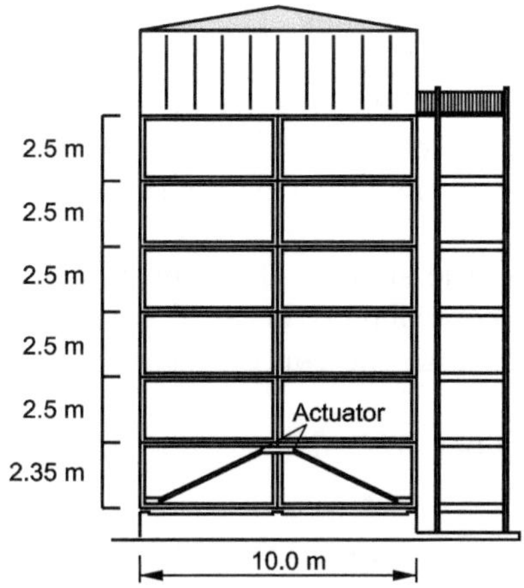

Fig. 8.57. Active tendon system

specifications: 735 mm length, 152.4 mm piston diameter, 63.5 mm rod diameter, ±50 mm stroke and 344 kN average capacity. Although the expected movement in the actuators was only ±12 mm, larger-sized actuators were chosen to enable length corrections during construction. In future applications, a much shorter actuator would be sufficient. The average capacity of the actuator was based on the working pressure (20.68 MPa) of the hydraulic oil and the average piston area. Two hydraulic actuators were coupled in series in each direction and are monitored by one servovalve and one servovalve controller of type MTS 458. The inner control loop for the hydraulic actuators is used for position feedback. The servovalve MTS 252.2x can supply up to 55 l/min at a pressure drop of 6.89 MPa.

The final design of the hydraulic system allows the active system to remain ready for full power controlled operation, while requiring the hydraulic pump to operate for only a few seconds each hour to keep the system full charged. An analog-digital converter was chosen based on the requirements that the analog controller must be compatible with the hydraulic service manifold and with the servovalves, and be capable of simultaneously controlling the two sets of servovalves.

The microcomputer executed the control algorithm, monitored the status of operation of various hydraulic components and monitored the status of the structural system. The system consisted of a PC computer with an Intel processor equipped with a math coprocessor. Two analog-to-digital and digital to analog conversion boards provided interface for up to 16 channels of differential inputs from sensors and four channels of analog outputs to controllers. In addition, 16 digital logic channels were available on the computer boards.

The control system had four servovelocity seismometers of type Tokyo Sokushin VSE 11 for each principal direction of the structure, with an output range of ±100 cm/s. The velocity sensors were located on the ground, at the first, at the third and at the sixth floors of the building. The same sensors could provide acceleration information up to $\pm 1000\,\mathrm{cm/s^2}$. Additional transducers were mounted at each floor to monitor building behaviour. Each actuator was equipped with a displacement transducer (LVDT) having a range of $\pm 12\,\mathrm{mm}$, which is used to adjust the length of the brace via the servovalve loop.

Some recorded samples show in Fig. 8.58 the structural response under 32% El Centro earthquake (uncontrolled and controlled structure). Also from the top-floor deflection (Fig. 8.59) one learns that the control measure will be effective with time delay.

The results presented in the Reinhorn paper [195] demonstrate the following:

- The concept of an active tendon or bracing system, originated almost 30 years ago, has led to the successful development of the device for civil engineering structural control.
- The success of the full-scale active brace system performance is the culmination of numerous analytical studies and carefully planned laboratory experiments involving model structures.
- The control system can be implemented with existing technology under practical constraints such as power requirements and under stringent demand of reliability.
- The use of the control measure in existing structures can be a practical solution for retrofit, as demonstrated by this full-scale experiment. Note that the active braces were added only after the structure was completed.
- The experience gained through the development of this system can serve as an invaluable resource for the development of active structural control systems in the future.

A new type of tuned active damper (Mitsubishi) for high-rise buildings is shown in Fig. 8.60. As high-rise buildings are being put up everywhere, and

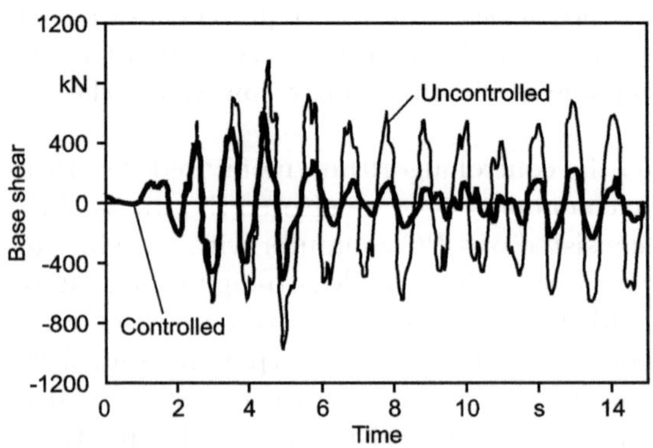

Fig. 8.58. Base shear response of structure

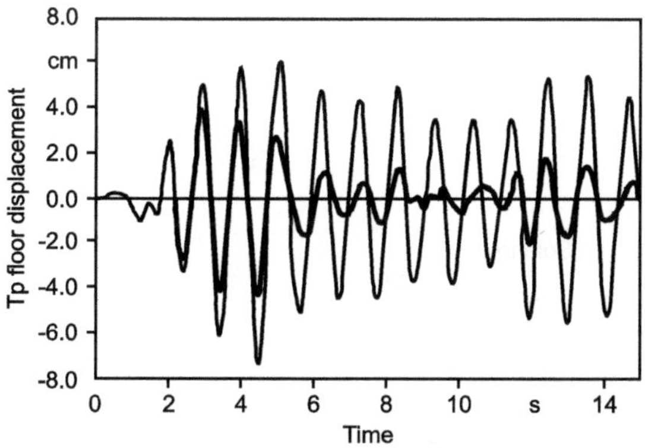

Fig. 8.59. Top-floor displacement

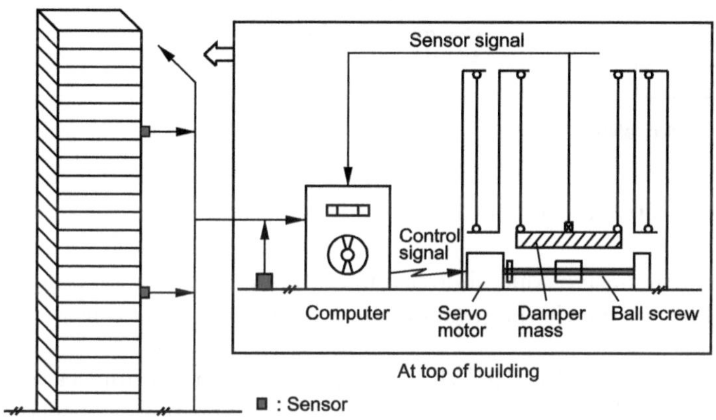

Fig. 8.60. Pendulum-type active damper

are put to many uses (high-class hotels, offices, apartments), whatever the purpose, one of the more unpleasant characteristics of such buildings is the degree of sway which occurs in high winds or earthquakes. The pendulum-type tuned active damper is a result of experiences with tuned mass dampers in towers, stacks and bridges.

The pendulum has the same natural period as the building. It has a multi stage suspended damper mass that can be housed in a single storey. When the building sways, the damper counteracts the buildings movement and tends to absorb it. The sensors detect the movement of the building and the computer controls servo motors driving ball screws to position the damper so that it makes the optimum use of its natural tendency to absorb the motion. In case of winds, the damper immediately goes into operation. It is effective against vibrations of only 1 Gal, but it works equally to counteract strong sway. Concerning earthquake, it cannot eliminate the effects of a major earthquake but it can bring residual vibrations under control.

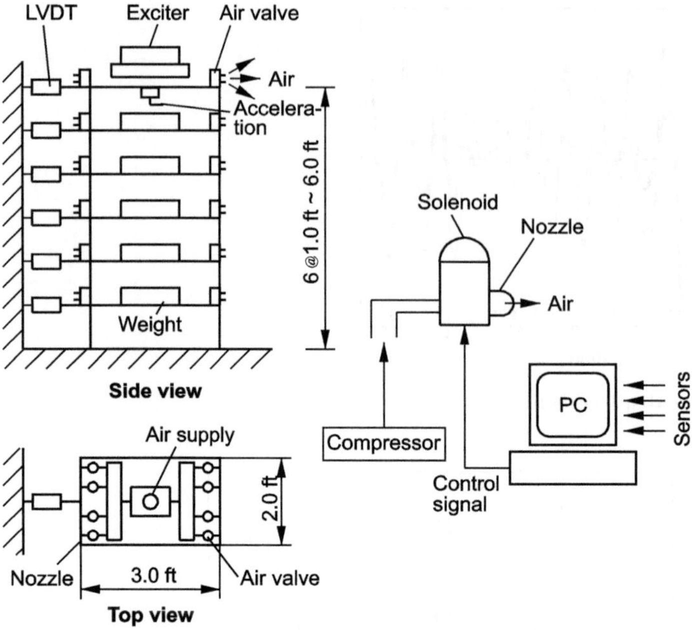

Fig. 8.61. Pneumatic active control device

Finally a pneumatic active control device may be an interesting example of active control. The schematic diagram of control configuration is shown in Fig. 8.61.

It has been demonstrated by [196] that a sequence of force impulses applied to a structure can be selected in such a way that the power spectral density of the displacement at a particular point within the structure matches the spectral density produced by earthquake ground motions as closely as desired. This result naturally suggests the possibility of using the force pulses to counteract or reduce displacements produced by earthquakes.

Some experimental results on control pulse and top-floor relative displacement are shown in Fig. 8.62.

8.4.5 Conclusions

The placement of large masses at the top of a structure (a tall building, for example) appears to be an expensive solution, and the approach may be limited by the size and stroke of the active mass damper. In both the active and passive state, the active member is a functional structural component. This is important relating to the realization of adaptronics in civil engineering structures (tall buildings, industrial plants, bridges etc.). The placement of the active members at locations of maximum strain energy using colocated sensors and actuators provides a redundant and robust control system and appears very attractive for civil engineering structures. Since only the first

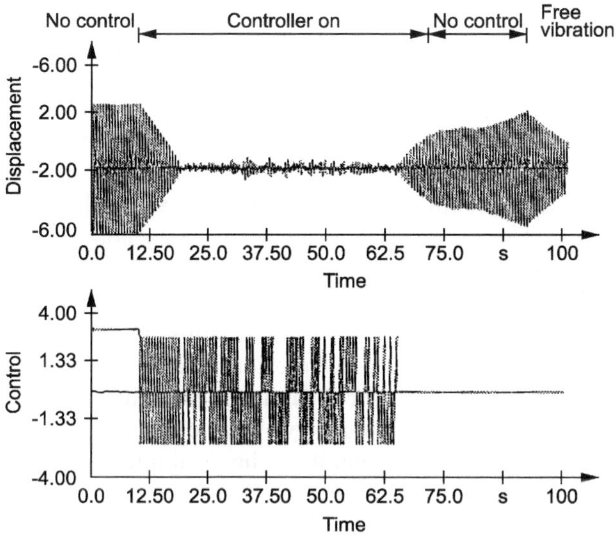

Fig. 8.62. Experimental results on pulse-control

few vibration modes are of interest, the locations of the maximum strain energy will be near the base of the structure. The measurement of the forces at the base is similar to using accelerometer data at the top of the structure. The sensors used for control can be member forces, the direct measurement of a parameter related to structure collapse. As Wada and Das (1992) showed [197], the load-carrying member can become the actuator itself; it has the requirement of a large load-carrying system with the potential to dissipate energy when placed at the locations of maximum strain energy. The actuator would be very inexpensive, located near a more attractive location near the base (or pylon in case of bridge control), and will not require additional functional physical space. Actuators developed for space systems are not appropriate because of their force and stroke limitations. An active member may include a hydraulic system that triggers the member force to introduce phase delay and stiffness changes to attenuate the loads in the structure. Adaptive structures may present attractive options for the control civil engineering structures in future.

8.5 Adaptronic Vibration Absorbers for Ropeway Gondolas

H. Matsuhisa

Dynamic vibration absorbers are widely used for mitigating adverse vibration in mechanical system such as machinery and tall buildings because they are simple, durable, and inexpensive. They do not require sensors or controllers and yet they behave as automatically adaptable devices. Recently, a new theory was developed that is vital to understanding their fundamental behaviour and establishing guidelines for optimizing their performance. Although vibration absorbers are typically installed on the structure at a location with the largest amplitude, the new theory explains that for pendulum-type structures such as a ropeway gondola, the absorber must be located far from the center of oscillation which is a small distance below the center of gravity. This means that an absorber at the fulcrum (a location where the translational amplitude is zero) is superior to one located near the gondola floor (large amplitude). This theory is based on the moment given by the absorber to mitigate the rolling motion of the host structure. Then, it can be applied to various structures which have rolling motions such as ships. A gyroscopic moment also can be used to mitigate the rolling motion. When the motion of the axis of the gyro rotor is attached to a rotational spring and damper, a single degree of freedom system is formed and a gyroscopic moment can be used to suppress the gondola swing. In this chapter, the dynamic absorbers described are explained in terms of actual application examples.

8.5.1 Dynamic Vibration Absorbers

Conventional dynamic vibration absorbers are composed of a mass, spring, and damper. Although dynamic vibration absorbers do not have sensors or controllers, they can provide vibration mitigation similar to that of actively controlled systems with a complicated sensor, control, and actuator system. Since an absorbers mass/spring/damper forms a single degree of freedom (DOF) vibration system, it consequently has a single resonant frequency and can exhibit an amplified response at this frequency. Dynamic absorbers behave similar to a system with a sensor to detect the specific frequency and a controller to amplify the vibration. Therefore, the absorbers natural frequency should be carefully tuned to a specific frequency for which the vibration amplitude of the host structure is to be reduced. The tuned frequency usually corresponds to natural modes and harmonically excited vibrations of a system.

In a fundamental theoretical model, the dynamic absorber $(m_{\mathrm{a}}, k_{\mathrm{a}}, c_{\mathrm{a}})$ is attached to a single degree of freedom system (m_1, k_1) as shown in Fig. 8.63. The equations of motion are

$$m_1\ddot{x}_1 + c_{\mathrm{a}}(\dot{x}_1 - \dot{x}_{\mathrm{a}}) + k_1 x_1 + k_{\mathrm{a}}(x_1 - x_{\mathrm{a}}) = F \tag{8.1}$$

$$m_{\mathrm{a}}\ddot{x}_{\mathrm{a}} + c_{\mathrm{a}}(-\dot{x}_1 + \dot{x}_{\mathrm{a}}) + k_{\mathrm{a}}(-x_1 + x_{\mathrm{a}}) = 0\,. \tag{8.2}$$

When the damping of the absorber is zero ($c_{\mathrm{a}} = 0$) and the disturbance force is sinusoidal ($F = \hat{F}\sin\omega t$), the responses of the structure and absorber, respectively, are

$$x_1 = \frac{(-m_{\mathrm{a}}\omega^2 + k_{\mathrm{a}})}{(-m_1\omega^2 + k_1)(-m_{\mathrm{a}}\omega^2 + k_{\mathrm{a}}) - m_{\mathrm{a}}k_{\mathrm{a}}\omega^2}\hat{F}\sin\omega t \tag{8.3}$$

$$x_{\mathrm{a}} = \frac{k_{\mathrm{a}}}{(-m_1\omega^2 + k_1)(-m_{\mathrm{a}}\omega^2 + k_{\mathrm{a}}) - m_{\mathrm{a}}k_{\mathrm{a}}\omega^2}\hat{F}\sin\omega t\,. \tag{8.4}$$

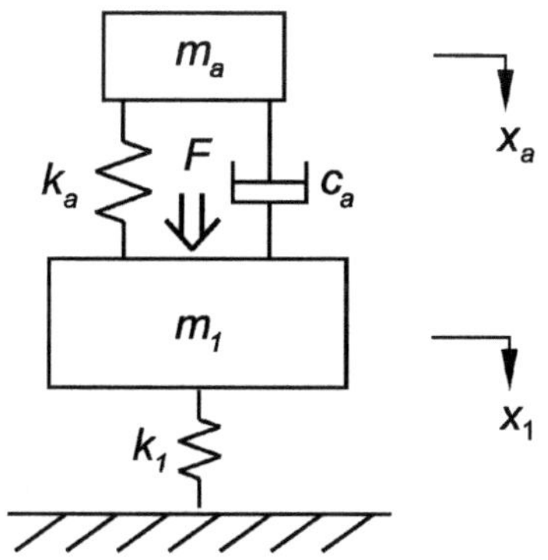

Fig. 8.63. Standard model of a dynamic absorber on a single DOF system

The numerator of (8.3) suggests that the vibration of the main system can be reduced by tuning the natural frequency of the absorber $\omega_a = \sqrt{k_a/m_a}$ to the disturbance frequency ω. Under these conditions, the force F that is imparted to the main structural mass by the motion of the absorber via the spring k_a is

$$F_a = k_a x_a = -\hat{F} \sin \omega t = -F \ . \tag{8.5}$$

This means that the absorber automatically moves such that the disturbance force is cancelled, which in turn minimizes the main mass movement. This represents a perfectly controlled system and it is called an anti-resonance dynamic absorber or anti-resonance dynamic damper.

When the disturbance is harmonic and its frequency is constant, the anti-resonance dynamic absorber is most effective. However, when the disturbance frequency varies, the tuning condition ($\omega_a = \omega$) is not satisfied and the absorber works poorly. Furthermore, it is possible that the disturbance frequency coincides with the resonance frequency. This would have the adverse effect of amplifying the main mass vibration. There are two methods to avoid the resonance. The first is adaptive control of the spring constant k_a. This method can be used for a case when the disturbance force is harmonic and the frequency changes slowly. This can be implemented using a simple control algorithm. When $\omega_a > \omega$, the phase difference between x_a and x_1 is 0 degrees and when $\omega_a < \omega$, the phase difference is 180 degrees. Based on this concept, it is possible to maintain $\omega_a = \omega$ by changing k_a by monitoring the phase difference. The second method for avoiding resonance involves damping. This method is commonly used to avoid the resonance peak. In general, the disturbance force is not harmonic but random, similar to that resulting from wind loading or an earthquake. The resonance peak can be reduced by increasing the damping constant c_a, at the expense of making the anti-resonance shallow. Figure 8.64 shows the frequency responses of the system shown in Fig. 8.63.

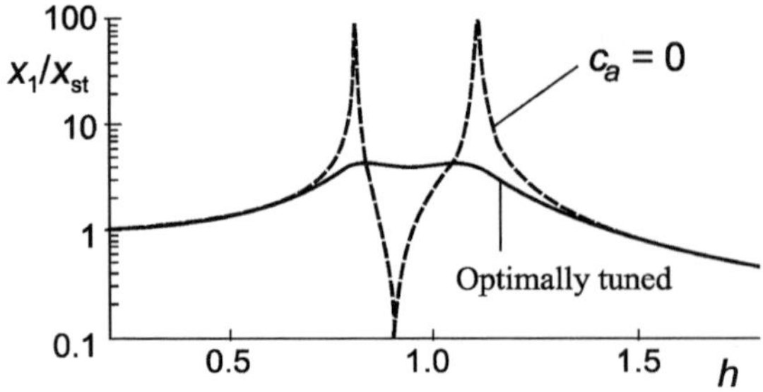

Fig. 8.64. Frequency responses of a single DOF system with a dynamic absorber

The horizontal axis is the non-dimensional exciting frequency $h = \omega/\omega_1$, ($\omega_1 = \sqrt{k_1/m_1}$) and the vertical axis is the non-dimensional amplitude X_1/X_{st}, ($X_{\mathrm{st}} = \hat{F}/k_1$). Since the efficiency of the absorber is proportional to the absorber mass m_{a}, it is selected first with consideration of cost and design. Subsequently, the spring constant k_{a} and damping coefficient c_{a} are determined such that the maximum value of the frequency response is minimized as shown in Fig. 8.64. This procedure is explained in detail by Den Hartog [198].

8.5.2 Dynamic Vibration Absorbers for Gondola

Ropeways are commonly used for skiing and sightseeing venues and urban transportation. The swing of ropeway carriers is easily induced by wind loading, rendering them inoperable for wind speeds in excess of about 15 m/s. This problem has attracted much research attention in the past few decades. The research work has primarily focused on two methods to reduce swing that involve a dynamic vibration absorber and a gyroscope.

In the case of tall buildings, the dynamic absorber is typically installed near the top of the building because the effectiveness of the absorber is proportional to the square of the vibration amplitude. Building on this idea, various researchers installed dynamic absorbers at locations of large motion on a gondola (near the bottom) and demonstrated that they work poorly; their results led other researchers to incorrectly conclude that it was impossible to reduce vibration of pendulum-type structures using dynamic absorbers. However, in 1993, Matsuhisa showed that the swing of a ropeway carrier could be reduced by using a dynamic absorber if it was located far above or below the center of oscillation which is a small distance below the center of gravity [199]. Based on this finding, a dynamic absorber composed of a moving mass on an arced track was designed for practical implementation. This type of dynamic absorber was installed on ski chair-lifts in Japan in 1995 for the first time in the world. Following the successful application of the new dynamic absorber on the ropeway carrier, they have been installed on many chair-lift-type and gondola-type carriers in Japan. It was shown that a dynamic absorber mass weighing one tenth of that of the carrier can half the swing amplitude [200].

As shown in Fig. 8.65, a ropeway carrier can be regarded as a rigid-body pendulum of mass m_1, inertia moment of I, and with negligible damping. Assuming that the distance between the center of gravity G and the fulcrum O is l_1, the equivalent length of pendulum OG′ is $l_1' = I/m_1 l_1$. In this case the point G′ is called the center of oscillation. There are several types of dynamic absorbers: a linear mass and spring type, a pendulum type, and so on. Considering factors such as easy of tuning the natural frequency, reliability, and cost, an arc-track type of dynamic absorber is chosen for a ropeway carrier. For this system, assume that the mass of the dynamic absorber is given by m_2, the radius of the arc track is l_2, the damping constant is c, and the distance between the fulcrum O and the mass of the dynamic absorber is l. By letting θ_1 and θ_2 represent the angular displacements of the carrier and the absorber, respectively, and q be the disturbance moment acting on the carrier, the linearized equations of motions can be expressed as

$$I\ddot{\theta}_1 + m_1 l_1 g\theta_1 - cl_2 l\dot{\theta}_2 + m_2(l_2 - l)g\theta_2 = T \tag{8.6}$$

$$m_2 l_2\ddot{\theta}_2 + cl_2\dot{\theta}_2 + m_2 g\theta_2 + m_2 l\ddot{\theta}_1 + m_2 g\theta_1 = 0\,. \tag{8.7}$$

From these equations, the frequency responses are obtained using a standard procedure to give

$$\frac{\hat{\theta}_1}{\hat{T}} = \frac{}{(-I\omega^2 + m_1 l_1 g)} \cdots$$

$$\cdots \frac{-m_2 l_2\omega^2 + m_2 g - \mathrm{i}cl_2\omega}{(-m_2 l_2\omega^2 + m_2 g + \mathrm{i}cl_2\omega) - \{m_2(l_2 - l)g - \mathrm{i}cl_2 l\omega\}(-m_2 l\omega^2 + m_2 g)}\,, \tag{8.8}$$

where i is the imaginary unit, $\theta = \hat{\theta}\mathrm{e}^{\mathrm{i}\omega t}$ and $T = \hat{T}\mathrm{e}^{\mathrm{i}\omega t}$.

In the case of a normal absorber as shown in Fig. 8.63, one could evaluate the efficiency by considering the mass ratio m_a/m_1. Clearly, in most applications the efficiency of the absorber is a critical parameter and should therefore be maximized. In the case of an absorber installed in a building, efficiency is proportional to the mass ratio and square of the displacement amplitude of the building. Therefore, the absorber must be located on a high floor. In the case of the gondola, the efficiency is given by

$$\mu_\mathrm{e} = \frac{m_2}{m_1}\left(1 - \frac{l}{l_1'}\right)^2\,. \tag{8.9}$$

This expression is referred to as the equivalent mass ratio. The equivalent mass ratio is thus proportional to the nominal mass ratio m_2/m_1 and the square of the distance between the absorber and the center of oscillation G'. This means that if the dynamic absorber is attached to the center of

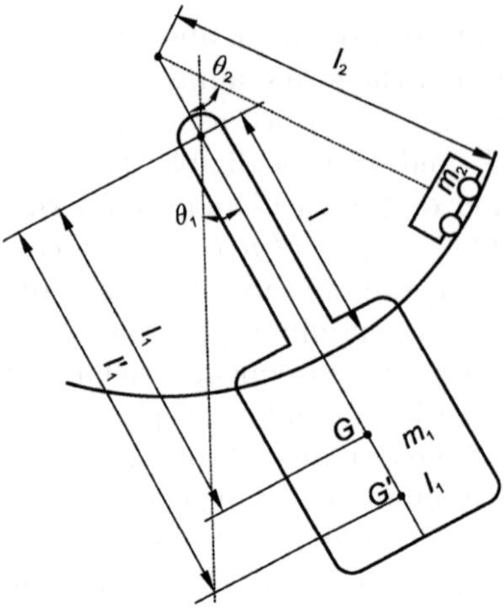

Fig. 8.65. Schematic diagram of a gondola with an arc-track type dynamic absorber

oscillation G', the dynamic absorber does not work, and therefore must be attached at a point as far from the center of oscillation as possible.

Equqtion (8.7) governs the relative motion of the absorber, and the fourth and fifth terms can be regarded as exciting forces. When the absorber is tuned such that its natural frequency is the same as that of the gondola and the absorber is located at the center of oscillation G', the fourth term (inertia force due to the acceleration of the gondola) and the fifth term (gravity force due to the decline of the gondola) cancel each other. Consequently, the dynamic absorber would not be excited. Equation (8.6) governs the swing of the gondola. When the absorber is located at the center of oscillation G' ($l_2 = l$), the motion of the absorber does not affect the motion of the gondola. From the above discussion, it is clear why the efficiency is proportional to the square of the distance between the absorber and the center of oscillation G'. Even if the dynamic absorber is attached to the fulcrum, the dynamic absorber can reduce the vibration remarkably. Furthermore, it is better to attach the dynamic absorber above the fulcrum because, in this case, $l < 0$, and the inertia force and the gravity force have the same direction, which increases the motion of the absorber and thus reduces the amplitude of the gondola swing. Alternatively, for a conventional absorber, the gravity force does not appear in the equation of motion and the efficiency is proportional to the square of amplitude which is proportional to the inertial force.

Figure 8.66 shows the theoretical prediction of the frequency response of a ropeway gondola for six passengers ($m_1 = 1000\,\text{kg}$, $l_1 = l'_1 = 4\,\text{m}$) with an optimally tuned dynamic absorber. In this calculation, the damping ratio ζ of the gondola is assumed to be 1%. In the case of a gondola without an absorber, the maximum value of the normalized amplitude (θ_1/θ_{st}) is 50.

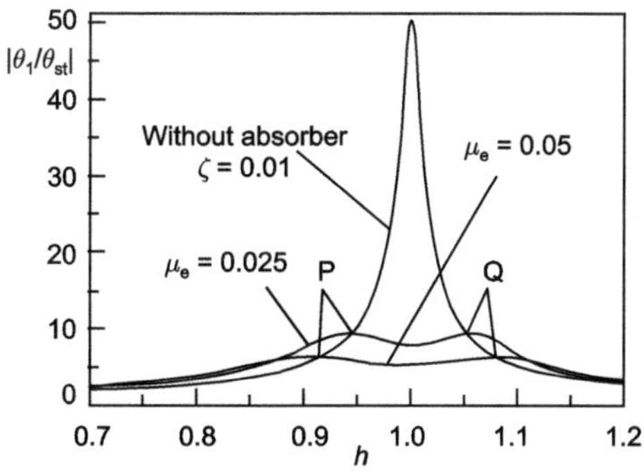

Fig. 8.66. Compliance curves for various equivalent mass ratios

When the gondola has a dynamic absorber with the equivalent mass ratio $\mu_e = 0.025$, the maximum value of the normalized amplitude is 9, and when $\mu_e = 0.05$, the maximum value is 6.4. Figure 8.67 shows the free vibration of the system. Figure 8.68 shows the random response of the system due to artificially generated wind force. These figures show that the dynamic absorber is very effective in reducing the swing of the gondola and the effectiveness is represented by the equivalent mass ratio μ_e.

It is possible to use another type of dynamic absorber for a gondola; for example, a pendulum-type dynamic absorber or a typical dynamic absorber composed of a mass and a spring on a straight track. In the case of a pendulum-type absorber, the equations of motions are the same as those for the arc-track type absorber. In the case of the mass-spring type absorber, the equations of motions are slightly different; however, the same equivalent mass ratio is obtained. For all types of dynamic absorbers, the equivalent mass ratio is given by (8.9). Thus the theory described above appears to be universal and can be applied to all kinds of dynamic absorbers attached to pendulum-type structures.

Figure 8.69 shows a prototype dynamic absorber ($m_2 = 48$ kg) installed on a gondola ($m_1 = 790$ kg) for 10 passengers. The experimental results

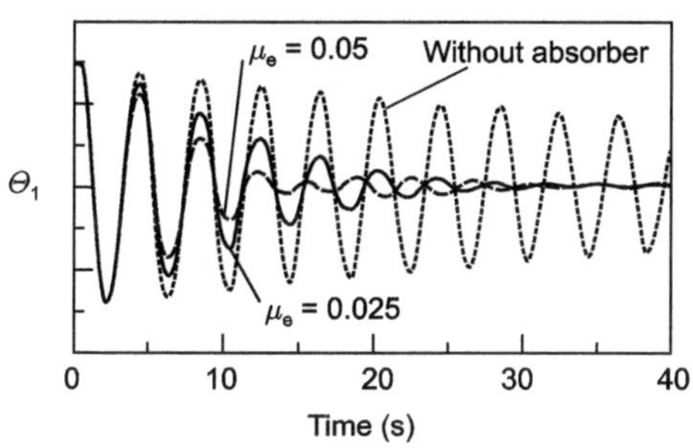

Fig. 8.67. Free vibrations for various equivalent mass ratios

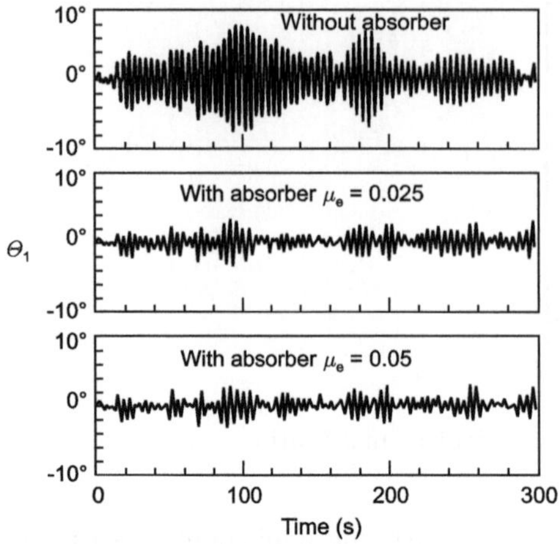

Fig. 8.68. Random responses for various equivalent mass ratios

Fig. 8.69. Prototype dynamic absorber on a gondola for ten passengers

for its free vibration are shown in Fig. 8.70. In this case, the absorber was attached to a very high position and the swing was attenuated rapidly. Since this absorber did not employ a damping device, the time response shown in Fig. 8.70 has a small beat.

Based on the theoretical predictions and the experiments described above, in 1995 dynamic absorbers were installed on chair lifts in Japan. This is the first application of dynamic absorbers for ropeway carriers in the world. An

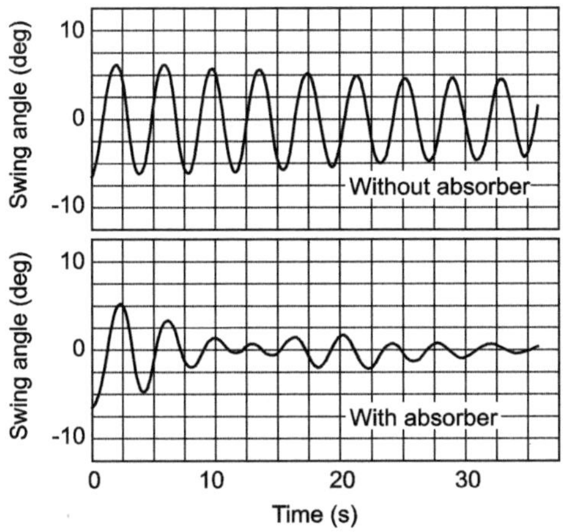

Fig. 8.70. Free vibrations of a gondola with the prototype dynamic absorber

Fig. 8.71. The dynamic absorber Libra was attached to chairlifts for the first such use in the world (1995)

image of the ropeway carrier and its free vibration is shown in Figs. 8.71 and 8.72. The weight of the carrier was 156 kg and the weight of the moving mass on the arc-shaped aluminum pipe was 11 kg. The radius of the arc-shaped track was 2300 mm and its length was 1400 mm. The damping was induced by electromagnetic force caused by a permanent magnet attached to a moving mass on the aluminum track. The swing of the lift was decreased from 10 degrees to 1 degree in six periods.

In 1996, the new dynamic absorber was also installed in 15 passenger gondolas in Japan, as shown in Fig. 8.73. The weight of the gondola was 830 kg and the weight of the moving mass was 17.6 kg. In this case, since the rope span between the neighboring towers was long, the natural frequency of

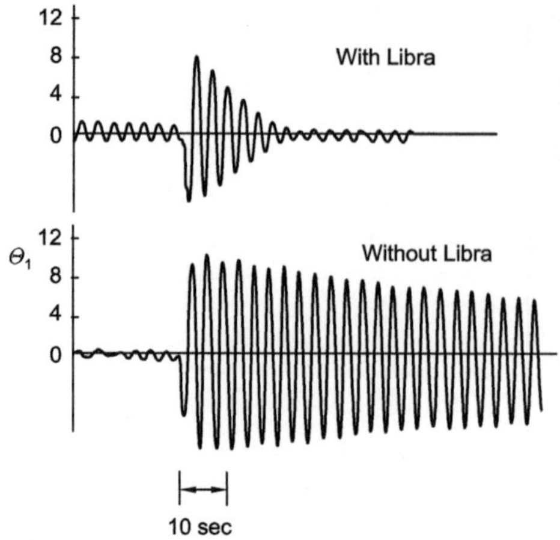

Fig. 8.72. Free vibrations of the chairlift with the dynamic absorber Libra

Fig. 8.73. Gondolas for 15 passengers with double dynamic absorbers Libra

the swing of the gondola varied depending on the location. When the gondola was located at the middle of the span, the natural period was 4.8 seconds, and when it was near the tower, the natural period was 4.3 seconds. Thus the gondola had two dynamic absorbers whose natural periods were 4.3 seconds and 4.8 seconds. The free vibration is shown in Fig. 8.74. The swing was remarkably decreased by the double dynamic absorber. While the gondola without the dynamic absorber could not operate for wind velocities in excess of 15 m/s, the new gondola with the dynamic absorbers was able to operate in wind velocities as high as 20 m/s.

8.5.3 Gyroscopic Moment Absorber for Gondola

It is well known that the tilt of a rotor axis induces gyroscopic moment, and the gyroscopic moment can be used to control the rolling motion of the

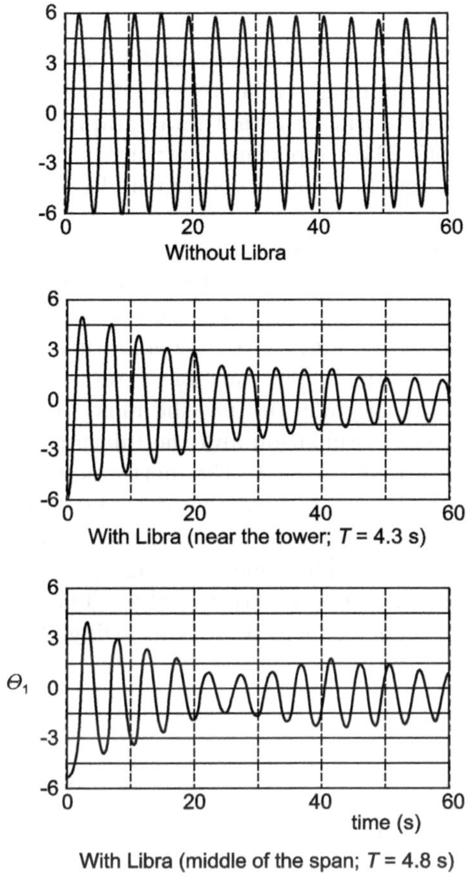

Fig. 8.74. Free vibrations of the gondola with the double dynamic absorbers Libra

primary system such as ships. In fact, such devices were installed in several aircraft carrier battleships before World War II to reduce rolling. It is also possible to use gyroscopic moment to control the swing of the gondola; however, they have not been used in practice because it is difficult to supply ropeway carriers with electrical power. To address this issue, a gyroscopic absorber powered by battery power could be developed. One method to obtain the moment is active control of the gyroscope axis tilt, known as CMG (control moment gyroscope). As this method consumes a large amount of energy to control, it is not suitable for ropeway gondola.

The second method is a passive gyroscope in which the rotor axis is connected to a rotary spring and damper and the tilt of the rotor axis forms a single degree of freedom of vibration system. Swing in the rolling direction of the gondola applies a moment to the rotor axis in the pitching direction. The tilt provides a reactive moment to the gondola to reduce swing. This system forms a two degree of freedom of vibration system and its function is similar to the dynamic absorber. This system is referred to as a passive gyroscopic absorber. The passive gyroscopic absorber is applicable for practical use because the energy to keep the rotor spinning is relatively small. Several design methods for determining the rotary spring constant and damping coefficient were proposed by Nishihara [200–204]. Experiments with prototype passive gyroscopes on a six-passenger gondola were carried out and reasonable results were obtained.

A schematic diagram of a gondola with a passive gyroscopic absorber is shown in Fig. 8.75. Assume for this theoretical analysis that the gondola swings about the rope; the rotor axis z is vertical; the rotor tilts about the x axis (which is parallel to the pitching axis); and the gyroscopic torque is exerted about the y axis (parallel to the rope). The variable I represents the inertial moment of the gondola about the fulcrum; m_1 the mass of the gondola; l_1 the length between the center of oscillation and the fulcrum; I_{R} the polar moment of inertia of the rotor about z axis; I_{G} the inertia moment of the rotor and the frame about x axis; Ω the rotational speed; c and k the rotary damping coefficient and spring constant, respectively; g the gravity acceleration; and θ_{x} the tilt angle of gyro rotor. The gondola swing generates a gyro moment $I_{\mathrm{R}}\Omega\dot{\theta}_1$ around the x axis, causing a rotor tilt of θ_{x}. The tilt generates a gyro moment $I_{\mathrm{R}}\Omega\dot{\theta}_{\mathrm{x}}$ around the y axis which prevents swing. The equations of motion are

$$I\ddot{\theta}_1 + m_1 l_1 g\theta_1 - I_{\mathrm{R}}\Omega\dot{\theta}_{\mathrm{x}}\cos\theta_{\mathrm{x}} = T \tag{8.10}$$

$$I_{\mathrm{G}}\ddot{\theta}_{\mathrm{x}} + c\dot{\theta}_{\mathrm{x}} + k\theta_{\mathrm{x}} - I_{\mathrm{R}}\Omega\dot{\theta}_1\cos\theta_{\mathrm{x}} = 0\,. \tag{8.11}$$

Assuming $\cos\theta_{\mathrm{x}}=1$, the frequency response is

$$\frac{\hat{\theta}_1}{\hat{T}} = \frac{-I_{\mathrm{G}}\omega^2 + k - \mathrm{i}c\omega}{(-I\omega^2 + m_1 l_1 g)(-I_{\mathrm{G}}\omega^2 + k + \mathrm{i}c\omega) - I_{\mathrm{R}}^2\Omega^2\omega^2}\,. \tag{8.12}$$

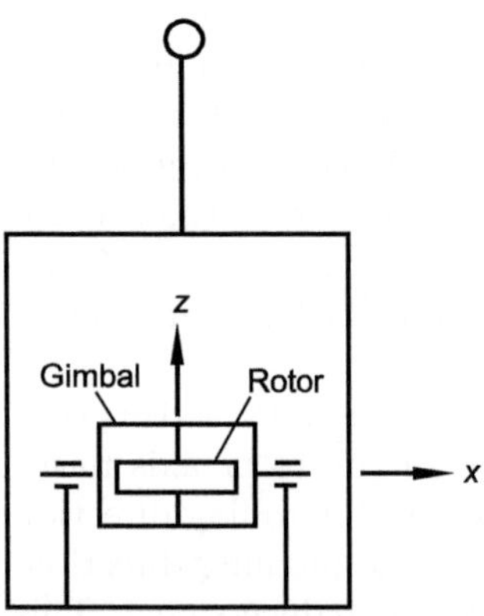

Fig. 8.75. Configuration of the passive gyroscopic damper

Fig. 8.76. The passive gyroscopic damper

The equivalent mass ratio is given by

$$\mu_e = \frac{I_R \Omega^2}{I\omega^2} . \tag{8.13}$$

The numerator of (8.12) suggests that the gondola swing can be reduced by tuning the natural frequency of the rotor tilt $\sqrt{k/I_G}$ to the exciting frequency ω. Equation (8.13) shows that the efficiency of the absorber can be increased by increasing the rotor speed Ω. The optimum values of damping and spring stiffness of the gyroscopic absorber are determined based on the frequency response or time response [201–205].

Two prototype gyroscopes shown in Fig. 8.76 were installed on an actual six-passenger gondola to carry out the experiment. It is possible to diminish the torque about the x axis and the z axis which may cause pitching and yawing by using two gyroscopes with opposing rotation directions. The weight of the gondola was 500 kg and its natural period was 3.5 seconds. The

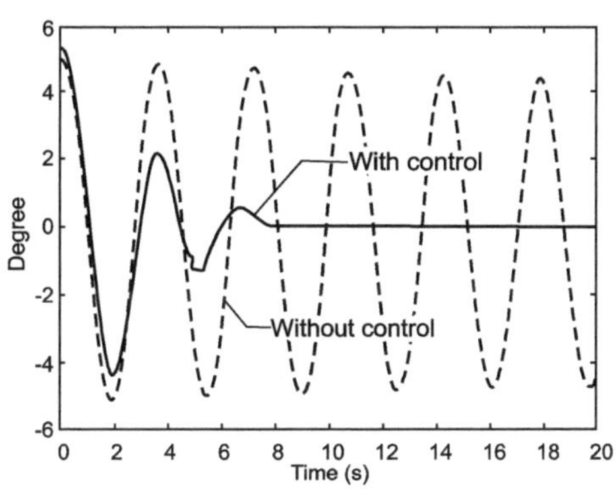

Fig. 8.77. Free vibrations of the gondola with passive gyroscopic dampers

dimension of the gyroscope was 700 mm (height) and 450 mm (width and depth). The moment of inertia of the rotor was 0.434 kg/m^2, its diameter was 390 mm, and its weight was 17.6 kg. The gimbals axis was connected to a rotary-type viscous damper and a spring with a pulley and wire. The rotor was driven by a 24 volt DC motor rotating at 2400 rpm. The gyroscopes were set such that the gimbals axes were vertical to the floor of the gondola and parallel to the pitching axis. The free vibration of the gondola displacement is shown in Fig. 8.77. The swing was reduced from 5 degree to zero in two periods.

8.5.4 Conclusions and Outlook on Future Research

Wind-induced vibration of ropeway carriers is an inevitable problem that has limited their use during windy conditions. However, in 1993 it was found that their vibration can be reduced easily by carefully locating dynamic vibration absorbers on the carriers. The use of dynamic vibration absorbers for ropeway carriers has been well received, and such absorbers have been used in Japan since 1995. Since the theory regarding the location of the dynamic absorber can be applied to many rolling structures (such as ships), research involving dynamic absorbers will continue to be significant and other types of dynamic absorbers are likely to be developed. Moreover, the passive and active-type gyroscope absorbers are very effective in reducing rotary vibration and they could also be applied to many structures such as rope-suspended bridges, ships, cranes, and robot arms.

References

1. Kudva, J.N.; Sanders, B.; Pinkerton-Florance, J. and Garcia, E.: *The DARPA/AFRL/NASA Smart Wing Program – Final Overview.* Proc. SPIE Vol. 4698 (2002), pp. 37–43
2. Martin, C.A.; Hallam, B.J.; Flanagan, J.S. and Bartley-Cho, J.: *Design, Fabrication and Testing of Scaled Wind Tunnel Model for the Smart Wing Phase 2 Program.* Proc. SPIE Vol. 4698 (2002), pp. 44–52
3. Bartley-Cho, J.D.; Wang, D.P. and West, M.N.: *Development, Control, and Test Results of High-Rate, Hingeless Trailing Edge Control Surface for the Smart Wing Phase 2 Wind Tunnel Model.* Proc. SPIE Vol. 4698 (2002), pp. 53–63
4. Scherer, L.B.; Martin, C.A.; Sanders, B.; West, M.; Florance, J.; Wiesemann, C.; Burner, A. and Fleming, G.: *DARPA/AFRL Smart Wing Phase 2 Wind Tunnel Test Results.* Proc. SPIE Vol. 4698 (2002), pp. 64–75
5. McGowan, A.-M. R.; Washburn, A.E.; Horta, L.G.; Bryant, R.G.; Cox, D.E.; Siochi, E.J.; Padula, S.L. and Holloway, N.M.: *Recent Results from NASAs Morphing Project.* Proc. SPIE Vol. 4698 (2002), pp. 97–111
6. Pitt, D.M.; Dunne, J.P.; White, E.V.: *SAMPSON smart inlet design overview and wind tunnel test; Part 1 – Design overview. Part II – Wind tunnel test.* Proc. SPIE Vol. 4698 (2002), pp. 13–36

7. *Proc. Smart Materials Symp. 2001.* R&D Institute Metals and Composites for Future Industries (RIMCOF), Tokyo/Japan
8. Mangalgiri, P.D.; Upadhya, A.R. and Selvaradjan, A. (Ed.s): *Proc. Internat. Conf. on Smart Materials, Structures & Systems.* Bangalore/India; Allied Publishers (1999)
9. Dattaguru, B.; Gopalakrishnan, S. and Mohan, S. (Ed.s): *Proc. Internat. Conf. on Smart Materials, Structures & Systems (ISSS-SPIE 2002).* Microart Multimedia Solutions, Bangalore/India (2002)
10. Galea, S.C. and Rajic, N.: *Structural Health Monitoring and Life Extension of Military Aerostructures using Smart Materials.* Proc. 15th Internat. Conf. on Adaptive Structures (ICAST), Bar Harbour, Maine, USA (2004)
11. McAdam, G; Newman, P.J.; McKenzie, I.; Davis, C. and Hinton, B.R.W.: *Fibre optic sensors for detection of corrosion within aircraft.* Structural Health Monitoring – Int. J., 4, 1 (2005) pp. 47–56
12. Walley, A.P. and Rajic, N.: *In-situ structural health monitoring of an impact damaged F/A-18 horizontal stabilator.* Proc. 2nd Australasian Workshop on Structural Health Monitoring, Monash Univ., Melbourne, Dec. 16–17 (2004)
13. http://www.spie.org/info/ss-nde
14. http://www.cansamrt.com
15. http://structure.stanford.edu/workshop
16. http://www.shm-europe.net
17. Bronowicki, A. J.; Das, A. and Wada, B. K. (Ed.s): *Special Issue on Smart Structures for Space.* Smart Mater. & Struct., 8, Nr. 6 (1999)
18. Böhringer, K. F. (Ed.): *Special Issue on Space Applications for MEMS.* Smart Mater. & Struct., 10, Nr. 6 (2001)
19. Chopra, I. (Ed.): *Special Issue on Application of Smart Structures Technology to Rotorcraft Systems.* Smart Mater. & Struct., 5, Nr. 1 (1996)
20. Gandhi, F. (Ed.): *Special Issue on Rotorcraft Applications.* Smart Mater. & Struct., 10, Nr. 1 (2001)
21. Garcia, E.: *Smart Structures and Actuators: Past, Present, and Future.* SPIE Vol. 4698, pp. 1–12 (2002)
22. Loewy, R.G.: *Recent Developments in Smart Structures with Aeronautical Applications.* Smart Mater. & Structures, Vol. 6, No. 5 (1997) pp. R11–R42
23. Giurgiutiu, V.: *Review of Smart-Materials Actuation Solutions for Aeroelastic and Vibration Control.* J. Int. Mat. Syst. & Struct., 11 (2000) pp. 525–544
24. Staszewski, W.J.; Boller, C. and Tomlinson, G.R.: *Health Monitoring of Aerospace Structures.* Wiley (2003)
25. http://aar400.tc.faa.gov/Programs/agingaircraft/rotorcraft/index.htm
26. Fisher; C.: *Gas Turbine Condition Monitoring Systems – An Integrated Approach.* Power Engineer, Vol. 1 Nr. 4 (1997), pp. 499ff.
27. Read, I.J. and Foote, P.D.: *Sea and flight trials of optical fibre Bragg grating strain sensing system.* Smart Mater. Struct., 10 (2001), pp. 1085–1094
28. Betz, D.: *Acousto-ultrasonic sensing using fibre Bragg gratings.* PhD thesis, Department of Mechanical Engineering, Sheffield University, UK (2004)
29. http://www.analatom.com
30. Trego, A.: *Installation of the Autonomous Structural Integrity Monitoring System.* 4th Int. Workshop on Structural Health Monitoring, Stanford/CA, USA; DEStech Publ. (2003), pp. 863–870
31. Akdeniz, A.: *Structural Health Management Technology Implementation Commercial Airplanes.* Boeing internal white paper (2004)

32. Akdeniz, A.; Trego, A. and Haugse, E.: *Structural Health Management Technology Implementation Commercial Airplanes.* Presentation material at 7th DoD/FAA/NASA Conf. on Ageing Aircraft, New Orleans, LO, USA (2003)
33. www.acellent.com
34. Ihn, J.-B. and Chang, F.-K.: *Detection and monitoring of hidden fatigue crack growth using built-in piezoelectric sensor/actuator network: Part I. Diagnostics; Part II. Validation using riveted joints and repair patches.* Smart Mater. Struct., 13 (2004), pp. 609–630
35. Washabaugh, A.; Zilberstein, V.; Schlicker, D.; Shay, I.; Grundy, D. and Goldfine, N.: *Shaped-Field Eddy-Current Sensors and Arrays.* SPIE Vol. 4702 (2002), pp. 63–75
36. Lemistre, M.B. and Balageas, D.L.: *A New Concept for Structural Health Monitoring Applied to Composite Materials. Part I: Theoretical Considerations; Part II: Experimental Validation.* Proc. 1st Eur. Workshop on Structural Health Monitoring, DesTech Publications (2002), pp. 493–507
37. Schmidt, H.-J. and Schmidt-Brandecker, B.: *Structure Design and Maintenance Benefits from Health Monitoring Systems.* In: Structural Health Monitoring, (Ed. F.-K. Chang), CRC Press (2001), pp. 80–101
38. Kishi, T. and Takeda, N. (Ed.s): *Special Issue on Japanese Smart Materials Demonstrator Program and Structures System Project.* Adv. Composite Mater., 13, No. 1 (2004)
39. DeCamp, R.W.; Hardy, R. and Gould, D.K.: *SAE Internat. Pacific Air and Space Technology Conf.* Melbourne/Australia (1987)
40. Kudva, J.N.; Lockyer, A.J. and Appa, K.: *Adaptive Aircraft Wing.* AGARD LS-205, Paper10 (1996)
41. Kudva, J.N.; et al.: *Overview of the DARPA/AFRL/NASA Smart Wing Program.* SPIE Vol. 3674 (1999), pp. 230–236
42. Martin, C.A.; et al.: *Design and Fabrication of Smart Wing Model and SMA Control Surfaces.* SPIE Vol. 3674 (1999), pp. 237–248
43. Jardine, A.P.; et al.: *Improved Design and Performance of the SMA Torque Tube for the Smart Wing Program.* SPIE Vol. 3674 (1999), pp. 260–269
44. AFRL-ML-WP-TR-1999–4162, Northrop Grumman Corp.: *Smart Materials and Structures – Smart Wing Phase 1 Final Report.* Air Vehicles Directorate, AFRL, Wright-Patterson Air Force Base (1999)
45. Kudva, J.N.; Sanders, B.; Pinkerton-Florance, J. and Garcia, E.: *Overview of the DARPA/AFRL/NASA Smart Wing Phase 2 Program.* SPIE Vol. 4332 (2001), pp. 383–389
46. Scherer, L.B.; Martin, C.A.; Sanders, B.; West, M.; Florance, J.; Wieseman, C.; Burner, A. and Fleming, G.: *DARPA/AFRL Smart Wing Phase 2 Wind Tunnel Test Results.* SPIE Vol. 4698 (2002), pp. 64–75
47. Cabell, R.H.; Schiller, N.; Mabe, J.H.; Ruggeri, R.T.; Butler, G.W.: *Feedback Control of a Morphing Chevron for Takeoff and Cruise Noise Reduction.* Proc. ACTIVE 04; Williamsburg/VA, USA (2004)
48. Bein, Th.; Hanselka, H. and Breitbach, E.: *An adaptive spoiler to control the transonic shock.* Smart Mater. Struct., 9 (2000), pp. 141–148
49. Seifert, A. and Pack, L.G.: *Separation Control at Flight Reynolds Numbers: Lessons Learned and Future Directions.* AIAA Paper 2000–2542; Fluids (2000)
50. Wilkinson, S.P.: *Investigation of the Effect of an Oscillating Surface Plasma on Turbulent Skin Friction.* AIAA-Paper, 1st Flow Control Conf. (2002)

51. Roth, J.R.; Sherman, D.M. and Wilkinson, S.P.: *Electrohydrodynamic Flow Control with a Glow-Discharge Surface Plasma.* AIAA J., Vol. 38, No. 7 (2000), pp. 1166–1172
52. Jacot, D.; Calkins, T. and Smith, J.: *Boeing Active Flow Control System (BAFCS)-III.* Proc. SPIE Vol. 4698 (2002); pp. 76–84
53. Calkins, F.T. and Clingman, D.J.: *Vibrating Surface Actuators for Active Flow Control.* Proc. SPIE Vol. 4698 (2002); pp. 85–96
54. Anhalt, C.; Breitbach, E. and Monner, H.P.: *Adaptronics in Airliner Design – A New Structural Approach.* Proc. SPIE Vol. 4698 (2002); pp. 364–375
55. Breitbach, E.J.: *Research Status on Adaptive Structures in Europe.* Proc. 2nd Joint Japan/US Conf. on Adaptive Structures; Technomic Sc. Publ. (1991)
56. Wada, B.K. and Garba, J.A.: *Advances in Adaptive Structures at Jet Propulsion Laboratory.* AGARD-CP 531, Paper 28 (1992)
57. Dürr, J.K.; Herold-Schmidt, U.; Zaglauer, H. and Arendts, F.J.: *Integration of piezoceramic actuators in fibre-reinforced structures for aerospace applications.* SPIE Vol. 3326 (1998), pp. 81–92
58. Dürr, J.K.; Honke, R.; von Alberti, M. and Sippel, R.: *Development and manufacture of an adaptive lightweight mirror for space application.* Smart Mater. Struct., 12 (2003), pp. 1005–1016
59. Dürr, J.K.; Honke, R.; von Alberti, M. and Sippel, R.: *Analysis and Design of an Adaptive Lightweight Satellite Mirror.* Proc. SPIE Vol. 4698 (2002); pp. 351–363
60. Misra, M.S.; Carpenter, B. and Maclean, B.: *Adaptive Structure design Employing Shape Memory Actuators.* AGARD-CP 531, Paper 15 (1992)
61. Barrett, R.: *Active Aeroelastic Tailoring of an Adaptive Flexspar Stabilator.* Smart. Mater. & Struct., 5 (1996), pp. 723–730
62. Lazarus, K.B. and Crawley, E.F.: *Multivariable High-Authority Control of Plate-like Active Structures.* AIAA-Paper No. 92–2529 (1992)
63. Moses, R.W: *Vertical Tail Buffeting Alleviation Using Piezoelectric Actuators – Some Results of the Actively Controlled Response of Buffet-Affected Tails (ACROBAT).* NASA Technical Memorandum 110336 (1997)
64. Fuller, C.R.; et al.: *Active Control of Interior Noise in Model Aircraft Fuselages Using Piezoceramic Actuators.* AIAA-J., 30 (11) (1992), pp. 2613–2617
65. Borchers, I.U.; et al.: *Selected Flight Test Data and Control System Results of the CEC BRITE/EURAM ASANCA Study.* In Proc. Internoise 93 (1993), pp. 59–64
66. Simpson, J. and Schweiger, J.: *An Industrial Approach to Piezo Electric Damping of Large Fighter Aircraft Components.* 5th Annual Int. Symp. on Smart Struct. & Mat., San Diego/CA (1998)
67. Becker, J. and Luber, L.: *Comparison of Piezoelectric Systems and Aerodynamic Systems for Aircraft Vibration Alleviation.* 5th Annual Int. Symp. on Smart Struct. & Mat., San Diego/CA (1998)
68. Dürr, J.K.; Herold-Schmidt, U.; Zaglauer, H.W. and Arendts, F.J.: *Integration of Piezoceramic Actuators in Fiber-Reinforced Structures for Aerospace Applications.* SPIE Vol. 3326 (1998), pp. 81–92
69. Becker, J.: *Active Buffeting Vibration Alleviation – Demonstration of Intelligent Aircraft Structure for Vibration and Dynamic Load Alleviation.* Proc. the ESF-NSF Workshop on Sensor Technol. and Intelligent Struct (2002)
70. Simpson, J. and Boller, C.: *Performance of SMA-reinforced composites in an aerodynamic profile.* Proc. SPIE Vol. 4698 (2002); pp. 416–426

71. Ben-Zeev, O. and Chopra, I.: *Advances in the Development of an Intelligent Helicopter Rotor Employing Smart Trailing-Edge Flaps.* Smart Mater. Struct., 5 (1996), pp. 11–25
72. Roglin, R.L. and Hanagud, S.V.: *A Helicopter with Adaptive Rotor Blades for Collective Control.* Smart Mater. Struct., 5 (1996), pp. 76–88
73. Fenn, R.C.; et al.: *Terfenol-D Driven Flaps for Helicopter Vibration Reduction.* Smart Mater. Struct., 5 (1996), pp. 49–57
74. Konstanzer, P.: *Decentralized Vibration Control for Active Helicopter Rotor Blades.* Doctoral thesis Univ. Stuttgart, VDI Fortschrittsberichte, Reihe 8, Nr. 923, VDI-Verlag Düsseldorf/Germany (2001)
75. Clark R.L. and C.R. Fuller, 1991: *Control of Sound Radiation with Adaptive Structures.* J. Intell. Mater. Syst. and Struct., 2, pp. 431–452
76. Barrett, R.: *Intelligent rotor blade structures development using directionally attached piezoelectric crystals.* MSc-thesis, Univ. of Maryland, College Park, MD (1990)
77. Chen, P.C.; Baeder, J.D.; Evans, R.A.D. and Niemczuk, J.: *Blade-vortex interaction noise reduction with active twist smart rotor technology.* Smart Mater. Struct., 10 (2001), pp. 77–85
78. Liu, Q.; Chattopadhyay, A.; Gu, H. and Zhou, X.: *Use of segmented constrained layer damping treatment for improved helicopter aeromechanical stability.* Smart Mater. Struct., 9 (2000), pp. 523–532
79. Jänker, P.; Hermle, F.; Lorkowski, T.; Storm, S.; Wettemann, M.; Gerle, M.: *Actuator Technology based on smart materials for adaptive systems in aerospace.* Proc.: ICAS 2000, Harrogate/UK (2000)
80. Bernhard, A.P.F. and Chopra, I.: *Analysis of a bending-torsion coupled actuator for a smart rotor with active blade tips.* Smart Mater. Struct., 10 (2001), pp. 35–52
81. Cesnik; C.E.S. and Shin, S.J.: *On the twist performance of a multiple-cell active helicopter blade.* Smart Mater. Struct., 10 (2001), pp. 53–61
82. Cesnik, C.E.S.; Shin, S.J. and Wilbur, M.L.: *Dynamic response of active twist rotor blades.* Smart Mater. Struct., 10 (2001), pp. 62–76
83. Kube, R. and Klöppel, V.: *On the role of prediction tools for adaptive rotor system developments.* Smart Mater. Struct., 10 (2001), pp. 137–144
84. Grohmann, B.A.; Maucher, C.K.; Jänker, P.; Altmikus, A. and Schimke, D.: *Aero-servo-elastic predesign of a smart trailing edge tab for an adaptive helicopter rotor blade.* Proc. Int. Forum on Aeroelasticity and Struct. Dynamics (IFASD); DGLR Bonn-Bad Godesberg/Germany (2005)
85. Lee, T. and Chopra, I.: *Design of piezostack-driven trailing-edge flap actuator for helicopter rotors.* Smart Mater. Struct., 10 (2001), pp. 15–24
86. Straub, F.K.; Ngo, H.T.; Anand, V. and Domzalski, D.B.: *Development of a piezoelectric actuator for trailing edge flap control of full scale rotor blades.* Smart Mater. Struct., 10 (2001), pp. 25–34
87. Koratkar, N.A. and Chopra, I.: *Wind tunnel testing of a Mach-scaled rotor model with trailing-edge flaps.* Smart Mater. Struct., 10 (2001), pp. 1–14
88. Centolanza, L.R.; Smith, E.C. and Munsky, B.: *Induced-shear piezoelectric actuators for rotor blade trailing edge flaps.* Smart Mater. Struct., 11 (2002), pp. 24–35
89. Gandhi, F.; Wang, K.W. and Xia, L.: *Magnethorheological fluid damper feedback linearization control for helicopter rotor application.* Smart Mater. Struct., 10 (2001), pp. 96–103

90. Epps, J.J. and Chopra, I.: *In-flight tracking of helicopter rotor blades using shape memory alloy actuators.* Smart Mater. Struct., 10 (2001), pp. 104–111
91. Bebesel, M.; Maier, R.; Hoffmann, F.: *Reduction of Interior Noise in Helicopters by using Active Gearbox Struts – Flight Test Results.* Proc. 27th European Rotorcraft Forum, Moscow/Russia (2001)
92. Mahapatra, R.D.; Gopalakrishnan, S. and Balachandran, B.: *Active feedback control of multiple waves in helicopter gearbox support struts.* Smart Mater. Struct., 10 (2001), pp. 1046–1058
93. Bronowicki, A.J.; Abhyanka, N.S. and Griffin, S.F.: *Active vibration control of large optical space structures.* Smart Mater. Struct., 8 (1999), pp. 740–752
94. Davis, L.; Hyland, D.; Yen, G. and Das, A.: *Adaptive neural control for space structure vibration suppression.* Smart Mater. Struct., 8(1999), pp. 753–766
95. Nye, T.W.; Manning, R.A. and Qassim, K.: *Performance of active vibration control technology: the ACTEX flight experiments.* Smart Mater. Struct., 8 (1999), pp. 767–780
96. Meyer, J.L.; Harrington, W.B.; Agrawal, B.N. and Song, G.: *Vibration suppression of a spacecraft flexible appendage using smart material.* Smart Mater. Struct., 7 (1998), pp. 95–104
97. Döngi, F.; Johann, U. and Szerdahelyi, L.: *Active structural subsystem of the OISI interferometry testbed.* Smart Mater. Struct., 8 (1999), pp. 709–718
98. Vaillon, L. and Philippe, C.: *Passive and active microvibration control for very high pointing accuracy space systems.* Smart Mater. Struct., 8 (1999), pp. 719–728
99. Vaillon, L.; Petitjean, B.; Frapard, B. and Lebihan, D.: *Active isolation in space truss structures: from concept to implementation.* Smart Mater. Struct., 8 (1999), pp. 781–790
100. Bushnell, G.S.; Becraft, M.D.: *Flight test of an international space active rack isolation prototype system.* Smart Mater. Struct., 8 (1999), pp. 791–797
101. Cobb, R.G.; Sullivan, J.M.; Das, A.; Porter Davis, L.; Tupper Hyde, T.; Davis, T.; Rahman, Z.H. and Spanos, J.T.: *Vibration isolation and suppression system for precision payloads in space.* Smart Mater. Struct., 8 (1999), pp. 798–812
102. Friebele, E.J.; Askins, C.G.; Bosse, A.B.; Kersey, A.D.; Patrick, H.J.; Pogue, W.R.; Putnam, M.A.; Simon, W.R.; Tasker, F.A.; Vincent, W.S. and Vohra, S.T.: *Optical fibre sensors for spacecraft applications.* Smart Mater. Struct., 8 (1999), pp. 813–838
103. O'Regan, S.D.; Burkewitz, B.; Fuller, C.R.; Lane, S. and Johnson, M.: *Payload noise suppression using distributed active vibration absorbers.* Proc. SPIE Vol. 4698 (2002); pp. 150–159
104. Priou, A.: *Electromagnetic Antenna and Smart Structures.* AGARD LS-205, Paper 11 (1996)
105. Lockyer, A.J.; et al.: *Development of a Conformal Load Carrying Smart Skin Antenna for Military Aircraft.* SPIE 2448/53 (1995)
106. Altandal, K.H.: *Smart Skin Structure Technology Demonstration.* SPIE Meeting on Smart Structures and Mater.; San Diego/CA (1996)
107. Howard, B.M.; et al.: *Thermoadaptive Antennas.* SPIE North American Conference on Smart Structures and Mater., San Diego/CA (1996)
108. Howard, B.M.; et al.: *Electrochromic Adaptive Antennas.* ibid (1996)
109. Varadan, V.K.: *Design and Development of a Conformal Spiral Antenna.* ibid (1996)

110. Yoon, H.-S. and Washington, G.: *Piezoceramic actuated aperture antennae.* Smart Mater. Struct., 7 (1998), pp. 537–542
111. Kiely, E.; Washington, G. and Bernhard, J.: *Design and development of smart microstrip patch antennas.* Smart Mater. Struct., 7 (1998), pp. 792–800
112. Varadan, V.K.; Jose, K.A. and Varadan, V.V.: *Design and development of electronically tunable microstrip antennas.* Smart Mater. Struct., 8 (1999), pp. 238–242
113. Barnes, A.; Despotakis, A.; Wong, T.C.P.; Anderson, A.P.; Chambers, B. and Wright, P.V.: *Towards a smart window for microwave applications.* Smart Mater. Struct., 7 (1998), pp. 752–758
114. Chambers, B.: *A smart radar absorber.* Smart Mater. Struct., 8 (1999), pp. 64–72
115. Wright, P.V.; Chambers, B.; Barnes, A.; Lees, K. and Despotakis, A.: *Progress in smart microwave materials and structures.* Smart Mater. Struct., 9 (2000), pp. 273–279
116. Tennant, A. and Chambers, B.: *Adaptive radar absorbing structure with PIN diode controlled active frequency selective surface.* Smart Mater. Struct., 13 (2004), pp. 122–125
117. Hunt, S.; Rudge, A.; Carey, M.; Parfitt, M.; Chase, J.G. and Huntsman, I.: *Micro-electro-mechanical-systems direct fluid shear stress sensor arrays for flow control.* Smart Mater. Struct., 11 (2002), pp. 617–621
118. Huang, A.; Ho, C.M.; Jiang, F. and Tai, Y.-C.: *MEMS Transducers for Aerodynamics – A Paradym Shift.* AIAA-Paper 00–0249 (2000)
119. Suhonen, M.; Graeffe, Y.-C.; Sillanpää, T.; Sipola, H. and Eiden, M.: *Scanning micromechanical mirror for fine-pointing units of intersatellite optical links.* Smart Mater. Struct., 10 (2001), pp. 1204–1210
120. Vinoy, K.J. and Varadan, V.K.: *Design of reconfigurable fractal antennas and RF-MEMS for space-based systems.* Smart Mater. Struct., 10(2001), pp. 1211–1223
121. Barrett, R.; Frye, P. and Schliesman, M.: *Design, construction and characterization of a flightworthy piezoelectric solid state adaptive rotor.* Smart Mater. Struct., 7 (1998), pp. 422–431
122. Barrett, R.; Gross, R.S. and Brozoski, F.: *Missile flight control using active flexspar actuators.* Smart Mater. Struct., 5 (1996), pp. 121–128
123. Barrett, R.: *Developmental History of a New Family of Subscale, Convertible, High Performance UAVs.* Presentation given at the MAV Workshop held at Schloss Elmau (2003)
124. Barrett, R. M.; Burger, C.; Melian, J. P. and Fidler, K.: *Recent advances in uninhabited aerial vehicle (UAV) flight control with adaptive aerostructures.* Proc. Eur. Conf. on Smart Technol. Demonstrators and Devices; Edinburgh/Scotland (2001)
125. Null, W.; Wagner, M.; Shkarayev, S.; Jouse, W. and Brock, K.: *Utilizing Adaptive Wing Technology in the Control of a Micro Air Vehicle.* Proc. SPIE Vol. 4698 (2002), pp. 112–120
126. Thomas, J.P.; Qidwai, M.A.; Matic, P.; Everett, R.K.; Gozdz, A.S.; Keennon, M.T. and Grasmeyer, J.M.: *Structure-Power Multifunctional Materials for UAVs.* Proc. SPIE Vol. 4698 (2002), pp. 160–170
127. Kornbluh, R.; et al.: *Application of Dielectric Elastomer EAP Actuators.* In: Bar-Cohen J. (Ed.), 2004: Electroactive Polymer (EAP) Actuators as Artificial Muscles; SPIE Press, Bellingham/USA (2004), pp. 529–581

128. Pornsin-Sirirak, T. N.; Lee, S. W.; Nassef, H.; Grasmeyer, J.; Tai, Y. C.; Ho, C. M.; Keennon, M.: *MEMS Wing Technology for a Battery-Powered Ornithopter.* Proc. the 13th IEEE Annual Int. Conf. on MEMS, Miyazaki/Japan (2000)
129. Ho, S.; Nassef, H.; Pornsin-Sirirak, N.; Tai, Y. C. and. Ho, C. M: *Unsteady Aerodynamics and Flow Control for Flapping Wing Flyers.* Progress in Aerospace Sciences, 39 (2003), pp. 635–681
130. Pornsin-Sirirak, T. N.; Tai, Y. C.; Ho, C. M. and Keennon, M.: Microbat: *A Palm-Sized Electrically Powered Ornithopter.* ISM 2000 Int. Symp. on Smart Structures and Microsystems, Hong-Kong/China (2000)
131. Pornsin-Sirirak, T. N.; Liger, M.; Tai, Y. C.; Ho, S. and Ho, C. M.: *Flexible Parylene-Valved Skin for Adaptive Flow Control.* Proc. 15th IEEE Annual Int. Conf. on MEMS, Las Vegas/USA (2002)
132. Hitt, D.L.; Zakrzwski, C.M. and Thomas, M.A.: *MEMS-based satellite micropropulsion via catalyzed hydrogen peroxide decomposition.* Smart Mater. Struct., 10 (2001), pp. 1163–1175
133. Terry, M.; Reiter, J.; Böhringer, K.F.; Suh, J.W. and Kovacs, G.T.A.: *A docking system for microsatellites based on MEMS actuator arrays.* Smart Mater. Struct., 10 (2001), pp. 1176–1184
134. Yao, J.J.; Chien, C.; Mihailovich, R.; Panov, V.; DeNatale, J.; Studer, J.; Li, X.; Wang, A. and Park, S.: *Microelectromechanical system radio frequency switches in a picosatellite mission.* Smart Mater. Struct., 10 (2001), pp. 1196–1203
135. Tung, S.; Witherspoon, S.R.; Roe, L.A.; Silano, A.; Maynard, D.P. and Ferraro, N.: *A MEMS-based flexible sensor and actuator system for space inflatable structures.* Smart Mater. Struct., 10 (2001), pp. 1230–1239
136. *Adaptives Fahrwerk im Vectra.* ATZ 9/2004 Jahrgang 106, Vieweg, Wiesbaden (2004), p. 751
137. Sonderheft ATZ/MTZ: *Der neue Audi TT.* Vieweg, Wiesbaden (2006), pp. 72–74
138. Böcking, F.; Sugg, B.: *Piezo Actuators: A Technology prevails with injection valves for combustion engines.* Proc. 10th Int. Conf. New Actuators, Bremen, Germany (14–16 June 2006), pp. 171–176
139. Carlson, J. D.; Sheng, P.; Wen, W.: *Magnetorheological and Electrorheological Fluid Highlights – 2006.* Proc. 10th Int. Conf. New Actuators, Bremen, Germany (14–16 June 2006), pp. 235–240
140. *Piezoelectric Materials – Global Technology Developments* (Technical Insights), Frost & Sullivan, D619 (March 2006)
141. Seipel, B.: *Reversible Verriegelungsaktorik zur Verbesserung der Crashsicherheit.* Haus der Technik, Veranstaltung 'Fahrzeugtüren', Essen (2006)
142. Thomaier, M.; Atzrodt, H.; Herold, S.; Mayer, D.; Melz, D.: *Simulation of a Complete System Using the Example of an Active Interface for Vibration Reduction.* Proc. Virtual Product Development in Automotive Eng., Mürzzuschlag / Austria (2005)
143. Elliott, S.J.: *Signal Processing for Active Control.* Academic (2001)
144. Mayer, D.; Atzrodt, H.; Herold, S.; Thomaier, M.: *An approach for the model based monitoring of piezoelectric actuators.* II ECCOMAS Thematic Conf. on Smart Structures and Mater., Lisbon (2005)
145. Uchino, K.; (et. al.): *High-Power Piezoelectric Transformers.* Proc. 13 Int. Conf. Adaptive Structures Tech. (ICAST) (2002)

146. Spencer Jr., B.F.; Ruiz-Sandoval, M.E.; Kurata, N.: *Smart sensing technology: opportunities and challenges.* Structural Control and Health Monitoring 11 (2004), pp. 349–368
147. Hagood, N.W.; von Flotow, A.: *Damping of Structural Vibrations with Piezoelectric Materials and Passive Electrical Networks.* J. Sound and Vibration 146 (2) (1991), pp. 243–268
148. European Automotive Research Partners Association: *Future Road Vehicle Research FURORE.* R&D Technology Roadmap, a contribution to the identification of key technologies for a sustainable development of European road transport, Thematic Network supported by the European Commission under the 5th. Framework Programme, Contract Number: G3RT-CT-2002–05089 (2003)
149. Preumont, A.: *Vibration Control of Active Structures.* Kluwer Academic, Dordrecht, The Netherlands, 2nd Edition (2001)
150. Kalinke, P.; Gnauert, U.; Fehren, H.: *Einsatz eines aktiven Schwingungsreduktionssystems zur Verbesserung des Schwingungskomforts bei Cabriolets.* Proc. Adaptronic Congress 2001, Berlin (4–5 April 2001)
151. Kalinke, P.; Gnauert, U.: *ATC: Active Torsion Control zur Optimierung des Schwingungskomforts bei Cabriolets.* Proc. Adaptronic Congress 2002, Potsdam (23–24 April 2002)
152. Melz, T.: *Entwicklung und Qualifikation modularer Satellitensysteme zur adaptiven Vibrationskompensation an mechanischen Kryokühlern.* PhD Dissertation Darmstadt 2001
153. Krix, P.: *Mehr als nur Schmuckstück – der Audi TT wird erwachsen.* Automobilwoche edition, Crain Communications, Oberpfaffenhofen (2006), pp. 20–21
154. Backfisch, K. P.: *Hightech als Ladenhüter.* Automobil Industrie 06/2006, Vogel Auto Nedien, Würzburg (2006), pp. 58–60
155. Marienfeld, P.M.; Bohn, C.; Karkosch, H.-J.; Svaricek, F.: *Reduzierung des motorseitig eingeleiteten Körperschalls durch Einsatz adaptiver und aktiver Lagersysteme.* Global Chassis Control, Haus der Technik, Essen (2002)
156. Janocha, H.: *Steuerbares Motorlager mit magnetorheologischer Flüssigkeit – Controllable engine mounting with MRF.* AUTOREG 2006, VDI-Berichte Nr. 1931 (2006), pp. 313–326
157. Fursdon, P.M.T.; Harrison, A.J.; Stoten, D.P.: *The Design and Development of a Self-Tuning Active Engine Mount.* IMechE (2000)
158. Goroncy, J.: *Hier federt der Strom.* Automobil Industrie 1–2/2005, Vogel Auto Medien, Würzburg (2005), pp. 54–56
159. Matsuoka, H.; Mikasa, T.; Nemoto, H.: *NV Countermeasure Technology for a Cylinder-On-Demand Engine Mount.* Proc. SAE World Congress (Paper 2004–01-0423), Detroit (2004)
160. Gruber; Winner, H.; Härtel, V.; Holst, M.: *Beeinflussung des Fahrverhaltens durch adaptive Fahrwerklager.* VDI-Tagung Reifen-Fahrwerk-Fahrbahn, Hannover (10/2003)
161. Matthias, M.; Thomaier, M.; Melz, T.: *Entwicklung, Bau und Test eines multiaxialen, modularen Inferfaces zur aktiven Schwingungsreduktion für automotive Anwendungen.* Proc. Adaptronic Congress, Göttingen (2005)
162. Atzrodt, H.; Herold, S.; Mayer, D.; Thomaier, M.; Melz, T.: *Gesamtsystemsimulation aktiver Strukturen am Beispiel eines aktiven Interfaces.* IFM, Int. Forum Mechatronik, Augsburg (2005)

163. Schmidt, K.; Thörmann, V.; Weyer, T.; Mayer, D.; Herold, S.; Krajenski, V.: *Aktive Schwingungskompensation an einer PKW-Dachstruktur.* Proc. Adaptronic Congress 2003, Wolfsburg (2003)
164. Melz, T.; Matthias, M.: *The Fraunhofer MaVo FASPAS for smart system design for automotive and machine tool engineering.* 12th SPIE Int. Symp., San Diego, California, USA (6. - 10.03.2005)
165. *Leitprojekt Adaptronik.* Förder-Nr.: 03N8500, http://www.lp-adaptronik.de/
166. European Commission: *Manufacturing Visions Report No. 3: Integrating Diverse Perspectives into Pan-European Foresight.* Delphi Interpretation Report, Contract No. NMP2-CT-2003–507139-MANVIS (11/2005)
167. *Eindeutige Zielvorgaben.* Interview with Dr. Alois Seewald, head of preliminary development at TRW, Automobil Industrie 6/2006, Vogel Auto Medien, Würzburg (2006), pp. 66–67
168. Dr. Klose: *Wirtschaftliche Fahrzeug-Leichtbaukonzepte für verkürzte Entwicklungszeiten großer Baureihen.* Darmstadt, LM Consulting, Sindelfingen (2006)
169. Busse, M.; Wöstmann, F.-J.: *Intelligent Cast Parts – Application of Adaptronic Components with Cast Parts.* Proc. Adaptronic Congress, Göttingen (May 2006)
170. Bräutigam, V.: *Integration of Piezoceramic Modules.* In: Die Castings – A New Producion Technology, Proc. Adaptronic Congress, Göttingen (May 2006)
171. Gosebruch, H.: *Rundschleifen im geschlossenen Regelkreis.* PhD thesis, Universität Hannover, VDI-Verlag GmbH, Düsseldorf (1990)
172. Zinngrebe, M.: *Adaptive Prozessführung beim Innenrundschleifen mit digitalen Grenzregelungen.* PhD thesis, Universität Hannover, Fortschritt-Berichte VDI, Düsseldorf (1990)
173. Großmann, K. (ed.): *Intelligente Funktionsmodule der Maschinentechnik.* TU Dresden, Lehrstuhl für Werkzeugmaschinen (1999)
174. Denkena, B.; Will, J.C. and Sellmeier, V.: *Prediction of process stability and dynamic forces of an adaptronic spindle system.* Proc. Adaptronic Congress 2006, 3.–4. Mai, Göttingen, Germany (2006)
175. Großmann, K.; Müller, J. and Schween, A.: *Mikro-Achse als Zusatzaggregat für Großdrehmaschinen.* ZWF Zeitschrift für wirtschaftlichen Fabrikbetrieb, Jahrg. 96 (2001) 9, pp. 470–473
176. Kemmerling-Lamparsky, M.: *Dynamische Stabilisierung spanender Fertigungsprozesse mit aktiven Zusatzsystemen.* PhD thesis, Universität Hannover, Fortschritt-Berichte VDI, Düsseldorf (1987)
177. Fleischer, J.; Knödel, A.; Munzinger, Ch. and Weis, M.: *Designing Adaptronical Components for Compensation of Static and Quasi-Static Loads.* Proc. ASME 2006 Int. Des. Eng. Tech. Conf. & Comp. and Inform. in Eng. Conf., Philadelphia, Penns. USA (DETC2006–99461), Sept 10–13 (2006)
178. Munzinger, Ch.: *Adaptronische Strebe zur Steifigkeitssteigerung von Werkzeugmaschienen.* PhD thesis, Universität Karlsruhe (TH) (2006)
179. Hesselbach, J.; Abel-Keilhack, C.: *Active hydrostatic bearing with magnetorheological fluid.* J. Appl. Phys., Vol. 93, No. 10 (2003), pp. 8441–8443
180. *Adaptronische Transportsysteme mit elektrorheologischen Flüssigkeiten (ERFs) zur Beförderung sensibler Güter.* Joint project funded by the German Federal Ministry of Education and Research (BMBF), 13N6986, www.tib.uni-hannover.de

181. Nitsche, R.; Gaul, L.: *Lyapunov design of damping controllers.* Archive of Applied Mechanics, 72 (2003), pp. 865–874
182. NN: *Ruhig gebettet.* MM Das Industrie-Magazin, 31/32 (2005), pp. 22–23
183. Kobori, T.; Minai, R.: *Analytical Study on Active Seismic Response Control.* Trans. Arch. Inst. Japan. 66 (1960), pp. 257–260
184. Yao, J.T.B.: *Concept of Structural Control.* ASCE. J. Structural Div. 98 (7) (1972), pp. 1567–1574
185. Leipholz, H.H.E.: *Structural Control.* North-Holland, Amsterdam, New York, Oxford (1980)
186. Leipholz, H.H.E.: *Structural Control.* Martinus Nijhoff, Amsterdam (1985)
187. Soong, T.T.: *Active Structural Control.* Longman Scientific & Technical, Essex (1990)
188. Housner, G.W.; Masri, S.F.: *International Workshop on Structural Control.* Hawaii USC Publ. Number CE-9311, Los Angeles (1993)
189. Housner, G.W.; Masri, S.F. and Chassiakos, A.G.: *First World Conference on Structural Control.* Proc. Int. Association for Structural Control, USC, Los Angeles (1994)
190. Wada, B.K.; Fanson, J.; Crawley, E.: *Adaptive structures.* J. Spacecraft and Rockets 27 (3) (1990), pp. 157–174
191. Bachmann, H. et al.: *Vibration Problems in Structures.* Birkhäuser, Basel, Boston, Berlin (1995)
192. Sockel, H. et al.: *Wind-excited Vibrations of Structures.* CISM Courses and Lectures No. 335, Springer, Wien, New York (1994)
193. Koshika, N. et al.: *Research, development and application of active-passive composite tuned mass dampers.* Proc. 4th Int. Conf. on Adaptive Structures, Technomic (1993)
194. Tamura, K.: *Technology of active control systems for structural vibration.* Int. Post-SMiRT Conf. Seminar, Capri (1993)
195. Reinhorn, A.M. et al.: *Active bracing system – A full scale implementation of active control.* Tech. Report NCEER-92–0020 (1992)
196. Traina, M.I. et al.: *An experimental study of the earthquake response of building models provided with active damping devices.* Proc. 9th World Conf. on Earthquake Eng., VIII 447–4522 (1988)
197. Wada, B.K. and Das, S.: *Application of adaptive structures concepts to civil structures.* Intelligent Structures-2, Ed. Wen, Y.K., Elsevier (1992)
198. Den Hartog, J.P.: *Mechanical Vibration.* 4th ed., McGraw-Hill (1956), pp. 87–106
199. Matsuhisa, H.; Gu, R.; Wang, Y.; Nishihara, O. and Sato, S.: *Vibration Control of a Ropeway Carrier by Passive Dynamic Vibration Absorbers.* Jpn. Soc. Mech. Eng. Int. J. (C), 38 (4) (1995), pp. 657–662
200. Matsuhisa, H.; Nishihara, O.; Sato, K.; Otake, Y. and Yasuda, M.: *Design of a Dynamic Absorber for a Gondola Lift.* Proc. Asia-Pacific Vibration Conf. (1995), pp. 215–220
201. Nishihara, O.; Matsuhisa, H. and Sato, S.: *Methods for Designing Vibration Control Mechanisms with Gyroscopic Moment.* Proc. Asia-Pacific Vibration Conf. 1 (1991), pp. 3.56–3.61
202. Nishihara, O.; Matsuhisa, H. and Sato, S.: *Optimum Design of Vibration Control Mechanisms with Gyroscopic Moment for Harmonic and Stationary Random Excitations.* Proc. 1st Int. Conf. on Motion and Vibration Control 1 (1992), pp. 321–326

203. Nishihara, O.; Matsuhisa, H. and Sato, S.: *Passive Gyroscopic Damper for Stabilization of rigid Body Pendulum.* Proc. Asia-Pacific Vibration Conf. 3 (1993), pp. 889–894
204. Nishihara, O.; Ishihara, H.; Matsuhisa, H. and Sato, S.: *Design Optimization of Passive Gyroscopic Damper by Genetic Algorithms – Monte Carlo Optimization under Random Excitations.* (in Japanese), Trans. Jpn. Soc. Mech. Eng., No. 640–26(1) (1994), pp. 31–34
205. Nishihara, O.; Yasuda, M.; Kanki, H; Nekomoto, Y.; Sato, K.; Otake, Y. and Matsuhisa, H.: *Stability Maximization of Passive Gyroscopic Damper for Ropeway Gondola.* Proc. Asia-Pacific Vibration Conf. (1995), pp. 864–869

9 Adaptronic Systems in Biology and Medicine

9.1 The Muscle as a Biological Universal Actuator in the Animal Kingdom

W. Nachtigall

Active movement is one of the signs of life. The universal actuator in the animal kingdom is the muscle. The striated skeletal muscle – which is the focus of the following discussion – is the evolutionary highlight of biological actuators. Early stages functioning as contractile elements can even be seen in protozoa such as amoebas.

Actuators took shape on the molecular scale as thread-like protein molecules formed that they could attach themselves to other similar molecules or biological surfaces with a type of crossbridge, and that the angle of attachment of these crossbridges can be changed by applying chemical energy. The 'carrier molecules' of the crossbridges must move actively relative to the point of attachment. This discovery took place very early in biological evolution, surely more than 600 million years ago. The discovery proved so useful that it not only still forms the basis for movement and mobility today, but it also made way, with the appearance of multicellular creatures, for a highly specialised cellular differentiation: muscle fiber. Muscle fiber is the functional unit upon which all actuators in the animal kingdom are based.

Muscle types of a very different nature – from slow, smooth muscles, as found in the intestines of vertebrates, to extremely quick, oscillating fibrillary wing muscles of small insects – have developed in response to the various demands placed on such actuators (quick/slow contraction, large/small force, sustained contraction/brief twitching, etc.). The striated skeletal muscle exhibits the broadest range of application. Despite numerous modifications, this muscle type has maintained a uniform construction and functionality. The following summary concentrates on the striated skeletal muscle. Special biological and physiological details will be left out; instead, an attempt will be made to elaborate on the mechanics of contraction, the mechanical aspects of producing force and extension, the way in which such biological actuators interact with skeletal elements and, finally, their incorporation into complete systems with feedback control. This description should provide the engineer, technician and physicist with a direct analogy to technical actuators, perhaps spurring initiative for autonomous technological developments or improvements – i. e. aspects of 'bionics'. At its Bionics Conference in Düsseldorf, Germany in 1993, the VDI (German engineering association) and the author presented the following definition: as a scientific discipline, bionics deals with the technical realisation and application of principles involving construction, process and development found in biological systems. Actuators would fall into the categorisation constructional bionics [1–16].

9.1.1 Principles of Construction and Function

Coarse and Fine Structures – Crossections

Characteristic, categorizable elements can be seen upon cutting across any skeletal muscle, such as musculus biceps brachii (the biceps). A closer look reveals finer and finer structures. Each hatched or shaded element of Fig. 9.1 in the lower sketch of this figure, respectively, is presented in more detail in a hierarchical fashion.

The muscle is several centimeters in diameter. It works within a relatively stiff sheath of connective tissue, the fascia, from which the muscle is separated by a layer of loose connective tissue. The muscle is divided into bundles of muscle fibers by internal boundaries (perimysium).

Such a muscle fiber bundle has a diameter of about one centimeter. The fiber bundle consists of a series of muscle fibers, each enveloped in a fiber sheath and held apart from the others by a loose connective tissue, permitting relative motion during muscle contraction.

The muscle fiber is formed in ontogeny (individual development) as a fusion of single cells to form a sort of giant polynucleate cell. Its thickness is no greater than 100 µm. The partially aqueous interior medium contains bundles of myofibrils in addition to nuclei and mitochondria (cellular power plants).

The myofibril (approximately 1 µm in diameter) is composed in cross-section of a regular hexagonal arrangement of interfaces between molecular filaments, of which the myosin filament is about twice as thick as the actin filament. Each myosin filament is surrounded by six actin filaments, resulting in a submicroscopic regularity nearly resembling a crystalline structure.

Fine Structures – Longitudinal Sections

A longitudinal section of the muscle reveals that the thin actin filaments are connected from both sides to a common anchoring membrane (the Z-membrane; Fig. 9.2). The thick myosin elements are located in between, and their ends appear to maintain a certain distance from the Z-membranes when observed by the light microscope. The thick and thin elements glide along each other in a telescopic fashion until the ends of the myosin elements

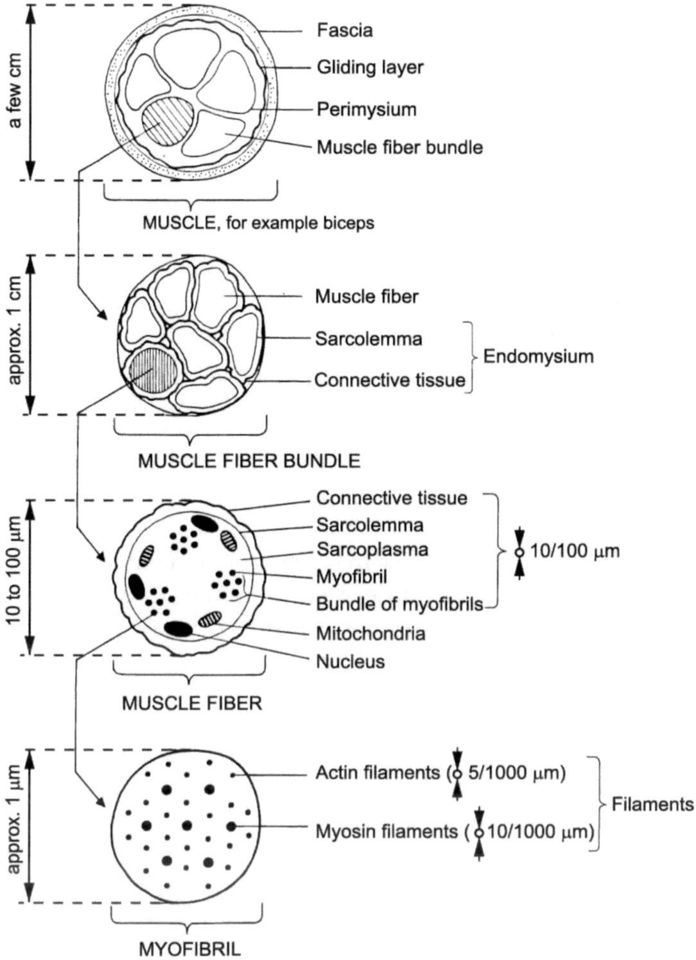

Fig. 9.1. Construction (crossections) of a striated mammalian muscle

hit the Z-membranes. (Holes in the Z-membranes of a very specialized muscle type, the so-called super contractile muscle, allow the myosin ends to travel a bit further.)

There are two additional proteins. Firstly sometimes a meshwork of scaffolding proteins can be recognized between the middle parts of the myosin filaments, probably stiffening them in their center regions. Electron microscopically they form the M-line (Fig. 9.2). Secondly a protein called titin is located between the ends of the Myosin filaments and the Z-membranes. This can only be seen by the electron microscope after special preparations. Due to its elasticity the system of myofibrils and therewith the total muscle becomes extensible. This is important for the muscles ability to store energy when passively stretched.

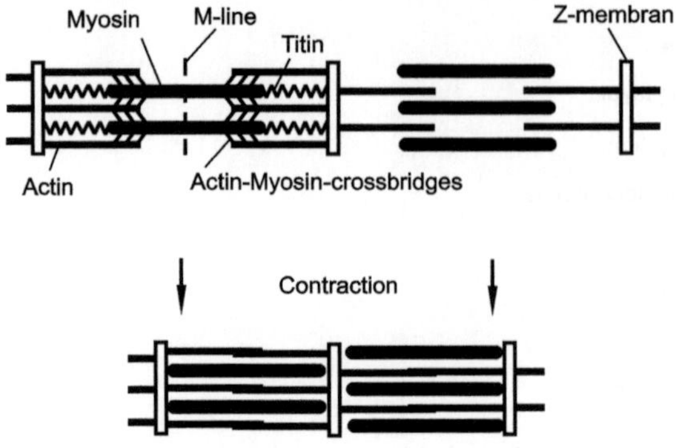

Fig. 9.2. Longitudinal section of a striated mammalian muscle

9.1.2 Analogies to Muscle Function and Fine Structure

The Boulder Analogy

Analogies often help to clarify complex structures. Let us assume that a stone-age man is trying to bring two boulders closer to one another (see Fig. 9.3). When he pulls on one rope, he is too weak and slips (Fig. 9.3a). Should he pull on two ropes connected to the boulders lying opposite one another, he won't slip, but he is still too weak (Fig. 9.3b). Several men pulling next to each other on long ropes (Fig. 9.3c) cause the system to cant, but they are still too weak. By adding more ropes and arranging themselves diagonally opposite one another (Fig. 9.3d), the system no longer cants – *summa summarum* – but there is not enough room for the necessary team of men.

The men now divide themselves into two groups, connect themselves by a bracing in the centre and, alternating, grip diagonally outward onto one of two ropes connected to each boulder (Fig. 9.3e). They can now slide the boulders a bit closer to each other by pulling strongly with both arms. By alternately releasing, gripping and pulling at other places along the two ropes, they manage to bring the boulders closer and closer together.

Submicroscopic Fine Structure of the Longitudinal Section

The following associations can be made to the biological model for the analogy presented. Boulders = Z-membranes; ropes connected to boulders = actin filaments; central bracing = myosin filaments; men = myosin heads; arms directed diagonally outward = actin-myosin crossbridges (Fig. 9.2 and Fig. 9.3e).

A photo taken with an electron microscope with the same orientation (longitudinal section) is displayed in Fig. 9.3f. The actin filaments run toward each other from the net-like Z-membrane structures; the crossbridges, radiating outward from the thick, centrally located myosin filaments to the actin filaments, are plainly visible.

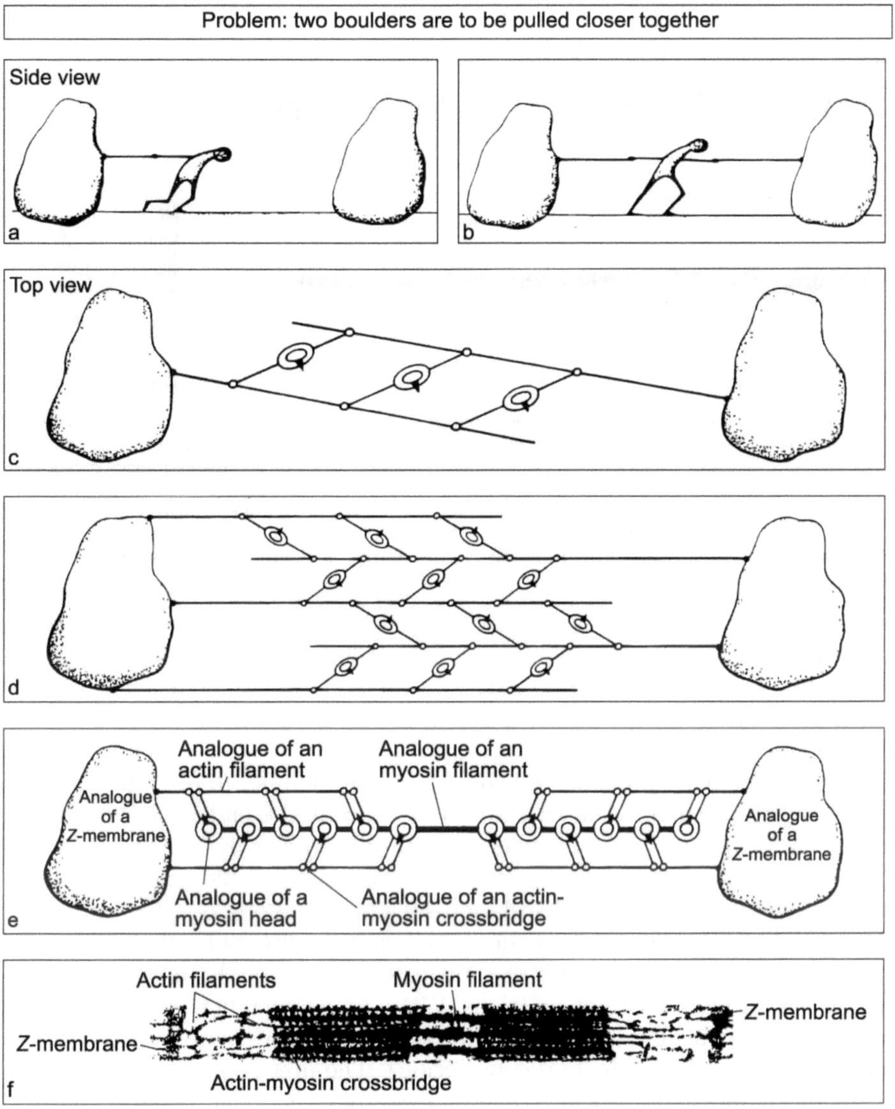

Fig. 9.3. An analogy to the muscle function

9.1.3 Muscle Contraction

Filaments and Elementary Contraction

An actin filament is about 1 µm long (see Fig. 9.4). It consists of two threads (F-actin) wound about each other and composed of globular monomers of actin maintaining their polarity. Fine tropomyosin threads, onto which troponin molecules are attached at regular intervals, run into the niches. The

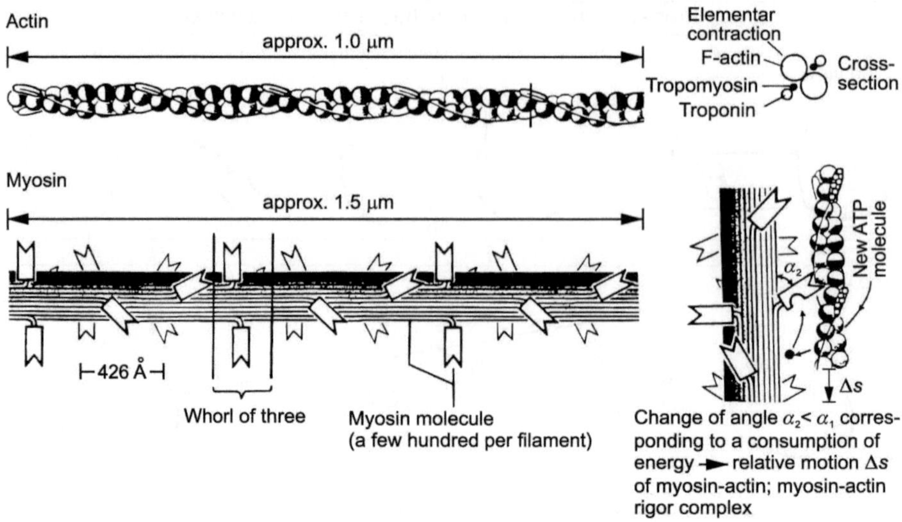

Fig. 9.4. Description of muscle contraction: actin, myosin and elemental contraction

entire configuration is functionally significant for the connection and detachment of crossbridges; their peculiarities cannot, however, be discussed in detail here.

The myosin filaments, which are approximately twice as thick as actin filaments, are about 1.5 µm long and consist of several hundred myosin molecules connected in parallel, each with a braced head at the end. These heads are located at a distance of 426 Å from one another in whorls of three.

Due to its charge and geometry, a myosin head is able to combine with a monomer from the thread-like F-actin. Upon supplying it with energy (reaction and disintegration of an adenosine triphosphate molecule (ATP)), the F-actin changes its angle with the longitudinal axis of the myosin filament through a complex series of reactions. When many such heads take hold, the myosin and actin filaments move relative to one another over a certain elemental distance Δs (elemental contraction). The precise processes (reaction, chemomechanical transduction, detachment, reattachment, aspects concerning energy) cannot be described in detail here. Fundamental aspects of the chemomechanical transduction are still unknown.

Increase of Stress with Progressing Extension

Excitation of a muscle fiber (or an entire muscle) under increasing extension results in a maximum of the relative stress (maximum stress set to 100%) when plotted against the relative length (unstimulated length set to 100%) in the vicinity of the unstimulated length (points 2 and 3 in Fig. 9.5). Under compression, the configuration is distorted and the stress that can be devel-

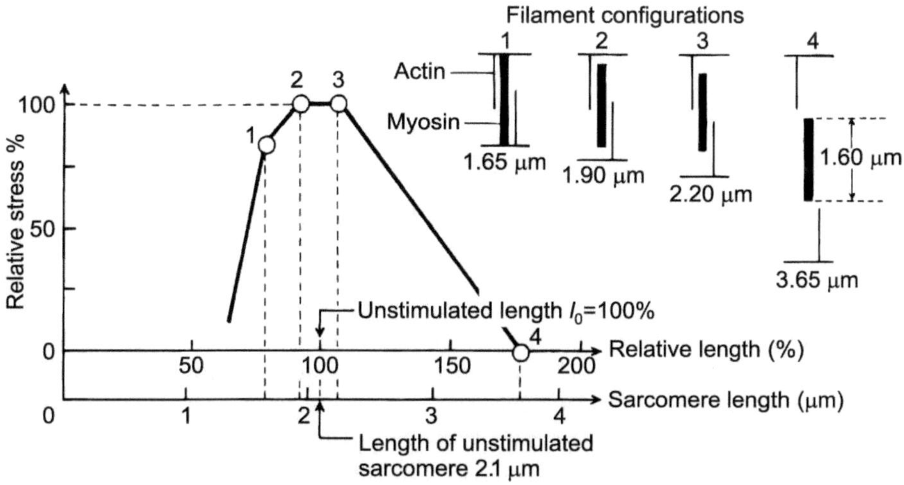

Fig. 9.5. Descriptions of a muscle contraction: length-stress relationship and filament configuration

oped reduces (point 1). During passive elongation, the number of crossbridge possibilities reduces and so also does the stress developed by excitation. The stress reduces to zero when no more crossbridges can take hold.

Single Twitches and Tetanus

A typical skeletal muscle reacts to an artificially induced electrical impulse with a brief contraction, resulting in a mechanical deflection under suitable experimental conditions (see Fig. 9.6). An increase in the excitation results in a superposition of contractions when the successive impulse takes place before the muscle has dilated completely: incomplete tetanus. Single twitches are no longer detectable at an excitation frequency of about 50 impulses per second: complete tetanus.

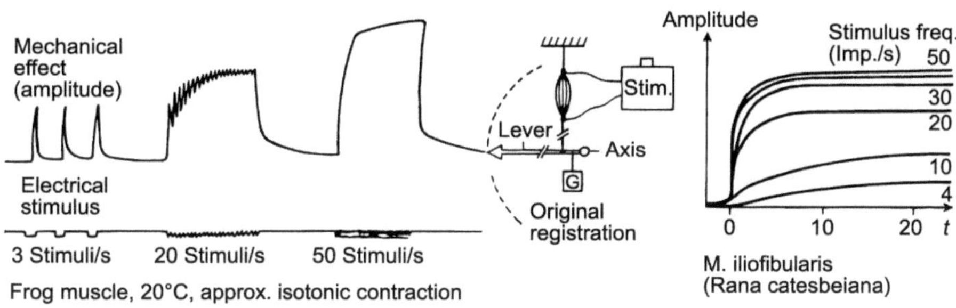

Fig. 9.6. Description of a muscle contraction: from single twitch to tetanus

Muscle operation often approximates to complete tetanus under natural conditions. The amplitude of contraction increases for increasing motor impulse frequency.

9.1.4 Aspects of Muscle Mechanics

Losses in Muscle Work During the Elongation and Contraction Cycle

Different curves (elongation and relaxation curves) result when plotting the load against the length during passive elongation and successive relaxation of muscle fiber (without electrical or neural excitation) – see Fig. 9.7. Since the area in a force-length diagram has the dimension 'work', the area enclosed by the elongation and relaxation branches of the curve corresponds to the losses per cycle of passive stretching. In a similar fashion losses also result per contraction cycle of an active muscle. The elastic efficiency of the system undergoing passive elongation can be determined as indicated in the diagram. The efficiency of skeletal muscles typically lies around 85%. The mechanical efficiency can be determined in a similar fashion for active contraction. In this case, the values vary due to a strong dependency on the boundary conditions.

Possibilities of Contraction

An isotonic type of contraction (reduction in length under constant stress) is exemplified by an isolated muscle excited to lift a weight hanging on one end – see Fig. 9.8. (The cross-sectional area experiencing loading is assumed to remain approximately constant.) By fixing both ends and applying an

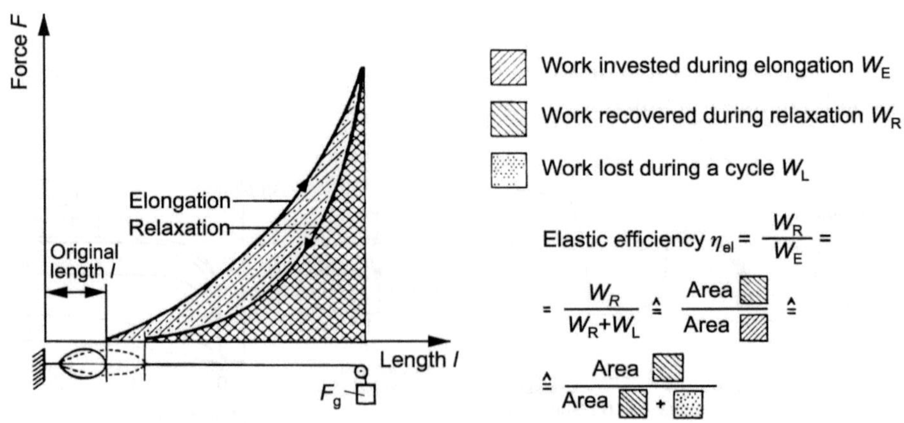

Fig. 9.7. Descriptions of muscle mechanics: passive stretch curve, energy loss and elastic efficiency

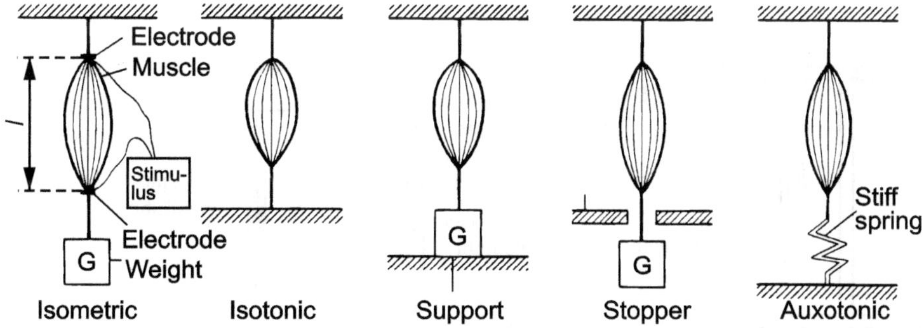

Fig. 9.8. Description of muscle mechanics: experimental contraction modes of a skeletal muscle

excitation, the stress increases without a change in the length: an isometric type of contraction.

A muscle works under isometric conditions followed by isotonic conditions when lifting a weight from a supporting surface (supporting contraction). Upon reaching a limit near completion of an isotonic contraction, the isotonic work of the muscle changes to isometric work (limited contraction). Isotonic and isometric behavior is combined when letting the muscle work, for example, against a strong elastic spring; this type of work is referred to in the field of physiology as auxotonic contraction and, among the experimental cases mentioned, exhibits the closest resemblance to contraction occurring in nature. If the spring is very stiff, a near isometric contraction occurs.

Pars pro toto the isometric contraction is discussed a bit more in detail (Fig. 9.9). By using a measuring device corresponding to Fig. 9.9a but not activating the muscle by an electric stimulus a certain lengthening force can be measured at a certain muscle length (prestress). From many such measurements force-length-curve of the not active muscle (curve A in Fig. 9.9c) can be derived.

When the same measuring device is used and the force transducer is rather stiff, as a consequence of a 'supramaximal' electric stimulus a (near) isometric contraction of the muscle under a certain prestress can be recorded. From many such measurements a force-length curve of the active muscle can be derived. It is known as the curve of the isometric maxima (curve B in Fig. 9.9c).

Assuming a certain-prestress resulting in a muscle length L_1 passive stress results in a passive force F_1. If the muscle is activated electrically in this position, it develops an active force $F_2 - F_1$ additionally, resulting in a total force F_2. One can device from Fig. 9.9c, that the active force is a function of the prestress, that means the length of the muscle under the experimental conditions discussed. The muscle can develop its maximum active force when prestressed to its length in-situ (i. e. when incorporated in the living

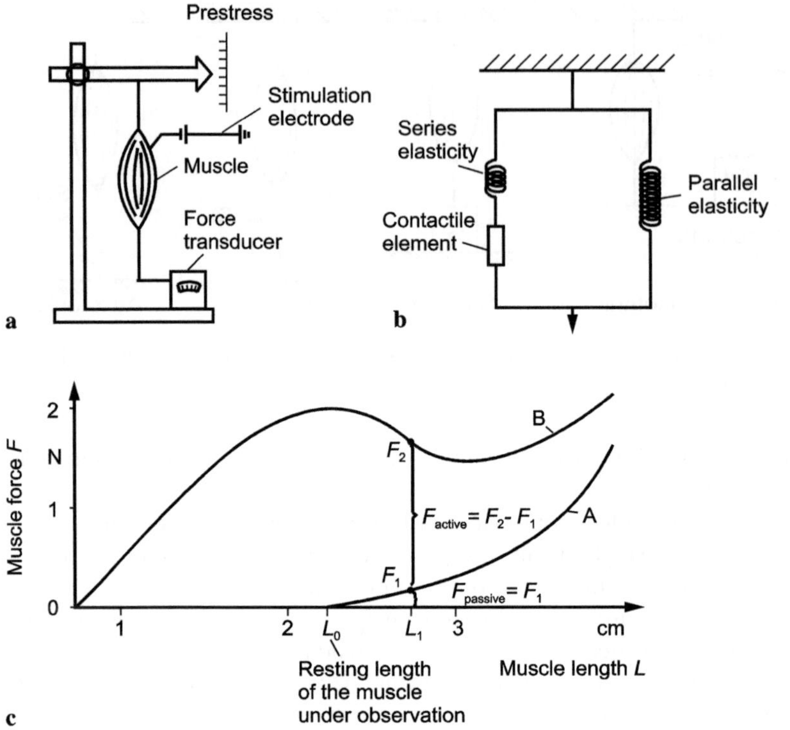

Fig. 9.9. Experimental setup, mechanical abstractions and force-length diagrams of mammalian muscle (comp. the text)

body). When overstressed experimentally its active force gets lower up to a point of zero force. This can easily be explained by Fig. 9.5: no active force any more, when no crossbridge elements between Myosin and Actin are in contact.

The contractile elements can only transduce their actively developed force to tendons (and further to bones or to measuring devices as in Fig. 9.9a) by elastic structures. When contracting they firstly extend structures as the crossbridges themselves, the active filaments, Z-Membranes and the bases of tendons. These structures can be modelled as series elasticities (Fig. 9.9b). Structures such as the sarcolemm, connective tissues between the fibres ant titin are extended, and modelled as parallel elasticities (Fig. 9.9c). At higher prestress of the not active fibre additionally more elastic elements come into working position. That means that the slope of curve A (Fig. 9.9c) is higher at higher prestress and this again means that the modulus of elasticity of a muscle is greater at higher extension.

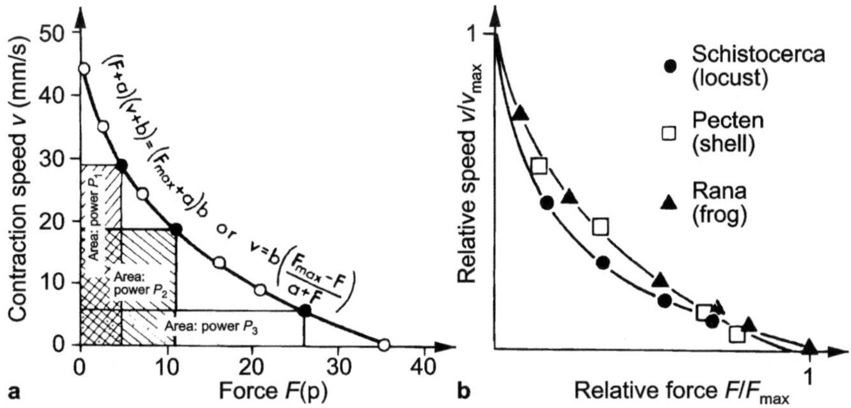

Fig. 9.10. Description of muscle mechanics: **a** force-speed diagram and **b** standardized force-speed diagram

Force-Speed Relationships

It is clear that an actuator cannot develop a large force and a high speed of contraction simultaneously; the two parameters counteract each other. A hyperbolic relationship (the Hills equation) results when plotting the speed of contraction of a muscle against the load (see Fig. 9.10a).

The area of a speed-force diagram has the dimension 'power'. As can be seen, a muscle is capable of producing its greatest power at medium values of speed and force; the expendable power sinks in the vicinity of the extremes (by either high contraction rates or large loads).

The characteristic curves of diverse muscles can in practice be represented by a single curve when normalised with respect to the maximum speed and maximum force. This applies to muscular systems functioning in different ways, including the wing muscles of insects, shell-closing muscles specialised to tonic contraction, or the leg muscles of cold-blooded animals. This indicates an inherent constructional principle among all of these greatly varying muscles (Fig. 9.10b).

9.1.5 Principal Types of Motion Achievable by a Muscle and its Antagonists

Muscles are always arranged so as to interact with an antagonist. This antagonist is typically an opposing muscle. However, mechanically elastic elements can also fulfil the function of an antagonist. In principle, an angle, a distance, an area or a volume can be changed by the contraction of a muscle.

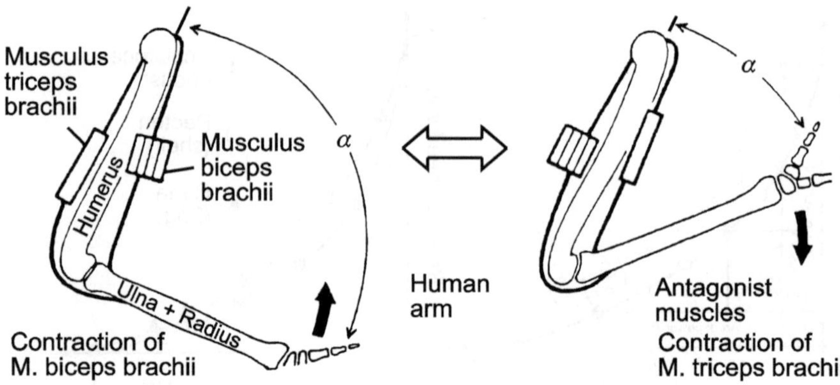

Fig. 9.11. Muscle: basic functions and antagonists: change of an angle between two skeletal elements

Change of Angle

A contraction of our *musculus biceps brachii* leads to a reduction of the angle α between the upper arm and the forearm (see Fig. 9.11). The opposing musculus triceps brachii undergoes a simultaneous extension. A contraction of the triceps leads to an increase in the angle α and to an extension of the biceps. Biceps and triceps are muscular antagonists.

Change of Distance

The downstroke and upstroke muscles that are used to drive the wings found in a dragonfly also operate as antagonistic muscle pairs. These muscles tilt the wing about its basal joint, causing the angle of the longitudinal wing axis to change relative to a reference axis (see Fig. 9.12a).

More highly developed flies and bees function in a different manner. Their thoracic capsule oscillates quickly (up to several hundred strokes per second!) through a more automatic indirect drive. In a sort of lid-and-pan system, the 'lid' (the upper side of the thoracic capsule, displayed a bit smaller in the oversimplified model) is pulled into the larger 'pan' (remaining portion of the capsule) by dorsoventral muscles fixed between the two thoracic pieces (see Fig. 9.12b). This action results in an upward motion of the connected wings. Dorsal muscles running longitudinally through the capsule deform it in the other way, causing the wings to effect a downward stroke. The dorsoventral muscles and their antagonists do not effect a change of an angle but rather the distance between their attachment points.

Changes in Area and Volume

The cuttlefish can change its lightness and colouring within fractions of a second. This is due to quickly contracting, fine radial muscles that are capable

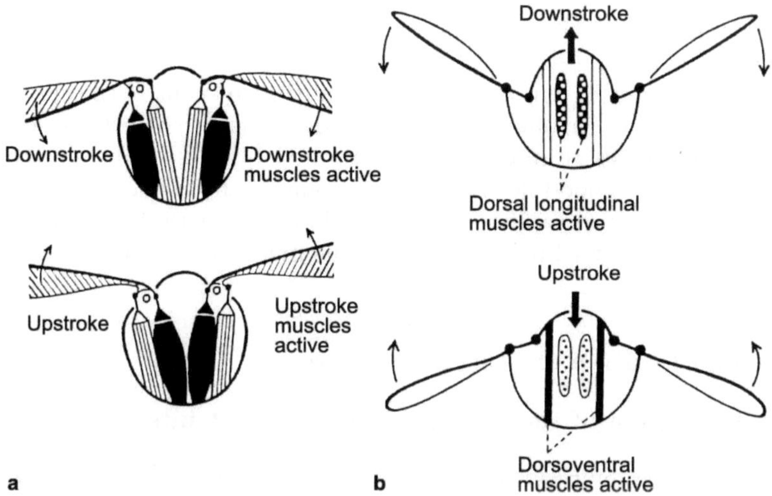

Fig. 9.12. Muscle: basic functions and antagonists. **a** Direct drive (dragonfly), **b** indirect drive (fly, bee)

of pulling pigment-filled cells apart: the result is dark colouring. When the muscle contraction releases, the cell pulls together due to its elasticity, the pigments are condensed point by point, and the light background appears: light colouring (Fig. 9.13a).

Segmentally arranged so-called wing muscles (having nothing to do with insect wings) work in a rhythmic sequence in an insect heart. Valve flaps on the inside prevent the circulatory fluid (hemolymph), sucked out of lateral openings, from flowing in the reverse direction. A unidirectional flow of circulatory fluid results. The elasticity of the entire system and, to some

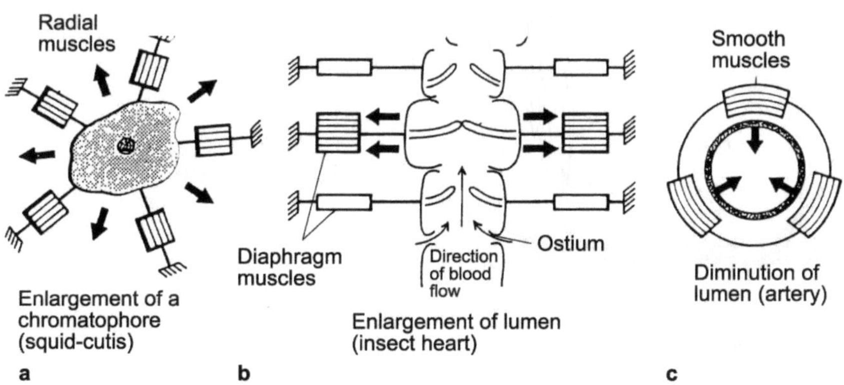

Fig. 9.13. Muscle, basic functions and antagonists. Change of areas and lumina: **a** enlargement of chromatophone, **b** enlargement of lumen, **c** diminution of lumen

degree, the muscles undergoing subsequent contraction function together as antagonists (Fig. 9.13b).

To reduce the volume of our small arteries and thereby control the speed of flow of our blood, we involuntarily place under tension smooth muscle fibers that enclose the vessel in a circular or spiral fashion. By reducing the free lumen, the speed of blood flow is changed dramatically, as described by the Hagen-Poiseuille law of capillary flow. The blood pressure functions as antagonist, causing the vessel to expand when the muscle contraction ceases (Fig. 9.13c).

9.1.6 Force and Position of Muscular Levers

Balance and Location of the Muscle Attachment Points

Of two conceivable possibilities – weak actuators that contract over great distances, or strong actuators that generate large displacements via translation (muscular levers) – nature generally chooses the latter. The biceps muscle is fixed relatively close to the joint. An angular motion of the forearm therefore requires large forces but small displacements. Figure 9.14 illustrates an additional concept for fixation, which could also be used to raise the hand, but an unseemly supply system would be necessary and the arm would hardly be a practical tool for daily activities. Working with powerful actuators, small displacements and large forces leave room for free motion of the lever arm.

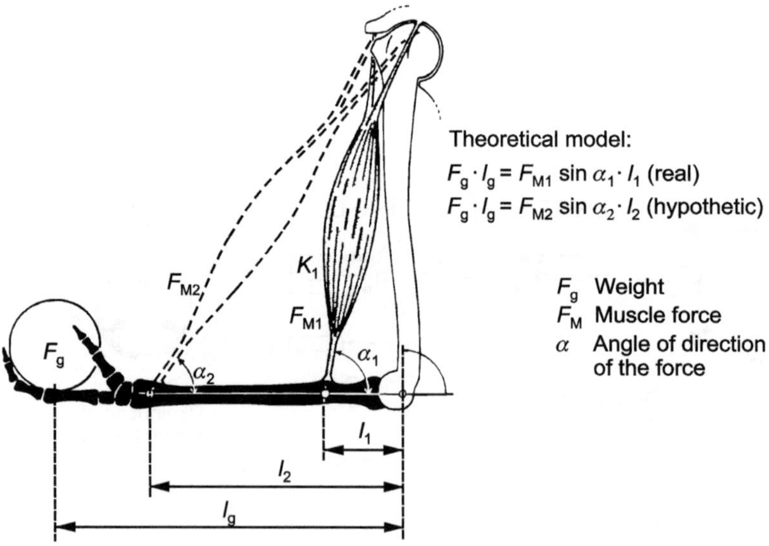

Fig. 9.14. Aspects of muscular levers. Balance of multilateral lever

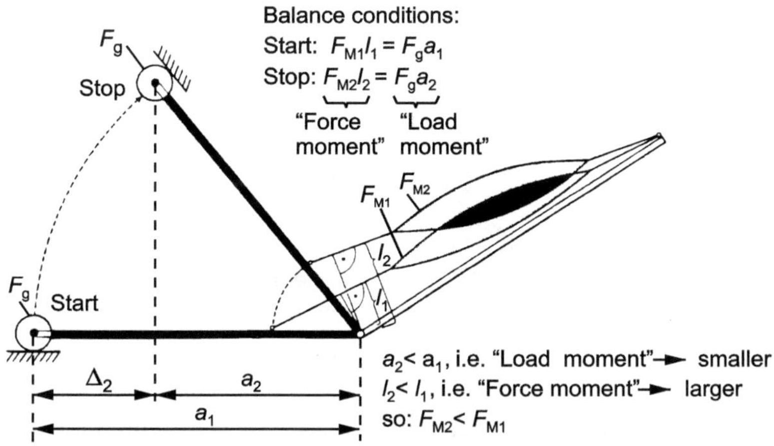

Fig. 9.15. Aspects of muscular levers. Reduction of force under continuous change of angle

Principle of Catapulting Twitch

Beginning with a wide angle between the upper arm and the forearm, the biceps works with a small leverage and therefore must generate a large force (see Fig. 9.15). Through its angular motion, the horizontal distance between the weight (in the hand) and the elbow joint becomes smaller; and so the moment created by the load becomes smaller. At the same time, the perpendicular distance between the muscle tendon and the joint increases; so (for a constant muscular stress) the force moment increases. Both tendencies are favorable and accommodate the characteristics of the muscular actuator: it can operate initially with a large force (nearly isometric) and the needed force decreases as the arm angle is reduced more and more. As soon as the weight has been set in motion, the muscle can stop contracting even before reaching the end position of the movement: 'a catapulting twitch'. (The opposing muscle must simultaneously begin its contraction before the end position of the movement has been reached in order to achieve early braking.)

Tensor Muscles

Not all muscular actuators function in the sense of inducing motion as presented here so far. Some muscles place mechanical systems under tension – systems that first become capable of motion upon being driven by other muscles. An example is found in the click mechanism of flies. A double leverage allows a central joint (over schematised in Fig. 9.16) to toggle after it has pressed the periphery joints apart. In this fashion, the wing is torn upwards or downwards (as applied in the cap of a can of shoe polish). However, the system only operates when the outer joints are drawn toward each other by

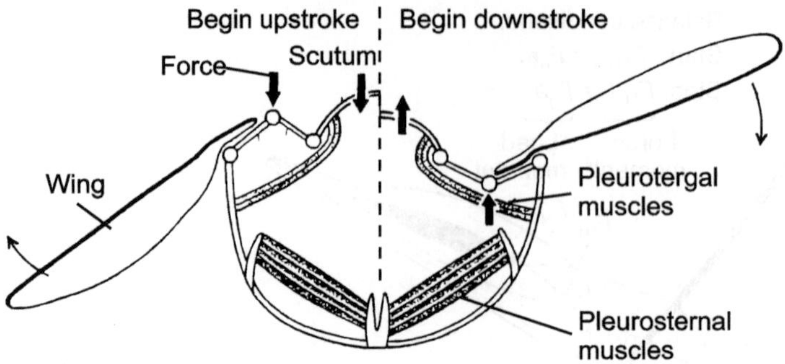

Fig. 9.16. Aspects of muscular levers. Thoracic stress and click mechanics in flies

a mechanical load. This loading is effected by tensor muscles represented in the drawing by the pleurotergal muscles (which pull between the sides and the top) and the pleurosternal muscles (which pull between the sides and the basal piece).

9.1.7 Cooperation of Unequal Actuators

Model calculations show that the jumping motion of animals can be considerably more efficient than their running motion (assuming the same average speed of locomotion). Several muscles work together to create jumps. One muscle performs the main drive and at least one secondary muscle assumes auxiliary functions. Jumping motion is a good example of the cooperation of unequal biological actuators.

Jump of a Locust

Figure 9.17a shows phases of jumping of a large locust. Phase 3 is presented in more detail in Fig. 9.17b. Clearly, the angle between the femur and the tibia is increased during the jump, while the tarsus located on the ground is unrolled until shortly before losing contact with the ground. The main driving muscle is the *musculus extensor tibiae* (which increases the angle between the tibia and the femur). As shown in Fig. 9.17b, this muscle works at a very small leverage distance from the joint, implying short contraction and large force. The instantaneous muscle force can be determined based on the geometric configuration, the mass of the body, and the acceleration during the jump. The force is about 5 N, corresponding to a mechanical stress in the muscle of $140\,\text{kNm}^{-2}$. For a cross-sectional area of the muscle apodem of $0.01\,\text{mm}^2$, the stress reaches $500\,\text{Nmm}^{-2}$. The experimentally determined ultimate tensile strength of the biological apodem is about $600\,\text{Nmm}^{-2}$.(This strength is comparable to that of structural steels, which starts at $450\,\text{Nmm}^{-2}$.) The factor of safety against breakage of the apodem is only 1.2, and the locust jumps near its biomechanical limit.

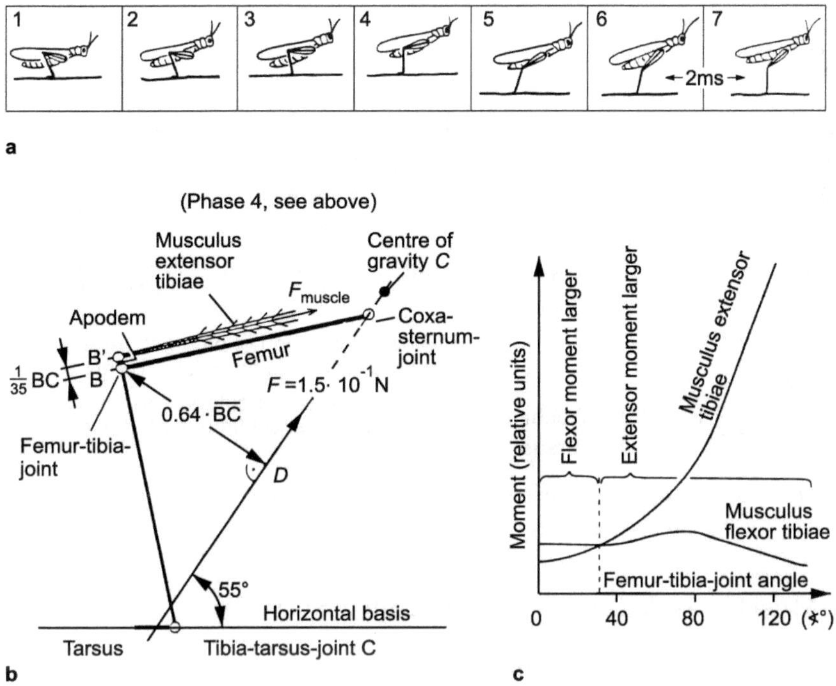

Fig. 9.17. Unequal actuators during jumping: **a** jumping phases of a locust, **b** jump mechanics, **c** flexor and extensor moments (From Brown)

As shown in Fig. 9.17c, a flexor muscle (*musculus flexor tibiae*) works against the jumping muscle (*musculus extensor tibiae*). At the beginning of a jump, the moment generated by the flexor exceeds that of the extensor until an angle of about 30° is reached. A sort of auxiliary catapult is prestressed and unloaded after reaching the 30° angle, thereby reducing the launch time.

Jump of a Flea

In a similar fashion, a flea jumps by contracting the main jump muscle (which is affixed to the 'reverse side') and deforming a highly elastic biological polymer (resilin) in compression. After storing this energy, an auxiliary muscle pulls the tendon-like apodem of the main jump muscle to the 'correct side' allowing the leg to spring out and catapult the flea, see Fig. 9.18.

This is a biological catapult, i. e. a translation of power, where the energy necessary for the jump is stored slowly and released over a shorter period than is possible with direct muscle contraction.

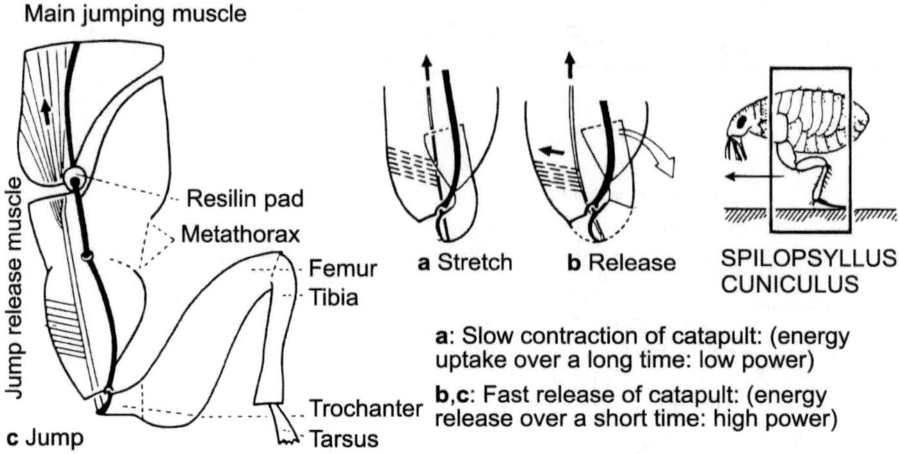

Fig. 9.18. Unequal actuators during jumping. Jump of a flea (From Bennet-Clarc): **a** slow contraction, **b** fast release, **c** the jump

9.1.8 Muscles as Actuators in Controlled Systems

Servo Assistance in the Control of Extension

Numerous control mechanisms take place in the arm musculature when we guide a glass of water to our mouth. Muscle contraction receptors (muscle spindles) effect servo assistance. The process can be described by four steps (Fig. 9.19).

1. The extrafusal fibers EF of the muscle M and the intrafusal fibers IF of the muscle spindles MS contract simultaneously when commanded by the central nervous system G.
2. The feedback signal sent via nerve fibers (so-called Ia-afference) remains constant for equivalent reduction in length of EF and IF; the spindle control loop is inactive.
3. Disturbances induced for example by unexpected increases or reductions in the load L cause the sensor ends SE in the muscle spindles to stretch or become compressed.
4. As a result, the Ia-feedback signal changes, effecting a corresponding change in the excitation of the so-called α-motoneurons. The spindle control loop is active. This process can be described as conditioned feedback. The disturbance is essentially compensated (servo assistance within the γ-loop).

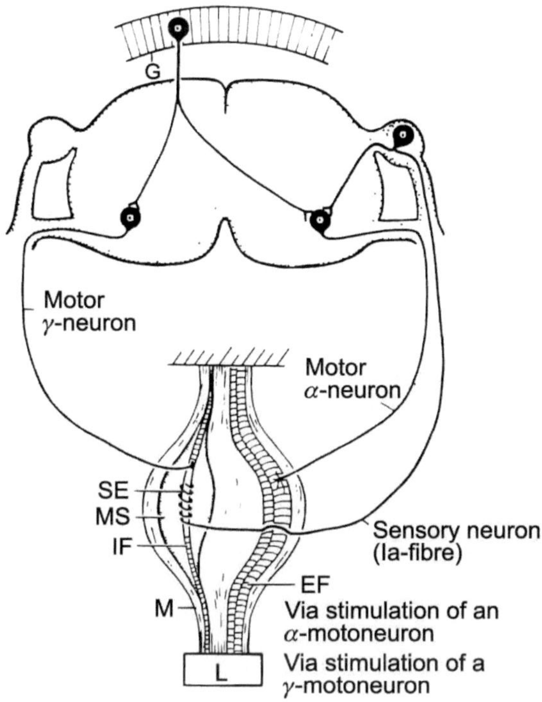

Fig. 9.19. Muscular actuators in feedback systems. Servo support of movement via muscle spindles (length regulation)

Final Control of the Stress

1. The extrafusal fibers EF of the muscle M contract on command from the central nervous system G, as shown in Fig. 9.20. (The muscle spindle control loop is ignored here.)
2. Tendon and tendon organs experience increased stretching.
3. Due to the geometry of the tendon organs (see insert in Fig. 9.20), the sensitive nerve endings are squeezed and thereby excited.
4. This excitation is fed back negatively to the α-motoneuron via the so-called Ib-fibers and an inhibitory interneuron I.
5. The resulting effect is a reduction of muscle tension. The tendon organ practically functions as a limit switch that shuts off muscle activity before the threat of tearing the tendon becomes real.

Crossed Extensor Reflex

Figure 9.21 is self-explanatory: the dilemma of having to react extremely quickly by reflex without upsetting a stable position of the body is accomplished in nature through positive and negative control of the bending and

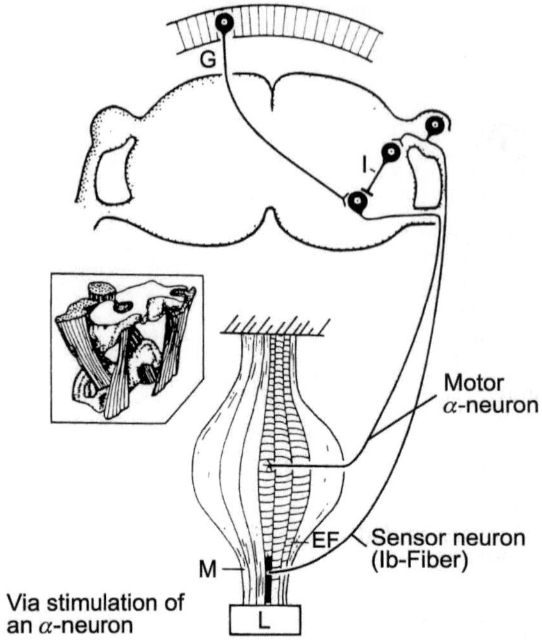

Fig. 9.20. Muscular actuators in feedback systems. Control of muscle activity via the tendon organ stress regulation in muscles

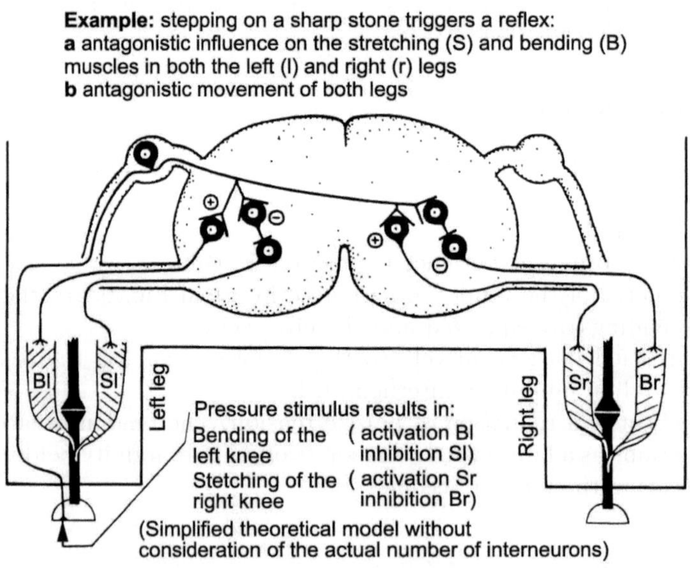

Fig. 9.21. Muscular actuators in feedback systems. Crossed extensor reflex – a more complicated reflex

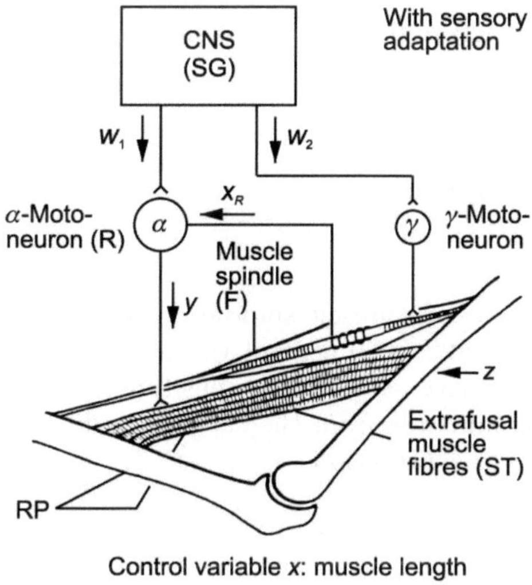

Fig. 9.22. Muscular-cybernetic analogy. Feedback control of muscle length

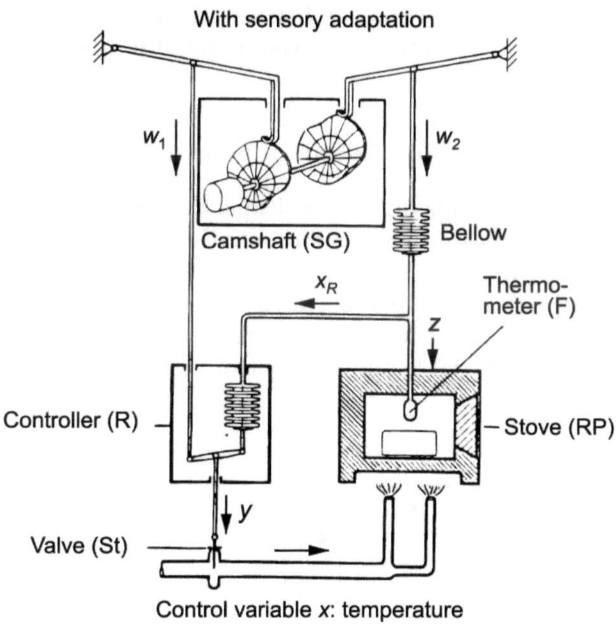

Fig. 9.23. Muscular-cybernetic analogy. Feedback control of furnace temperature

stretching muscle groups located in the left and right legs. A negative interneuron is necessary to switch from positive to negative control.

9.1.9 Control Loops in Biology: Similarities Within Biology and Engineering

Control of Muscle Length

The control of length with servo assistance via muscle spindles, as described in Sect. 9.1.8, is displayed again in Fig. 9.22. The control variable x is the muscle length. The control involves adaptive sensors: the sensitivity of the muscle spindle as a sensor is set by commands from the set-point adjuster before or during the preliminary phase of contraction.

Control of a Furnace

The control variable in the example in Fig. 9.23 is the temperature. A camshaft can influence the sensitivity (via a second cam in the model) of the thermoactuator (by changing the fill volume). The analogy between the two examples (control of length and control of temperature) is given in detail in Figs. 9.22 and 9.23.

Common Control Loop: Control with Adaptive Sensing

The control loop presented in Fig. 9.24 is just as applicable to biological as to technical control. The set-point adjuster informs not only the controller (via reference input 1) but also the sensor (via reference input 2) of its intentions. The sensor is therefore more capable of reacting with the proper signal at the right time. The transient response can be accelerated thereby for small differences between the command variable and the control variable.

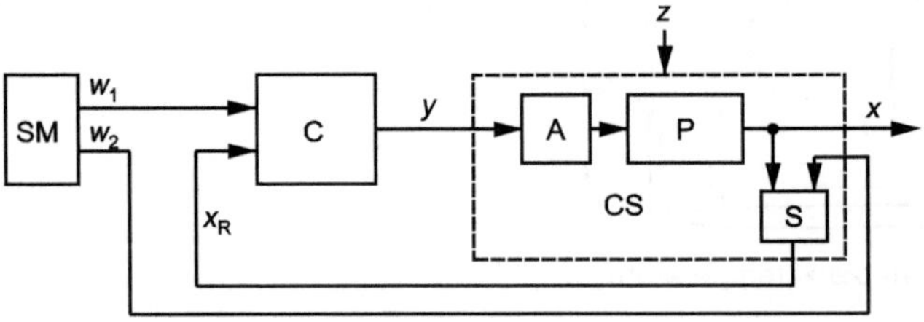

SM: Setting means	S: Sensing device	y: Manipulated variable
C: Controller	CS: Controlled system	x: Output quantity
A: Actuator	w_1: Reference input 1	x_R: Feedback variable
P: Process or system to be controlled	w_2: Reference input 2	z: Disturbance

Fig. 9.24. Muscular-cybernetic analogy. Cybernetic control scheme, valid for Figs. 9.22 and 9.23

Outlook

The similarity of processing within biological and engineering systems is apparent although the morphological designs of their actuators differ. Biology and engineering can be viewed as final components within a joint continuum governed by the laws of nature. By applying this viewpoint, boundaries that have developed between scientific disciplines through traditional dividing strategies could be dissolved.

9.2 Adaptronic Systems in Medicine and Medical Technology

J.-U. Meyer, T. Stieglitz

9.2.1 Introduction

Adaptronic technical systems and structures in medicine are characterized by monitoring the biological system with different sensor modalities (physical, electrical, chemical) and adjusting their performance using multifunctional elements in order to accomplish a beneficial effect for the subject. Medical devices demand adaptive features since most biological systems exhibit time-variant and non-linear properties often with metabolism related reaction time. The controller design has to adapt to the varying system dynamics caused by the variability of the biological system and the lack of a complete description of state-variables. Beyond classic estimation methods in control theory, incomplete description of the system is often compensated by employing artificial neural networks (ANN) or fuzzy based control algorithms. Open-loop feed-forward and closed-loop feedback strategies and adaptive systems are applied.

Adaptronic systems imply versatile sensor-actuator interfaces with possible hierarchical control systems. They are realized with adaptive technical systems that rely heavily on microelectronics and microelectromechanical components integrated in one element for improved performance and enhanced functionality per device size. Low level and high level control circuits are perfectly interlocked to drive the actuator pathways and transmit the right amount of information to our consciousness. Our body is a perfect example of a highly complex, well tuned and adaptive system – for example, the proper moment to apply torque in our motor systems depends on a continuous inflow of sensory information. The peripheral motor system and the motor cortex of the brain extract the information necessary to guide the movements and transmit the signals by the brain stem and spinal cord to the skeletal muscles including all the knowledge about the metabolic background that influences the reaction time and the performance of the skeletal muscles.

Three possible adaptronic systems for biomedical applications are conceived in general (Fig. 9.25a–c).

- An open-loop adaptronic system with closed-loop options (Fig. 9.25a) is an example for applications in the neural/muscular control of a mechanical prosthesis (open-loop) or for a drug delivery system (closed-loop) [17]. In the latter, the biological system as a plant is altered in response to the effector performance. It closes the loop in the control structure and adapts the function of the controller. The signal controller is decoding the information of the biological signals and generates an actuation paradigm for the effector.

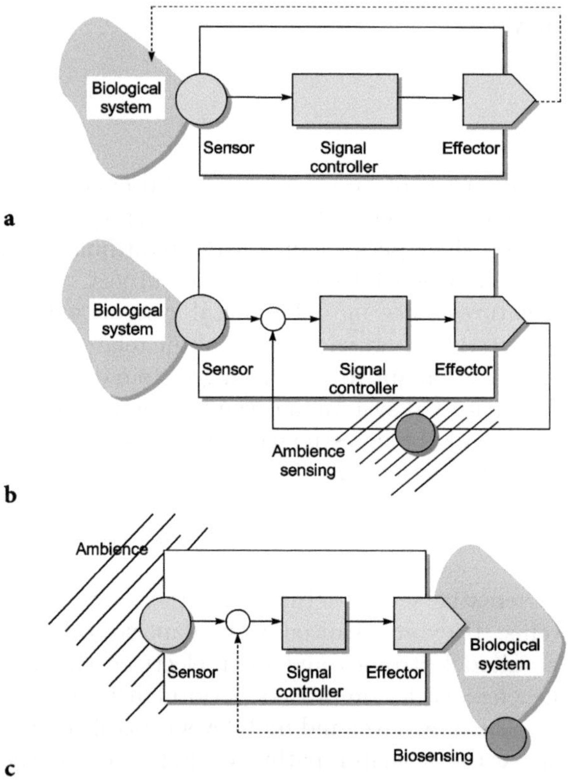

Fig. 9.25. Three possible adaptronic control systems for biomedical applications. **a** Open-loop (*solid line*) adaptronic system with closed-loop option (*dashed line*), as e. g. applied in the neural/muscular control of a mechanical prosthesis (open-loop) or a drug delivery system (closed-loop), **b** control depends on two variables, namely the biological system and the ambiance sensor input, **c** signal is encoded from an ambiance input to achieve an effector response in the biological system. Biosensing delivers state variables for adaptive control

- Incorporation of further ambient signals (Fig. 9.25b) in addition to the biological ones leads to another control scheme of adaptronic systems. Signals from technical and biological sensors and actuators are combined in a single system to handle complex tasks, e. g. in a sensorized prosthesis where the ambient sensor is placed outside the body.
- If neural signals are to be used to control prostheses (Fig. 9.27), additional encoding between the technical and the biological system has to be done to transfer the information in an appropriate code. The sensor has to be placed inside the body to record signals that are used as command variables in the control task (Fig. 9.25c). One example for this adaptronic system is an implantable neural stimulator for grasp in paralyzed people with feedback response from an implanted sensor to control grasp force [18].

In the following, adaptronic systems for biomedical applications have been selected that illustrate recent activities in the field. An emphasis is given on systems that employ microtechnologies. The described microsystems comprise electronic implants, as advanced pacemakers and neural prostheses as well as adapting diagnostic devices, as tactile sensors and self-adjusting blood flow monitors.

9.2.2 Adaptive Implants

Microelectronic implants have gained increasing interest in the bioengineering research community [19] and the medical device industry. Heart pacemakers are the most prominent example of an implantable microsystem that exhibits adaptive properties [20]. In the following section, adaptive properties of advanced pacemaker systems are described. Experience gained from heart stimulation devices has led to the development of a new class of implantable neural stimulators and sensor systems, namely neural prosthetic devices. More than a hundred thousand devices have been implanted in clinical practice, even though many applications are unknown to the general public [21, 22]. Fabrication and the envisioned adaptive control mechanisms are described here upon.

Advanced Pacemakers and Implantable Defibrillators

Common pacemakers stimulate the heart with a fixed rate. For the benefit of the patient, it is desirable to adjust the cardiac output to the physiological state of the patient. Among other factors, the physiological condition of a patient is influenced by body motion, posture, metabolism, ambient conditions, and emotional states. It is the objective of advanced pacemaker systems to adjust cardiac output by either altering the cardiac stroke volume or the cardiac rate. The regulation of stroke volume is refined by its inherent limited range. Current activities focus on rate-adaptive heart pacing systems. Cardiac and non-cardiac state variables can be used for the regulation of cardiac output. Variables include central venous oxygen saturation, oxygen uptake and ventilation, mean arterial blood pressure, parasympathetic and sympathetic activity, which are difficult to monitor in an integrative manner [20, 23, 24]. Applications are corporeal control parameters, such as motion and central venous temperature, as well as cardiac control parameters that include intracardiac impedance measurements and transfer the cardiac pacemaker into an adaptronic system (Fig. 9.26).

Sensory inputs may comprise parameters for body motion, central venous temperature, and intracardiac impedance. Body motion is measured using piezoelectric or micromachined accelerometers that are located in the housing of the pacemaker. Implantable electrodes are used for sensing intracardiac impedance and for heart pacing. More physiological control paradigms gain information by recording the autonomous nervous system in the heart and extracting the conduction velocity between the atrium and the ventricle of the heart (atrio-ventricular conduction time, AVCT). Under load, the AVCT is changed in the working heart. Stroke volume and conduction time have been taken into account in the latest implant concepts. Adaptronic closed-loop systems for dromotropic and inotropic heart control lead to increased activities of daily living for the patient due to better load adaptability [25].

Even more severe than the diseases leading to cardiac pacemaker implantation is the sudden cardiac death (SCD). It is the most often reason for death in western industrial countries and accounts for 1200 deaths per day in the USA. SCD is caused by ventricular tachycardia or fibrillation and death occurs within minutes. The only possibility to overcome the tachycardia and to induce a regular heart beat is to defibrillate the heart. In these cases, instead of a 'normal' cardiac pacemaker, a different implant has to be applied. Implantable cardioverter-defibrillators (ICD) have been developed in the 1980s and more than 25 000 devices have been implanted worldwide up to the mid

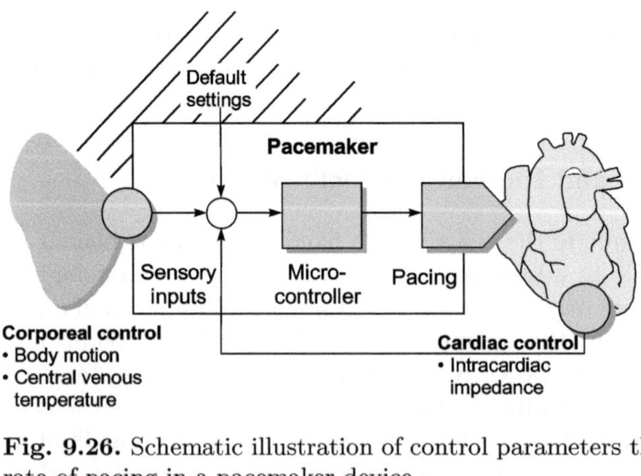

Fig. 9.26. Schematic illustration of control parameters that are used to adjust the rate of pacing in a pacemaker device

1990s. The ICD are similar to a standard pacemaker but have more powerful batteries and circuitry for stimulation. They detect fibrillation of the myocard and defibrillate immediately. In the case of tachycardia in the atrium, the method of choice is called cardioversion. The implant detects the R-spike of the electrocardiogram (ECG) and the stimulation is triggered with a delay of 20 ms after the R-spike to prevent a short fibrillation of the ventricle myocard.

Neural Prostheses

Neuroprosthetic microimplants are considered an emerging field in rehabilitation engineering [26, 27] even though many applications are unknown to a wide public [21] or even to most of general physicians. The neural protheses are designed to compensate for lost or impaired nervous functions or to modulate the nervous system as therapy for incontinence, chronic pain or in degenerative diseases like Parkinsons disease [22]. Biocompatible, long-term functional interfaces have to be established to the neural tissue for sensor applications to record bioelectrical signals or as actuators applying functional electrical stimulation. Detection and decoding of peripheral and central neural signals, respectively, are needed when decoding neural information, e. g. to control an artificial limb. Microimplants with neural interfaces [28] for recoding and stimulating of neural structures are applied when restoring motion in paralyzed limbs or to mimic sensory functions [27]. The same implantable electrode elements serve as sensors for nerve signal registrations and as actuators for nerve stimulation.

Neural Interfaces for Amputees. Artificial limbs after amputation trauma are mainly crude and simple technical systems, even in the 21st century. Only a few systems like the upper limb prostheses Myohand or the lower limb prosthesis c-leg (Otto Bock HealthCare, Duderstadt, Germany) include intelligence that can be summarized under the term adaptronics. In principle, a simplified open-loop adaptronic system for controlling an artificial limb with muscular or nervous signals has to detect the command signals in the amputation stump and has to transfer them into suitable control signals to actuate the prosthesis (Fig. 9.27). A robust proportional control with a myoelectric interface is commercially available with the Myohand (Otto Bock HealthCare, Duderstadt, Germany). However, functionality is limited by the two input channels of the device and the mechatronic design that only allows hand rotation and a cylinder grip. More sophisticated prostheses need different hand designs and actuation strategies and a higher number of (muscle or nerve) interface channels.

The main challenge of the approach (Fig. 9.27) in terms of adaptive signal processing is the design of the controller. The controllers task is to decode the extracellular nerve potentials and identify the ones that are related to efferent signals containing information about motion. Different paradigms postulate

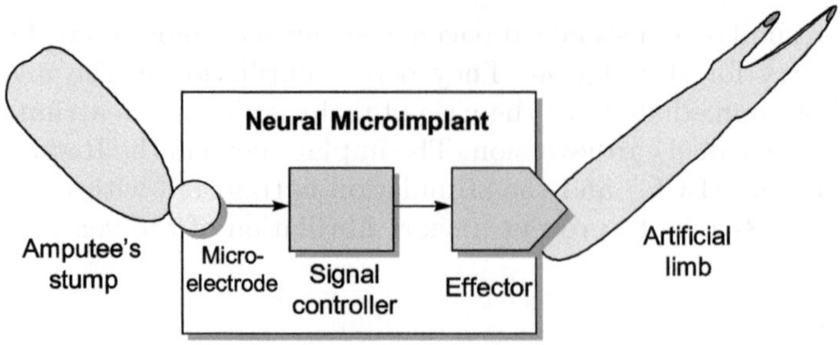

Fig. 9.27. Open-loop adaptronic system for controlling the movements of a mechanical prosthesis

the use of single nerve fiber action potentials or the use of compound signals, respectively. Highly sophisticated signal processing techniques including wavelet analysis, support vector machines, fuzzy logic, artificial neural networks or others are applied for separating the various signal components of the neural activity. Subsequently, meaningful information about the neural activity is converted into electric signals that control the artificial limb. However, only a few electrodes are sufficient to obtain motor control and sensory feedback within an arm prosthesis, if electrodes are properly placed in the corresponding nerves [29].

Micromachined flexible, neural electrodes have been designed and manufactured for contacting the tightly packed neural elements in the nerve trunk [30,31]. Polyimide was chosen as the substrate and insulation material for a design of multichannel sieve electrodes with integrated interconnects

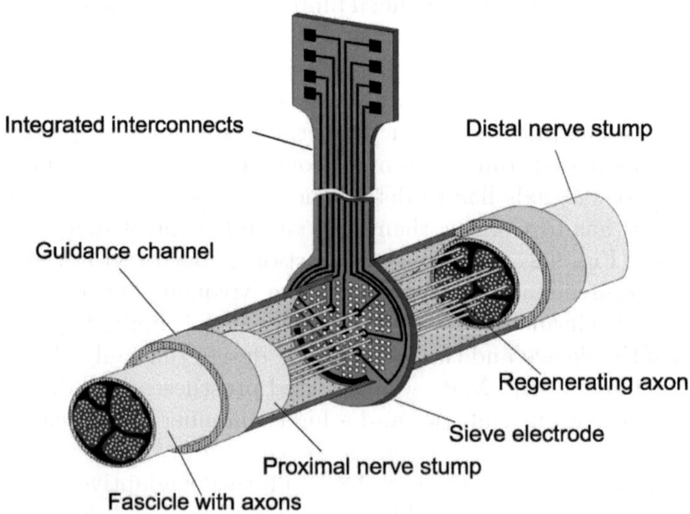

Fig. 9.28. Flexible, polyimide based sieve electrodes interfacing regenerating axons in peripheral nerves

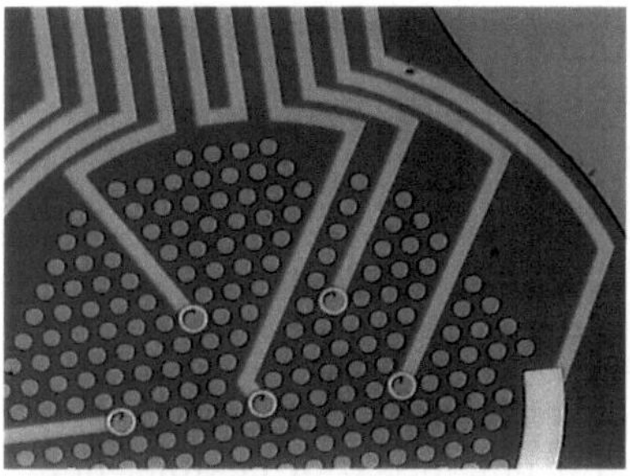

Fig. 9.29. Micrograph of the micromachined sieve structure with circular platinum electrodes. The diameter of the through holes is about 40 µm

(Fig. 9.28) that are interfacing regenerating nerve axons of the stump. While early designs (Fig. 9.29) started with seven platinum rings as the electrode sites, more than 20 electrode sites have been distributed uniformly over the sieve to interface a larger portion of the nerve fiber population. Bidirectional coupling, i. e. recording of nerve signals and neural stimulation, is feasible [32], although the signal-to-noise ratio is quite small and stability over time still has to be improved.

Neural Implants. Neural stimulation is applied to restore motor and sensor functions and to modulate central nervous system malfunctions for medically intractable diseases, so called neuromodulation. The latter field is dominated by spinal cord stimulation to treat incontinence [33] and to suppress chronic pain and by vagal nerve stimulation with respect to epilepsy, obesity and severe depression. The field of functional electrical stimulation (FES) classically addresses the neuromuscular apparatus to restore motion in paralyzed extremities. Stimulation is achieved through rather bulky surface electrodes that are part of an orthosis or individually attached to the skin. Investigators have applied self adaptive neuro-fuzzy algorithms to control an actuated orthosis [35]. The controller utilizes a closed-loop supervised learning adaptive network controller. Inputs comprise knee and hip angles from which hip torque and the pulse width of the stimulation are generated as outputs. Models for hand-free standing of paralyzed subjects are developed [36,37] but no implant is commercially available due to low patient numbers and little performance on stance and gait.

However, secondary benefits like decubitus and osteoporosis prevention as well as improvement of circulation regulation have been recently under discussion. The generation of appropriate stimulation parameters to overcome inverse recruitment and fatigue is the major challenge for controlling motor activity. A major drawback of functional electrical stimulation with surface electrode is the rather inconvenient handling of the electrodes and the adjustment of supporting structures. Therefore, recent research in rehabilitation engineering has focused on implantable neuromuscular stimulators with adaptronic functions. However, normal recruitment cannot be obtained with neuromuscular stimulation due to physical reasons. Spinal cord stimulation in animal models has recently achieved weight bearing force for gait pattern with correct muscle recruitment and without fatigue [34] but the transfer of these results into spinal cord injured humans still has to be done. It is mainly a question of evolutionary brain developments from cat to man, if it can be done at all and is not the subject of any control theory or technological development.

Various microelectrode designs have been realized for contacting nerve trunks and neural tissue. We have developed a new breed of flexible, lightweight multi-channel electrodes for interfacing nerves using micromachining technologies [38]. The electrodes are designed to allow the integration of microchips on the electrode substrate in order to obtain an adaptronic microsystem particularly suited for nerve stimulation and recording [39]. An adaptronic microsystem of the future (Fig. 9.30) for restoring grasping in paralyzed arms and hands may comprise motion sensors as inputs for control variables and to support hand-eye coordination via gaze control, a neural network for encoding the information into stimulation pulses, several telemetric units, a transcutaneous signal and energy transmitter, the subcutaneous microelectrodes with integrated chips, and an implantable force sensor or an neural interface to record neural force information for tactile feedback. Mate-

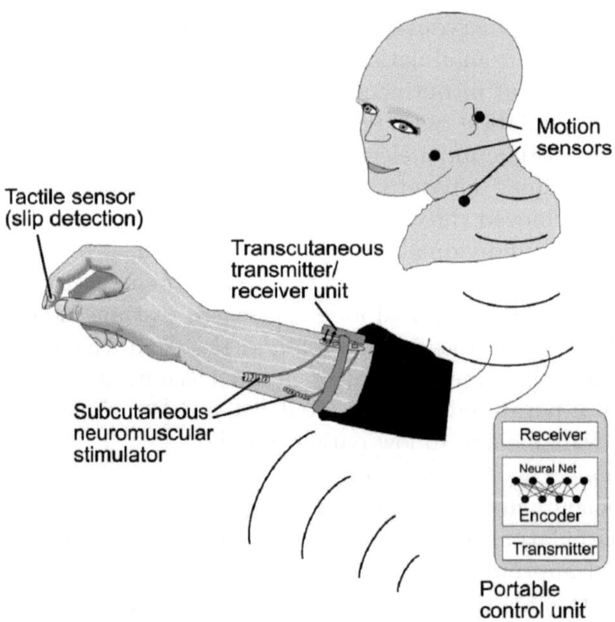

Fig. 9.30. Sophisticated multicomponent control system with adaptronic elements (slip detector, motion sensors for gaze control) for controlling grasping in a paralyzed arm. Brain-Machine-Interfaces are currently under invention for direct thought control

rials as piezoelectric foils or conducting polymers, can be used as smart skins for slip detection.

The latest developments use human control signals recorded from the motor cortex of the brain with so-called brain-machine-interfaces via multiple electrode sites on the skull, on the brain (i. e. epicortical) or with intracortical with penetrating microelectrodes. The electrical signals from the cortical areas represent vectors of motion and allow a trajectory control of robotic arms [40–42]. Studies on primates showed stable and reliable performance after training periods thereby proving the adaptability of the brain to setup transfer functions without knowing all the state variables.

A major challenge in biomedical engineering has been the development of microelectronic systems for substituting impaired or lost sensory functions. Cochlear implants have demonstrated that only eight to twenty electrodes are sufficient to stimulate sensory nerves in the cochlea with the effect of regaining a hearing perception. Cochlea systems consist of a holter device sized speech processor and a transcutaneous inductive link to a microprocessor implant which generates the appropriate signals for the stimulating electrodes in the cochlea (for details see e. g. [22]). Their function corresponds to an open-loop control system (Fig. 9.25a). During the training phase, parameters of the speech processor are adjusted for gaining an optimal performance of hearing. In general, cochlea systems are open-loop regulating systems that perceive feedback of perception only during training and readjustment of the system.

Even more ambitious are research programs in the USA and in Europe aiming to develop a neural prosthesis for the blind. Different research groups work on interfacing the retina, the optic nerve or the visual cortex, respectively [43]. In Germany, an epiretinal vision prosthesis has been jointly developed by our team of twelve researchers, who are experts in ophthalmology, neuro-informatics, and microtechnology, and are funded from the German research ministry (BMBF). Our work within the team focused on the design of flexible, multiple channel microelectrodes for stimulating retinal nervous tissue and on biocompatible system design and assembling and packaging techniques. Retinal contact microstructures have been developed (Fig. 9.31) to investigate chronic compatibility in contact with the retinal tissue and spatial selective excitation of the cortex after retinal stimulation. Pilot experiments with a wireless powered implant proved the hypothesis of selectivity in the visual cortex after electrical stimulation of electrode pairs on the retina [44].

Another emphasis has been given to the design of optimal and appropriate adaptive stimulation with respect to the interindividual retinal degeneration in retinitis pigmentosa and the implantation site. An adaptive retinal encoder has been under development [45]. Adaptive visual fields are employed which are adjusted to the function of retinal ganglion cells. The arrangement of electrodes allows sophisticated stimulation procedures. The described retinal prosthesis is considered to be one of the most challenging adaptronic systems of the future. The first patients have been implanted in Europe and the USA

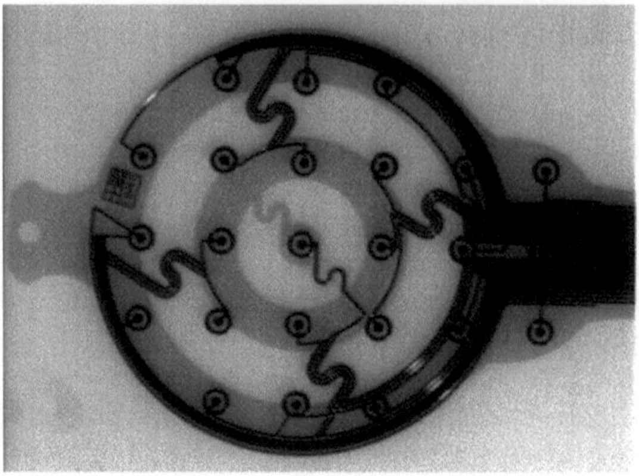

Fig. 9.31. Micrograph of the polyimide based retina contact structure

with precision mechanics implants. In 2005, German research groups started clinical human trials with microtechnical retinal vision implants. Adaptronic solutions still have to be integrated but initial results from implant stability in patients are promising (for details see [27]).

9.2.3 Adaptive Diagnostic Systems

Adaptive diagnostic systems are characterized by their enhanced performance when adapting to the varying dynamics of the biological system under multimodal inspection and control. Enhanced performance include additional feedback information received from sensory front-ends or is expressed by adaptive tracking of the biological system. The devices described below represent two implementations of adaptronic systems that have been developed in the recent years.

Tactile Sensing and Feedback

Endoscopic diagnostics and therapy has gained worldwide recognition as a method for minimal invasive interventions. However, endoscopic procedures are restricted to instrumentation that does not allow for a direct tactile contact or palpation of the operators hand or fingers with the tissue under observation. Our research has investigated sensor and actuator principles that have provided tactile information for the surgeon. Endoscopic forceps has been developed with an integrated array of piezoresistive silicon pressure sensors [46]. The pressure sensors are covered by a thin steel foil to withstand rugged handling and sterilization of the endoscopic instrument. In this version, the pressure signals have been displayed on a monitor. Since robotic assisted minimal invasive surgery has entered the surgical theaters with the option of tele-operation over large distances, haptic feedback has become more important. Arthroscopy, laparoscopy and needle insertion techniques, for example, can be performed with highest precision. Tele-surgery, robotic-assisted interventions and of course training and education benefit from various sensors and visualization techniques and finally lead to steeper learning curves. The latest research is conducted on multimodal feedback including the sense of smell to increase the information content for the surgeon [47].

Neonatal Blood Flow Monitoring

Ultrasonic monitoring of blood flow is a well established diagnostic tool in medicine. Despite recent advances in Doppler blood flow measurements, long-term, operator-independent monitoring of blood flow is still facing the problem of tracking the blood vessel under observation, particularly, when movements occur. Developments in electronics and smart materials resulted in electrical ultrasonic beam steering to accomplish continuous tracking of blood vessels under investigation. Electrical beam steering is achieved utilizing phased arrays. Phased arrays comprise an assembly of single piezoelectric transducers arranged in a line (1-D array) or in rows and columns (2-D array) and electronics for properly delaying the signals going to the elements for transmission or signals arriving at the elements for receiving [48]. Appropriate, adaptive algorithms are needed to apply ultrasonic beam steering to a moving target, e. g. a cerebral blood vessel. As an example, a cerebral blood flow monitor for pre-term infants with automatic tracking of a sample volume under observation (Fig. 9.32) is described in detail.

Pre-term infants face the risk of intraventricular hemorrhage that may lead to neurological disorders. Most commonly, the hemorrhage is a consequence of disturbed cerebral blood flow. The Fraunhofer-Institute for Biomedical Engineering (St. Ingbert, Germany) has developed a Doppler flow system

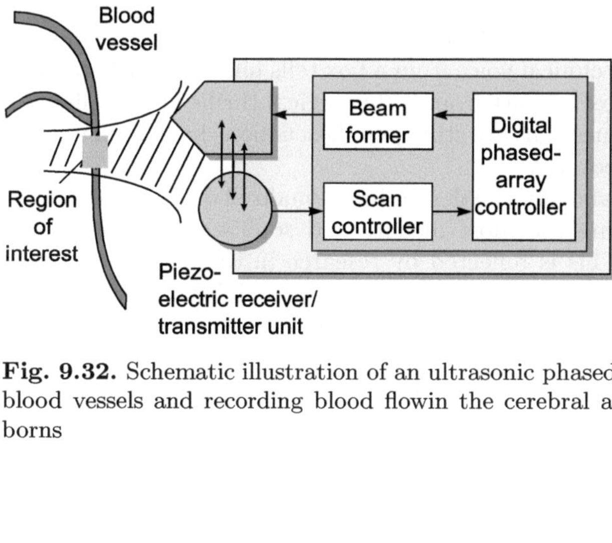

Fig. 9.32. Schematic illustration of an ultrasonic phased array system for tracking blood vessels and recording blood flowin the cerebral arteries of premature newborns

that is capable of long-term cerebral blood flow tracking in premature newborns. Major components of the system are board-based plug-ins, namely a scan converter, a microprocessor based controller board, and up to sixteen beam formers. Each beam former contains twelve piezoelectric elements which perform the functions of beam steering and Doppler signal detection. The system works with a dynamic, adaptive aperture control performed in the scan converter. Signals are received from a center point and four points where positions rotate around the center. The recordings at multiple sites are achieved without spatial readjustment of the array position. A new center position is determined as a function of a vector that points to an expected new signal maximum.

9.2.4 Conclusions and Outlook

Adaptronics in medicine is an evolving field that is presently in its infant stage. The examples above have given a brief insight into the possibilities of adaptronic systems in medicine. The application field of minimal invasive surgery has highlighted some potentials of adaptronic systems. Shape memory alloy (SMA) actuated microgrippers have been demonstrated whose dimensions are smaller than one half of a millimeter [49]. Several research teams are working on SMA actuated endoscopes that can be steered through cavities, lumen, and blood vessels inside the human body. Implantable drug delivery systems will address glucose control by insulin hopefully solving the stability problem of implantable biosensors. Promising research is directed towards the design of glucose sensitive gels that actuates valve and pumping systems in a self-regulating manner.

Interfaces with the nervous system will adapt to the output of the technical system, and the input of the biological system. Even taking over of cognitive functions by technical systems is under discussion. However, the human perception of unwanted help from a technical cognitive system, e. g. as support for elderly persons suffering from degenerative diseases such as Alzheimers is completely unknown. What does a person with cognitive deficits feel and experience, if a technical voice from a box tells him or her where to go or to drink a glass of water? Apart from technological thrills and challenges, many ethical and social questions are still open. Discussions have been started to address these issues recently.

Despite technical advances in the design functionalized or biomimetic materials, nature is providing the most mature and sophisticated adaptronic materials and systems. This is reflected by research in the field of bioartificial organs. For example, pancreas cells possess the capability of glucose level dependent production of insulin. The insulin production is inherently regulated inside the cell. It will be a major challenge of the future to employ the adaptability and multifunctionality of living cells to design the bioadaptronic systems of the future.

References

1. Bourne, G.H. (ed.): *The structure and function of muscle.* 2nd. ed., 4 vols., Academic, New York, London (1973)
2. Gordon, A.M.; Huxley, A.F. and Julian F.J.: *The variation in isometric tension with sarcomere length in vertebrate muscle fibres.* J. Physiol. (Lond) 184:170 (1966)
3. Huddart, H.: *Comparative structure and function of muscles.* Pergamon, Oxford, London, New York, Toronto (1974)
4. Huxley, H.E.: *The mechanism of muscular contraction.* Science 164:1356 (1969)
5. Huxley, A.F.: *Muscular contraction.* J. Physiol. (Lond) 243:1–43 (1974)
6. Irving, M.; Lombardi, V.; Piazzesi, G. and Ferenczi, M.A.: *Myosin head movements are synchronous with the elementary force generating process in muscle.* Nature (Lond) 357:156–158 (1992)
7. Jewell, B.R.; Wilkie, D.R.: *The mechanical properties of relaxing muscle.* J. Physiol. (Lond) 152:30–47 (1960)
8. Milligan, R.A.; Flicker, P.F.: *Structural relationships of actin and tropomyosin revealed by cryo-electronmicroscopy.* J. Cell. Biol. 105:29–39 (1987)
9. Nachtigall, W.; Wilson, D.M.: *Neuro-muscular control of Dipterian flight.* J. Exp. Biol. 47 (1967) pp. 77–97
10. Nachtigall, W.: *Insektenflug. Konstruktionsmorphologie, Biomechanik, Flugverhalten.* Springer, Berlin (2003)
11. Phillips, G.N.; Fillers, J.P. and Cohen, C.: *Tropomyosin crystal structure and muscle regulation.* J. Mol. Biol. 192:111–131 (1986)
12. Reedy, M.K.: *Myosin-actin motors. The partnership goes atomic.* Structure 1:1–15 (1993)
13. Rüegg, J.C.: *Calcium in muscle contraction.* 2nd. ed., Springer, Berlin Heidelberg New York (1992)
14. Usherwood, P.N.R.: *Insect muscles.* Academic, New York, London, San Francisco (1975)
15. Wilkie, D.R.: *The relation between force and velocity in human muscle.* J. Physiol. (Lond) 110:249–280 (1950)
16. Wilkie, D.R.: *Muskel. Struktur und Funktion.* B.G. Teubner, Stuttgart (1983)
17. Woodruff, E. A.: *Clinical care of patients with closed-loop drug delivery systems.* In: Biomed. Eng. Handbook, Bronzino, J. D. (ed.), Boca Raton, FL (USA), CRC and IEEE (1995), pp. 2447–2458
18. Sinkjaer, T.; Haugland, M.; Inmann, A.; Hansen, M.; Nielsen, K.D.: *Biopotentials as Command and Feedback Signals in Functional Electrical Stimulation Systems.* Med. Eng. Phys. 25(1) (2003), pp. 29–40
19. Edell, D. J.: *Tomorrows implantable electronic systems.* Edell, D. J.; Kuzma, J. and Petraitis, D. (eds), 18th Annual Int. Conf. IEEE EMBS, Minisymposia. IEEE. Amsterdam (1996), pp. 1–3
20. Schaldach, M.: *Reestablishment of physiological regulation, a challenge to technology.* In: Electrotherapy of the Heart, M. Schaldach (ed.), Berlin, Heidelberg: Springer-Verlag (1992), pp. 209–221
21. Cavuoto, J.: *Neural Engineerings Image Problem.* IEEE Spectrum, vol. 41, no. 4 (April 2004), pp. 20–25
22. Stieglitz, T.; Meyer, J.-U.: *Neural Implants in Clinical Practice.* In: Urban, G. A. (ed.) BIOMEMS, Dordrecht: Springer-Verlag (2006), pp. 41–70

23. Bolz, A.; Urbaszek, W.: *Technik in der Kardiologie: eine interdisziplinäre Darstellung für Ingenieure und Mediziner.* Berlin, Heidelberg, New York, Springer-Verlag (2002)
24. Webster, J.G. (ed.): *Design of cardiac pacemakers.* Piscataway, NJ: IEEE (1995)
25. Hexamer, M.; Drewes, C.; Meine, M.; Kloppe, A.; Weckmüller, J.; Mügge, A.; Werner, J.: *Rate-Responsive Pacing Using the Atrio-Ventricular Conduction Time: Design and Test of a New Algorithm.* Med. Biol. Comp. 42 (2004), pp. 688–697
26. Horch, K.; Dhillon, G. (eds.): *Neuroprosthetics: Theory and Practice.* Series on Bioengineering and Biomedical Eng., Vol. 2, River Edge, London, Singapore: World Scientific (2004)
27. Stieglitz, T.; Meyer, J.-U.: *Biomedical Microdevices for Neural Implants.* In: Urban, G. A. (Ed.), BIOMEMS, Dordrecht, Springer-Verlag (2006), pp. 71–138
28. Navarro, X.; Krueger, T.B.; Lago, N.; Micera, S.; Stieglitz, T.; Dario, P.: *A Critical Review of Interfaces with the Peripheral Nervous System for the Control of Neuroprostheses ad Hybrid Bionic Systems.* J. Periph. Nerv. Sys., vol. 10 (2005), pp. 229–258
29. Dhillon, G.S.; Horch, K.W.: *Direct neural sensory feedback and control of a prosthetic arm.* IEEE Trans. Neural. Sys. Rehabil. Eng., 13(4) (2005), pp. 468–472
30. Stieglitz, T.; Beutel, H.; and Meyer, J.-U.: *A flexible, light-weight multichannel sieve electrode with integrated cables for interfacing regenerating peripheral nerves.* Sensors and Actuators A 60 (1997), pp. 240–243
31. Lago, N.; Ceballos, D.; Rodríguez, F.J.; Stieglitz, T.; Navarro, X.: *Long Term Assessment of Axonal Regeneration through Polyimide Regenerative Electrodes to Interface the Peripheral Nerve.* Biomaterials, 26 (2005), pp. 2021–2031
32. Navarro, X.; Calvet, S.; Rodríguez, F. J.; Stieglitz, T.; Blau, C.; Butí, M.; Valderrama, E.; Meyer, J.-U.: *Stimulation and Recording from Regenerated Peripheral Nerves through Polyimide Sieve Electrodes.* J. Peripheral Nervous Sys. (3) 2 (1998), pp. 91–101
33. Rijkhoff, N.J.M.: *Neuroprostheses to treat neurogenic bladder dysfunction: current status and future perspectives.* Childs Nerv. Sys. 20 (2004), pp. 75–86
34. Saigal, R.; Rwnzi, C.; Mushahwar, V.K.: *Intraspinal Microstimulation generates functional movements after spinal-cord injury.* IEEE Trans. Neural Sys. Rehab. Eng. 12 (4) (2004), pp. 430–440
35. Trasher, A.; Wang, F. and Andrews, B.: *Self adaptive neuro-fuzzy control of neural prostheses using reinforcement learning pp.* (CD-ROM version), 18. IEEE EMBS Conf.. IEEE. Amsterdam (1996)
36. Davoodi, R.; Andrews, B.J.: *Computer Simulation of FES Standing up in Paraplegia: a Self-Adaptive Fuzzy Controller with Reinforcement Learning.* IEEE Trans Rehab. Eng. 6(2) (1998), pp. 151–161
37. Jacobs, R.: *Control model of human stance using fuzzy logic.* Biol Cybern. 77(1) (1997), pp. 63–70
38. Stieglitz, T.; Beutel, H.; Schüttler, M.; Meyer, J.-U.: *Micromachined, Polyimide-based Devices for Flexible Neural Interfaces.* Biomedical Microdevices 2 (4) (2000), pp. 283–294
39. Stieglitz, T.; Koch, K.P.; Schüttler, M: *Flexible, Polyimide-Based Modular Implantable Biomedical Microsystems for Neural Prostheses.* IEEE Eng. in Med. and Biology Magazine, vol. 24, no. 5 (2005), pp. 58–65

40. Serruya, M.D.; Hatsopoulos, N.G.; Paninski, L.; Fellows, M.R.; Donoghue, J.P.: *Instant neural control of a movement signal.* Nature 416 (2002), pp. 141–142
41. Patil, P.G.; Carmena, J.M.; Nicolelis, M.A.; Turner, D.A.: *Ensemble recordings of human subcortical neurons as a source of motor control signals for a brain-machine interface.* Neurosurgery 55(1) (2004), pp. 27–38
42. Schwartz, A.B.: *Cortical Neural Prosthetics.* Annu. Rev. Neurosci. 27 (2004), pp. 487–507
43. Veraart, C.; Duret, F.; Brelén, M.; Oozeer, M.; Delbeke, J.: *Vision rehabilitation in the case of blindness.* Expert Rev. Medical Devices 1(1) (2004), pp. 139–153
44. Walter, P.; Kisvárday, Z.F.; Görtz, M.; Alteheld, N.; Rössler, G.; Stieglitz, T.; Eysel, U.T.: *Cortical Activation with a Completely Implanted Wireless Retinal Prosthesis.* Investigative Ophthalmology and Visual Science, vol. 46, no. 5 (2005), pp. 1780–1785
45. Eckmiller, R.: *Learning retina implants with epiretinal contacts.* Ophthalmic Res., 29 (1997), pp. 281–289
46. Flemming, E.: *Tactile Sense in Minimal Invasive Surgery: The TAMIC-Project.* mst news, vol. 19, VDI/VDE (February 1997), p. 13
47. Spencer, B.S.: *Incorporating the sense of smell into patient and haptic surgical simulators.* IEEE Trans. Inf. Technol. Biomed. 10(1) (2006), pp. 168–173
48. Shung, K. K.; Zipparo, M.: *Ultrasonic transducers and arrays.* IEEE Eng. in Medicine and Biology Magazine, vol. 15, no. 6, IEEE. New York (1996), pp. 20–30
49. Heyn, S. P.: *Microsystem technologies and allied technologies in medicine: spotlights on recent activities in the USA, Canada, and Europe.* mst news, Special Issue, VDI/VDE (June 96), pp. 134–139

10 Future Perspectives: Opportunities, Risks and Requirements in Adaptronics

B. Culshaw

10.1 What's in a Name?

Adaptronics remains an enigma for much of the technical community and for virtually everyone whose professional life does not involve technology. Typing adaptronics into Google gives around 10 000 hits most of which concern smart structures and materials. Inserting this latter term into Google produces well over 5 million hits so perhaps this is better understood. Of course many of us view nanotechnology as a key enabler for smart materials. This gives well over 40 million hits. It is rough and it is crude but it could be convincingly argued that the number of hits on Google is a broad indicator of the understanding which the world thinks it has of a particular topic.

The inevitable implication of these observations is that adaptronics remains highly technical and culturally insular. This is surely not the intention. For adaptronics or even smart structures and materials to make its mark it must become understood as the vital enabler which it most definitely is within a community which extends far beyond its practitioners. Or should it really simply recognise that is a part of smart structures and materials and accede to popular demand? Personally I remain unconvinced of the technological content of this latter term but we must acknowledge that it has gained some community currency and this is at least as important as the technology itself. Names, whether we wish it so otherwise, are important as communication aids.

Why this discussion? Well, in order to make a mark, the potential which any technique offers must be recognised by the prospective users. We have certainly seen during recent years a plethora of society and community problems into which adaptronics could make a substantial contribution. Measuring the state of, responding to this state measurement and predicting the future performance of a physical or social system promises to contribute very significantly within our future lifestyle. Over the past decade many of the enabling tools which promise to release this potential have become much more impressive. The storage, accessing and manipulation of data in particular have progressed substantially, becoming simpler and more accessible to the non-specialist user.

There have been countless attempts to define the subject area. With many others, I too have contributed to this intriguing, arguably futile, debate. I concluded by observing that adaptronics, smart structures or whatever we call it is nothing other than a synonym for good engineering and surely good engineering (Fig. 10.1) is what the profession aspires towards? Adaptronics is arguably a universal panacea. It applies the tools available from material science, mathematics, computing and numerical analysis to the optimisation of material artefacts. The history of spectacular failure and less spectacular but far more costly decay in the infrastructure which uses these artefacts provides compelling evidence that any potentially useful optimisation process, whether applied to the design, to the use, to the maintenance or preferably all three must offer very substantial benefit. These benefits can be quantified – if the highway had not needed so much repairing the cost differentials can be estimated and the cost of the social disruption can also be estimated. The same logic may be applied to grounded aircraft, automobiles called back for design faults, potential savings in energy consumption, in transport, industrial processes and heating and cooling – indeed across the entire sector. But the take up is, at best, leisurely. I think the principal reason is that the universal enabler (Fig. 10.2) has to cope with not only engineering integration but also (and especially for major infrastructure) the conflicts between the political and economic aspirations of the various sectors within the society in which our adaptronic structure must operate.

It is however far from a totally gloomy picture. At one extreme civil and military aircraft and extraterrestrial vehicles epitomise the intelligence and adaptability which could be built into current mechanical engineering concepts. They are also an excellent model from which to expose the needs in the future. The global telecommunications network remains a remarkable example of the fact that an extremely complex system can be made remarkably reliable and deceptively easy to use. The civil engineering infrastructure on the other hand presents the paradox of an extremely conservative design to code targeted at building nominally permanent structures with almost totally unproven materials – most notably concrete. It seems that the extrapolation that the survival of medieval masterpieces was evidence that anything derived from stone was automatically permanent. What is more if

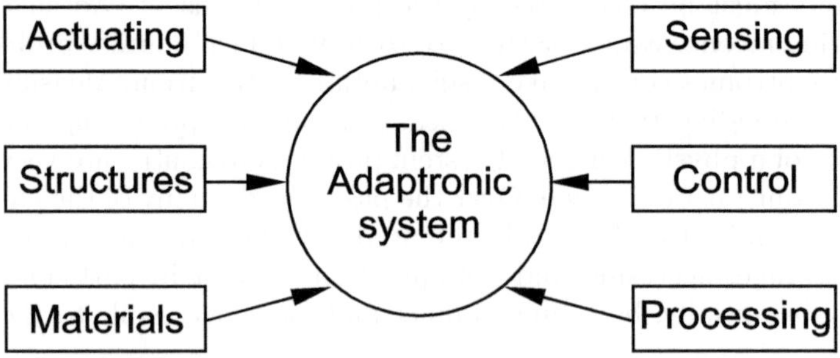

Fig. 10.1. Adaptronics, smart structures, call it what you will – the net effect is essentially good integrated engineering

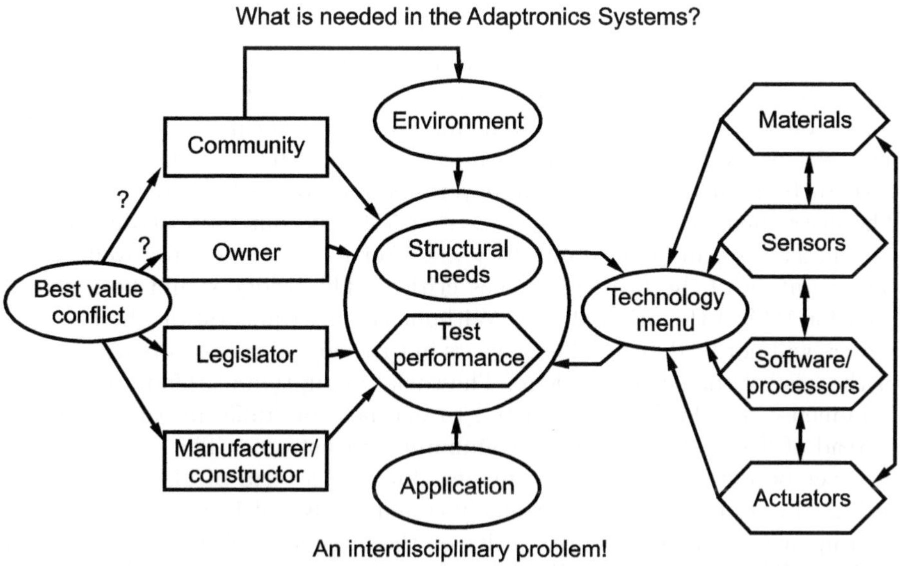

Fig. 10.2. Implementing adaptronics – who works out its real value? And against what criteria, over what timescales...?

you reinforced it with steel it would be more permanent still. The potential for adaptronics to cope with anomalous loads, to detect deterioration and decay, to identify tests and prove new materials and new material systems is only now becoming appreciated within a civil infrastructure which accounts typically (Table 10.1) for 5% to 10% of GDP in most developed economies.

Of course any additional complexity in any of these systems implies inevitably additional cost. This additional cost is only tolerated if either, there is a demonstrable benefit to the customer within the timescale which is rele-

Table 10.1. How Countries Spend Their Money

Item / Country	GNP ($B)	Agriculture %	Mining %	Manufacture %	Construction %	Transport and Communication %
US	6738 (2)	1.9	1.1	17.6	3.7	5.9
Japan	4693 (2)	2.2	0.3	30.4	8.8	6.4
UK	1042 (1)	2.0	2.2	20.9	5.4	8.5
India	81 (1)	30.3	2.4	17.3	5.8	8.1
Brazil	471 (1)	12.4	1.8	24.9	7.4	6.2

(1) 1993 (2) 1994 [Current figures demonstrate similar trends]

vant to the customer, or society puts in place legislature which compels complexity. The aerospace industry is an excellent example of the latter. The dilemma in the former may be exemplified by a hypothetical discussion of a new concrete bridge structure to be purchased by a local or national government driven by politicians who wish to continue to be elected and therefore wish to be seen to be saving taxpayers money during their term of office. The instrumented bridge costs an extra £5 000 000 but the instrumentation will direct a repair programme 15 years after it is installed and well before the condition of the bridge becomes critical. The bridge would be repaired then for 10% of the original cost. Without the instrumentation the need for repairs will not be detected for 25 years but the cost of the repairs will be half of the cost of the original bridge. The social disruption of this major repair is immense, the bridge is restricted to one lane of traffic in each direction instead of five and a city full of commuters is frustrated. The politician will however be determined to save money during his term of office unless the technical arguments can be extremely convincingly put forward and agreed.

Engineers and technologists thrive on the satisfaction of the application of their art. The design code philosophy which dominates most branches of large project engineering does however mitigate against the risk taking adventurer, but this conservatism has produced a whole catalogue of spectacular mistakes. There is then a need to take risks, to be adventurous, to gamble on the prospects of spectacular successes to demonstrate that the integration of the diverse engineering professions can produce hitherto unrealised benefits.

Even now it is far from impossible to cite some examples of both the risks of conservatism and the benefits stemming from the adventurous spirit.

10.2 Where Could Adaptronics Contribute: the Future?

The basic components of adaptronic systems encompass an integrated engineering philosophy which has much to offer well beyond the currently accepted wisdom that most should be led by aerospace and similar high-tech and infrastructure support industries. In recent times the immense potential offered by technology into addressing major social, environmental and cultural issues has seen ever increasing exposure [note 1]. In September 2005 Scientific American devoted a whole issue to scientific and technological approaches to defining solutions to cultural, social and environmental problems. In September 2006, the focus was on the post carbon energy economy. The 2005 IEEE President has toured the world with his lecture 'On the role and challenges of the engineer in the prosperity and well being of the world'. The Royal Society of Arts in the UK has defined, also in 2005, the fifteen greatest challenges to mankind and many of these embrace technological solutions.

Abraham Maslow, a psychologist based at Brandeis University proposed his now famous 'Hierarchy of Needs' in the 1960s. The first four: physiological needs, safety needs, the need to belong and the need for esteem are, in ascending order of necessity, fairly self explanatory but the fifth and ultimate need – self actualisation, the art of being at one with yourself – is more obscure. This according to Maslow comes but rarely and cannot be realised without the first four being achieved (Fig. 10.3 and note 2).

The reason for mentioning this is that whilst western societies worry about their cottage in the country, seventy percent of the worlds populace has no access to the basic resources for those most fundamental of physiological and safety requirements. Responsible application of adaptronics could begin to address the creation of the basic infrastructure which is fundamental to human needs.

Clean water, whilst by no means entirely furnished through the application of adaptronics, is one such need to which the technology could substantially contribute. The modelling, monitoring and control of flow rates through purification plant, the simplification of the purification process especially for use in remote areas, the careful monitoring of water supplies and their purity through even very simple measurement systems. All these – and many others–could begin to make the cost effective difference. Improved agriculture is another prospect where much can be done through the simple means of optimising irrigation processes, monitoring the quality of soils and providing readily available information on climate predictions particularly through the use of the Internet.

Waste generation and the management thereof is another example where adaptronic systems could contribute significantly. Possibilities here include accurate wide scale monitoring of environmental ground borne and water borne toxins and the definition of well controlled mechanisms through which the generation and localisation of these toxins can be suitably controlled. Whilst western consumer products contribute much to the generation of mountains of waste, there are also more basic needs, essentially sanitation systems, which continue to threaten much of the worlds population.

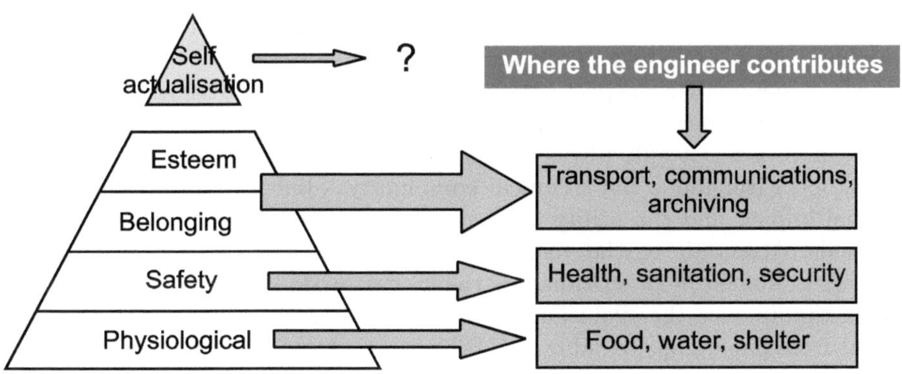

Fig. 10.3. Addressing our hierarchy of needs using broad based engineering: adaptronic is at the heart of much of this, but is engineering essential at the pinnacle?

And of course the most obvious direct area for adaptronic systems within the social environmental context lies in the control of combustion processes and emissions from the consumption of hydrocarbon fuels. Despite continuing protestations that alternative energy sources must be found, the relatively low key approach indicates either dogged complacency or convincing knowledge in the minds of those who know that hydrocarbon fuels are with us for some considerable time to come. Remember though that every liter of hydrocarbon fuel which burns in your car, oil fired power station, ship or aeroplane inevitably produces around 3 kilograms of carbon dioxide. Whilst the debate continues the current scientific view is that this carbon dioxide is a significant threat to our future safety and prosperity and measured steps are needed to combine our thirst for energy with the need for survival [note 4]. The principles of adaptronics, which apply not only to the combustion process but to the management of the entire system, could evolve a viable and technically sound approach to this pressing issue.

Much could also be done in enhancing the lifetime, reliability and recyclability of both the manmade infrastructure and the consumer products which operate within it. This is probably the natural domain for adaptronics, the one in which the technology first began to be defined.

So there is much that this technology can contribute, not only in its traditional arena but also in the far more diverse domains which could impact significantly on our collective future as a human society.

10.3 But it is More Than Technology

A very great deal can be achieved within the confines of currently available technology and natural resources – though as ever-new technological tools which in particular improve the efficiency of processes would always be welcome. The biosciences appear to offer much here but our corner of adaptronics has relatively mature tools already available. Much can be improved by optimising the utility of what we have.

Design is a critical aspect of this. If consumers are to be content with product for longer then design for longevity becomes a new engineering problem. This embraces not only physical ruggedness – a concept which is readily understood throughout the technical community – but also ensuring that the visual impact and the utility of the product retain their value over lengthy periods. Attending to the visual is reasonably well recognised. For passive products numerous practitioners – Henry Petrowski is a personal favourite – have championed the evolution of the passive, examining in the process artefacts from paper clips to cable stay bridges, combining function, elegance and cost in context. To date though very little has happened in the more complex area of optimising the intellectual and emotional interface between an active product – mobile phone, portable GPS receiver, even a video recorder – and the people who will use it. There is a little on the topic of emotional design but this interaction needs far greater recognition [note 3].

Critical too is communicating the viability of the technology to address major issues into a group of societies which have become obsessed by single criterion, simple, numerical, fiscal accounting – namely the books must show a profit [note 7]. Many of the concepts discussed here cannot be profitable in the direct and traditional sense of the word. However for the past couple of centuries or so advances in technology – engineers continually enhancing their skill in incorporating natural resources into every day life – have continued to increase the net worth of society. There is no reason why this trend should not continue so that even after using some of the resource to address the basic needs elsewhere there will still be more than adequate left for the originators to enjoy. This process of communication and education, highlighting the role of the engineer in society, is an issue which all of us in technology should take very seriously. Currently politicians and economists make the major strategic decisions, the former responding to the whims of a largely non-technical electorate, the latter frequently, but thankfully not always, drawn to the simplistic criteria encapsulated in the bottom line. Hence we scramble for land hungry visually intrusive wind farms in remote wild places and forget the impressive contributions and safety record of the nuclear option (except of course in France...). More than ever we need as a profession to speak fluently to the community [note 4].

A few nations and institutions are beginning to recognise these needs. Within the past year the US National Academy of Sciences published its 'The Engineer in 2020' report [note 5]. Its hundred or so pages are very well worth the attention of any educationalist or policy maker concerned with science and technology. The basic thesis is that communities will continue to wish to develop, notably at the moment China and India, and coupled to this the issues which society faces which can be addressed using technology, will continue to expand. The major themes that emerge concern the need for interdisciplinarity and the need for the engineer to communicate not simply within his own technological comfort zone, but to venture out into the community and the political and economic process. The UK Engineering Council has ventured some way along this route in its recently issued UK-Spec, which offers guidance for undergraduate education. This of course leads inevitably into the educational domain especially at the higher levels in universities. How should the education system evolve? (Fig. 10.4)

By far the vast majority of respectable technological universities have enhanced their degree portfolio over the past decade or so by offering ever-increasing specialisation. In parallel academic tenure and departmental and institutional assessment processes encourage more and more of learning more and more about less and less. It is after all relatively straightforward to recognise the (almost always) incremental advances in research in a highly specialised area. Meanwhile serious pleas emerge from the community (and the 2020 report is only one example) for the interdisciplinary generalist who re-

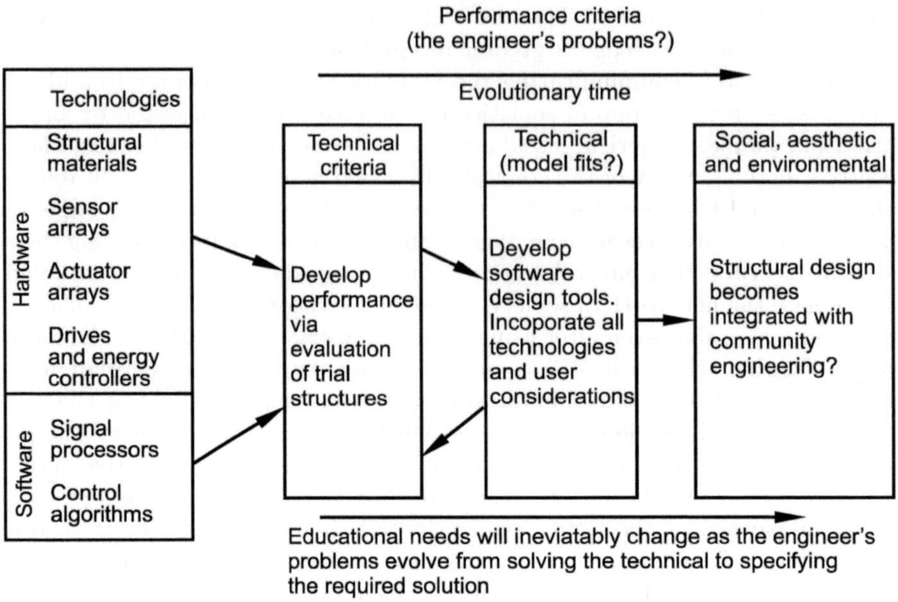

Fig. 10.4. Evolving the engineers role – the integrator of the problem in the round, will become increasingly important as manufacture becomes more centralised, but localised custom solutions become paramount

ally understands society and business as well as technology. The needs for the specialist, which undoubtedly continue, are addressed, whilst the needs for the generalist rely on in career evolution. There are a few exceptions, of which the pioneering course at the Technical University of Eindhoven in Intelligent Products is probably the best example [note 6]. Sceptical academics though, thanks to a career of advancement in ever narrowing channels, instinctively subscribe to the need for more technology and less context. Hopefully we shall see a gradual acceptance of the need for the interdisciplinary engineer to work alongside and integrate the contributions of specialists into a more effective engineered system measured against both technological and social/environmental criteria.

10.4 Educating the Public

Raising public awareness of solutions and problems associated with science and technology also becomes more pressing as engineering artefacts pervade more and more into our daily lives. A fine example of this linked into the recent European Union directive on managing waste associated with electronic and electrical equipment (WEEE). The legal aspects of this directive are profound making manufacturers responsible for the eventual disposal of

Fig. 10.5. The WEEE man... shown here on Londons South Bank

their product. Conveying its impact on our everyday existence is difficult to do in terms of chemical composition and numerical tallies. The WEEE directive did however stimulate the UK Royal Society of Arts to commission the WEEE man (Fig. 10.5 and note 7). The sculpture, standing over seven meters high and weighing more than three tons, was unveiled on Londons South Bank and has since started on a comprehensive tour. The sculpture is made from the electronic and electrical kit which the average western European uses over a lifetime. It also brought into sharp focus not only the launch of the directive but also the personal responsibilities associated with the ever-increasing presence of domestic appliances. Of course what it did do was highlight a problem. What is needed is similar initiatives to highlight the solutions which an interdisciplinary engineer can bring to bear within society.

10.5 The International Dimension: And Musings on Technology Transfer

Education though must go much further than simply raising awareness in the broad context to the need to synthesize technology into socially and economically relevant solutions. We hear much of technology transfer, we see international aid budgets predominantly project driven and often constrained towards contracting from the nation from which the aid originates. Indeed there seems to be an attitude that providing the technological solution provides the entire solution.

So if our adaptronic system finds itself usefully purifying water, maintaining and monitoring the civil infrastructure, producing efficient energy conversion – indeed playing a useful role in society – then this is undoubtedly beneficial. These systems though are often complex and nothing invented by engineers of whatever persuasion is infinitely reliable particularly in remote or ill-characterised environments. Consequently educating engineers internationally must be another important responsibility for the academic community to pursue as technologies spread and contribute to worldwide well-being [note 8].

10.6 And What About Technology?

Our discourse here has focussed on context above technology. The technical and scientific community is well adept at recognising and enthusing over technical and scientific advances. Whilst we muse over context, the scientific world is undoubtedly changing, perhaps more rapidly than ever. We see the sequenced genome and genetic manipulation; unprecedented skills in nano-scale materials engineering. Technology facilitates ever longer bridges and ever taller buildings as national egos make their mark; greater demands on transportation systems, especially as global warming becomes more convincingly linked to carbon emissions, the prospects of controlled ageing (at least for western societies where soon more will be over 60 than under 30...). The remaining chapters in this book have explored technology and its prospects for the adaptive systems, so little more on the topic is needed here.

Apart from to observe that adaptronics, whatever it may be, encapsulates the concepts of precision, control, responsiveness, indeed design for purpose with all that this simple statement implies. The contributions into other scientific and technological disciplines are absolutely critical. Without the precision control, the ability to interpret data, the integration of the mechanical and the electrical most – arguably all – of the current scientific and technical evolution would be severely hampered. Part of our context then is the scientific world beyond, which few if any of us understand but which needs the basic ideas in this book to progress. We should also recognise that as a genuinely interdisciplinary field of activity, we shall see inroads – new tools – emerging from surprising quarters, of which the biosciences and nano engineering are perhaps the most obvious. Consequently, an awareness of the world outside and its potential is more than interesting – it is the source of new and yet unimagined capacity to realise new prospects.

10.7 Some Concluding Thoughts

These are all simple observations but the implementation and the necessary actions are complex and intertwine politics, business, education and technology – an interacting web well beyond the capabilities of current or even projected simulation and modelling. We shall simply have to be stumbling experimentalists.

I do believe though that it is critical that a part of our interdisciplinary engineering community – exemplified in the adaptronics label – become aware of and broaden their ambitions into these far wider and far more challenging dimensions.

The environmental social, political and economic issues associated with the responsible use of technology have become more apparent and more important since the first edition of this book was published. Meanwhile though the enabling technologies with the exception of the immense and continuing increases in computer power and data access have advanced, dare I say it, at a relatively modest rate. But should this be a source of real concern? What does remain is the continuing conundrum of how to really make the best use of the technological tools which are already available.

Perhaps this highlights the need for more of the interdisciplinarity which this book encapsulates and indeed perhaps the need to venture even further into the numerous other spheres of activity which engineering touches upon.

Whilst preparing this short contribution a plethora of events – the principal ones of which are highlighted in the notes below – conspired to make similar largely political and social points concerning the future of engineering. We were fortunate to host Cleon Andersons lecture here in Glasgow during his tenure as President of the IEEE and his persuasive vision of engineering as an all pervasive enabler for social change reinforced and clarified some of the thoughts presented here. Authoritative discourse by the RSA and through Science, Nature and Scientific American reinforced this yet further. These reflections are of course personal and one objective of presenting this discussion has been to highlight the prospects that the interdisciplinary engineer has to contribute in a broader sense to improving the society in which we live. Adaptronics, smart structures and intelligent materials have a central role to play and in areas well beyond those which feature in this volume. Going beyond the technological domain, by far the majority of the contributors to this book find their homes in universities where not only is there a responsibility to advance the technology, but there is also a responsibility to seriously examine our role as interdisciplinary practitioners and encourage the art of communication among and outside our disciplines. There is much to be done.

Notes

Note 1: In September 2005 Scientific American devoted a special edition to the topic 'Crossroads for planet earth'.
The August 2005 journal of the RSA lists the fifteen challenges (Jerome Glen, Future Gazing, pages 16–21). Much of this concerns the social and political impact of scientific and technological development. Also in 2005 current IEEE President Cleon Anderson toured with his lecture: the role and challenges of the engineer in the prosperity and wellbeing of the world, highlighting the responsible application of engineering to the use of available resources.

Note 2: Abraham Maslow became noted for these needs which, whilst they still raise some scepticism in the philosopher, are generally accepted as being a fair reflection of the human condition. Much is available on the web and his book, *The Further Reaches Of Human Nature*, New York, Viking Press 1971, also presents these ideas in more detail.

Note 3: There are many books on visual design but Henry Petrowskis contributions are one of a few written from the perspective of a mainstream engineer. These are very readable including 'Invention by Design' (1996) and 'Design Paradigms' (1994) Harvard University.
There is however relatively little on human interface as a design issue. Donald Norman's book, *Emotional Design*, (2004 – Basic Books) is a useful example. Whilst on the topic of the role of technology in society, Thomas Hughes book *The human built world: how to think about technology and culture*, University of Chicago Press 2004 takes a broader view and examines the history of the engineer in working in society.

Note 4: Science Volume 309 no 5738, 19 August 2005, presented a news focus on rethinking nuclear power (pages 1168 to 1179).
The Stern Review (2006) (www.hm-treasury.gov.uk/independent_reviews/stem) presents an economists view of climate change, complementing and augmenting the scientific debate.
Scientific American, September 2006, explores the impact of carbon, the evolution of the carbon economy and the need for urgent coordinated response.

Note 5: The full report 'The engineer in 2020' is available online from the National Academy Press at www.nap.edu/catalogue/10999. UK-SPEC is obtainable free from Engineering Council UK, 10 Maltravers Street, London WC2R 3ER or at www.uk-spec.org.uk.

Note 6: Technical University of Eindhoven at www.tue.nl describes the Masters programme in Intelligent Products and Systems within the Department of Industrial Design.

Note 7: Much has been written on the WEEE man.
www.theRSA.org/project/WEEE_man.asp
and also at www.weeeman.org present a plethora of information on the WEEE man and its implications.
Al Gore, former US presidential candidate, has discussed these issues of accounting and climate change in many forums. See for example 'Lobbying for Earth', RSA Journal p33 October 2006 and www.theRSA.org/events.

Note 8: David A King, (UK Chief Scientific Advisor) *Aid to enhance Africa's Skills*, editorial, Science 314 p385 October 2006 makes this point very eloquently.

Index

About the Authors

Horst Baier was born in 1950 near Frankfurt. He gained his degree in mechanical engineering at Technische Universität Darmstadt in 1972. From 1972 until 1977 he was a research assistant at the Institute of Lightweight Structures in Darmstadt, with emphasis on finite element techniques and multicriteria optimization methods. In 1977 he joined Dornier Satellitensysteme where he soon became responsible for structural analysis and technology of mechanical systems. He became increasingly involved in the interaction of mechanical and control systems, and adaptive structures became one of his research interests. In the summer of 1997 he was appointed professor and head of the Institute of Lightweight Structures at Technische Universität München. His main research activities are in adaptive structures, fiber composite structures, as well as multidisciplinary structural and design optimization.

Christian Boller studied Structural Engineering at the Technical University of Darmstadt/Germany and received a Dipl.-Ing. degree in 1980 and a Dr.-Ing. degree in 1987. He was awarded a Japanese government scholarship for a stay with the Fatigue Testing Division of the National Research Institute for Metals in Tokyo/Japan in 1984/85. His involvement in smart technologies dates back to 1990 when he joined the Military Aircraft Division of MBB (now EADS) in Ottobrunn/Germany. He was appointed on the newly established chair in Smart Structural Design at the University of Sheffield/UK in 2003 which includes research work in structural health monitoring and micro aerial vehicles. He is the European editor of 'Smart Materials and Structures', a major international journal and has more than 100 papers and books published.

Peter Boltryk graduated in 2000 from the University of Southampton, UK, with an MEng in mechanical engineering. In 2004 he successfully completed a PhD which specialised in developing certain aspects of signal processing and data-based model techniques for application in a novel ultrasonics-based navigational aid for underwater submersibles. His current post-doctorial research interests at the University of Southampton, include non-contact surface metrology, audio recovery from mechanical sound recordings, signal processing, intelligent sensing and condition monitoring, and data analysis. He is an associate member of the IMechE.

William A. Bullough was born in Westhoughton, Lancashire in 1939. He was a trade apprentice, student engineer and assistant hydraulics engineer with De Havilland Aircraft. Following his M.Sc. in thermo fluid mechanics at the University of Birmingham he worked as a systems analyst with Hawker Siddely Dynamics and was latterly research associate in the Osborne Reynolds Hydraulics Laboratory, University of Manchester. From 1967 he has worked as lecturer, senior lecturer and reader at the University of Sheffield, UK where he is currently Honorary Reader. He is, among other things, a chartered engineer, a fellow of the Institution of Mechanical Engineers and a member of the Royal Institution.

Wenwu Cao received his PhD degree in condensed matter physics from the Pennsylvania State University, USA in 1987. After working briefly at the Laboratory of Atomic and Solid State Physics, Cornell University, he joined Penn. State University to become a faculty member in 1990. Currently, he is a professor of mathematics and materials science, a joint appointment between the Department of Mathematics and the Materials Research Institute of Penn. State. He conducts both theoretical and experimental research mainly on functional materials and their applications. To date, he has authored and co-authored more than 230 scientific journal articles. He is a member of the American Physical Society.

J. David Carlson, born in Erie, Pennsylvania in 1946, earned degrees in physics from Case Western Reserve University (B.S.) in 1968 and the University of Colorado (PhD) in 1972. Since 1976 he has been with the Lord Corporation, a global manufacturer of vibration and motion control systems and specialty materials, where he is senior engineering fellow. Since the mid 1980s he has provided technical leadership that has transformed magneto-rheological (MR) fluid technology from an interesting concept to a successful business. He holds 59 US patents and is the author of over 130 technical papers and books. He is the inventor of the 'MR fluid syringe' that has been used worldwide to effectively demonstrate the marvel of smart MR fluids. He is an adjunct professor in mechanical engineering at Virginia Tech University and a fellow of the American Physical Society.

Federico Carpi was born in Italy in 1975. He received the Laurea degree in electronic engineering in 2001 from the University of Pisa, Pisa, Italy. He received the PhD degree in bioengineering in 2005 from the University of Pisa. Since 2000 he has carried on his research work at the Interdepartmental Research Centre Piaggio, University of Pisa, where he currently has a post-doctoral position. His main research activities are related to polymer based materials and devices for biomedical engineering and robotics. He is author of several technical and scientific publications.

Frank Claeyssen, born in 1962, received in 1985 an engineering diploma from ISEN, Lille. He achieved a PhD Thesis at INSA Lyon in 1989 from his work at DCN (French Navy) on low frequency electro-acoustic transducers (design & construction of magnetostrictive sonar transducers). In 1989, he joined Cedrat to fund the active materials applications (AMA) activities dealing with smart materials and applications. Since 2001, he has been the technical & marketing director of Cedrat Technologies S.A., covering AMA and electrical engineering activities. He has invented several patented devices such as piezoelectric, magnetostrictive and magnetic actuators, motors and sensors. Some such as the famous APAs are manufactured and successfully marketed by Cedrat Technologies S.A.

Brian Culshaw is professor of optoelectronics at the University of Strathclyde, where he has acted as head of department and as vice dean of the engineering faculty. His research, spanning over 30 years has encompassed microwaves, optics and ultrasonics, both at device and system level, with applications in communications and sensing. He has published seven research texts in microwave semi-conductors, fibre sensing and smart structures and over 400 journal and conference contributions including many invited. He has been active in professional societies including two periods as a director of SPIE, of which he is currently president elect and as an editor of Applied Optics. He is a founder director of OptoSci limited and of Solus Sensors. He has chaired numerous technical conferences in the UK and abroad in optical fibre sensors and smart structures

Danilo De Rossi received the Laurea degree in chemical engineering in 1976 from the University of Genova, Genova, Italy. From 1976 to 1981, he was a researcher with the Institute of Clinical Physiology of CNR. Since 1982, he has been with the School of Engineering, University of Pisa, Pisa, Italy, where presently he is a full professor of bioengineering. He has been president of the Biomedical Engineering Teaching Track of the University of Pisa. Since 1999, he has also been an adjunct professor of material science with Wollongong University, Wollongong, Australia. His scientific activities are related to the physics of organic and polymeric materials, and to the design of sensors and actuators for bioengineering and robotics. He is the author of over 150 technical and scientific publications, and is co-author of several books.

Frank Döngi was born in Worms, Germany in 1966. He studied aerospace engineering at Universität Stuttgart and at the Cranfield Institute of Technology, England, where he was awarded an engineering diploma in 1991 and M. Phil. in 1993, respectively. In 1996 he received his doctorate in engineering from the Universität Stuttgart for a dissertation on active flutter suppression by means of smart structures. He then joined Daimler-Benz Aerospace/Dornier Satellitensysteme GmbH as a structural mechanics engineer where he was responsible for smart structures and active structural control. From 1999 to 2005, he was responsible for system design, engineering, and projects within Jena-Optronik GmbH, a company producing optoelectronic sensors and instruments for aerospace and defence applications in Jena. In January 2006, he joined EADS Astrium Satellites in Toulouse, France, where he is operations and process improvement manager within central engineering.

Göran Engdahl Doctor in Electricity in 1981, Uppsala University, Uppsala. Professor in electrotechnical design in 2001, Kungliga Tekniska Högskolan, Stockholm. Research field: theory, methods, and tools for design of electrotechnical systems and components based on data from material characterisations, and new phenomenological modelling algorithms with special interest in hysteresis and magnetoelastic phenomena in magnetic materials. Research interests: applied electro-mechanics and electromagnetism and modelling of electrotechnical components and systems comprising materials or phenomena involving functional, nonlinear or hysteretic behaviour. Publications: more than 70 international journals and conference publications.

Victor Giurgiutiu is professor of mechanical engineering and director of the Laboratory for Adaptive Materials and Smart Structures at the University of South Carolina, USA. He received his aeronautical PhD. (1977) and B.S. (1972) from the Imperial College, London, UK. In 1992–1996, he worked as research professor in the Center for Intelligent Materials Systems and Structures in Virginia Polytechnic Institute and State University, USA. From 1977 until 1992 he worked in the Aviation Research Institute, Bucharest, Romania. His research interests include adaptive materials and smart structures, structural health monitoring, mechatronics; embedded ultrasonic with piezoelectric wafer active sensors (pwas); active biomedical sensors, integrated nano sensors. He serves as associate editor of the Journal of Structural Health Monitoring; he has been guest editor for several journals. Dr. Giurgiutiu is a fellow of the Royal Aeronautical Society and of the American Society of Mechanical Engineers.

Martin Gurka was born in Öhringen, Germany in 1967. He received his diploma in experimental physics from the Ruprecht-Karls-Universitiy of Heidelberg in 1994. From 1994 to 1998 he was research associate at the Institute of Applied Physical Chemistry at the University of Heidelberg, where he received his PhD in physical chemistry in 1998. From 1998 to 2000 he was project manager R&D in the automotive industry. Since September 2000 he has been the managing director of Neue Materialien Würzburg (NMW) GmbH, an R&D-Service-Provider in the field of multifunctional materials. NMW also manufactures custom made PZT-Composites for application as sensors or actuators in adaptive structures.

Wolfgang R. Habel, born in 1949, received the diploma degree from the Technical University of Ilmenau, and the doctoral degree in engineering from the Technical University of Berlin. Since 1984 he has been engaged in research and development on fibre optic sensors for monitoring of structures and characterization of materials. In 1997, he joined the Federal Institute for Materials Research and Testing (BAM), where he has been head of the research group for fibre optic sensors since 2003. His main research activities concern strain transfer of embedded and surface-applied fibre optic deformation sensors as well as reliability and validation aspects for their practical use. He is a member of several German and international societies, head of the study group 'fibre-optic measurement technology' within the German VDI/VDE Association, and author or co-author of numerous national and international invited publications.

Jürgen Hesselbach was born in Stuttgart in 1949. From 1968 to 1975 he studied mechanical engineering at the University of Stuttgart and received his PhD degree from the Institute of Control Engineering of Machine Tools in 1980. Afterwards he joined the department of 'Industrial Equipment' of the Robert Bosch Company. From 1990 to 1998 he headed the Institute of Production Automation and Handling Technology (IFH) at the Technical University (TU) of Braunschweig. Since 1998 he has been the head of the Institute of Machine Tools and Production Technology (IWF) which was incorporated with the IFH. The institute is researching robotics, microassembly, product and life cycle management, fine and wood machining, and new actuators. Besides the institute leadership he was elected for President of the TU Braunschweig in 2005.

Gerhard Hirsch† was born in Pogegen, Lithuania, in 1924. He received his Engineering diploma from the Aachen University of Technology 1954. From 1954 to 1972 he was assistant and senior assistant lecturer. Lecturing and research work on structural dynamics and vibration control of lightweight structures were the basis for the senior lectureship that he held from 1972 onwards. He retired in 1996 and worked as a consultant for TMM Ltd., Eschweiler. G. Hirsch died in October 2003.

Hartmut Janocha was born in 1944 and studied electrical and mechanical engineering at the Technical University (today: Leibniz-University) in Hannover, Germany. In 1969 he graduated in the field of high-frequency technology, finished his doctorate in 1973 and his habilitation in 1979. Since 1989 he has been professor at Saarland University, where he holds the chair of the Laboratory of Process Automation (LPA). Between 1992 and 1994 he served as president of the German Assembly of Electrical Engineering Departments, between 1995 and 1997 as vice president of Saarland University, and between 2002 and 2004 as dean of the faculty of physics and mechatronics. His main fields of work include unconventional actuators, 'intelligent' structures, new signal processing concepts, and machine or robot vision.

Klaus Kuhnen, born in 1967, received the Dipl.-Ing. degree in electrical engineering at the University of Saarland in 1994. Following graduation, he has been working there as a scientific collaborator at the Laboratory of Process Automation (LPA) in the fields of solid-state actuators and control of systems with hysteresis and creep. Since 2000 he has worked as scientific assistant at the LPA in the fields of self-sensing solid-state actuators, active vibration damping with active material systems and adaptive identification and compensation with convex constraints. He received the Dr.-Ing. degree in engineering sciences at the University of Saarland in 2001. His research interests are in mechatronic systems with active materials, and the control of systems with hysteresis and creep and adaptive control theory.

Ronan Le Letty was born in Pont l'Abbé, France in 1967. He graduated with an engineering degree from the Institut Supérieur d'Electronique et du Numérique (ISEN Lile) in 1990. He got his PhD from the Institut National des Sciences Appliquées (INSA Lyon) in 1994, working on modelling of piezoelectric motors. He has been with Cedrat Technologies since 1991, working on piezoelectric and magnetic actuators, and adaptronics applications, especially for the aerospace industry. He is now technical director in charge of production.

Hiroshi Matsuhisa was born in Osaka, Japan in 1947. He received his B.S. in mechanical engineering from Kyoto University in 1970, and his MS in industrial engineering from Georgia Institute of Technology 1972 and his doctoral degree from Kyoto University for the study of vibration and noise of train wheel in 1982. Since 1976 he has been an instructor, associate professor and professor at Kyoto University. His main fields of research are vibration control, noise control and human dynamics.

Dirk Mayer was born in 1973 in Bochum, Germany. He studied electrical engineering at Ruhr-Universität Bochum from 1993 to 1998. From 1998 to 2003 he worked as a research assistant at Otto-von-Guericke-Universität Magdeburg, Institute of Mechanics. The focus of his research work within several public funded and industrial projects was the active control of vibrations. In 2003, he received a doctoral degree at the Technische Universität Darmstadt for his thesis on the application of adaptive filters for identification and control of smart structures. Since 2003, he has been working at the Fraunhofer Institute for Structural Durability and System Reliability (LBF) in the mechatronics/adaptronics department, being concerned with research on system integration and control.

Alberto Mazzoldi† graduated in Chemical Engineering from the University of Pisa in 1993. From then to 1995 he worked on biosensors based on conducting polymers and enzymes. From 1995 to 1998 he performed his PhD studies in Bioengineering at University of Pisa, working on conducting polymer fibres as dry actuators. Following post-doctoral fellowship, from 2000 to 2005 he was a researcher at the school of engineering of the University of Pisa. His main scientific interests concerned sensors and actuators for biomedical and robotic applications. He has been author of several technical and scientific publications. He died prematurely in May 2006.

Tobias Melz was born in Hildesheim, Germany in 1968. He studied mechanical engineering at the Technical University Braunschweig and received his diploma in 1996. He then joined the Institute of Structural Mechanics at the German Aerospace Center (DLR) and was responsible for several R&D aerospace projects involving smart structures. He received his PhD at TU Darmstadt in 2001 for active vibration reduction of mechanical cryocoolers for small satellites, and joined the Fraunhofer Gesellschaft in the same year. Since then he has been leading a research department in smart structure technology at Fraunhofer LBF (Institute for Structural Durability and System Reliability) in Darmstadt which is directed by Prof. H. Hanselka. He is also the managing director of the Fraunhofer Alliance Adaptronics (FVA).

Jörg-Uwe Meyer received his engineering degree from the Applied Sciences University in Giessen, Germany, in 1981 and his PhD degree in Biomedical Engineering from the University of California, San Diego, USA, in 1988. He worked as a principal investigator at the NASA-Ames Research Center, Moffett Field, USA. In the years 1990–2002 he was heading the sensor systems and microsystems department at the Fraunhofer Institute for Biomedical Engineering, St. Ingbert, Germany. Since August 2002, he has been leading the research unit at the Draegerwerk AG, a human safety and medical device company (www.draeger.com). Jörg-Uwe Meyer is a member of the technical and natural science faculty at the University of Lübeck. His expertise is on biomedical and industrial sensors and on integrated systems at the acute point of care (APOC). He is a member of the board of directors of the German Biomedical Engineering Society (DGBMT).

Uwe Müller was born in 1975 in Neu-Ulm, Germany. From 1996–1998 he completed the first part of his mechanical engineering studies ('Vordiplom') at Universität Stuttgart. From 1998–2002 he completed the second part of his mechanical engineering studies ('Hauptdiplom') at Technische Universität München with a focus on the simulation of aerospace structures and materials. In 2000 he spent six months under the supervision of professor Nesbitt Hagood as a visiting student at the active materials and structures lab. (AMSL) at the Massachusetts Institute of Technology (MIT) in Cambridge, USA. Since receiving his degree in 2002 he has been working as a research assistant under the supervision of Professor Horst Baier at the institute of lightweight structures at Technische Universität München. His areas of interest are smart structures, monitoring and sensor systems.

Werner Nachtigall, born 1934 in Saaz, studied biology and technical physics in Munich. After working as an assistant lecturer in the zoology and radiobiology departments of the Munich University and as research associate at the University of California, Berkeley, he was appointed director of the Zoology Department at the University of Saarland, Saarbrücken. His main fields of research were movement physiology and biomechanics. For several years he has offered a training in Technical Biology and Bionics. Combining biology and physics is an important subject of interest to him. He retired in 2002.

Dieter Neumann was born in 1955 in Gelnhausen, Germany. After finishing his studies in physics and sports he earned the degree of doctor at the University of Göttingen. He started his career in Göttingen at the Max-Planck-Institute for Fluid Mechanics. From 1991 to 1998 he worked as a technology consultant at the VDI Technology Centre and directed VDIs IT department in Düsseldorf (Germany). During the following years Dr. Neumann held several leading positions, being employed by major well-known IT-companies as area manager or divisional head. Since September 2005, Dr. Neumann has been Managing Director of Acteos GmbH in Gilching (Germany). Acteos provides solutions for implementing advanced mobile technologies to synchronize processes, material and information flow along the entire supply chain.

Raino Petricevic 1988–1995 study of physics at University of Würzburg and State University of NY at Buffalo; 1995 diploma; 2000 conferral of doctorate; project manager at the Bavarian Centre of Applied Energy Research. Working fields: material development for advanced energy storage technologies (fuel cells, super capacitors, lithium ion cells). Since 2000: project manager at Neue Materialien Wüzburg GmbH. Working fields: development/processing of adaptive materials, composites and structures, unconventional piezo actuators and sensors, polymer actuators, functional material, electro-rheological fluids.

Helmut Seidel was born in Munich, Germany in 1954. He received his diploma in physics in 1980 from Ludwig-Maximilians-University in Munich and his PhD in 1986 from the Free University in Berlin, Germany. In 1980 he joined the Fraunhofer-Institute for Solid-State Technology in Munich, Germany as a research scientist. His research focuses on micro electro mechanical systems (MEMS) ever since that time. In 1986 he joined the aerospace company Messerschmitt-Bölkow-Blohm (MBB), which later became part of Daimler-Benz Corporate Research. In 1996 he joined Temic, managing the department for microsensor development. His main focus was on developing sensors for automotive applications, particularly airbag accelerometers, gyroscopes and pressure sensors. In 2002 he became a full professor at Saarland University, Saarbrücken, Germany, holding the chair for micromechanics, microfluidics/microactuators. He has obtained more than 30 patents and has co-authored a large number of publications.

Thomas Stieglitz was born in Goslar, Germany, in 1965. He received the Dipl.-Ing. degree in electrical engineering from the University of Technology Karlsruhe (Germany) in 1993. From 1993 until 2004, he was with the Fraunhofer-Institute for Biomedical Engineering, St. Ingbert (Germany). He received a Dr.-Ing. degree (summa cum laude) in electrical engineering in 1998 and qualified as a university lecturer in 2002, both at the University of Saarland (Germany). In 2000 Dr. Stieglitz received the science award of the Saarland state for his work on flexible, neural prostheses. Since 2004 he has been a full professor of biomedical microtechnology in the department of microsystems technology (IMTEK), University of Freiburg. His research interests include biomedical microdevices, neural prostheses, neuromonitoring, and functional electrical stimulation.

Martin Thomaier was born in 1975 in Groß-Gerau, Germany. He gained his degree in mechanical and process engineering at Technische Universität Darmstadt in 2003. From 1992 until 1995 he did an industrial apprenticeship at Carl Schenck AG, Darmstadt as a technical drawer/mechanical engineer. Subsequently he worked as a technical drawer at Schenck PPSystems GmbH, Darmstadt. Until January 2004 he was a research assistant at the Fraunhofer-Institute for Structural Durability and System Reliability (LBF) in Darmstadt within the department of mechatronics/adaptronics. His work is focused on the design, simulation and development of active systems for vibration and noise reduction especially for automotive applications.

Neil White holds a personal chair in the school of electronics and computer science, University of Southampton. He has been active in sensor development since 1985. In 1988 he was awarded a PhD from the University of Southampton. He has considerable experience in the design and fabrication of a wide variety of sensors, formulation of novel thick-film sensing materials and intelligent sensor systems. Professor White is director and co-founder of the University of Southampton spin-out-company Perpetuum Ltd. He has over 200 publications in the area of instrumentation and advanced sensor technology. His professional qualifications include chartered engineer, fellow of the IET, fellow of the IOP, chartered physicist and senior member of the IEEE.

Thomas Würtz attended the Hochschule für Technik und Wirtschaft in Saarbrücken, Germany (Saarland University of Applied Sciences). Since 1993, at the Laboratory of Process Automation (Saarland University), he has been responsible for the area of power electronics for driving unconventional actuators. In addition to developing product lines mainly for driving piezo actuators used in fuel injection, he has contributed, in some cases as task leader, to various multi-lateral research projects funded by the EU and the Federal Ministry of Education and Research. Also in these cases the focus was on the development of power electronics and signal processing, tailored to the technical problem and the selected actuator.

MIX
Papier aus verantwortungsvollen Quellen
Paper from responsible sources
FSC® C105338

If you have any concerns about our products,
you can contact us on
ProductSafety@springernature.com

In case Publisher is established outside the EU,
the EU authorized representative is:
Springer Nature Customer Service Center GmbH
Europaplatz 3, 69115 Heidelberg, Germany

Printed by Libri Plureos GmbH
in Hamburg, Germany